Membrane Filtration

POLLUTION ENGINEERING AND TECHNOLOGY

A Series of Reference Books and Textbooks

EDITOR

PAUL N. CHEREMISINOFF

Associate Professor
of Environmental Engineering
New Jersey Institute of Technology
Newark, New Jersey

1. Energy from Solid Wastes, *Paul N. Cheremisinoff and Angelo C. Morresi*
2. Air Pollution Control and Design Handbook (in two parts), *edited by Paul N Cheremisinoff and Richard A. Young*
3. Wastewater Renovation and Reuse, *edited by Frank M. D'Itri*
4. Water and Wastewater Treatment: Calculations for Chemical and Physical Processes, *Michael J. Humenick, Jr.*
5. Biofouling Control Procedures, *edited by Loren D. Jensen*
6. Managing the Heavy Metals on the Land, *G. W. Leeper*
7. Combustion and Incineration Processes: Applications in Environmental Engineering, *Walter R. Niessen*
8. Electrostatic Precipitation, *Sabert Oglesby, Jr. and Grady B. Nichols*
9. Benzene: Basic and Hazardous Properties, *Paul N. Cheremisinoff and Angelo C. Morresi*
10. Air Pollution Control Engineering: Basic Calculations for Particulate Collection, *William Licht*
11. Solid Waste Conversion to Energy: Current European and U.S. Practice, *Harvey Alter and J. J. Dunn, Jr.*
12. Biological Wastewater Treatment: Theory and Applications, *C. P. Leslie Grady, Jr. and Henry C. Lim*
13. Chemicals in the Environment: Distribution · Transport · Fate · Analysis, *W. Brock Neely*
14. Sludge Treatment, *edited by W. Wesley Eckenfelder, Jr. and Chakra J. Santhanam*
15. Wastewater Treatment and Disposal: Engineering and Ecology in Pollution, *S. J. Arceivala*
16. Carcinogens in Industry and the Environment, *edited by James M. Sontag*
17. Membrane Filtration: Applications, Techniques, and Problems, *edited by Bernard J. Dutka*

Additional Volumes in Preparation

Membrane Filtration

APPLICATIONS, TECHNIQUES, AND PROBLEMS

edited by Bernard J. Dutka
National Water Research Institute
Canada Centre for Inland Waters
Burlington, Ontario, Canada

MARCEL DEKKER, INC. New York and Basel

Library of Congress Cataloging in Publication Data
Main entry under title:

Membrane filtration.

(Pollution engineering and technology; 17)
Includes index.
1. Sanitary microbiology--Technique. 2. Membrane filters. 3. Indicators (Biology) 4. Water--Microbiology--Technique. I. Dutka, B. J. II. Series
[DNLM: 1. Filtration--Methods. 2. Water microbiology. 3. Membranes, Artificial. 4. Water supply. WA 690 M533]
QR48.M45 628.1'61 81-4103
ISBN 0-8247-1164-5 AACR2

MARCEL DEKKER, INC.
270 Madison Avenue, New York, New York 10016

Current printing (last digit):
10 9 8 7 6 5 4 3 2 1

PRINTED IN THE UNITED STATES OF AMERICA

PREFACE

In this book, an attempt has been made to present an international review of the developments and the use of membrane filters in the field of water microbiology. The initial chapters describe the development of the membrane filter and the needs for standardization. The realization that all membranes are not equally able to resuscitate and grow all bacteria has helped explain some of the discrepancies noted in media comparison studies. Thus, the need for membrane filter standardization has been realized; various organizations such as the American Society for Testing and Materials (ASTM) and the International Standards Organization (ISO) are trying to develop techniques to help users evaluate membranes.

Several chapters are devoted to the use of membrane filters in the isolation and enumeration of most of the common indicator organisms tested for in water quality studies. In addition, detailed information is presented on the use of membranes in fungi and virus enumeration procedures. In many instances, specific media and procedures are detailed, along with details of potential problem areas.

To many readers, some of the most informative and important chapters will be the five devoted to the philosophy toward membrane filters and procedures used by European countries, South Africa,

Latin America, the Caribbean countries, Japan, and New Zealand. These chapters will provide North American readers with a better understanding of problems and accomplishments around the world.

Dr. W. A. Hoadley has prepared a very comprehensive and illuminating chapter on the effects of stress on the recovery of bacteria from water. This chapter will, no doubt, help the reader (perhaps for the first time) understand some of the reasons some procedures work on one sample and not on another. Along this same vein, the reader will also find a chapter dealing with structural differences between membrane filters and how these differences may account for growth and resuscitation variations.

Finally, there are several chapters dealing with innovative and possible future uses of the membrane filter in water and industrial microbiology.

The editor firmly believes that the reader of this book will obtain a thorough and up-to-date review of membrane filter technology as it relates to water.

Bernard J. Dutka

CONTRIBUTORS

G. I. BARROW Environmental Hygiene Reference Laboratory, Public Health Laboratory Service, Centre for Applied Microbiology and Research, Porton Down, Salisbury, Wiltshire, England

ELIZABETH C. BEAUDOIN* Department of Environmental Sciences, University of Massachusetts, Amherst, Massachusetts

NORMAN P. BURMAN† Metropolitan Water Services, Scientific Services, Thames Water Authority, London, England

VERA G. COLLINS† Bacteriological Department, Freshwater Biological Association, Ambleside, Cumbria, England

R. R. COLWELL Department of Microbiology, University of Maryland, College Park, Maryland

J. WILLIAM COSTERTON Department of Biology, University of Calgary, Calgary, Alberta, Canada

RICHARD A. COTTON Technical Liaison, Millipore Corporation, Bedford, Massachusetts

BERNARD J. DUTKA Microbiology Laboratories Section, Analytical Methods Division, National Water Research Institute, Canada Centre for Inland Waters, Department of the Environment, Burlington, Ontario, Canada

---

Present affiliation:
*Department of Microbiology, University of Massachusetts, Amherst, Massachusetts
†Dr. Berman and Dr. Collins are retired.

CHARLES W. FIFIELD Millipore Corporation, Bedford, Massachusetts

CLIFFORD F. FRITH, Jr.* Quality Control, Millipore Corporation, Bedford, Massachusetts

GILL G. GEESEY Department of Microbiology, California State University, Long Beach, California

EDWIN E. GELDREICH Microbiological Treatment Branch, Drinking Water Research Division, Research Laboratory, U.S. Environmental Protection Agency, Cincinnati, Ohio

CHARLES P. GERBA Department of Virology and Epidemiology, Baylor College of Medicine, Houston, Texas

SAGAR M. GOYAL Department of Virology and Epidemiology, Baylor College of Medicine, Houston, Texas

W. O. K. GRABOW Water Quality Division, National Institute for Water Research, Council for Scientific and Industrial Research, Pretoria, South Africa

ALFRED W. HOADLEY† School of Civil Engineering, Georgia Institute of Technology, Atlanta, Georgia

JAMES B. KAPER‡ Department of Microbiology, University of Maryland, College Park, Maryland

MORRIS A. LEVIN§ Office of Exploratory Research, U.S. Environmental Protection Agency, Washington, D.C.

WARREN LITSKY Department of Environmental Sciences, University of Massachusetts, Amherst, Massachusetts

MARIA THEREZINHA MARTINS Department of Biological Analysis, Companhia de Tecnologia de Saneamento Ambiental (CETESB), São Paulo, Brazil

GORDON A. McFETERS Department of Microbiology, Montana State University, Bozeman, Montana

MINORU MORITA Laboratory of Pharmacology, Suntory Institute for Biomedical Research, Suntory Limited, Osaka, Japan

---

Present affiliation:
*Vaponics, Inc., Plymouth, Massachusetts
†World Health Organization, Dacca, Bangladesh
‡Department of Microbiology, University of Washington, Seattle, Washington
§U.S. Army, Dugway Proving Ground, Dugway, Utah

WILLIAM G. PRESSWOOD* Gelman Sciences, Inc., Ann Arbor, Michigan

B. H. PYLE† Hamilton Science Centre, Water and Soil Division, Ministry of Works and Development, Hamilton, New Zealand

ANSAR A. QURESHI‡ Microbiology Section, Laboratory Services Branch, Ministry of the Environment, Rexdale, Ontario, Canada

I. GARY RESNICK§ Health Effects Research Laboratory, U.S. Environmental Protection Agency, Cincinnati, Ohio

D. A. SCHIEMANN Environmental Bacteriology Laboratory, Ontario Ministry of Health, Toronto, Ontario, Canada

ANTHONY N. SHARPE Microbiology Automation Section, Bureau of Microbial Hazards, Food Directorate, Health Protection Branch, Health and Welfare Canada, Ottawa, Ontario, Canada

JAMES P. SHERRY Microbiology Laboratories Section, Analytical Methods Division, National Water Research Institute, Canada Centre for Inland Waters, Department of the Environment, Burlington, Ontario, Canada

DAVID G. STUART¶ Department of Microbiology, Montana State University, Bozeman, Montana

RICHARD S. TOBIN Bureau of Chemical Hazards, Health Protection Branch, Health and Welfare Canada, Ottawa, Ontario, Canada

---

Present affiliation:
*Nuclepore Corporation, Pleasanton, California
†Department of Agricultural Microbiology, Lincoln University College of Agriculture, Canterbury, New Zealand
‡Microbiology Section, Alberta Environmental Centre, Vegreville, Alberta, Canada
§Department of Microbiology, University of Texas, Austin, Texas
¶The Baker Company, Inc., Sanford, Maine

CONTENTS

Membrane Filtration

1

THE MEMBRANE FILTER: ITS HISTORY AND CHARACTERISTICS

*WILLIAM G. PRESSWOOD** *Gelman Sciences, Inc., Ann Arbor, Michigan*

I. INTRODUCTION

Membrane filter technology has expanded more since 1957 than in the more than 100 years of its history preceding that date. The science of membrane filter technology, as we know it today, had its beginnings in the development of a membrane filter technique for the bacteriological analysis of drinking water. Immediately after World War II, the U.S. government, concerned with biological warfare and bacterial contamination of water supplies, commissioned a task group to study German membrane technology. Within three years, 1947—1950, American scientists had developed successful membrane manufacturing procedures. In 1957, as a result of many research reports on the membrane filter, the United States officially accepted the use of the membrane filter for bacterial analysis of drinking water. Acceptance of the membrane filter procedure by public health authorities led to its expansion and growth in other applications, such as industrial production of sterile products.

The use of membrane filters for sanitary analysis of water by growing bacteria upon the filter's surface was a novel application

**Present affiliation:* Nuclepore Corporation, Pleasanton, California

to historical uses of membranes. During the first century of the development of the membrane filter, investigators used membranes for dialysis and osmotic studies and for ultrafiltration. Investigators made their membranes for these experiments from nitrocellulose. The contributions by individual investigators of procedures for making membranes is a record of the history of the membrane filter.

The history of the cellulosic membrane filter may be divided into four periods: discovery, experimental, developmental, and modern. Before the discovery of the cellulosic membrane, natural membranes from animals and plants were used for diffusion and ultrafiltration experiments. Researchers of diffusion phenomena in the early 1800s used the pericardia of cows, the bladders of pigs and fish, the skins of frogs, and the skins of onions for dialysis experiments, osmotic studies, and ultrafiltration applications. Matteucci and Cima in 1845, using animal membranes for osmotic studies, were the first researchers to report on permeability differences related to the asymmetry of pores in membranes. They observed an acceleration in the flow of water through the membrane when they reversed the sides of the membrane to the direction of flow.

A similar observation was made by Schmidt (1856), who first performed what is now called ultrafiltration. Schmidt found a marked difference in the flow of water through animal membranes when he reversed the interior side of the membrane in the direction of the flow. Without understanding the reasons, these investigators had made a significant observation regarding the pore structure of membranes. Later it was found that artificial collodion membranes have this same anisomorphous pore structure, which influences the retentive and bacterial recovery properties of membranes.

II. DISCOVERY PERIOD

Scientists had long searched for a synthetic membrane to duplicate the diffusion and ultrafiltration processes accomplished in nature. Progress was made toward that goal in 1845 when Schoenbein accidentally synthesized nitrocellulose, and in 1855 when Fick made the first nitrocellulose membrane which he substituted for vegetable and animal membranes for his dialysis experiments. Despite meager success with these membranes, Fick had discovered what would be recognized a century later as a valuable analytical tool for a multitude of applications.

III. EXPERIMENTAL PERIOD

The years 1855–1918 are now seen as the experimental period of membrane filter technology. During this period, investigators repeatedly attempted to make membranes for their own work. The first practical uses of artificial membranes were in physiological studies for sizing macromolecules and for diffusion studies. During the six decades of this original experimentation, at least 53 investigators tried to devise methods for producing collodion or other types of membranes. Their early experiments centered almost exclusively on cellulosic membranes, chiefly nitrocellulose.

Nitrocellulose is a generic name for nitration products of cellulose. Collodion is a more commonly used term for cellulosic polymers and is a solution of nitrocellulose in an ether-alcohol mixture or an acetic acid-acetone mixture. All the early efforts to make collodion membranes involved independent applications of similar basic procedures with comparable results. The basic procedure used was to dissolve nitrocellulose in a solvent, either alcohol-ether or acetic acid. The mixture was poured to a thin film on a flat surface and allowed to gel by the evaporation of volatile solvents. Pore structure was controlled by the time allowed for evaporation of solvents, the time for evaporation being controlled by washing the film with water. Important differences in methods of preparation were use of solvents and nonsolvents in different ways. Scientists of the period tended to view membranes as attractive and promising objects of curiosity but with little practical value.

In 1860, Schumacher, using Fick's original procedure for making collodion films, experimented with those films in a closed sac or thimble design for dialysis studies. Baranetzky (1872) introduced the concept of preparing membranes in sheets. He made thin films of collodion on filter paper supports by pouring the collodion onto glass plates. He noted that water stopped the action of solvents, which evaporated from the mixture, forming pores in the membrane. However, Baranetzsky's collodion films failed to interest other scientists. The method was obscured until revived by Bechhold, who expanded it to produce the first successful series of membranes with different sizes of pores.

The collodion sac designed by Schumacher was used by Sanarelli in 1891 for ultrafiltration of blood plasma *in vivo*. Metchnikoff et al. employed these sacs for their 1896 research on *Vibrio cholera* by separating bacterial toxins from the cells. The collodion sacs or thimbles were easily prepared by dipping a test tube into a collodion solution, allowing it to dry, and then immersing it in water. Water loosened the film, which was then easily removed. Filling test tubes with collodion and pouring off the excess, leaving a thin collodion film on the glass, was another method of preparing collodion films. The film was permitted to dry, immersed in water, and removed from the test tube.

Collodion sacs could not be uniformly produced, gave uneven permeability, and varied in thickness from top to bottom. To overcome these obstacles, Bechhold (1907) made flat sheet membrane filters by impregnating filter paper with collodion dissolved in acetic acid. His was the first successful attempt to produce a graded series of membranes with varying permeability.

Bechhold found that permeability varied inversely with the concentration of collodion, high collodion concentrations producing less permeable membranes than lower collodion concentrations. Bechhold was the first researcher to successfully estimate the diameters of membrane pores. He performed well-defined permeability measurements by determining the pressure required to blow air through water-filled pores. Bechhold found that the air pressure required to force water through the pores was related to the surface tension of the water and to pore diameter.

This procedure is now commonly referred to as the bubble point measurement for determining the diameter of pores in membranes. Bechhold produced membranes with diameters of pores ranging from 1 to 5 μm. He developed ultrafilters of formalized gelatin with pore diameters of less than 10 nm. Bechhold initiated the use of air pressure in ultrafiltration and is given credit for the term "ultrafiltration," which has become standard.

Bechhold's method of utilizing a filter paper support was the first practical procedure proposed for making membranes. The reason he used supports for his membranes was because acetic acid mixtures of collodion produced fragile collodion films. Bechhold explored other uses for membranes by coating porcelain supports with collodion. His membranes were useful for many applications, such as ultrafiltration experiments, but the requirement of a rigid support was a hinderance in manipulation.

Bigelow and Gemberling (1907) introduced flat disks of collodion membranes without a supporting matrix. They substituted an ether-alcohol solution of collodion for the acetic acid procedure used by Bechhold and successfully produced membranes sufficiently strong to be self-supporting. They noted that variations in the drying time of collodion before immersion into water affected permeability. They also introduced the process of pouring thin layers of ether-alcohol solutions of collodion onto a leveled glass plate and regulated pore size formation by controlling the evaporation of solvents during the drying stage. This is the basic method later developed by Zsigmondy and Bachmann (1918) in their commercial production process of membrane filters.

Schoep in 1911 contributed to the experimental advance of membrane filter technology with his discovery that glycerol increased the diameters of pores in ether-alcohol collodion membranes. Concomitantly, he introduced the idea of adding plasticizers to membranes by using castor oil in the collodion mix. Schoep found that castor oil and glycerol made membranes more flexible and less fragile.

Walpole (1915) produced experimental membranes by varying the drying time of collodion and evaporation of solvents before replacing the solvents with water. He allowed the alcohol-ether solvents to evaporate for a definite time and then immersed the membrane in water, a process similar to that of Bigelow and Gemberling. He defined permeability as the wetness of the membrane, which is a ratio of solvent to water remaining in the membrane. Brown (1915, 1917) introduced another variation in membrane manufacture when he controlled the permeability of membranes by drying the collodion films to a specific weight before immersing them in alcohol-water solutions. Alcohol caused swelling of the nitrocellulose, the degree of swelling depending upon the percent of alcohol in the water. Permeability was expressed in relation to the alcohol content of the alcohol-water mixture. Brown was first to use cellulose acetate for preparing membranes, and made the notable observation of asymmetry in pore depth, or the anisomorphous structure of artificial membranes, referred to earlier in animal membranes.

The forerunner to Goetz's membrane filter was developed by Zsigmondy and Bachmann, who patented a graded series of membranes in 1918. Zsigmondy, while Director of the Institute of Colloid Chemistry, the University of Göttingen, used cellulose and esters of cellulose in a method proposed for commercial production of membrane filters.

Zsigmondy used the same basic ingredients for producing collodion membranes as had been used previously by other workers. His contribution to the development of membrane filter technology was his procedure for preparing filters. The dilution of cellulose derivatives was mixed in a solvent to which water, insoluble in the cellulose dilution, was added. The mixture was cast on glass plates and evaporation of the solvents was regulated by a slow passage of air in known quantities over the mixture. Humidity and temperature were also controlled. Evaporation was continued to a specified time and ended by suddenly covering the collodion film with water. The remaining solvents were washed free. The film was cut into small disks and stored in water as hydrated gels until used.

Zsigmondy related the diameters of the pores of his membranes to a Z factor, defined as the time Z required to pass 100 ml of distilled water at ambient temperature through a 100 cm^2 disk at a differential pressure of 1 atm.

As with Brown's membranes, Zsigmondy's method produced membranes with conical pores. The shape of the pore is produced by the differential rate of evaporation of volatile solvents between the upper and lower sides of the films. This procedure creates larger openings on the upper side of the membrane. The Zsigmondy method proved to be an important historical contribution and led to the developmental period of membrane filter technology.

IV. DEVELOPMENTAL PERIOD

The Sartorius-Werke Aktiengesellschaft and Company, Göttingen, Germany, refined the Zsigmondy process and in 1927 began the commercial production of membrane filters on a small scale. A division of this company manufactures membrane filters today.

Although available since 1927, commercial membrane filters continued for some time to be merely interesting objects of study among scientists, who continued to make their own membrane filters. The potential usefulness of the membrane had not been fully recognized, and at first, this remarkable tool was exploited in only a limited fashion.

A number of early investigators manufactured their own membrane filters and sometimes contributed to advances in the technology. Eggerth (1921), who produced membranes of high permeability, varied the amounts of absolute ethyl alcohol and anhydrous ether in the collodion mixture while holding the drying time constant and thus was able to grade the diameters of pores of membranes. He increased membrane pore diameters by adding lactic acid to his mixture.

Asheshov (1925) dissolved nitrocellulose in a solution of alcohol-ether and regulated the permeability of his membranes by adding acetone to increase and amyl alcohol to decrease pore diameters. He thus initiated the use of volatile reagents in membrane manufacturing for controlling pore diameter and contributed to the development of the modern membrane filter.

Pierce (1927) found that ethylene glycol, a nonvolatile compound, increased the diameters of the pores of membranes. He compared Brown's (1915), Walpole's (1915), and Bechhold's (1907) methods, and concluded that controlling the solvent/nitrocellulose ratio before exposure to water was vital in controlling permeability. Duclaux (1928) was the first investigator to use a cloth support in membrane manufacture. He received a patent (1928) for making cellulose acetate membranes by this method. Recently, the Duclaux method has been commercially used by Gelman Sciences to produce membranes of excellent strength and flexibility.

In the early 1930s, Elford made significant developments in membrane filter technology. He found that acetone and amyl alcohol, recommended by Asheshov, were antagonistic in their solvent action upon nitrocellulose solutions. Either reagent in an ether-alcohol mixture would dissolve nitrocellulose; together, they caused the nitrocellulose to coagulate. Elford (1931) took advantage of this principle to prepare highly permeable membranes with good tensile strength. He capitalized on Bechhold's discovery of using acetic acid to dissolve nitrocellulose. He found that adding acetic acid to the nitrocellulose mixture decreased the pore size of collodion membranes and that adding water increased permeability. Based on these concepts, Elford achieved membranes offering membrane pore sizes from 3 to 10 μm in diamter. He evaluated the permeability or diameters of the pores of his membranes by employing a technique originated by Guérout in 1872.

Guérout (1872) envisioned the pores in a membrane as a bundle of capillary tubes at right angles to the surface and equal in length to the thickness of the membrane. Applying Poiseuille's law, derived to express the flow rate of liquids throughout one capillary tube of a uniform diameter, Guérout calculated the diameter of pores in animal membranes. Elford and other investigators of the period indicated that passage of fluids through collodion membranes was governed by Poiseuille's law. Although this concept was accepted by other investigators, the theory is given little credence in modern membrane filter technology. To apply Poiseuille's law to the flow of fluids through the pores of a cellulosic membrane would imply that the pores are uniform in diameter. The pores in cellulosic membranes are arranged in a tortuous pattern and are not uniform in either diameter or length. Another innovation by Elford was a process for preparing sterile membranes by using ultraviolet light during manufacturing, thus avoiding the use of preservatives, which were then necessary for membrane manufacturing.

The literature is sparse on significant developments of the collodion membrane filter from the time of Elford's work until the end of World War II. After the war, scientists began to benefit from the advances in membrane filter technology achieved by Germany during the war.

Before World War II, membrane filters were used primarily to remove microorganisms and particles from liquids, in diffusion studies, and for sizing of macromolecules. The art of using the membrane filter for growing bacterial colonies upon its surface was undeveloped and the idea obscure. The Russians (Rasumov, 1933) are reported to have used the membrane filter technique for bacteriological analysis of water supplies as early as 1933, but scientists in Germany are credited with bringing the membrane filter to a position of prominence in bacterial analysis of water.

During World War II, German water supplies and laboratories were often devastated and their water contaminated by bombing raids. German scientists were forced to seek an efficient and expedient method for determining the safety of their drinking water. Practical applications utilizing the membrane filter for bacteriological tests of water were developed at German hygiene institutes. Working at the Hygiene Institute of the University of Hamburg during the years 1943–1945, Gertrud Müller and others developed membrane filter techniques suitable for the successful culturing of coliform bacteria upon the membrane filter.

In 1947, Müller published her methods for culturing coliform bacteria by the membrane filter technique (Mueller, 1947a). The same year she reported on the isolation and growth of *Salmonella typhosa* using the membrane filter technique with a bismuth sulfite medium (Mueller, 1947b).

After the war, U.S. military strategists recognized the need for bacteriological methods to provide an early warning of microorganisms in bacteriological warfare. In 1946, the Joint Intelligence Objective Agency of the U.S. armed services commissioned Alexander Goetz to visit Germany and obtain information on membrane filter production methods. Goetz gathered details on the Zsigmondy process from personnel of the Membranfiltergesellschaft Sartorius Werke, Göttingen, Germany, where membrane filters were commercially manufactured.

The method involved preparation of a dilute cellulose ester solution in a solvent to which water and other liquids, insoluble in the cellulose solution, were added to form an emulsion. In this process the components are homogeneously distributed. The emulsion is then cast onto glass plates and dried under meticulously regulated humidity and temperature conditions. The completed membrane sheets are then cut into disks and calibrated to determine the diameter of the pores.

V. MODERN PERIOD

Because rapid methods for detecting bacteria in water supplies was a concern at the Biological Department of the Chemical Corps of the U.S. Army, they awarded research contracts to the California Institute of Technology, Pasadena, California, in 1947, for Goetz to study the development and improvement of the membrane filter. Within three years, Geotz developed new production methods that improved flow rates and uniformity of pores. His methods obviated the necessity of storing membranes in water and the practice of boiling them before use, an inconvenient requirement with Zsigmondy's membrane.

Goetz (1951) imprinted grid lines on filters to facilitate counting of bacterial colonies. His grid line design is still used by commercial manufacturers of membrane filters. Goetz recognized the usefulness of colored membranes and was the first investigator to use dyes in the collodion mix to produce colored membranes. The filter equipment designs he introduced, as well as his filter disk sizes, are still employed. His work gave the world an eminently successful membrane filter suitable for commercial production. This filter inaugurated the modern era of membrane filter manufacture and application.

The first practical application of the membrane filter technique was in the analysis of potable water for coliform bacteria. During the late 1940s, concurrently with Goetz's developmental work, personnel of the U.S. Public Health Service at the Environmental Health Center, Cincinnati, Ohio, conducted research on the culturing of coliform and other bacteria with the membrane filter technique. Initially, they were supplied with membranes and filtering apparatus by Goetz. This research (Clark et al., 1951) established methods for culturing coliform bacteria and *Salmonella* upon the membrane filter. Their investigation was perhaps the first systematic study of the membrane filter technique for drinking water analysis.

Because of the bacteriological warfare concerns, the military classified as secret the early stages of membrane filter development and applications. It was not until 1951, with the publication of the U.S. Public Health Service report, that membrane filter technology was declassified and the filter made available to the public health field.

Following Goetz's ground-laying research, further research was performed by the Lovell Chemical Company, Watertown, Massachusetts, under a U.S. Chemical Corps contract. Initially, they sought to establish process controls for the commercial production of membranes with pores of diameters of 0.45 and 0.8 μm. Then came the design and construction of production equipment for large-scale membrane filter manufacturing.

In 1954, the Lovell Chemical Company sold their membrane-manufacturing enterprise to the newly organized Millipore Corporation, which continued research and development of membrane filters. Manufacturing procedures were refined to produce membranes with larger pore sizes than the standard 0.45 μm and, conversely, membranes capable of retaining viruses. Eight different pore sizes, from submicrometer to 10 μm, were made available. Despite this progress, the science of membrane filter technology had not yet reached its full potential. Many further technical advances were still ahead. Research continued in the United States and in Europe on comparisons of the membrane filter technique using the most probable number (MPN) method. Millipore Corporation, then the sole U.S. commercial company producing membrane filters, was instrumental in promoting adoption of the membrane filter technique by state health laboratories.

In 1957, the U.S. Public Health Service and the American Water Works Association officially accepted the membrane filter procedure for recovery of coliform bacteria from water as an alternative method to the completed test of the standard MPN method. After 102 years of experimentation and development, the membrane filter had reached official recognition in the United States and Europe as a valuable laboratory analytical tool.

Soon after adoption of the membrane filter into the public health field, methods for enumerating a wide variety of microrganisms were developed. The literature gives techniques and procedures using membrane filters for culturing or isolating bacteria, yeasts, molds, algae, protozoans, and viruses. The many uses of membrane filters are discussed in the succeeding chapters.

VI. PHYSICAL CHARACTERISTICS OF MEMBRANE FILTERS

Membrane filters are classified as surface or screen filters. Particles are retained on the surface of the filter or within a depth of 10-15 µm. It is this characteristic that distinguishes membrane filters from depth filters or filter aids, which trap particles within the filter matrix. Cellulosic membrane filters are approximately 150 µm thick with a myriad of conically shaped pores arranged in a tortuous pattern. A typical membrane filter has 400–500 million pores per square centimeter of surface area. The pores in cellulosic filters may be envisioned as a spongelike construction and comprise about 80% of the filter's volume. The extreme thinness of membrane filters compared to depth filters gives them a distinct advantage. Because microorganisms are retained on or near the membrane surface, they can absorb nutrients through the filter and produce visible colonies.

It should be noted that membrane filters retain microorganisms and particles by mechanisms other than a simple sieving action. Microbes and particles may be adsorbed by the filter, react with the membrane itself, or may be retained on the membrane by coagulation. The fundamental mechanisms of filtration may be considered as sieving modified by adsorption and blocking arising from the large ratio of pore length to pore diameter. The ratio of pore length to width in a 0.2 µm pore size rated filter is approximately 750. This ratio, in combination with the geometric configuration of a tortuous pore structure, heavily influences the retention of particles by membranes. Variation in cellulosic esters produces membrane filters that selectively adsorb viruses when the pore diameter greatly exceeds the size of the virus, whereas a different combination of cellulosic esters allows viruses to pass through. The basic guidelines are: nitrocellulose adsorbs viruses; cellulose triacetate repels viruses.

When used to retain viruses, membrane filters, with a thickness of about 5000 virus diameters, function as depth filters. Virus particles trapped in the matrix of a nitrocellulosic filter can be eluted from the filter.

Introduction of the membrane filter into microbiology revolutionized the analysis of water and other liquids for microorganisms. By the use of membrane filters, it became possible to filter large volumes of liquids containing low concentration of microorganisms and to cultivate them *in situ*.

Initial applications of the membrane filter, after Goetz's work, was for analysis of water for coliform bacteria. It was because of the need for an easier method and a shorter time delay between bacteriological analysis and final results that interest was created in developing the membrane filter. For over 50 years, the MPN technique had been the standard method for detecting coliform bacteria in water. The membrane filter made it possible to obtain results of analysis of water for coliform bacteria in less than 24 hours versus a maximum of 96 hours by the MPN method. Results of the MF method are a direct colony count of coliform bacteria. MPN results are a statistical estimate of coliform concentrations. Other advantages of the MF technique are less manipulation of samples, requirement of less laboratory space, and thus greater economy compared to the MPN method. Colonies on the MF may be kept for further study without additional isolating procedures. The filter may be preserved as a permanent record of the analysis.

The past century has witnessed dynamic progress for membrane filter technology and applications, resulting in the diversity of filters and uses common today. What were once viewed as scientific curiosities are now recognized as invaluable scientific aids. Considering this progress, it would be naive to assume that either the technology or applications of membrane filters have reached an end point. Membrane filters continue to surprise the experts with their versatility. The modern era of membrane filtration is well begun, but it is just beginning.

VII. TYPES OF MEMBRANE FILTERS

Until 1963, membrane filters were predominantly, but not exclusively, composed of nitrocellulose and esters of cellulose acetate. Filters of other types developed during the late 1920s and early 1930s reveal an interesting variation of types. Nickel was plated on nickel and bronze wire mesh to make filters with pore diameters from 50 to 300 nm. Zinc was distilled out of brass leaving a porous copper membrane. Silver membranes were made using a similar process. These methods were novel ways of making microporous structures suitable for specific applications for filtering liquids. However, none could equal the collodion membrane filter in performance and ease of production

Gelman (1965) has written a comprehensive review of types of membrane filters developed before 1965 and their industrial applications. Some of the more interesting highlights of new developments in membrane filters are listed in his report.

In 1963, scientists began studying the use of a variety of aromatic polymers for making membrane filters. A polyethylene membrane was marketed in 1963, but lack of quality forced it off the market. At about the same time, cellulose acetate membranes, devoid of nitrocellulose, were developed with good thermotolerance. Soon afterward, cellulose triacetate membranes were perfected by the Gelman Instrument Company. Cellulose triacetate has good tensile strength, inertness, and is not subject to hydrolytic clevage of the polymer during autoclaving. Regenerated cellulose membranes introduced in the mid-1960s were adaptable for filtering most organic solvents. Membranes made from nylon were developed at about the same time. Polyvinyl chloride was used to make a strong, nonsupported membrane highly resistant to chemical attack.

In the early 1970s, Nuclepore introduced a membrane with straight-through cylindrical pores (as opposed to the tortuous pore configuration of conventional membranes). Initially, the membrane was made from polycarbonate; more recently a polyester polymer has been added to the capillary-pore line. These capillary-pore membranes have some unique advantages for certain applications because of their thinness (nominally 10 μm), transparency, and sharp fractionation retention capability. A membrane made from Tungsten was developed by the Hughes Research Laboratories in the early 1960s. Selas Flotronics in 1964 introduced a metallic membrane of pure silver. Polymers recently used for making membranes are polyamide and polysulfone. Research is continuing on new polymers, suitable for casting into membrane filters, that will certainly revolutionize membrane filter technology.

REFERENCES

Asheshov, I. (1925). Preparation of graded collodion membranes. *C.R. Soc. Biol 92*: 362–363.

Baranetzky, J. (1872). Diosmotishe untersuchungen. *Pogg. Ann. 147*: 195–245.

Bechhold, H. (1907). Kolloidstudien mit der Filterationesmethode. *Z. Phys. Chem. 60*: 257–318.

Bigelow, S. L., and Gemberling, A. (1907). Collodion membranes. *J. Am. Chem. Soc. 29*: 1576–1589.

Brown, W. (1915). On the preparation of collodion membranes of differential permeability. *Biochem. J. 9*: 591–617.

Brown, W. (1917). Further contributions to the technique of preparing membranes for dialysis. *Biochem J. 11*: 40–57.

Clark, H. F., Geldreich, E. E., Jeter, H. L., and Kabler, P. W. (1951). The membrane filter in sanitary bacteriology. *Public Health Rep. 66*: 951–977.

Duclaux, J. (1928). United States patent 14472.

Eggerth, A. H. (1921). The preparation and standardization of collodion membranes. *J. Biol. Chem. 48*: 203–221.

Elford, W. J. (1931). A new series of graded collodion membranes suitable for general bacteriological use, especially in filterable virus studies. *J. Pathol. Bacteriol. 34:* 505–521.

Fick, A. (1855). Ueber diffusion. *Pogg. Ann. 94: 59-86*.

Gelman, C. (1965). Microporous membrane technology: Part 1. Historical development and applications. *Anal. Chem. 87*: 29–34.

Goetz, A. and Tsuneishi, N. (1951). Application of molecular filter membranes to the bacteriological analysis of water. *J. Am. Water Works Assoc. 43*: 943–967.

Guérout, M. (1872). *Compt. rend Acad. 75*: 1809–1812.

Matteucci, O. H., and Cima, A. (1845). Mémoire sur l'endosmose. *Ann. Chim. Phys. 13*: 63–86.

Metchnikoff, E., Roux, E., and Taurelli-Salimbeni (1896). Membranes in biological studies. *Ann. Inst. Pasteur, 10*: 261–264.

Müller, G. (1947a). Lactose-fuchsin plate for detection of coli in drinking water by means of membrane filters. *Z. Hyg. Infektionskr. 127*: 187–190.

Müller, G. (1947b). Eine trinkwassergebundene Ruhrepidemie. *Zentralbl. Bakteriol. Parasitenkd. Infekionskr. Hyg. Abt. I Orig. 152*: 133–135.

Pierce, H. F. (1927). Nitrocellulose membranes of graded permeability. *J. Biol. Chem. 75*: 795–815.

Rasumov, A. S. (1933). Novye metody i puti kachestvennogo i kolichestvennogo izucheniia mikroflory vody. *Mikrobiologiya 2*: 346–352.

Sanarelli, G. (1891). *Zentralbl. Bakteriol. Parasitenkd. Infektionskr. Hyg. Abt. I 9*: 457.

Schmidt, W. (1856). Verusche über filtrationsgeschwindigkeit verschiedener flussigkeiten durch thierische membranen. *Pogg. Ann. 89*: 337–340.

Schoenbein, B. (1845). British patent 11407.

Schoep, A. (1911). Ueber ein neues ultrafilter. *Kolloid Z. 8*: 80–87.

Schumacher, W. (1860). Collodion membranes. *Ann. Phys. Chem. 47*: 337–342.

Walpole, G. S. (1915). XXVI notes on collodion membranes for ultrafiltration and pressure dialysis. *Biochem. J. 9*: 284 297.

Zsigmondy, R., and Bachmann, W. (1918). Ueber neue filter. *Z. Anorg. Allgem. Chem. 103*: 119–128.

2

STANDARDIZATION OF MEMBRANE FILTERS FOR MICROBIOLOGICAL APPLICATIONS

RICHARD A. COTTON and CHARLES W. FIFIELD Millipore Corporation, Bedford, Massachusetts

I. INTRODUCTION

The membrane filter is more than just a particle separation medium and is used extensively in various technical and analytical applications. As a result, the precise control of structure, pore size, and physical characteristics is critical. The controls that are applied in the production and testing of membrane filters are discussed in this chapter.

A. Membrane Filter

Because of its unique structure, the membrane filter has become an important analytical and process tool in microbiological applications. The structure comprises a matrix with a multiplicity of closely packed pores that have considerable uniformity of size. The membrane filter can quantitatively remove organisms from liquids and gas and retain those organisms on its surface for analysis by either colony culture or direct microscopic observation. However, all membrane filters of similar pore size rating do not give identical results in sensitive performance tests (Brodsky and Schiemann, 1977; Schillinger et al., 1977; Green et al., 1975). Therefore, standardization of membranes for specific sensitive microbiological applications is needed.

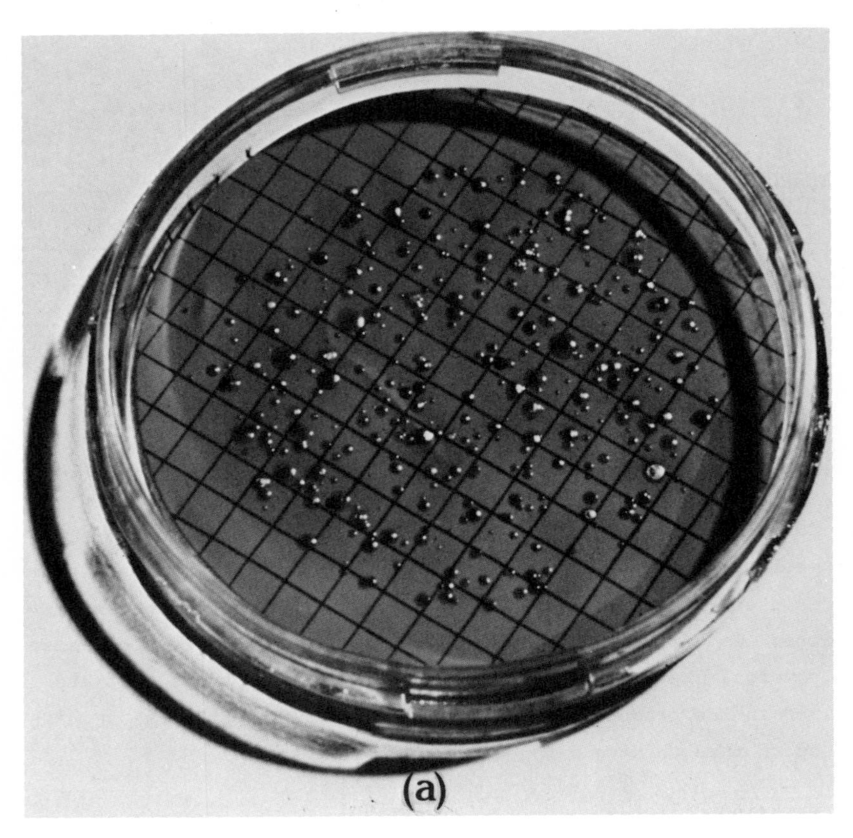

FIGURE 1 Colony growth: bacterial colonies cultured on the membrane filter: (a) total membrane; (b) close-up (magnification 30x). (Courtesy of Millipore Corporation, Bedford, Mass.)

The inert structure of a properly produced membrane filter does not affect or alter the bacterial cell retained by the membranes and should not affect the viability of the organism. The high porosity created by close packing of the pores results in a high rate of filtration, which makes possible a rapid concentration of organisms from large volumes of liquids. The capillary pore structure of the membrane filter enables bacteriological nutrient to be supplied to the retained organisms in a controlled manner. This is accomplished by feeding the nutrient through the membrane filter from a reservoir

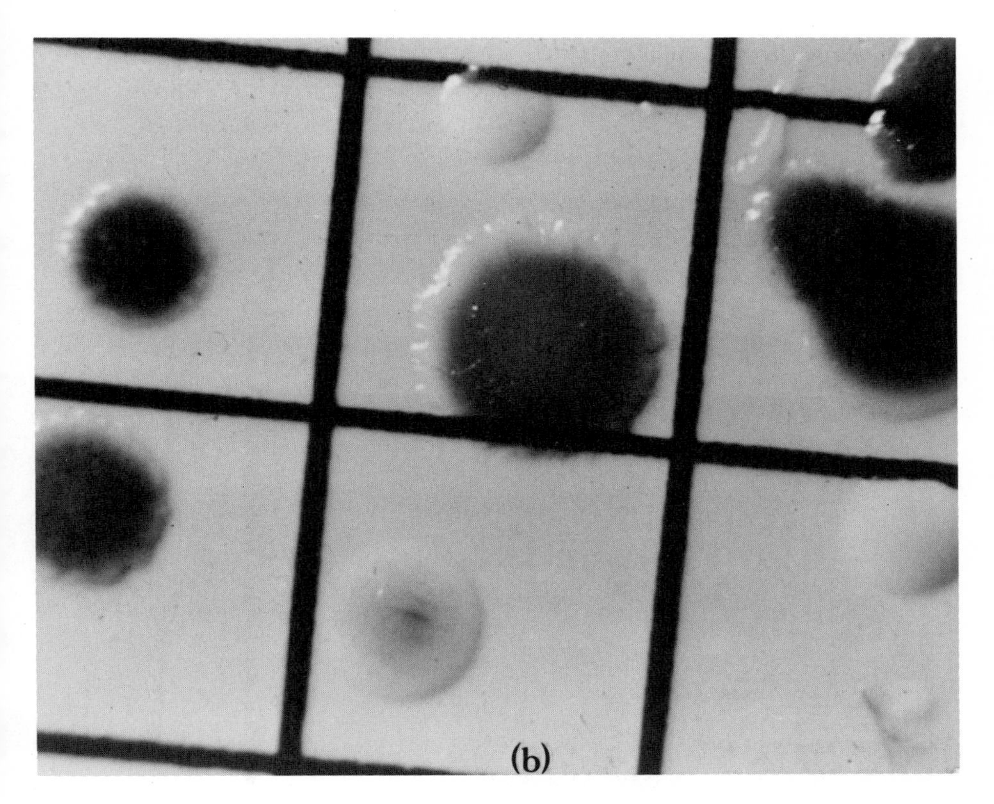

FIGURE 1 (Continued)

of medium that is contained in an agar gel or a nutrient pad beneath the membrane. Colony growth on the membrane filter is illustrated in Fig. 1(a) and (b).

The smooth, flat surface of the membrane facilitates direct microscopic examination of organisms retained by filtration. Staining techniques to improve contrast and visibility are normally employed for direct microscopic observation of the organisms (Schaufus and Krabek, 1955). Since transmitted light microscopy is advantageous for this purpose, the membrane filter can be rendered transparent by filling its pores with an immersion oil or other nonvolatile fluid which has a refractive index identical to that of the material comprising the membrane matrix. The technique is illustrated in Fig. 2.

Before describing the membrane filter parameters that require standardization and critical evaluation for microbiological applications, the membrane filter itself should be defined. The American Society for Testing and Materials (ASTM) has developed the following

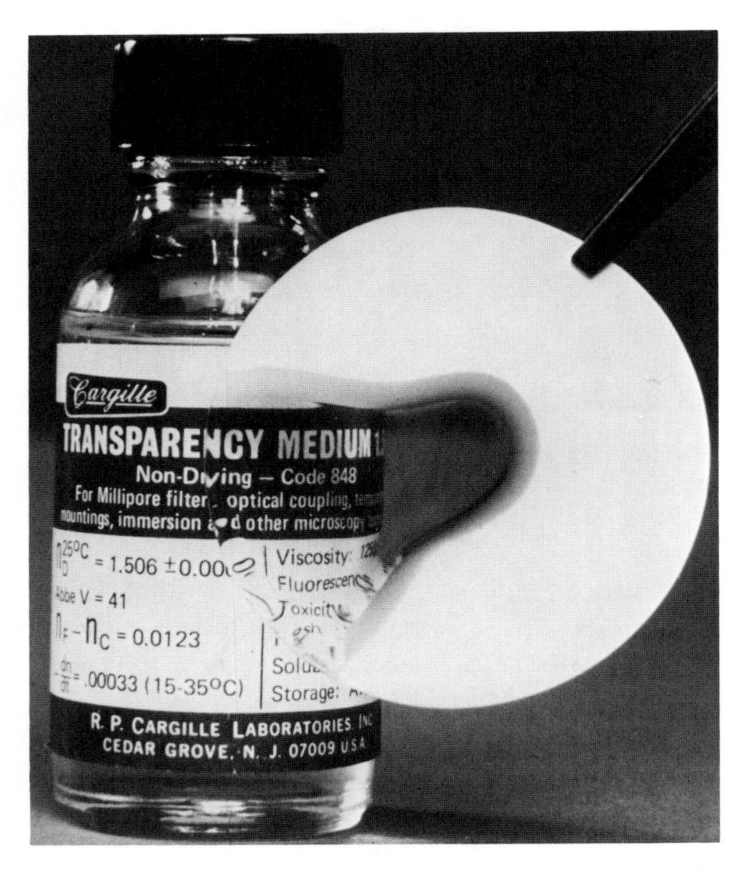

FIGURE 2 Membrane filter rendered transparent: membrane saturated with immersion oil $N_r = 1.515$. (Photo courtesy of Millipore Corporation, Bedford, Mass.)

definition (Subsection D19.08.04): "A thin nonfibrous filtration medium for fluids, having a mean pore size larger than 0.01 micrometer in diameter, on which particles larger than the rated pore size are retained at or near the delivery surface."

While the membrane filter, by virtue of its certified controlled pore size, is responsible for the sterilization of liquids by filtration, the bubble point test method can be employed to provide additional assurance of complete bacteria removal. The bubble point test method determines maximum pore size by measuring the minimum air pressure required to force a wetting liquid through the filter pores.

B. Pore Size

The pore size of porous media used in filtration refers to the effective openings in the media through which liquids or gases flow. Suspended particles that are larger than the pore size are unable to pass through. Since these openings are not usually circular in shape, the use of pore "radius" or "diameter" to describe the opening is misleading. Pore size distribution refers to the variety of pore sizes usually present in a filtration medium. The pore size distribution is best measured by physical structure determination using the mercury intrusion technique (Honold and Skau, 1954) and the mean flow pore size measurement.

II. METHODS OF MEASURING PORE SIZE

Pore size may be measured directly by various physical tests provided that contour irregularity is taken into consideration and accounted for in calculations and measurements. Pore size can also be determined when particles of known size have been retained by the membrane filter. The pore size determination by function measurement may be affected by factors other than direct screening. Random entrapment, adsorption, and electrostatic attraction tend to increase retention efficiency and, therefore, indicate a pore size smaller than that measured by the mercury intrusion or bubble point test.

A. Mercury Intrusion Method

The mercury intrusion porosimeter method (Honold and Skau, 1954) is based on determining the external pressure required to overcome the forces of surface tension, which tend to prevent the entrance of the nonwetting liquid mercury into the pores of a filter medium. Although individual pores cannot be measured by this technique, a pore size distribution can be obtained by immersing the test filter medium in a reservoir of mercury and applying pressure to the mercury in increments. The pressures at which mercury is forced into the pores of a filter can be determined by observing the apparent volume change of the mercury as pressure increases. Pore size distribution can then be calculated as follows:

$$\text{P.S.} = K\left(\frac{4\gamma \cos \theta}{p}\right)$$

where P.S. = pore size, μm

K = pore radius morphology constant

γ = surface tension of mercury at test temperature (25°C)

θ = contact angle between mercury and filter media

p = pressure

The morphology constant K would be 1 for a circular pore and less than 1 for an elliptical or irregular-shaped pore. This pore shape factor dictates the need for other measurements to supplement the intrusion data. Furthermore, a few large pores in a membrane filter would not be detected using the intrusion measurement. One additional problem occurs with certain types of filter media: If the medium is very compressible, the apparent change of volume will be larger than it should be and an erroneous pore size distribution curve will result.

B. Bubble Point Method

The bubble point method provides a means for measuring the largest pore of a porous media. By measuring air flow vs. pressure with a prewet filter, a pore size distribution may be obtained. ASTM Method F316-70 presents procedures for determining maximum pore size and pore size distribution of membrane filters. The method described can also be applied to a wide variety of porous media other than membrane filters. The apparatus employed for bubble point testing of membrane filter cartridges is shown in Fig. 3.

To determine maximum pore size, the filter is completely prewet with a liquid of known surface tension. By measuring the minimum air pressure required to overcome capillary attraction and force

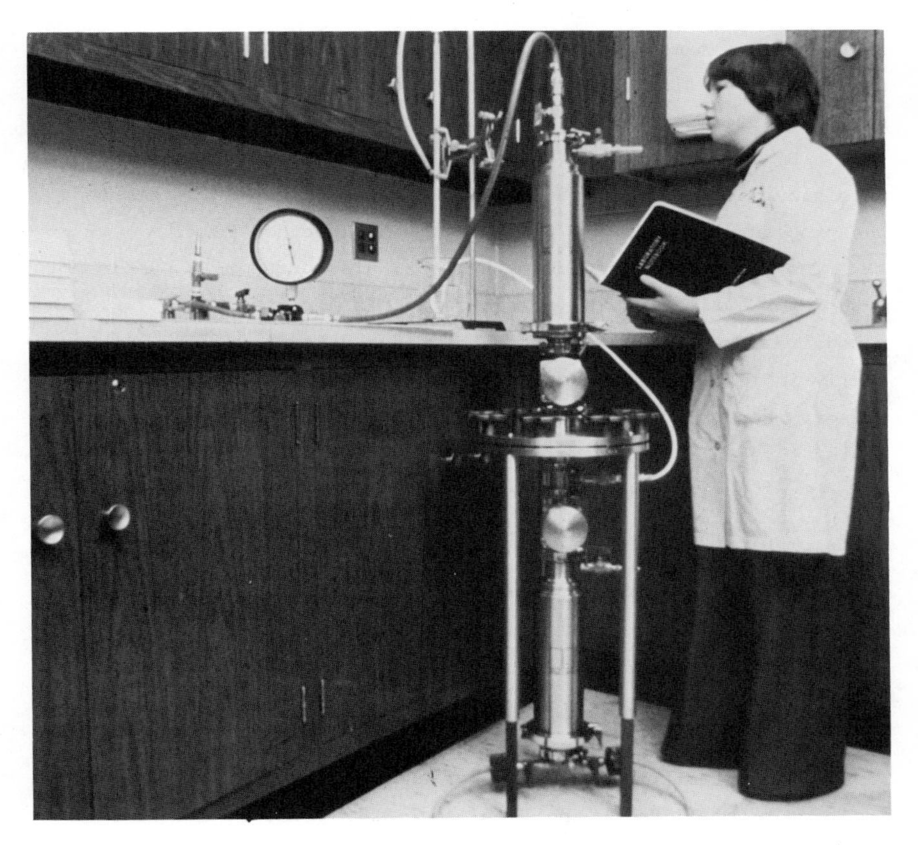

FIGURE 3 Bubble point testing, showing apparatus used for testing maximum pore size of cartridge membrane filters used for sterilizing filtration. (Courtesy of Millipore Corporation, Bedford, Mass.)

this liquid from the pores, maximum pore size can be calculated from the capillary rise equation:

$$p = \frac{2\gamma \cos \theta}{r}$$

However, this equation assumes circular pores; since this is not the case with most filter media, a capillary morphology correction factor must be applied.

The mean pore flow is determined in the second part of this procedure by comparing the air flow rates of a wet filter and a dry filter of the same composition at the same pressures. The percentage of flow passing through filter pores larger than or equal to a specific size may then be calculated from the capillary rise equation. By measuring change in air flow at small pressure increments, pore size distribution and mean flow pore size may be determined. The procedure details are presented in ASTM Method F316.

The older ASTM procedure, E128-61, also presents a bubble point method for evaluating the pore size of rigid porous media. E128 recognizes that the maximum pore diameter determined does not necessarily indicate the physical dimensions of the largest pore in the filter. Because of pore shape irregularity, the filter may be expected to retain particles much smaller than the maximum diameter determined by the bubble point.

C. Passage Testing

Because of the irregular pore shapes, neither ASTM standard allows for exact pore size calculation. However, data from test methods can provide a general description of filtration characteristics for the medium, and the integrity of the filter in its housing can be assured.

Passage testing with viable organisms provides the optimum method for establishing the absolute cutoff point of a filter medium. Pure bacteria cultures are easily prepared, and the size of the organism may be readily determined (Rogers and Rossmoore, 1969). The test provides data on absolute retention of organisms larger than the maximum pore size as well as nominal percent passage data for viable "particles" that are smaller than the rated pore size.

III. STANDARDIZATION REQUIREMENTS

A. Filter Composition and Structure

Numerous membrane filters in a wide variety of materials are currently available. To assure consistent performance in analytical and process applications, standardizing membrane filters by employing standard test methods has become increasingly important.

Various polymeric materials are used to produce synthetic membrane filters. Of these materials, cellulose nitrate, polytetrafluoroethylene, and polyvinylidene fluoride are among the more widely known. In using the membrane filter for microbiological analysis, many aspects of its structure and properties are significant for proper microbial retention, growth, and identification (Geldreich, 1977). Methods of evaluating the important properties of membrane filters and their significance are presented in Table 1.

TABLE 1 Properties of Membrane Filters Requiring Standardization or Qualification

Property	Method of measurement	Significance
Pore size distribution	Hg intrusion porometer, bubble point, microscopy, electron microscopy	Most critical property of the membrane filter
Surface pore size	Electron microscopy (see Fig. 4)	Bacteria cultivation
Retentivity	Passage testing	Sterile filtration
Bacteria growth	Fecal coliform culture	Bateria recovery
Bacteria inhibition	Fecal coliform culture	Measuring inhibition
Sterility	*U.S. Pharmacopeia* sterility test	Sterile filtration
Flow rate	Water flow--air flow test	Practical filtration

TABLE 1 (Continued)

Property	Method of measurement	Significance
Thickness	Thickness Gage--micrometer	Affects filter strength and bacteria growth
Strength	Microtensile tester	Physical stability
Temperature stability	Autoclave test	Heat stability during sterilization
Wettability	Water-wetting pattern	Media feed
Refractive index	Comparative Refractometer	Rendering filter transparent
Percent porosity	Thickness and weight measurement	Filter life and flow rate
Solvent resistance	Compatibility testing	Filter stability
Extractable components	Extraction and instrument analysis	Purity--filtrate and residue
Residual ash	Combustion and gravimetric analysis	Chemical and instrumentation analytical methods
Grid marking	Microscopy and coliform growth	Noninhibitory and proper dimensions

Pore size distribution, maximum pore size, and surface pore size are important characteristics in respect to completely removing bacterial organisms from the suspending fluid for bacterial analysis and enumeration and for sterilizing filtration. For direct microscopic examination (Ehrlich, 1955) or for culture analysis, the surface retention of organisms is imperative. For culture techniques such as fecal coliform growth, semisurface retention is necessary to assure optimum colony growth (Sladek et al., 1977). Specific pore size is also critical to assure proper nutrient feed to the retained organisms by capillarity. Whereas pores that are too small will inhibit nutrient flow to the organism, pores that are too large will cause surface flooding, which results in spreading colonies and overgrowth. Retention is a function of

FIGURE 4 Electron photomicrograph of membrane filter, showing membrane structure at 50,000x magnification. (Courtesy of Millipore Corporation, Bedford, Mass.)

maximum pore size and is the determining factor in using filtration as a means of sterilizing fluids. Passage testing to determine retentivity of membrane filter cartridges is shown in Fig. 5.

Bacterial growth and inhibitory characteristics of membrane filters are significant factors. The presence of inhibitory substances in the membrane material will retard or prevent growth of sensitive or injured organisms. A hydrophobic structure will retard media feed, and proper colony growth will be prevented.

To ensure that the organisms on the filter came from the test sample and not from the filter itself, membrane filter sterility is

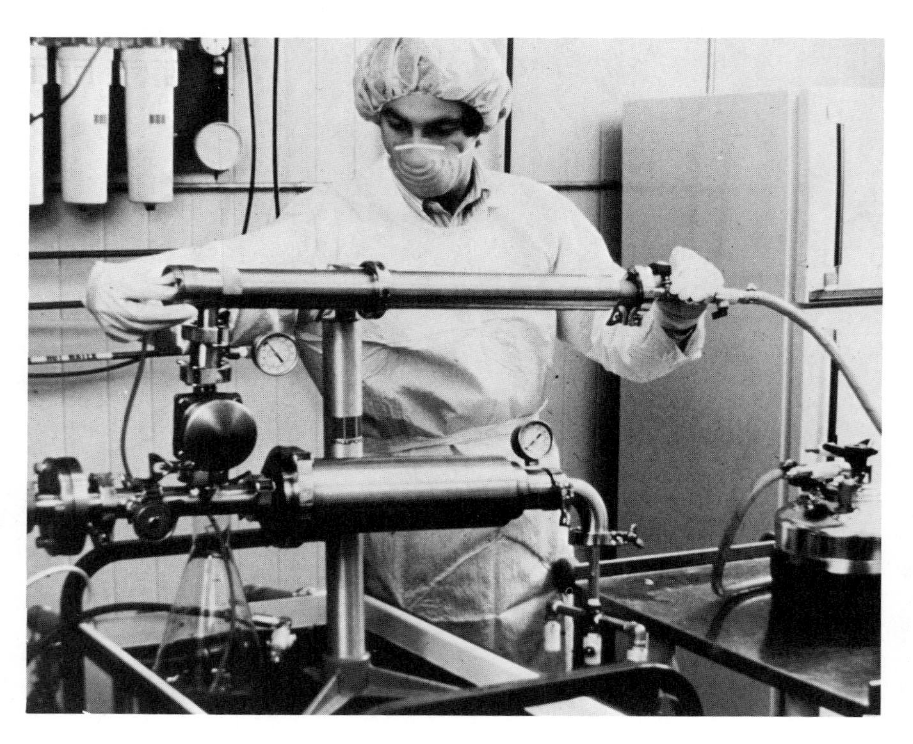

FIGURE 5 Passage testing, showing apparatus used for conducting bacterial passage tests on membrane filter cartridges. (Courtesy of Millipore Corporation, Bedford, Mass.)

important. When sterilizing fluids by filtration, sterility must also be assured on the downstream portion of the filter. Since filters are often sterilized by autoclaving, they must be capable of withstanding autoclave temperatures and conditions. The ability to withstand gas sterilization and radiation sterilization without damaging the filter structure and without permanently adsorbing the sterilizing gas is also important.

B. Physical Properties

The physical characteristics, flow rate, thickness, and strength are important in providing a satisfactory filtration device. If the flow rate is too slow, the tests will take too long and become impractical for routine sample analysis. Flow rates that are higher than normal indicate either filter rupture or a mean pore size that is too large for proper retention of bacteria. Thickness specifications are critical: A membrane that is too thin may have insufficient strength, and a membrane that is too thick will retard nutrient feed and inhibit bacteria growth. A filter with insufficient strength may break or rupture during handling or filtration.

Temperature stability is required for membranes to withstand steam sterilization and the filtration of hot fluids. For nutrient media to feed through the membranes to all the retained organisms, uniform wettability must be maintained. Nonwetting spots on the membranes will prevent proper bubble point testing, and nonwetting filters will not permit filtration of aqueous fluid unless high filtration pressures are employed.

Control of the refractive index is important when membrane filters must be rendered transparent in order to examine organisms on the filter surface directly using transmitted light and straining techniques. The percent porosity has an effect on flow rates, filter clogging rate, and nutrient media feed. Too high a porosity will result in a filter structure that is too weak to withstand routine handling and filtration procedures.

C. Chemical Properties

Solvent resistance, extractables, and residual ash are important chemical characteristics of membrane filters. Of these characteristics the most important is the extractables component, which can either inhibit or enhance bacteria growth and give false-negative or false-positive results. When culturing microorganisms with selective media as in the coliform test, membrane additives such as glycerol can produce an apparent sheen in a noncoliform colony, which will result in a false-positive count. The membrane structure should be inert so that the membrane itself neither suppresses nor enhances bacterial growth. The resistance or nonresistance to solvents is important in certain procedures where the membranes must either resist damage by the solvent used or be dissolved by the solvent. In certain gravimetric analytical procedures, the ash weight of the membrane must be negligible to prevent interfering with analytical results.

Finally, the grid marking on analytical membrane filters must be of proper dimensions to provide correct statistical data, must be noninhibitory to bacterial growth, and must not leach materials into the body of the membrane.

IV. MICROBIOLOGICAL APPLICATIONS

As membrane filters are widely used for various microbiological applications, there is a definite need for standardization and control.

A. Analytical Microbiology

1. Detecting indicator organisms of sanitary significance
2. Diagnostic microbiology
    a. Identification of disease-producing organisms
    b. Antibiotic sensitivity testing (Millipore Corporation, 1970)
3. Research microbiology--studying cell growth and propagation
4. Sterility testing (Frediani, 1964; Bowman, 1966)
5. General microbiological research

B. Process Biological Filtration

1. Sterilization by filtration (Schaufus, 1962)
2. Process microfiltration--water purification (Dwyer, 1968)
3. Beverage stabilization (cold pasteurization of beer) (Bush, 1956)
4. Purification of clinical laboratory fluids (Russell, 1969)

V. STANDARDS ORGANIZATIONS

There are many national and international organizations involved in the standardization of membrane filters. These organizations include the following:

ASTM	American Society for Testing and Materials
ANSI	American National Standards Institute
NCCLS	National Committee for Clinical Laboratory Standards
USP	*United States Pharmacopeia*
ISO	International Standards Organization
US-EPA	Environmental Protection Agency
DMSC	U.S. Armed Forces Medical Procurement Agency Defense Medical Supply Center
APHA-AWWA	Standard Methods Committee of American Public Health Association and American Water Works Association

These organizations are responsible for establishing standard test methods for determining the presence and type of microorganisms in a wide variety of fluids. As the membrane filter technique is widely used for these analyses, the organizations have established and are working on a number of test methods for standardizing and controlling the quality of membrane filters used in these test procedures.

Copies of the test methods established by the standardization organizations are available from the organizations listed above. A list of titles and committee code number designations that will enable you to obtain copies of membrane filter tests follows. Although the methods listed are not presented in complete form, the scope of each test method is described briefly.

A. ASTM Test Methods

1. D3508-78: *Evaluating water testing membrane filters for fecal coliform recovery.* The membrane filters are compared with spread plate cultures for recovery and growth of fecal coliform from untreated water and sewage samples using mFC agar medium at 44.5 ±0.2°C incubation temperature.

2. F316-70: *Pore size characteristics of membrane filters for use with aerospace fluids.* This method employs a procedure for determining the maximum pore size and the mean pore size of a membrane filter by measuring the initial bubble point and air flow vs. pressure through a liquid wet filter.

3. F317-72: *Liquid flow rate of membrane filters.* The liquid flow capability of membrane filters is determined by measuring the time required to filter a known volume of liquid at a specified temperature and differential pressure.

4. D3863-80: *Determining the retention characteristics of 0.4-0.45 μm membrane filters used in routine filtration procedures for the evaluation of microbiological water quality:* This method involves filtration of a suspension of *Serratia marcescens* organisms. A routine sterility test is then used to determine the presence of the test organism in the filtrate.

5. D3862-80: *Test for the retention characteristics of 0.2 μm membrane filters used in routine filtration procedures for the evaluation of microbiological water quality:* The method parallels the procedure described above for the 0.4–0.45 μm membrane. However, a suspension of *Pseudomonas diminuta* that is slightly larger than the 0.2 μm pore size rating is used as the test organism.

6. D3862-79: *Test method of determining the quantity of extractable matter in membrane filters:* The test involves gravimetric analysis of the water-extractable materials in the membrane filter. The preweighed filter is extracted in boiling reagent-grade water for 30 min, dried, and reweighed.

B. ASTM Proposed Membrane Test Method

1. *Evaluating inhibitory effects of membrane filters using the test organism Escherichia coli at 44.5°C:* In this test a comparison is made between the recoveries obtained by membrane filtration vs. spread plate culture of a known suspension of *E. coli* at 44.5 ±0.2°C.

C. ANSI Test Methods

ANSI has established a working group, TC 147-N82, to provide test methods and specifications for standardization of membrane filters used in water quality studies that are related primarily to microbiological applications.

D. ISO--Test Methods

A new working group WG9 has been formed within ISO TC 147-SC4. The Secretariat of this working group, "Membrane Filters," is held by Canada and chaired by B. J. Dutka. Two methods are at the Draft Proposal stage: (1) a method for evaluating membrane filters used for microbiological examination of water quality, and (2) a test method for confirming the sterility of membrane filters.

E. US-EPA

Laboratory certification requirements specify that membranes must be certified for microbiological analysis. Certification for each manufactured lot includes testing for pore size, flow rates, extractables, trace metals, phosphates, ammonium salts, organics, sterility, retention, and recovery of coliform organisms.

F. DMSC--Defense Medical Supply Center

Federal Specifications on Membrane Filter, Bacteriological/Particulate, NNN-D-00370 (DSA-DM)

The specification covers classification and quality control of microporous membrane filters in 12 pore size categories ranging from 0.01 to 8.0 μm. In addition to specifying pore size and membrane filter thickness, the following parameters are covered:

1. Percent porosity
2. Refractive index
3. Percent extractables

4. Bubble point
5. Water flow rate
6. Air flow rate
7. Retention of specific microorganisms
8. Diffusibility
9. Toxicity to bacteria
10. Autoclave stability
11. Resistance to ethylene oxide (ETO) gas sterilization
12. Sterility

The membrane filter qualification tests described in Federal Specification NNN-D-00370 have become the most relied upon evaluation parameters for qualifying membrane filters used in microbiological testing and research. Specification NNN-D-00370 is available from the U.S. Defense Medical Supply Center, 3rd Avenue and 29th Street, Brooklyn, New York 11232. The following is a brief description of its contents:

1. Percent porosity is determined by measuring the volume and weight of the filter and calculating as follows:

$$\text{\% Porosity} = \frac{\text{theoret. wt. of solid disk - actual filter wt.}}{\text{theoretical wt. of solid disk}} \times 100$$

   (Theoretical wt. of solid disk is calculated from thickness, diameter, and density of the filter.)

2. Refractive index is determined by testing filters with immersion oils of varying refractive indices. If the refractive indices of the oil and filter are different, a line called a Becke line will appear at the interface. Observing the movement of the Becke line under a microscope at 100X magnification aids in determining the refractive index of the filter.

3. Percent extractables are measured by determining weight loss of the filter after a 20-min immersion in boiling reagent-grade water and drying at 70°C for 60 min.

4. Bubble point and maximum pore size are determined in accordance with ASTM Method E128-61. Method E128-61 is similar to the F316 bubble point method previously described.

5. Water flow rate is determined with a 47 mm stainless steel vacuum filtration apparatus at 25°C and 70 cm Hg pressure differential.

6. Air flow rate is determined at 25°C and 70 cm mercury differential pressure with an air flow meter.

7. Organism retention is determined using five different pore size filters (listed in the table) by passing a saline suspension of the specified organism through the sterilized test filter held in a sterilized membrane holder.

Pore size (μm)	Organism	Culture medium
0.22	*Vibrio percolans*	Tryptone glucose extract agar
0.30	*Serratia marcescens*	Wilson's peptone agar
0.45	*Serratia macescens*	Wilson's peptone agar
0.65	*Bacellus subtilus var globigie*	Tryptone glucose extract agar
1.2	*Saccharomyces cerevisiae*	Tryptone glucose extract agar

The filtrate is then passed through a reference filter of the same pore size and cultured on tryptone glucose extract (TGE) agar or Wilson's peptone agar as indicated. Both the test filter and reference filter are incubated at 32°C for 24 hr (48 hr for the 1.2 μm filters) and then examined for passage or edge leakage.

8. Diffusibility is determined by placing the test filter flat on clean water and measuring the wetting time while checking for nonwetting areas.

9. Toxicity test. The following tests are performed on the different pore size filters:

   a. *8.0 μm, 5.0 μm, 3.0 μm, 0.10 μm, 0.05 μm, 0.01 μm:* Prepare a spread plate culture of *Serratia marcescens* on solidified Wilson's peptone agar in a 100 mm petri dish. Place the test filter section on a seeded agar surface, incubate for 24 hr at 32°C, and examine for a zone of inhibition around the filter. Filters are considered nontoxic if the filter recovers at least 90% of the spread plate counts.
   b. *1.2 μm, 0.8 μm, 0.65 μm:* Compare the recovery of *Saccharomyces cerevisiai* retained on the filter with a spread plate culture of the same test sample, both cultured on TGE agar at 32°C for 48 hr.
   c. *0.45 μm, 0.3 μm, 0.22 μm:* Compare the recovery of *S. marcescens* retained on the filter with a spread plate culture of the same organism test sample, both cultured on Wilson's peptone agar at 32°C for 24 hr.

10. Autoclave stability. Filters in protective autoclave packaging are autoclaved for 10 min at 121°C and then examined for cracking or dimensional changes as specified in NNN-D-00370.

11. Resistance to ethylene oxide gas sterilization. Expose the test filters to 20% ETO/80% CO_2 gas mixture for 4 hr at 18 psi (124 kPa) and 50°C with the relative humidity at 35%. They should then be examined for structural stability and post-treatment performance.
12. Sterility. Asceptically place the test filter on an absorbent pad saturated with TGE broth in a petri dish. Invert the test filter and incubate it for 24 hr at 35°C. Absence of colony growth on the filter indicates sterility.

VI. SUMMARY

The membrane filter characteristics described in this chapter relate to critical requirements for microbiological applications. Performance criteria have been established that relate to these properties for various filter uses. Membrane filters that have the same rated pore size designation are not necessarily the same in technical performance.

The standardization of filter parameters will result in fewer false positives and negatives and will ultimately lead to more precise microbiological test results. The parameters that are important to filter performance have been individually described, and test methods for evaluating filter properties have been detailed. Since specifications on various critical parameters are usually given with the specific application, they have not been included in this chapter.

REFERENCES

ASTM (1961). Standard method of test for maximum pore diameter and permeability of rigid porous filters for laboratory use. E128-61. Philadelphia.

ASTM (1970). Standard method of test for pore size characteristics of membrane filters. F316-70. Philadelphia.

Bowman, F. W. (1966). Application of membrane filtration of antibiotic quality control sterility testing. *J. Pharm. Sci.* 55 (8):

Brodsky, M., and Schiemann, D. (1977). A comparison of membrane filters and media used to recover coliforms from water. EPA-600/9-77-024 and ASTM *Symposium on the Recovery of Indicator Organisms Employing Membrane Filters*, September, pp. 58–62.

Bush, J. H. (1956). Beer stability and controlled filtration. *Brew. Dig.*, February, pp. 48–51.

Dwyer, J. L. (1968). Control techniques in high purity water production. *Bull. Parenteral Drug Assoc. 22*(6): 267–275.

Ehrlich, R. (1955). Technique for microscopic count of microorganisms directly on membrane filters. *J. Bacteriol. 70*(3): 265–268.

Frediani, H. (1964). Membrane filter sterility testing. *Bull. Parenteral Drug Assoc. 18*(1): 25–27.

Geldreich, E.E. (1977). Performance variability of membrane filter procedures. EPA-600/9-77-024 and ASTM *Symposium on the Recovery of Indicator Organisms Employing Membrane Filters,* September, pp. 12–19.

Green, B., Clausen, E., and Litsky, W. (1975). Comparisons of the new Millipore HC with conventional membrane filters for the enumeration of fecal coliform bacteria. *Appl. Microbiol. 30* (4): 697–699.

Honold, E., and Skau, E. L. (1954). Application of mercury intrusion method for determination of pore size distribution of membrane filters. *Science* 120 (3124): 805–806.

Millipore Corporation (1970). Antibiotic sensitivity testing with the Millipore Zone Analyzer. Millipore AR 140. Bedford, Mass.

Rogers, B., and Rossmoore, H. (1969). Determination of membrane filter porosity by microbiological methods. *Proc. Soc. Ind. Microbiol.*, vol. II, August, pp. 453–459.

Russell, J. H. (1969). Millipore techniques for the clinical laboratory. *World Med. Instrum.*, December, pp. 6–13.

Schaufus, C. P. (1962). Further developments in Millipore sterile filtration techniques. *Bull. Parenteral Drug Assoc. 16*(5): 1–5.

Schaufus, C. P., and Krabek, W. (1955). Direct staining of microorganisms collected on the Millipore filter. *Bacteriol. Proc.*, p. 20.

Schillinger, J., Mcfeters, G., and Stuart, D. (1977). Comparison of membrane filters in recovery of naturally injured coliforms. EPA-600/9-77-024 and ASTM *Symposium on the Recovery of Indicator Organisms Employing Membrane Filters,* September, pp. 64–66.

Sladek, K., Suslavich, R., Sohn, B., and Dawson, F. (1977). Optimum structures for growth of fecal coliform organisms. EPA-600/9-77-024 and ASTM *Symposium on the Recovery of Indicator Organisms Employing Membrane Filters,* September, pp. 46–67.

3

MEMBRANE FILTER TECHNIQUES FOR TOTAL COLIFORM AND FECAL COLIFORM POPULATION IN WATER

EDWIN E. GELDREICH *U.S. Environmental Protection Agency, Cincinnati, Ohio*

I. INTRODUCTION

A major objective in water microbiology has always been the development of rapid, accurate, and uncomplicated techniques for the isolation, identification, and enumeration of primary indicator bacteria in water. With the advent of the membrane filter (MF) and associated microbiological techniques, many of these idealistic goals are being realized. The uniqueness of the membrane filter is that it permits the entrapment of particles, including microorganisms, on or near the surface of the filter during rapid passage of the sample. Following filtration, the membrane filter with entrapped bacteria may either be stained and examined microscopically or placed on a selective nutrient substrate, incubated at the desired temperature, and then examined hours later for the development of colonies from viable cells. Microcolonies developing after 2 hr of incubation may become visible with the aid of magnification in a few hours on an appropriate medium and in response to an optimum incubation temperature.

Cultivation of bacteria on the membrane filter permits the rapid transfer of the isolates from one medium substrate to another without creating an additional lag time in the growth curve. Thus, bacterial cultivation may be programmed to start with an enriched medium for optimum recovery of stressed organisms prior to transfer to a selective medium for differentiation of colonies. This sequential cultivation could also include the transfer from a differential medium to a selected biochemical substrate for further reaction and identification. Cultivation may also be controlled during a 24–72 hr transport of membrane filter cultures to the laboratory for final processing by use of a minimal growth medium. Upon arrival at the laboratory the culture is then placed on a growth substrate that allows differentiation of the indicator population. These properties of the membrane filter technique plus its adaptability to field use requiring limited space, equipment, and expendable materials have made membrane filter methodology a very useful tool for total coliform and fecal coliform analyses of samples from a variety of water and effluent sources.

II. TECHNICAL CONSIDERATIONS

Entrapment of bacteria during filtration followed by cultivation on the membrane filter surface results in some marked advantages over the multiple-tube procedure and the most probable number calculation. Flexibility in choosing an appropriately large test sample portion is a primary advantage when using the membrane filter technique. For example, properly treated potable water, protected groundwater, and swimming pools are expected to contain less than one indicator organism per 100 ml. The search for pathogen contamination of any high-quality waters and the large dilution factor involved would also necessitate the examination of large sample portions ranging from 1 to several liters. By contrast, examination of potable water by the multiple-tube procedure is generally restricted to testing a 50 ml sample that is divided into a series of five tubes of presumptive lactose fermentation medium, each containing 10 ml portions of the water. Few laboratories routinely inoculate 100 ml test portions in the multiple-tube procedure for drinking water samples because of the problems of preparing, handling, and incubating bottles large enough to culture 100 ml sample test portions. Consequently, the search for low-level coliform densities in high-quality waters is limited to detecting two organisms per 100 ml when the multiple-tube procedure is used. By using the MF procedure, the minimal detection level can be extended to one organism per liter or per 5 liters provided that sample turbidity does not exceed one nephelometric turbidity unit (NTU). Although test volume of this magnitude is desirable to measure the efficiency of treatment for optimum removal of bacteria of sanitary significance (Akin et al., 1975), practical constraints related to the cost of transporting samples of this size have dictated the acceptance of a 100 ml drinking water sample size for analysis by the membrane filter procedure.

The membrane filter procedure provides a direct count of organisms that are entrapped on or near the filter surface (Fig. 1) and are capable of growth on the culture medium provided. Direct counts of organisms may also be obtained by use of pour plates and spread plates, but their usefulness is limited to examination of a maximum of 2 ml sample (pour plate) or 0.5 ml (spread plate) and are therefore unacceptable for measurements of indicator bacteria at very low concentrations. The only alternative test procedure would be the multiple tube method and the population estimate provided by the most probable number (MPN) concept. In this approach, a statistical estimation of the true bacterial density in the sample is based on the combination of positive and negative tube results occurring in the test and is subject to a variability ranging from 31 to 289% of the absolute value, when using the five-tube test. Furthermore, the table of most probable numbers for the five-tube test, when developed included a 23% positive bias to compensate for statistical probabilities below the true sample density. Obviously, the MPN value is not a precise measurement and will often yield results 20–25% higher than parallel MF values determined from a direct count.

Rapid detection of changes in microbial quality of water is a major objective in every monitoring program. Although instantaneous

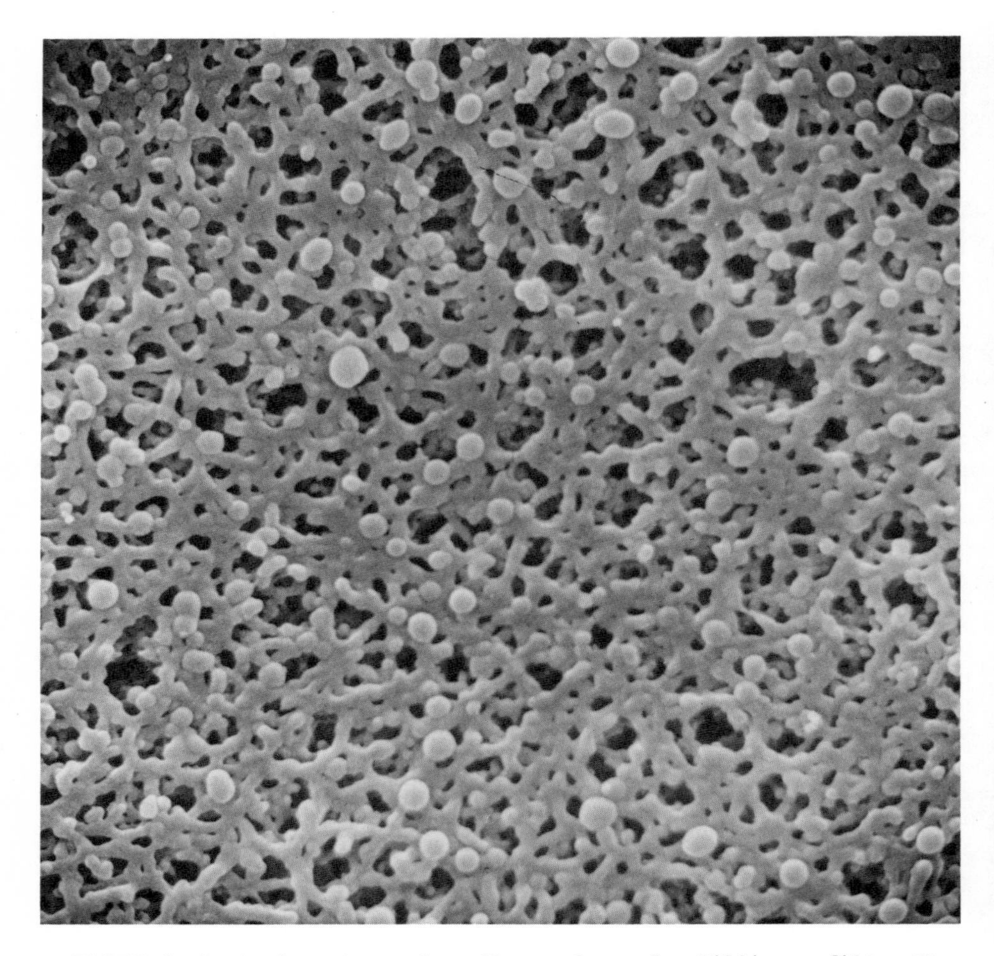

FIGURE 1 Bacteria entrapped on the surface of a Millipore filter HA type (0.22 μm) membrane filter. (Magnification approximately 4000X).

quantitative measurements for selected microbiological indicators are not yet a reality, a substantial breakthrough in sample processing time can be achieved through application of conventional membrane filter procedures as compared to equivalent multiple-tube procedures. For example, the routine total coliform analysis required in potable water monitoring may be concluded within 22 hr when using the MF technique but the multiple-tube procedure requires a minimum of 48 hr to demonstrate no coliform occurrence. Even more rapid detection (7 hr) of fecal coliforms in contaminated potable water and recreation waters is possible using conventional membrane filter equipment, incubation at 41.5°C and a lightly buffered lactose medium (Reasoner et al., 1979).

A. Membrane Filter Limitations and Solutions

Although the membrane filter offers substantial advantages for examination of the microbial quality of water, some sample characteristics may adversely affect the use of any membrane filter procedure and necessitate the use of the less desirable multiple-tube procedure.

Particulates in water can place a limitation on the size of the sample that can be analyzed by the MF procedure. Water samples from treated public water supplies will rarely present a problem of excessive turbidities until the sample size exceeds 1 liter. These turbidities may result from sediment movement in the distribution system and in some instances by passage of inadequately treated source water containing suspended particulates introduced by stormwater runoff and algal blooms. Groundwater turbidities are frequently chemical in nature but may be caused by growth of iron and sulfur bacteria. Turbidity increases in polluted waters are frequently accompanied by large increases in the bacterial densities; thus, dilution of the sample to achieve a satisfactory bacterial density for enumeration also minimizes turbidity. However, waters with algal blooms, raw source waters stored in settling basins, and poor-quality effluents that have been disinfected may present some problems in MF analysis. In such waters, indicator bacteria densities are generally low, requiring 10–100 ml of sample to achieve a statistically meaningful bacteriological result. Unfortunately, filtration is often impossible because of the presence of algal cells, gelatinous detritis, and finely divided silt that clog the membrane filter pores and interfere with discrete colony formation. Experience has indicated that prefiltering a turbid water sample through a course filter or decanting the supernatant from settled turbidity in a water sample results in a loss of organisms. This loss ranges unpredictably from 20 to 80%. A guide to acceptable turbidity limits for MF filtration cannot be offered because this varies with the type of particulate (sand, clay, insoluble phosphates, iron precipitates, and algal masses). Therefore, the microbiologist must judge the suitability of the MF method for examination of a given water sample based on experience. In general, visible turbidity in a potable water sample will preclude filtration of 100 ml. In these instances, filter two 50 ml portions and total the results for a 100 ml quantity. Where turbidity problems are unsurmountable, the examination must be performed by the multiple-tube procedure.

Considerable emphasis must also be placed on the proper selection of test volume to avoid overcrowding the membrane filter surface with bacterial colonies. Dense growth of colonies on nonselective media may result in confluency and will prevent discrete colony formation. On a selective medium, dense growth will result in loss of discrete colony formation and poor differentation of colonies. Although membranes with an effective filtering diameter of 47 mm are standard, the use of membrane filters with larger effective filtering areas would be one way to disperse the bacterial growth and prevent confluent growth (Fig. 2). Unfortunately, this approach is not practical because special filtration apparatus would be required to accommodate a membrane filter of larger surface area. The most practical approach is to reduce sample size to obtain a countable range of discrete colonies and thereby improve the statistical accuracy of the result. In general, sample volumes that yield membrane filter cultures with 20—60 colonies provide acceptable, statistically accurate results. Whenever the average colony size is less than 2 mm, the acceptable count range may be extended to approximately 20—80 colonies per membrane. Another limiting factor is the magnitude of background growth (i.e., the number of nonindicator colonies appearing in addition to total coliforms or fecal

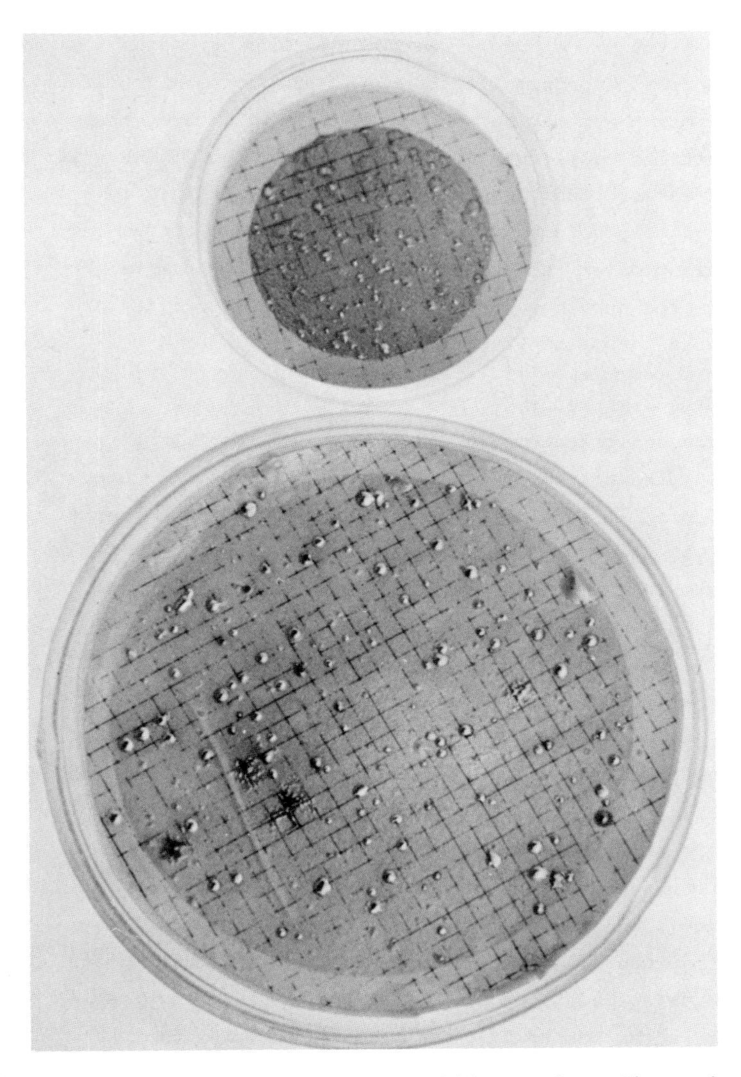

FIGURE 2 Turbidity and total coliform colony dispersion resulting from filtration of 1 ml of Ohio River water through the standard membrane filter (47 mm) and a larger membrane (76 mm diameter) used in specially designed filtration apparatus.

coliforms in a given MF culture). Although m-Endo MF medium was designed to suppress approximately 95—99% of the nontotal coliform bacteria, at times there may be sufficient breakthrough of the background growth to cause interference with coliform differentiation. In such cases, countable membranes must be obtained by filtering smaller sample volumes that still contain some total coliforms. This problem is less frequently encountered with the fecal coliform procedure because of the greather selectivity provided by elevated incubation temperature.

B. Stressed Organisms Recovery

Recovery of stressed organisms in chlorinated effluents, saline waters, and polluted natural waters (especially those with heavy metal ions or certain organic industrial wastes) can be a significant problem that is more readily apparent when using the membrane filter procedure than with either the multiple-tube test or pour plate method. Details on stressed organism recovery are given in Chap. 16.

Solutions to the problem of optimizing recovery of stressed organisms vary with the type of sample and state of stress. For chlorinated samples, sufficient dechlorinating agent must be present in the sample bottle to quickly stop disinfection action. Samples known to contain heavy metal pollutants should be collected in a sample bottle containing a chelating agent [372 mg of edetic acid (EDTA) per liter], and sample storage time must be kept to a minimum. Use peptone dilution water rather than phosphate-buffered water when preparing dilutions of samples containing heavy metal ions. However, after making dilution in 0.1% peptone dilution water, do not hold at room temperature for more than 30 min before making inoculations (Geldreich, 1975). This time limitation will prevent initiation of growth by cells in suspension.

III. BASIC MF FILTRATION PROTOCOL

Successful application of any MF procedure requires development of good laboratory operational practices (Fig. 3). Preanalysis activities include recording sample data in the laboratory log, disinfection of laboratory benchtop working area, and assembly of the necessary sterile filtration equipment and sterile materials (MFs, culture containers, pipettes, graduated cylinders, media, rinse water, etc.).

Prior to sample filtration, the appropriate medium is prepared either as a broth or agar. If broth is preferred, a sterile absorbent pad is placed in each culture dish, using sterile forceps.

FIGURE 3 Membrane filtration preanalysis preparation, which includes filtration assemblies with membrane filters in place on a manifold, UV sterilizer for filtration units, vacuum source, sterile culture dishes containing medium, 70% alcohol for forcep flaming, and sterile rinse water.

The amount of medium necessary to saturate an absorbent pad varies directly with pad thickness and degree of dryness, usually requiring from 1.8 to 2.2 ml. Pour any excess medium from the culture dish before rolling the membrane onto the absorbent pad. If the excess broth is not removed, flooding of the membrane may occur and cause confluent growth. Insufficient medium may result in development of small "starved" colonies. When an agar medium is employed, dispense 3–4 ml of the melted agar medium directly into each culture dish.

The filtration assembly should be sterile at the beginning of each filtration series. A filtration series is considered to be interrupted if an interval of 30 min or longer occurs between sample filtrations. Resumption of filtration after such an interruption requires that another set of sterile filtration units be used and is, therefore, a new filtration series. This protocol reduces the chance cross-contamination of a large series of samples from spills that may accidently occur during extended filtration operations. Rapid resterilization of the funnel by ultraviolet light exposure, flowing steam, or boiling water may be practiced between sample filtrations at the bench. To provide assurance of sterility of funnels and rinse water, a sterile rinse water blank (100 ml volume of in-use rinse water) should be filtered at the start and finish of each filtration series. If these controls indicate that contamination has occurred, all data on samples in the filtration series affected should be rejected and a request made for immediate resampling of those waters involved in the laboratory error.

A standard sample volume of 100 ml must be analyzed for all treated water supplies. Untreated water supplies (individual wells, springs, cisterns) may have excessive noncoliform bacterial populations that will necessitate the examination of two 50 ml portions per sample. Because the test is quantitative, all water sample volumes must be measured within a ±2.5% tolerance. The glass funnel 100 ml gradation mark may be used routinely to measure potable water sample volumes only after its accuracy has been verified. Although metal funnels may not have 100 ml marks impressed on the interior surface, marking the funnel interior wall at the 100 ml water level with a waterproof, heat-resistant ink or enamel provides a satisfactory substitute. The mark should not be etched into the metal wall with a diamond marker or other metal-cutting tool because this will prevent complete flushing of all residual bacteria from the stainless steel surface onto the membrane filter.

For the most accurate measurement of large sample volumes, use graduated cylinders. An individual, sterile, graduated cylinder or volumetric pipette should be assigned to each sample examined in the filtration series. Sample volumes can then be measured, poured into the funnel, and filtered. Rinse the graduated cylinder with a small portion of sterile rinse water (approximately 25 ml), then flush this water through the membrane filter being used for that sample. Follow this procedure with two separate short rinses (20—30 ml of sterile rinse water) to flush any residual bacteria from the funnel walls onto the MF surface. Rinsing the graduated cylinder and funnel before removing the MF ensures transfer of all bacteria in the sample to the membrane filter surface and prevents carryover of coliforms to the next sample.

Examination of natural bathing water, stormwater runoff, raw sewage, and treated sewage effluents for fecal coliforms requires the examination of a range of test volumes to obtain suitable fecal coliform densities (20—60 colonies) on the membrane. Therefore, appropriate sample volumes of 0.1 ml or smaller must be prepared from the original sample by use of serial dilutions in either phosphate-buffered water or peptone dilution water. The original sample is shaken vigorously to obtain an even distribution of bacteria in suspension, using care to secure the screw cap to prevent leakage during shaking. Immediately withdraw 1 ml of sample and add this quantity to a 99 ml sterile dilution blank (1:100 dilution). After shaking this dilution blank vigorously, 10 ml or 1 ml portions may be filtered for examination of 0.1 ml and 0.01 ml of the original sample. If necessary, withdraw 1 ml of this first dilution into a second dilution blank and shake vigorously. A 10 ml and 1 ml portion of the second dilution bottle then will correspondingly equate to 0.001 ml and 0.0001 ml of the original sample. To avoid uneven bacterial dispersion over the filtration surface and wide-ranging differences in replicate counts, do not attempt to filter original sample volumes or dilution portions smaller than 1 ml.

When the bacterial density of the sample is totally unknown, it is necessary to filter several decimal quantities of sample to obtain a countable density of coliforms. The best method is to estimate the ideal quantity expected to yield a countable membrane and use two additional quantities representing one-tenth and 10 times that quantity. Data in Table 1 may be used as a guide for selecting the appropriate volumes for various water and wastes.

Membrane filters are fragile and may be easily damaged by improper handling or storage. Grasp the outer part of the membrane filter (outside the effective filtering area) with sterile, smooth-tipped forceps. This procedure avoids smearing entrapped bacteria

TABLE 1 Suggested Guide for Fecal Coliform Filtration Quantities

	Quantities filtered (ml)						
Water source	100	50	10	1	0.1	0.01	0.001
Lakes, reservoirs	x	x	-	-	-	-	-
Wells, springs	x	x	x	-	-	-	-
Water supply, surface intake	-	x	x	x	-	-	-
Natural bathing waters	-	x	x	x	-	-	-
Sewage treatment; secondary effluent	-	-	x	x	x	-	-
Farm ponds, rivers	-	-	-	x	x	x	-
Stormwater runoff	-	-	-	x	x	x	-
Raw municipal sewage	-	-	-	-	x	x	x
Feedlot runoff	-	-	-	-	x	x	x

Source: Data from Geldreich (1975).

or the possibility of piercing the membrane filter surface and breaking its retention capabilities. Place the sterile membrane filter on the filter holder, grid side up, centered over the porous part of the filter support plate. To avoid damage to the membrane filter, the funnel should not be turned or twisted while it is being seated and locked to the lower element of the filter holder. Filter holding units featuring a bayonet joint and locking ring to join the upper element to the lower element require special care on the part of the operator. Turn this locking ring sufficiently to give a snug fit, but do not tighten excessively.

Immediately before filtering a measured sample, invert the sample and shake it vigorously. This vigorous shaking is needed to obtain a homogeneous distribution of suspended bacteria and is of particular concern with turbidity-laden waters. Turbidity in water settles rapidly, pulls suspended bacteria into the bottom sediment, and thereby creates an uneven distribution of the bacterial population.

After shaking the sample thoroughly, pour or pipette the measured sample volume into the funnel with the vacuum supply line connection turned off. To avoid uneven distribution of organisms over the effective filtering area, the vacuum should never be applied simultaneously with the addition of the sample test portion. When filtering 10 ml or less, add approximately 10 ml of sterile rinse water to the funnel before sample addition to ensure uniform dispersion of the bacterial suspension. Then apply the vacuum to rapidly draw the sample through the MF. Finally, rinse the funnel wall twice with 20–30 ml of sterile rinse water. Extensive tests have shown that with the proper rinsing technique, bacterial retention by the funnel walls is negligible.

Buffered dilution water or dilute peptone water (0.1%) used for rinse water in the MF procedure is often prepared in large flasks or carboys, autoclaved, and stored in the laboratory until needed. Because these containers may vary from 1 liter flasks to 20 liter carboys, the rinse water is generally dispensed by siphoning through glass, Teflon, or rubber tubing to the MF funnels or is poured into smaller, sterile wash bottles for ease of handling. Caution must be exercised that the siphoning devices and dispensing wash bottles do not become contaminated and, thereby, contribute microbial contamination to the filtration procedure. A single occurrence of heavy microbial growth in the rinse water can nullify the results of an entire days' water testing program by completely "masking" the membrane with noncoliform growths that interferes with coliform colony or sheen development. The use of a fresh sterile rinse water supply and dispensing system each day will avoid this contamination problem.

Upon completion of the rinse procedure, turn off the vacuum supply to the filtration assembly to avoid accidentally tearing the filter while transferring it in the next step. Disengage the filtration assembly and carefully transfer the MF, using sterile forceps, to a culture dish containing a medium-saturated absorbent pad or an agar preparation. Proper contact between the MF and the absorbent pad or the agar substrate requires the underside of the membrane to be uniformly wetted with culture medium without air bubble entrapment.

Air bubbles, easily recognized as colorless spots on the membrane surface or seen through the agar layer in the inverted culture dish, are produced when membranes are rolled too rapidly onto the substrate. Other causes of air bubble entrapment may relate to changes in the agar surface due to desiccation of the medium during storage or foaming of agar during the rapid ejection of medium from an automatic syringe or pipette into culture dishes. These entrapments of air preclude the diffusion of nutrients from either the medium-saturated absorbent pad or agar preparation to any bacteria on the MF surface directly above. This condition results in diminished potential for growth of the viable bacterial cells into differential colonies or hastens their death through desiccation. The net result would be an occasional reduction in the quantitative recovery of all indicator bacteria in a given water sample. This could create a serious error in the analysis of potable water with low levels of coliform contamination. Therefore, before incubation, inspect all MF cultures for any air bubble entrapment inside the effective filtration area. Air bubbles are easily removed by simply lifting the membrane with sterile forceps and gently rerolling it onto the membrane-saturated pad or agar substrate. Thereupon, close the culture container, invert it, and place it in the appropriate incubator, preferably within 10–15 min after filtration.

IV. TOTAL COLIFORM METHODS

Total coliform measurements are primarily restricted to the bacteriological examination of potable water for treatment effectiveness against fecal contamination. With adequate chemical treatment, filtration, and disinfection, total coliform occurrences in the processed water should be controlled consistently below a level of one organism per 100 ml.

A. Media Systems and m-Endo Medium Preparation

Differential media systems for total coliform analyses by the membrane filter technique are based primarily on biochemical reactions in the fermentation of lactose at 35 or 37°C. One approach is to use a pH indicator (or combination of indicators) such as bromcresol purple, phenol red, or aniline blue to detect a shift in pH caused by formation of organic acid end products of lactose fermentation. MacConkey medium, discussed in Chap. 11, incorporates a pH indicator system for total coliform differentiation. The major concern with use of any pH indicator system in a membrane filter medium is limitation of the zone of pH color change. In the presence of a large number of coliform colonies or several vigorously growing large coliform colonies, the pH color change may spread to noncoliform colony areas and confuse colony interpretation. This problem can be prevented by counting only membranes that have fewer than 60 total coliform colonies on the filter surface.

A second medium indicator system incorporates basic fuchsin and sodium sulfite in lactose medium (m-Endo-type media) for detecting the release of aldehyde as an intermediate produce in lactose fermentation (Table 2). Since basic fuchsin may differ in dye content from lot to lot and from manufacturer to manufacturer, media production must include a basic fuchsin standardization, which is more involved than adding only enough sodium sulfite to decolorize the fuchsin. In standardizing this medium, a number of tubes of the broth formulation without basic fuchsin are prepared and varying amounts of basic fuchsin are then added to each tube. Each of these medium variations is used to cultivate a known density

TABLE 2 m-Endo Medium[a]

Ingredients	Quantity
Tryptone or polypeptone	10 g
Thiopeptone or thiotone	5 g
Casitone or trypticase	5 g
Yeast extract	1.5 g
Lactose	12.5 g
Sodium chloride	5 g
Dipotassium hydrogen phosphate (K_2HPO_4)	4.375 g
Potassium dihydrogen phosphate (KH_2PO_4)	1.375 g
Sodium lauryl sulfate	0.05 g
Sodium deoxycholate	0.1 g
Sodium sulfite	2.1 g
Basic fuchsin	1.05 g
Distilled water containing 20 ml of ethanol	1.0 liter
Final pH 7.2 ±0.1 (*no* autoclaving)	
Single-strength dehydrated medium, 48g/liter	

[a]Commercial dehydrated medium is available from several media manufacturers. m-Endo agar may be prepared by adding 1.5% agar to the broth formulation.
Source: Fifield and Schaufus (1958).

of a coliform strain in the membrane filter procedure. From this experiment, the basic fuchsin concentration that provides optimum coliform recovery with development of a golden metallic colony luster is selected for that lot of dye.

The addition of pure grain ethanol to a final concentration of 2% (vol/vol) to form alcohol esters is essential for the development of coliform colonies with a maximum sheen and with less tendency toward confluent growth. These alcohol esters tend to suppress significant numbers of noncoliform organisms that could otherwise develop on the medium. Denatured ethanol, commonly available in the laboratory, must not be used because the denaturant commonly employed is either methanol or propanol, both of which are toxic to coliforms.

Where the laboratory directives severely restrict the availability of pure ethanol, a stock supply of ethanol for use in the MF procedure may be technically denatured by adding a few grains of m-Endo powder. The trace amount of basic fuchsin present in the small amount of dehydrated powder turns the ethanol pink; it does not adversely affect the m-Endo medium formulation and nullifies the illegal use of the product for human consumption.

Excessive heating of m-Endo medium destroys or reduces its specificity. Therefore, the medium is heated in a boiling-water bath just to the initial boiling point and is never autoclaved. As a general practice, only enough m-Endo medium is prepared to meet anticipated daily needs. However, surplus medium may be saved for use within a 96 hr period provided that the medium is stored in the dark at 2—10°C. Protected storage in the dark is essential because m-Endo medium is sensitive to strong artifical light or direct sunlight.

One formulation of Endo medium known as LES Endo agar (McCarthy et al., 1961) may be prepared by adding 1.5% agar to 75% of the recommended quantity of m-Endo powder per 100 ml of distilled water. The mixture is then heated in a boiling-water bath to completely dissolve the agar (described in Sec. III) and is poured into small culture dishes for use in the MF total coliform procedure.

B. Routine Total Coliform Procedure

Membrane filter examination for total coliform recovery requires 22-24 hr incubation at 35°C or 37°C for optimum growth and sheen development. For specific details on the filtration procedure, follow the recommendations described in Sec. III. Although many total coliform colonies can be differentiated within 18–20 hr, the extra 2–4 hr is especially important when examining treated drinking water samples because incomplete disinfection may have created stressed coliforms with damaged metabolic pathways. These coliforms are initially slow to develop the normal lactose fermentation end products that are the basis for differentiation.

In treated water supply, total coliform strains that survive some inadequate exposure to disinfectant are often in a stressed condition as they pass into the distribution system. Furthermore, the environment in distribution lines may also induce stress from exposure to slow-acting chloramines, marginal sources of essential nutrients in the water or pipe sediments, and competition from excessive populations of standard plate count organisms.

Recognition of stressed total coliforms in the membrane filter technique was noted by Clark et al. (1951), who proposed a 2 hr cultivation on an enrichment medium prior to transfer of the membrane filter culture to a differential medium. In that study, total coliform stress was possibly more related to medium toxicity than to environmental stresses present in natural waters. Although less toxic Endo-type media (m-Endo, LES Endo) are in use today, laboratories should be urged to consider the 2 hr general enrichment step in the total coliform procedure for optimum recovery of any stressed strains in waters examined from public water supplies.

The incubator should maintain a high level of humidity (approximately 90%). Reduced humidity often permits the surface of the membranes to lose moisture more rapidly than it is replenished by the diffusion of medium from an agar or absorbent-pad substrate. As a result, growth failure or small, poorly differentiated colonies may result. A conventional hot-air incubator may be used, but cultures in culture dishes with loose-fitting lids must be placed in a tightly closed container together with moist paper or cloth to maintain the necessary humid atmosphere. A vegetable crisper such as those used in home refrigerators is satisfactory for this purpose. Tight-fitting plastic culture dishes are preferred because the required humidity is established for each culture by the evaporation of some medium within the confines of the individual dish. No modification for higher humidity in the air incubator is necessary when tight-fitting plastic culture dishes are used.

Coliform colonies are best counted while in the moist state associated with their growth. Magnification of 10—15 diameters and illumination by a daylight fluorescent light source adjusted to an angle of 60—80° above the colonies are essential for optimum reflection of the golden metallic surface luster developed by all coliform strains on Endo-type media. Noncoliform colonies vary in appearance from colorless to a deep red color on Endo-type media. Inexperienced technicians often have great difficulty in interpreting confluent colonies and mirror reflections of fluorescent light on the colony surface, and mistake water condensate or particulate matter as colonies. In these instances, all doubtful colonies should be verified until the technician has gained a satisfactory proficiency in coliform colony recognition.

Total coliform results by the MF procedure are reported as "total coliforms per 100 ml" regardless of the size of test portion used. For example, if duplicate 50 ml portions of a well water are examined and the two membranes contain five and three coliform colonies, respectively, the count should be reported as eight total coliforms per 100 ml. The following rules (Water Supply Quality Assurance Work Group, 1978) for reporting any problem with MF results from potable water should be observed:

1. Confluent growth. Growth (with or without discrete sheen colonies) covering the entire filtration area of the membrane. These results are reported as "confluent growth per 100 ml, with (or without) coliform."

2. TNTC (too numerous to count). The total number of bacterial colonies on the membrane is too numerous (usually greater than 200 total colonies), not sufficiently distinct, or both. Since an accurate count cannot be made, results are reported as "TNTC per 100 ml with (or without) coliforms."
3. Response. In either result 1 or 2 above, a new sample must be requested and the sample volumes filtered must be adjusted to apply the MF procedure; otherwise, the multiple-tube procedure must be used.

C. Reaffirmation of Test Results

When coliforms are found in potable water samples, the proper authorities should be alerted promptly and repeat sampling at the same sites in the distribution network should be requested immediately. Retain the cultures, however, until the colonies have been verified, because synergistic false-positive reactions on Endo media may occur (Fifield and Schaufus, 1958; Schiff et al, 1970). This supplemental procedure consists of transferring each suspected coliform colony to lauryl tryptose broth (LTB) or lactose broth (LB) and incubation at 35°C for 24-48 hr. Gas-positive tubes are confirmed in brilliant green lactose broth (BGLB) for evidence of gas production at 35°C within the 48 hr limit. If all coliform-type colonies cannot be transferred, verify a random selection of at least five of these colonies. Avoid direct transfer of colonies to BGLB because of the inherent lower recovery of stressed coliform strains in this more selective medium. Omitting the BGLB step is undesirable because this medium eliminates most false-positive results from the lauryl tryptose broth.

In an effort to minimize the time delay resulting from verification of sheen colonies, it is permissible to transfer growth from each colony into pairs of LTB or LB and BGLB tubes. In this procedure, the verification is completed in 24 hr if both the LTB and BGLB culture pairs produce gas at 35°C. However, in those instances where the paired cultures are negative in 24 hr, the LTB or LB culture is reincubated for the second 24 hr period. If this tube is then positive, confirmation in a new tube of BGLB is necessary before verification is completed. This procedure of double inoculation from each sheen colony could reduce by 80–90% the test time for all coliform colony verification.

From the number of BGLB cultures that produce gas within 48 hr at 35°C, calculate the percent of colonies verified as coliforms. This percent figure is used to adjust the reported coliform count per 100 ml. For example, 10 sheen colonies from one MF culture are verified, but only 8 colonies produce gas after cultivation in LTB and BGLB. The percent verification is 80. The original coliform count was recorded as 20 organisms per 100 ml. Based on the verification of a random selection of 10 such colonies, the final coliform count recorded and reported would be 16 (80% X 2) organisms per 100 ml.

D. Delayed Incubation Procedure

Monitoring water supplies in remote areas of national parks and sparsely populated regions creates considerable difficulty in quickly transporting water samples to a laboratory for analysis within a 30–48 hr period. In these situations, the use of a delayed total coliform procedure may be desirable. This procedure involves field filtration of the water sample at the collection site and placement of the membrane filter on a holding medium (Geldreich et al., 1955). The holding medium contains a growth-suppressive agent to minimize colony development during the 24–72 hr time required to transport cultures to the laboratory for final processing.

To prepare the holding medium, add 0.384 g of sodium benzoate (U.S.P. grade) or 3.2 ml of a 12% (wt/vol) sodium benzoate solution to 100 ml of either m-Endo MF broth or LES Endo agar, which has previously been heated in a boiling-water bath to completely dissolve ingredients. Be sure to also include the 2 ml of pure ethanol per 100 ml of medium as normally prepared. Where overgrowth from fungus colonies causes problems, the addition of 50 mg of cycloheximide (actidione) per 100 ml of Endo holding medium is desirable.

Upon arrival in the laboratory, transfer the MF cultures to a fresh culture dish containing standard m-Endo MF broth or LES Endo agar, and incubate the plates at 35°C for 20–22 hr. If growth is visible at time of transfer, hold the cultures in a refrigerator until the end of the workday and incubate them at 35°C overnight (16–18 hr). Then examine cultures for sheen colonies (total coliforms) and record their density per 100 ml.

It is essential that the laboratory establish the validity of the delayed incubation test for total coliforms on those waters that are to be examined routinely by this procedure. Wide variations in ambient temperature and storage periods up to 72 hr before final processing of the MF cultures in the laboratory may stimulate the growth of some false-positive, noncoliform organisms that are capable of partial breakdown of lactose. Once the magnitude of these occurrences has been determined through sheen colony verification, data from the delayed incubation test for total coliform detection may be more accurately interpreted.

E. Portable Field Laboratory Procedure

By virtue of both the simplicity of operation and the compactness of essential apparatus, the MF procedure readily lends itself to field applications provided that the operator is adequately trained to use the technique effectively. Field uses include water supply monitoring on ships, overseas airlines, and national parks; water quality monitoring in sparsely populated regions; and emergency application following natural disasters. Recognizing the potential of this procedure for monitoring potability of water supplies used in military operations, Col. Thomas Sparks and his laboratory staff at Fort Sam Houston, Texas, designed and field-tested a MF portable laboratory package that had the general configuration of a suitcase, the size of a picnic cooler, and included standard laboratory filtration equipment. A commercially available version of this unit is now available (Millipore Portable Water Laboratory XX 63-001-50). Within the fiberglass carrying case is a filtration funnel, plastic culture dishes, ampuled or preweighed media vials, a hand-pumped vacuum source, a suitable electric incubator (35°C) designed to operate on 12/24 V dc or 115/230 V ac or dc, plus other necessary small items. With such a kit, the properly trained technician can test approximately 24 water samples for total coliforms.

Several different methods of medium preparation can be used in conjunction with the portable field laboratory. Ampuled m-Endo medium is the most convenient. Each ampul contains sufficient sterile, prepared medium to saturate one absorbent culture pad. Shelf life for ampuled m-Endo medium, stored at 4—10°C, is approximately 18 months. Ampuls that appear turbid or dark red in color may be contaminated and should not be used. Another approach to media supplies for field use is to prepare or purchase vials of preweighed dehydrated m-Endo medium. When needed, the desired number of preweighed vials of medium are reconstituted with the appropriate amounts of distilled water, 2% ethanol (not denatured) is added, and the medium is carefully heated to dissolve the ingredients. The finished medium preparation is then dispensed in 2 ml volumes into culture dishes containing absorbent culture pads. Poured agar plates of the appropriate MF medium normally used in the laboratory may also be used in the field kit provided that these agar preparations are made in tight-fitting dishes to reduce moisture loss in storage, are protected from light, and are not held beyond a 2 week refrigerated storage period. Dehydrated medium pads that are to be reconstituted by adding 2 ml of sterile distilled water are not recommended. Pads impregnated with m-Endo medium and dehydrated have a shelf life of approximately 3 months but when reconstituted do not produce a medium of uniform quality.

V. FECAL COLIFORM METHODS

Fecal coliforms are used as the primary indicator of fecal pollution in source waters for public water supplies, recreational waters, agricultural irrigation and livestock water, and in shellfish culture waters. This bacterial indicator is also of prime importance in determining sewage effluent quality discharged to receiving waters or applied to soil and for measurement of fecal content of landfill leachates and water surrounding ocean dumping sites for garbage and sludges. Following natural or human-caused disasters involving water supply systems, rapid detection of fecal contamination penetrating the water treatment barrier is of great urgency to public health officials and much research effort has been directed toward the development and application of rapid fecal coliform methods for this purpose.

A. Elevated Temperature Concept

The elevated temperature test for the separation of the fecal from nonfecal coliform groups was originally proposed by Eijkman (1904). It was based on his observations that coliform bacteria derived from the gut of warm-blooded animals produced gas from glucose at 46°C, whereas the coliform strains from nonfecal sources failed to produce growth. The Eijkman reaction or one of its many modifications was studied by an impressive list of investigators, but conclusions differed concerning sensitivity, specificity, and interpretation of data. Perry and Hajna (1944) proposed an elevated temperature test for *Escherichia coli* that gained little acceptance at that time by bacteriologists investigating pollution. Perry and Hajna, however, had improved the Eijkman concept by adding bile salts and a buffer system and by reducing the air incubation temperature to 44.5°C, which improved the medium sensitivity. The British proposed use of MacConkey's broth at 44°C for the confirmation of *E. coli* type I. Vaughn et al. (1951) recommended a buffered boric acid lactose broth for the enrichment and identification of *E. coli*. These investigators believed that the reduction of the incubation temperature to 43°C for 48 hr increased sensitivity and that the addition of the boric acid inhibited the growth of *Klebsiella-Enterobacter* strains. All these media and procedural modifications were designed for use in multiple-tube tests utilizing the most probable number (MPN) concept.

Following general acceptance of membrane filter procedures for total coliform detection in water analysis, methods for fecal coliform detection by this technique were soon proposed. Taylor et al

(1955) proposed a 2 hr enrichment on nutrient broth at 37°C for temperature acclimation followed by transferal of the membrane culture to a modified MacConkey broth for final incubation at 44°C for 16 hr. Geldreich et al. (1965) reported that excellent fecal coliform recoveries could be achieved using an mFC medium and direct incubation at 44.5°C for 24 hr. Recent concerns for improved recovery of stressed fecal coliforms from chlorinated sewage effluents, waters receiving industrial waste discharges of heavy metals, and coastal marine waters have prompted various procedural modifications involving temperature acclimation at 35°C prior to final incubation on mFC medium at 44.5°C.

A review of the literature on stressed fecal coliforms (Lin, 1977; Hartman et al., 1975; Tobin and Dutka, 1977; Burman et al., 1969; Rose et al., 1975; Green et al., 1977; Stuart et al., 1977; Presswood and Strong, 1978) suggests that chlorine inactivation of some coliform cells may be reversed without compromising test specificity provided that membrane filters with the largest pore diameter (Millipore HC) or with the widest range of pore diameters (Gelman GN-6) and consistent with optimum organism retention are used (Lin, 1977; Tobin and Dutka, 1977). This choice can be coupled with temperature acclimation (Vaughn et al., 1951; Taylor et al., 1955), or enrichment-temperature acclimation (Burman et al., 1969; Rose et al., 1975; Green et al., 1977; Stuart et al., 1977) if desirable for improved recovery. In temperature acclimation, cultures are preincubated on mFC medium for 5 hr at 35°C, followed by 18 hr at 44.5°C. The use of a temperature-programmed incubator to make the change from 35°C to 44.5°C after the 5 hr preincubation period would accommodate laboratory working schedules. To achieve automation in enrichment-temperature acclimation, a two-layer agar (mFC agar with an overlay of lauryl tryptose agar) can be incorporated with 2 hr incubation at 35°C followed by 22 hr at 44.5°C (Rose et al., 1975) in a temperature-programmed incubator. If the two-layer agar technique is used, prepare the mFC agar plate in advance but *do not* add the lauryl tryptose agar overlay more than 1 hr before use. This requirement controls the diffusion rate of the enrichment agar overlay so that MF cultures are exposed to this medium for 2 hr before diffusion of nutrients between the two-agar medium creates a slightly modified mFC agar substrate. Finally, deletion of the rosolic acid suppressive agent from mFC medium may improve recovery of stressed fecal coliforms (Presswood and Strong, 1978). Fecal coliform colonies are intense blue on the modified medium and are still distinguished easily from the cream, gray, and pale green colonies typically produced by nonfecal coliforms.

As a word of caution, any modifications of medium and procedures may decrease selectivity and differentiation of fecal coliform colonies. Therefore, before accepting any procedure modifications, verify at least 10% of the blue colonies from a variety of samples. Blue colonies to be verified are first transferred to tubes of lauryl tryptose broth or lactose broth for evidence of gas production within 48 hr at 35°C incubation. All gas-positive broth cultures are confirmed in EC broth for gas production at 44.5°C within 24 hr. All cultures that produce gas at the elevated temperature are fecal coliforms.

Precise control of the elevated temperature is essential because lower temperatures will permit the growth of many nonspecific organisms, and higher temperatures decrease the recovery of some strains of fecal coliforms. Once the sample volume is filtered and the membrane placed on the selected medium, the desired incubation temperature should be reached within 10—15 min and held precisely within the recommended range. For these reasons, incubation in a water bath or in a solid heat-sink incubator (such as aluminum) is desirable because precise temperature control in these systems is more easily attained than in conventional air incubators.

Accurate temperature measurements are essential for elevated-temperature tests. A continuous temperature recorder sensitive to ±0.2°C changes should be used for a permanent record. In addition, an accurate thermometer (with 0.1°C scale divisions) should be immersed in the water bath or inserted in the heat-sink aluminum block once each day to spot-check the precision of the recorder tracing. If the recorder tracing is inaccurate, the marking pen should be adjusted so that the temperature tracing agrees with comparative readings of the immersed thermometer. If a recording thermometer is not used to monitor water-bath temperatures, a daily record of temperature readings from an immersed or inserted thermometer or digital electronic thermometer probe must be made.

Large benchtop water baths with gabled covers can effectively maintain a temperature of 44.5°C within ±0.5°C. Temperature measurements in these noncirculating water baths may reveal that some are capable of temperature control within ±0.2°C; others exhibit a slightly greater deviation. Noncirculating water baths can generally be brought to within ±0.2°C temperature tolerances by adding a low-speed stirring motor to create a gentle circulation of water and prevent heat stratification. Coarse temperature controls, sediment accumulations on the bottom created from the corrosion of immersed wire racks or baskets, and inadequate heat diffuser plates may create more severe temperature control problems.

Circulating water baths or aluminum block heat-sink incubators may not have to be kept turned on during nonuse periods of 72 hr or longer provided that the laboratory has established (through adequate monitoring data) that the desired stable temperature can be achieved prior to time of use. Noncirculating water baths must be left on at all times because temperature stability in these units, at the recommended temperature tolerance of ±0.2°C, is marginal.

B. Media Systems and mFC Agar Preparation

Use of a basic fuchsin-sulfite indicator system in the elevated-temperature test medium has been found to be unsatisfactory because of the transient nature of aldehyde release at 44.5°C. Obeying a basic law of kinetics that states that for every 10°C rise in temperature, the biochemical reaction rate doubles, it is not surprising to find lactose fermentation completed with little aldehyde remaining as an intermediate residual by-product. When this condition occurs, fecal coliform colonies (in a vigorous stage of growth) on

Endo-type media rapidly lose their typical golden metallic luster and fade to a dirty-gray color. Thus, the differential characteristics of the indicator system deteriorate and become unreliable for routine analysis.

Therefore, differential media used for fecal coliform recovery by membrane filter methods generally incorporate pH indicator systems that allow detection of stable organic acid end products from the lactose fermentation. However, the choice of pH indicator is restricted to indicators that are basic or neutral in reaction, produce good color contrast for colony differentiation, do not develop large zones of color change beyond the colony area, are nontoxic to bacterial growth, and are nonreactive to membrane filter materials. These requirements have greatly restricted the choice of indicators useful in development of membrane filter differential media. One widely used fecal coliform medium is mFC agar (Geldreich et al., 1965).

mFC agar is an enriched growth medium containing lactose (Table 3). Bile salts and rosolic acid are included for suppression of a

TABLE 3 mFC Agar[a]

Ingredient	Quantity
Tryptose or biosate	10.0 g
Proteose peptone No. 3 or polypeptone	5.0 g
Yeast extract	3.0 g
Sodium chloride	5.0 g
Lactose	12.5 g
Bile salts No. 3 or bile salts mixture	1.5 g
Aniline blue	0.1 g
Agar	15.0 g
Distilled water containing 10 ml of 1% rosolic acid sale reagent	1.0 liter
Final pH 7.4 ±0.1 (*no* autoclaving)	
Single-strength dehydrated medium, 37 g/liter	

[a]Commercial dehydrated medium is available from several media manufacturers.
Source: Geldreich et al. (1965).

variety of nonfecal coliform organisms common to some source waters and the first flush of stormwater runoff. In this fecal coliform medium, analine blue is the indicator system used to detect lactose fermentation. Development of the blue colony color does not depend upon the addition of the rosolic acid salt reagent. After the medium ingredients are in solution, 1 ml of a 1% rosolic acid salt reagent is added and the medium is heated to the boiling point. As a general practice, only enough mFC agar is prepared to meet daily needs. However, surplus medium may be saved for use within a 96 hr period provided that the medium is stored in the dark at 2—10°C.

The 1% rosolic acid salt reagent is prepared by dissolving 1 g of rosolic acid in 100 ml of 0.2 N sodium hydroxide (0.8 g of NaOH in 100 ml of distilled water). Do not autoclave this solution. Rosolic acid salt reagent should be stored at 2—10°C in the dark and must be discarded after 2 weeks or sooner if the solution changes color from dark red to muddy brown. Background color in the MF will vary from a yellowish cream to faint blue, depending on the age of the reagent. When rosolic acid salt has been prepared within 1—2 hr of its addition to the medium, it does have a differential effect on some of the nonfecal coliform colonies. This phenomenon has been observed in the development of yellow and red nonfecal coliform colonies from samples on some canal waters. The identity of these organisms is not known, but the important point is that only blue colonies were verified as fecal coliforms.

C. Routine Fecal Coliform Procedure

For laboratories desiring to perform the routine fecal coliform test, the only special items necessary are a water bath that can be regulated at 44.5 ±0.2°C, mFC agar, and sealable plastic bags to protect the cultures during immersion in the water-bath incubator. For specific details on the filtration procedure, follow the recommendations described in Sec. III. After sample filtration, place the membrane filter on mFC agar contained in culture dishes with tight-fitting lids. If air bubbles have been entrapped between the medium and the membrane, the membrane must be reseated on the agar before the culture dish is inserted into a sealable plastic bag (Fig. 4). These waterproof plastic bags (Whirl-Pak or equivalent) may be used to hold three to eight culture dishes during submersion. These cultures must be placed in the incubator within 30 min of filtration because the elevated temperature (44.5°C) is critical to the fecal coliform test sensitivity. Where there is a concern for optimizing recovery of stressed fecal coliforms, variations in this protocol may involve use of Millipore HC or Gelman GN-6 membrane filters, preenrichment, temperature acclimation at 35°C, or deletion of rosolic acid reagent in the preparation of mFC medium, as previously discussed. Regardless of the initial stages of the test, final incubation or total incubation is done at 44.5°C and total incubation time is 24 ±2 hr. After cultivation, the membrane filter cultures are examined promptly (within 30 min) under low-power magnification. Fecal coliform colonies on mFC agar are blue in color. The fecal coliform density per 100 ml is counted and

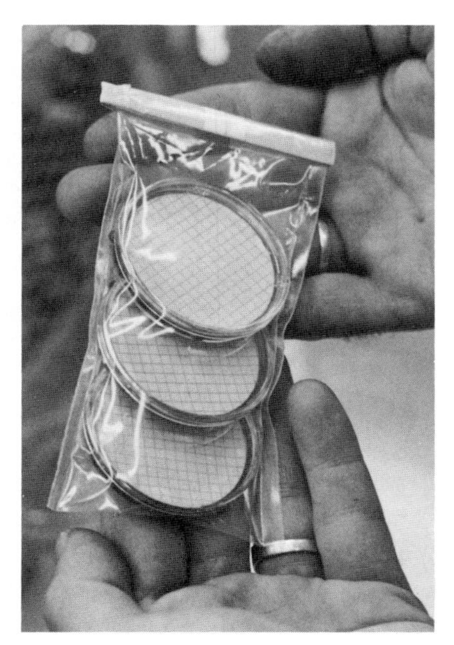

FIGURE 4 Fecal coliform cultures are inserted into a sealable plastic bag for waterproofing prior to submergence in the water bath.

calculated. When several different filtration quantities are used, the membrane filter culture with fecal coliform densities ranging between 20 to 60 colonies is selected, to provide the best statistical data.

Verification of fecal coliform colonies should be done on one complete set of samples at the start of a field study, basically as a quality control procedure to demonstrate the validity of the test for examination of those waters being investigated. Verification is also desirable to resolve any questions concerning borderline color development of colonies or to reassure a new technician as to the colony counts. In this procedure, growth from a fecal coliform colony is transferred to lauryl tryptose broth or lactose broth for evidence of gas production at 35°C within 48 hr. Positive cultures of broth are used to inoculate EC broth tubes; evidence of gas production at 44.5°C within 24 hr confirms the organism to be a fecal coliform.

D. Rapid Fecal Coliform Procedure

Rapid assessment of the sanitary quality of water is often needed for emergency testing of potable water supplies, bathing beach water whose quality may have deteriorated following storms, and shellfish growing waters subject to sewage pollution. One promising approach involves the use of a membrane filter procedure utilizing m-7HrFC agar (Reasoner et al., 1979; Geldreich, 1975). Detection of fecal coliforms in a rapid test (7 hr) requires the use of a lightly buffered enriched medium for early detection of lactose or mannitol fermentation (Table 4). The inclusion of mannitol in the lactose-based m-7 hr agar permits the fecal coliforms to utilize one of two pathways to rapid fermentation. This results in an early pH shift in the bromcresol purple/phenol red indicator system, creating a yellow color on fecal coliform colonies. To prepare the medium, rehydrate the ingredients and heat in a boiling-water bath to dissolve the agar. After solution is complete, heat for an additional 5 min and place in a 44.5°C water bath to temper the melted medium

TABLE 4 m-7HrFC Agar[a]

Ingredient	Quantity
Proteose peptone No. 3 or polypeptone	5.0 g
Yeast extract	3.0 g
Lactose	10.0 g
Mannitol	5.0 g
Sodium chloride	7.5 g
Sodium lauryl sulfate	0.2 g
Sodium deoxycholate	0.1 g
Bromcresol purple	0.35 g
Phenol red	0.3 g
Agar	15.0 g
Distilled water	1.0 liter
Adjust final pH to 7.3 ±0.1; approximately 0.35 ml of 0.1 N NaOH is required (*no* autoclaving)	
Single-strength dehydrated medium, 41.45 g/liter	

[a]Commercial dehydrated medium is available from Difco on special request.
Source: Reasoner et al. (1979) and Geldreich (1975).

before pouring the plates. Poured plates may be stored at 2–10°C for periods up to 30 days before use.

Following filtration, membranes are placed in culture dishes containing m-7HrFC agar. The plates are then placed in waterproof plastic bags and submerged in a 41.5 ±0.2°C water bath for 7 hr. Colonies must be examined within 15 min at 10–15X magnification using either a fluorescent light or an incandescent microscope light with a blue filter. Fecal coliform colonies appear yellow against a light purple background. Fecal coliform colonies are generally very bright and distinct, but all colonies having a yellow appearance should be counted. Initial verification of the colonies (as described for routine fecal coliform tests) on m-7HrFC agar is desirable to demonstrate the effectiveness of the medium to the technician using the procedure for the first time.

E. Delayed Incubation for Fecal Coliform

The delayed incubation concept can also be applied to fecal coliform measurements (Taylor et al., 1973). After the water sample has been filtered in the field, place the MF in contact with an absorbent pad saturated with vitamin-free casitone (VFC) holding medium and send by mail to the laboratory for final processing. The holding medium (m-VFC) for fecal coliforms does not require an indicator system because membrane filter cultures transported to the laboratory for final processing will be immediately placed on the differential mFC agar for final incubation at 44.5°C. The key to this medium (Table 5) is a vitamin-free minimal nutrient base containing sodium benzoate and ethanol to slow colony development and sulfanilamide to suppress many other organisms present in the water flora. Warm to dissolve all ingredients, then sterilize the medium by filtration through a 0.22 μm pore size membrane filter. If only a 100 ml quantity of the medium is prepared, it is easier to add the vitamin-free casitone as 2 ml of a 1:100 aqueous solution. Store the finished medium at 2–10°C, and discard any unused portions after 1 month of storage.

TABLE 5 mVFC Holding Medium

Ingredient	Quantity
Vitamin-free casitone	0.2 g
Sodium benzoate	4.0 g
Sulfanilamide	0.5 g
Ethanol (95%)	10.0 ml
Distilled water	1.0 liter
Final pH 6.7 ±0.1 (*no* autoclaving)	
Single-strength dehydrated medium, 4.7 g/liter	

Source: Taylor et al. (1973).

Although the delayed incubation procedure will hold for up to 72 hr, the holding period should be minimal, and for some distant locations, special priority mail service should be used. Upon receipt in the laboratory, transfer the membrane filter cultures to fresh culture dishes containing mFC agar, place in waterproof plastic bags, and incubate (submerged) in a 44.5°C water bath for 22 ±2 hr. Count the blue colonies (fecal coliforms) and calculate the count per 100 ml. Verification of fecal coliforms (as described for routine fecal coliform tests) is recommended as an initial control check of the delayed fecal coliform procedure when this technique is to be used for the examination of source waters on a continual monitoring basis. This check on the validity of delayed test results is useful for demonstration of the test accuracy when such laboratory data are to be used in enforcement actions.

ACKNOWLEDGMENTS

The author acknowledges the assistance given by Dr. Donald J. Reasoner, Dr. Martin J. Allen, and Mr. Raymond H. Taylor for technical and editorial review of the manuscript, and Dr. Allen for the electron micrographs of the membrane filter pore structure and bacterial entrapment on the membrane filter surface.

REFERENCES

Akin, E. W., Brashear, D. A., and Clarke, N. A. (1975). *A Virus-In-Water Study of Finished Water from Six Communities*, Environmental Protection Technology Series. EPA-600/1-75-003. U.S. Environmental Protection Agency, Cincinnati, Ohio.

Burman, N. P., Oliver, E. W., and Stevens, J. K. (1969). Membrane filtration techniques for the isolation from water, of coli-aerogenes, *Escherichia coli*, faecal streptococci, *Clostridium perfringens*, actinomycetes and microfungi. In *Isolation Methods for Microbiologists*, Technical Series No. 3, D. A. Shapton and G. W. Gould (Eds.). Academic Press, London.

Clark, H. F., Geldreich, E. E., Jeter, H. L., and Kabler, P. W. (1951). The membrane filter in sanitary bacteriology. *Public Health Rep. 66:* 951–977.

Fifield, C. W., and Schaufus, C. P. (1958). Improved membrane filter medium for the detection of coliform organisms. *J. Am. Water Works Assoc. 50:* 193–196.

Geldreich, E. E. (1975). *Handbook for Evaluating Water Bacteriological Laboratories*. EPA-670/9-75-006. U.S. Environmental Protection Agency, Cincinnati, Ohio.

Geldreich, E. E., Kabler, P. W., Jeter, H. L., and Clark, H. F. (1955). A delayed incubation membrane filter test for coliform bacteria in water. *Am. J. Public Health 45:* 1462–1474.

Geldreich, E. E., Clark, H. F., Huff, C. B., and Best, L. C. (1965). Fecal-coliform organism medium for the membrane filter technique. *J. Am. Water Works Assoc. 57:* 208–214.

Green, B. L., Clausen, E. M., and Litsky, W. (1977). Two-temperature membrane filter method for enumerating fecal coliform bacteria from chlorinated effluents. *Appl. Environ. Microbiol. 33:* 1259–1264.

Hartman, P. A., Hartman, P. S., and Lang, W. W. (1975). Violet red bile agar for stressed coliforms. *Appl. Microbiol. 29:* 537–539.

Lin, S. D. (1977). Comparison of membranes for fecal coliform recovery in chlorinated effluents. *J. Water Pollut. Control Fed. 49:* 2255–2264.

McCarthy, J. A., Delaney, J. E., and Grasso, R. J. (1961). Measuring coliforms in water. *Water Sewage Works 108:* 238–243.

Perry, C. A., and Hajna, A. A. (1944). Further evaluation of EC medium for the isolation of coliform bacteria and *Escherichia coli*. *Am. J. Public Health 34:* 735–738.

Presswood, W. G., and Strong, D. K. (1978). Modification of M-FC medium by eliminating rosolic acid. *Appl. Environ. Microbiol. 36:* 90–94.

Reasoner, D. J., Blannon, J. C., and Geldreich, E. E. (1979). Rapid seven-hour fecal coliform test. *Appl. Environ. Microbiol. 38:* 229–236.

Rose, R. E., Geldreich, E. E., and Litsky, W. (1975). Improved membrane filter method for fecal coliform analysis. *Appl. Microbiol. 29:* 532–536.

Schiff, L. J., Morrison, S. M., and Mayeux, J. V. (1970). Synergistic false-positive coliform reaction on M-Endo MF medium. *Appl. Microbiol. 20:* 778–781.

Stuart, D. S., McFeters, G. A., and Schillinger, J. E. (1977). Membrane filter technique for the quantification of stressed fecal coliforms in the aquatic environment. *Appl. Environ. Microbiol. 34:* 42–46.

Taylor, E. W., Burman, N. P., and Oliver, C. W. (1955). Membrane filtration technique applied to the routine bacteriological examination of water. *J. Inst. Water Eng. 9:* 248–263.

Taylor, R. H., Bordner, R. H., and Scarpino, P. V. (1973). Delayed-incubation membrane filter test for fecal coliforms. *Appl. Microbiol. 25:* 363–368.

Tobin, R. S., and Dutka, B. J. (1977). Comparison of the surface structure, metal binding, and fecal coliform recoveries of nine membrane filters. *Applied Environ. Microbiol. 34:* 69–79.

Vaughn, R. H., Levine, M., and Smith, H. A. (1951). A buffered boric acid lactose medium for enrichment and presumptive identification of *Escherichia coli*. *Food Res. 16:* 10–19.

Water Supply Quality Assurance Work Group (1978). *Manual for the Interim Certification of Laboratories Involved in Analyzing Public Drinking Water Supplies*. EPA-600/8-78-008. U.S. Environmental Protection Agency, Washington, D.C.

4

FECAL STREPTOCOCCI

ELIZABETH C. BEAUDOIN and WARREN LITSKY University of Massachusetts, Amherst, Massachusetts

I. INTRODUCTION

Controversy regarding the reliability of fecal streptococci as indicators of fecal pollution has persisted throughout the last 25 years. Within the United States this indicator is seldom used alone for water quality assessment but is commonly included as an adjunct to the coliform test, particularly when identification of a pollution source is desired. Proper interpretation of fecal streptococcus densities requires an understanding of the natural occurrence and survival of these organisms. In this chapter advantages and disadvantages of these bacteria as indicators are considered and methods for membrane filtration enumeration evaluated.

II. DEFINITION AND CLASSIFICATION

Terms such as fecal streptococci, enterococci, and group D must be interpreted with some caution in that no single definition for any of these categories has been universally adhered to. The following definitions, however, have gained widespread acceptance in recent years:

Fecal streptococci: The term "fecal streptococci" refers to all species of streptococci that normally occur in fecal matter.

This is therefore a broad group composed of *Streptococcus faecalis*,* *S. faecium*,† *S. bovis*, and *S. equinus*. In addition, *S. mitus* and *S. salivarius* are now frequently included. The latter organisms are buccal streptococci which appear in human feces in low numbers. *S. avium* (or group Q) is also often considered a member of this group because this species is recovered from the feces of fowl in significant quantities.

Enterococci: The enterococcus group is a more restricted category and includes only those streptococci capable of growth at 45°C and 10°C, in a salt concentration of 6.5%, and at pH 9.6. These four tests are commonly referred to as the Sherman criteria. Only *S. faecalis* and *S. faecium* are generally regarded as enterococci, although *S. avium* also conforms to the Sherman criteria.

Group D: Group D streptococci are defined as those species that possess the group D antigen. This antigen is a glycerol teichoic acid (Wicken and Baddiley, 1963; Wicken et al., 1963) located between the cell membrane and the cell wall (Smith and Shattock, 1964). It is routinely detected by the precipitin test. Group D includes *S. faecalis*, *S. faecium*, *S. bovis*, and *S. equinus*. *S. avium* also possesses the D antigen but is classified on the basis of the group Q antigen found in the cell wall of this species (Smith and Shattock, 1964).

The distribution of streptococcal species into each of the foregoing categories is shown in Fig. 1.

Speciation of the fecal streptococci is a matter of some importance in that the presence or predominance of certain species is indicative of the origin of fecal pollution. There are no simple

---

**S. faecalis* is composed of three subspecies: *S. faecalis* subspecies *faecalis*, *S. faecalis* subspecies *zymogenes*, and *S. faecalis* subspecies *liquefaciens*.

†*S. faecium* also includes those strains formerly called *S. durans*, which are distinguished by their inability to ferment mannitol and arabinose.

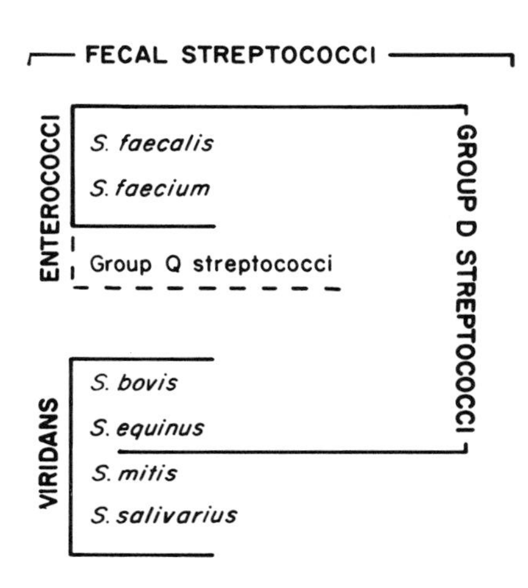

FIGURE 1 Definition of the terms "enterococci," "group D streptococci," and "fecal streptococci" based on *Streptococcus* species belonging to each group. (From Levin et al., 1975.)

tests by which speciation of these organisms is accomplished; a series of 10—20 biochemical tests is required. These tests are necessitated by the considerable variability among strains, which prevents any single reaction from providing an accurate identification of a particular species. It is not the intention of this chapter to dwell on the specific reactions by which speciation is accomplished. These tests are outlined in Fig. 2. Table 1 presents data obtained from the speciation of fecal isolates. The percentages of strains shown to produce atypical reactions to particular tests emphasize the necessity of considering the complete series of reactions. The reader is advised to consult *Microbiological Methods for Monitoring the Environment: Vol. I, Water and Wastes* (Bordner and Winter, Eds., 1977) and also Facklam (1972) for detailed information pertaining to the execution of these tests.

Even upon completion of the biochemical tests there remain a significant number of streptococci that fail to conform to the reaction pattern of any recognized species. These are referred to as biotypes. Several researchers have reported recovering high per-

TABLE 1 Positive Reactions of Group D Species Isolated from Human Sources

Test	Positive reactions (%) of streptococci from:				
	S. faecalis	*S. faecium*	*S. durans*	*S. bovis*	Variant
Group D reaction	100	96	69	96	100
Bile Esculin medium	99	100	100	100	100
Growth on 40% bile	100	100	100	100	100
Growth at 45°C	99	100	100	93	100
Acid in litmus milk	100	100	100	100	100
Growth in 6.5% NaCl broth	100	100	100	2	0
Growth in *S. faecalis* broth	98	96	100	2	0
Growth in pH 9.6 broth	98	86	100	39	67
Growth in methylene blue milk	98	100	100	61	43
Growth at 10°C	95	89	100	2	0
Reduction of tetrazolium	99	0	0	30	14
Resistant to tellurite	90	0	0	0	0
Clot in litmus milk	89	33	62	100	100

Hydrolysis of gelatine	28	0	0	0	0
Hydrolysis of starch	0	0	0	100	0
Beta hemolysis	17	0	0	0	0
Acid from:					
Esculin	99	100	100	100	100
Lactose	95	100	100	100	100
Sorbitol	97	7	0	2	0
Sucrose	97	89	77	100	100
Glycerol	78	4	0	0	0
Mannitol	98	100	0	91	0
Arabinose	4	96	0	2	0
Raffinose	6	67	39	98	29
Inulin	4	11	0	50	0
Production of extracellular polysaccharide on:					
5% Sucrose agar	0	0	0	84	0
5% Sucrose broth	0	0	0	57	0
Number of specimens tested	171	27	13	44	7

Source: Facklam (1972).

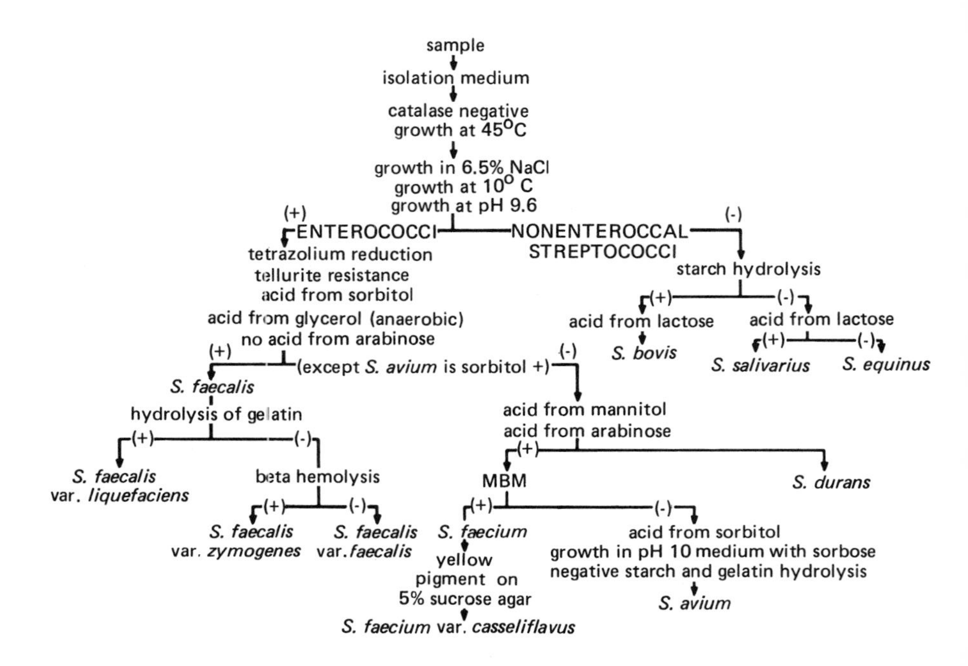

FIGURE 2 Biochemical identification of fecal streptococci.

centages of biotypes during speciation of streptococcal isolates obtained from a variety of sources. These data are summarized in Table 2. The importance of biotypes among plant isolates will be discussed in a subsequent section. Considerable numbers of biotypes were also isolated from polluted waters, sewage, and feces. The significance of these biotypes has not been established nor are they recognized by any taxonomic scheme. If the streptococci are to continue to be employed for characterization of pollutants, questions regarding the origin and identity of the biotypes merit more careful investigation.

TABLE 2 Percentages of Biotypes According to Various Studies

Source	Biotype						
	Enterococcal		*S. faecalis*			*S. faecalis* (starch +)	Unidentified streptococci
	Kenner et al., 1960	Tilton & Litsky, 1967	Cooper & Ramadan, 1955	Bartley & Slanetz, 1960	Kibbey et al., 1978	Geldreich and Kenner, 1969	Mundt, 1976
In feces							
Human	6.8	30.0	41.9				
Cow	12.4	6.5	57.9				
Sheep	28.8	18.8	75.8				
Pig	34.0						
Fowl	37.8						
In environmental samples							
Sewage or sludge				51	36		
River water				36			
Sea water				59			
Pond water				57			
Well water				42			
Plants					56	34.9	
Soil					72		
Dry and frozen foods							50.2

III. FECAL STREPTOCOCCI AS FECAL INDICATOR ORGANISMS

A. Detection of the Origin of Pollution

Fecal streptococci provide an indication of the origin of fecal pollution. This information can be important in locating pollution sources and cannot be obtained using coliform indicators alone. Whether pollution is of animal or human origin can often be determined using one of the approaches described next.

Fecal Coliform/Fecal Streptococcus Ratio

A number of studies have indicated that the proportion of fecal streptococci to fecal coliform bacteria is consistent for samples of similar origin. In sewage both total and fecal coliforms are predictably more numerous than fecal streptococci (Geldreich and Kenner, 1969; Cohen and Shuval, 1973; Burm and Vaughn, 1966). In stormwater runoff this relationship is reversed and fecal streptococci exceed fecal coliform bacteria (Geldreich and Kenner, 1969; Evans et al., 1968; Geldreich et al., 1968; Weibel et al., 1964). Actual densities of these bacteria are shown in Table 3. These findings are felt to be related to the fact that human feces contain a predominance of fecal coliform bacteria, whereas animal feces are

TABLE 3 Densities of Fecal Coliform and Fecal Streptococcus Bacteria Reported in Sewage, Stormwaters, and Feces

Sewage	Density/100 ml
Fecal coliforms	3.4×10^5-4.9×10^7
Fecal streptococci	6.4×10^4-4.5×10^6
Stormwater	
Fecal coliforms	2.7×10^3-1.3×10^4
Fecal streptococci	5.8×10^4-1.5×10^5
Human feces	Density/g
Fecal coliforms	1.3×10^7
Fecal streptococci	3.0×10^6
Livestock feces	
Fecal coliforms	2.5×10^5-1.6×10^7
Fecal streptococci	1.3×10^6-8.4×10^7

Source: Geldreich and Kenner (1969).

characterized by higher numbers of fecal streptococci (Table 3). Geldreich and Kenner (1969) have used these relationships to formulate a fecal coliform/fecal streptococcus ratio, which provides an indication as to whether pollution is of human or animal origin. If the ratio exceeds 4.0, pollution probably results from human contamination. However, if the ratio drops below 0.7, animal pollution is indicated. Ratios falling between 0.7 and 4.0 cannot be interpreted. The ratio is valid for only approximately the first 24 hr following discharge of fecal material into a body of water (Geldreich and Kenner, 1969; McFeters et al., 1974). Following this period, differing rates of die-off are found to alter the relationship. If pollution is of animal origin, rapid die-off of *S. bovis* will cause the ratio to rise with time. When human waste is the major pollutant, the ratio will tend to decrease because enterococci are generally more persistent than are fecal coliform bacteria (Faecham, 1975; McFeters et al., 1974).

Speciation

The isolation of particular streptococcal species from polluted waters may also aid in locating pollution sources. The streptococcal population of human feces is composed predominantly of enterococci, and *S. bovis* and *S. equinus* are rarely recovered. Certain species have been suggested to be unique to human fecal matter, and these include *S. faecalis* subspecies *faecalis* (Bartley and Slanetz, 1960), *S. durans* (Cooper and Ramadan, 1955), and *S. mitus* and *S. salivarius* (Kenner et al., 1960). Recovery of these organisms is therefore indicative of human contamination. Feces from livestock are distinguished by a high proportion of *S. bovis* and *S. equinus* (Geldreich and Kenner, 1969; Kenner et al., 1960), and fecal material from fowl has been reported to harbor significant numbers of *S. avium* (Nowlan and Deibel, 1967). It must be recognized, however, that survival of *S. bovis* and *S. equinus* in water is extremely limited (Geldreich and Kenner, 1969) and that *S. mitus* and *S. salivarius* are present in feces in very low numbers. It is therefore often difficult to recover these strains. Only a few attempts to isolate *S. avium* have been reported.

B. Survival

Fecal streptococci survive longer than fecal coliforms and are therefore sometimes a preferable indicator of pollution. The length of time that a fecal organism can persist in the environment is of great relevance to its usefulness as an indicator of pollution. An ideal indicator is expected to survive for as long as the most persistent fecal pathogen and must itself die off shortly following the destruction of such organisms. The actual number of days required for removal of both fecal indicators and pathogens is dependent on a variety of factors and is subject to considerable variation. Survival is known to be enhanced by low temperatures and high nutrient concentrations (Kibbey et al., 1978; Davenport et al., 1976; Geldreich et al., 1968; Allen et al., 1952). Numerous reports have compared the survival rate of the fecal streptococci with that of the fecal and total coliform bacteria. These survival studies have employed a variety of natural and polluted test waters, including sewage (Cohen and Shuval, 1973; Kjellander, 1960), stormwater (Geldreich et al., 1968), marine waters (Vasconcelos and Swartz, 1976), and well water (McFeters et al., 1974; Gyllenberg et al., 1960), as well as polluted and ice-covered rivers (Cohen and Shuval, 1973; Davenport et al., 1976). Although *S. bovis* and *S. equinus* die rapidly in the environment (Geldreich et al., 1968; Slanetz and Bartley, 1965; McFeters et al., 1974), there is almost universal agreement that the enterococci persist for a longer period than do the coliform group.

Proper interpretation of these findings is contingent on a knowledge of the persistence of pathogens in similar circumstances. A number of comparisons of the survival of indicator bacteria and *Salmonella* species have been conducted. Fecal coliform bacteria were included in most of these studies, but fecal streptococci were frequently omitted. Data indicated that the ability of *Salmonella* to survive in the environment sometimes exceeded that of the fecal coliform bacteria (Vasconcelos and Swartz, 1976; McFeters et al., 1974; Gallagher and Spino, 1968; Geldreich et al., 1968). Greater persistence would seem to give the enterococcus group an advantage under these circumstances, and studies that have compared die-off of *Salmonella*, fecal streptococci, and fecal coliforms lend support to this idea. Geldreich et al. (1968) compared the survival of these organisms in stormwaters. Some of these data are shown in Fig. 3.

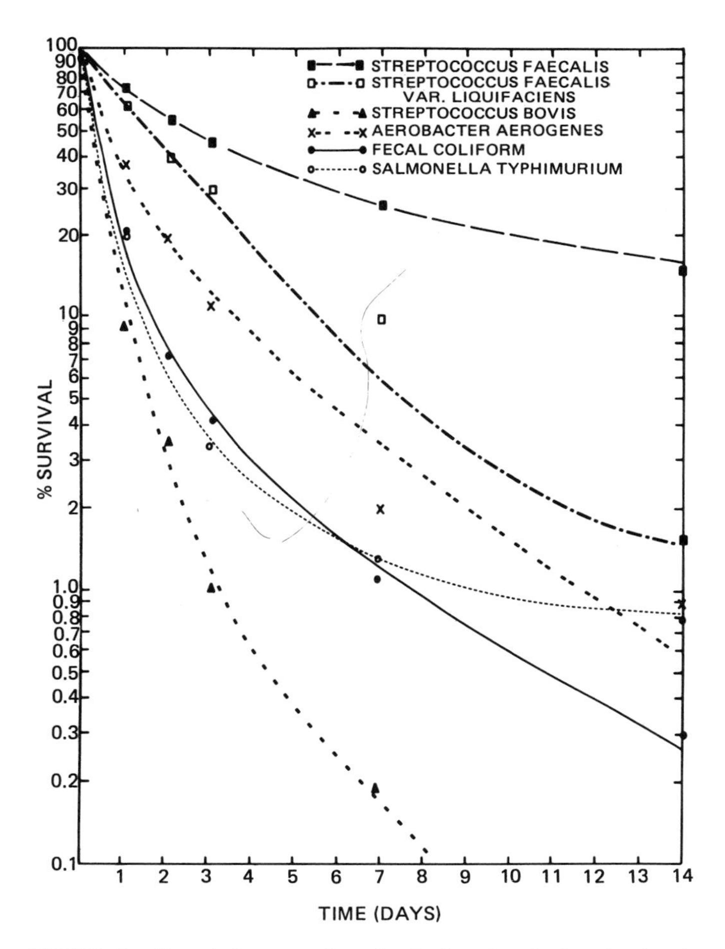

FIGURE 3 Persistence of selected enteric bacteria in stormwater stored at 20°C. (From Geldreich and Kenner, 1969).

Although the survival rate of the fecal coliforms best paralleled that of *Salmonella typhimurium*, die-off of these organisms was observed to exceed that of the *Salmonella* after a period of 7—12 days. The die-off of *S. faecalis* remained lower than that of the *Salmonella* throughout the duration of the experiment. McFeters et al (1974) studied the survival of a variety of fecal indicators and pathogenic organisms that were placed in membrane chambers and suspended in well water. The die-off of coliforms was slightly more rapid than that of enterococci. *Shigella* species appeared to survive longer than either indicator, and the persistence of *Salmonella* species varied but did not exceed the survival of the enterococci. *Salmonella* and *Shigella* are not the only bacterial pathogens that can outlive coliform indicators. *Vibrio parahaemolyticus* has also been found to be considerably more resistant than *Escherichia coli* in seawater at a mean temperature of 14.5°C (Vasconcelos and Swartz, 1976). Enteric viruses are also undesirable contaminants of water and are known to be removed very slowly. Although none of the fecal bacteria are suitable indicators of viral contamination, the possibility of viral pollution is better assessed by fecal streptococcal counts. Enterococcal die-off is less rapid than coliform removal and thus more nearly parallels that of viral destruction (Cohen and Shuval, 1973). This is particularly true for chlorinated sewage effluents. Viruses are relatively resistant to chlorination, and this persistence is better reflected by the enterococci, as they have a greater resistance to chlorine than do the coliform indicators (Cohen and Shuval, 1973; Kjellander, 1960; Silvery et al., 1974).

Factors other than survival must also be taken into account. The size of the initial population will affect the frequency with which an indicator organism is recovered. Thus, if the initial population of fecal coliforms is substantially larger than the number of streptococci present, it will be some time before the coliforms are reduced to a point at which they are recovered less frequently than the streptococci. Aftergrowth of *Salmonella* is best reflected by fecal coliforms, and this is considered in the following section. The presence or absence of fecal streptococci alone is not generally considered to provide a sufficient indication of water quality. It must be recognized, however, that fecal coliforms are not a satisfactory indicator under all conditions and that it is often advisable to consider both fecal coliform and fecal streptococcus densities when determining the sanitary quality of water.

C. Regrowth

Fecal streptococci do not multiply in sewage effluents, and therefore the density of these organisms reflects the degree of fecal contamination even in nutrient-rich waters. Regrowth of enteric pathogens and indicator bacteria can sometimes occur in the nutrient-rich effluents of sewage treatment plants and food processing plants. Regrowth of total coliforms in natural waters is a common phenomenon (Hendricks and Morrison, 1967; Silvery et al, 1974; Morris and Weaver, 1954). Aftergrowth of fecal coliforms has also been reported, although it appears to be unusual. Multiplication of fecal coliforms has been observed in chlorinated wastewater effluent (Shuval et al., 1973), in marine waters receiving sewage discharge* (Slanetz and Bartley, 1965), and in laboratory tests at a concentration of 0.28 ppm of organic matter (Allen et al., 1952). Other researchers have failed to detect any increase in these organisms below the outfall of sewage treatment plants (Silvery et al., 1974; Deaner and Kerri, 1969; Evans et al., 1968).

Salmonella occur in sewage effluents and are apparently capable of regrowth under certain conditions. Although Brezenski et al. (1965) determined that *Salmonella* were removed from sewage effluents by adequate chlorination, they have been recovered from chlorinated effluent (Silvery et al., 1974) and below outfalls of chlorinated effluent (Gallagher and Spino, 1968). Whether these organisms were resistant to chlorine or survived as a result of

---

*Effluent was contained in cellophane bags suspended in a polluted bay area.

inadequate chlorination was not determined. Aftergrowth of *Salmonella* below sewage outfalls has been reported in fresh (Hendricks and Morrison, 1967) and marine (Slanetz and Bartley, 1965) waters. Test organisms used in these studies were contained in dialysis tubing or cellophane bags and suspended in polluted waters. It is apparent that both *Salmonella* and fecal coliform bacteria are capable of multiplication in certain nutrient-rich effluents, although the frequency with which this occurs has not been established. It must presently be assumed that the regrowth of fecal coliforms is indicative of a potential for regrowth of *Salmonella* as well. Further work is required to verify this relationship. However, high densities of fecal coliform bacteria in nutrient-rich waters known to receive low levels of fecal pollution can not be overlooked but must be regarded as evidence of a potential health hazard.

Fecal streptococci have not been observed to multiply in sewage effluents (Slanetz and Bartley, 1965). The only waters in which regrowth of these organisms is reported to take place are the effluents of vegetable processing plants (Mundt et al., 1966). The streptococci present in these effluents are probably not of fecal origin. Fecal streptococci provide a more accurate estimation of the amount of fecal contamination received by other nutrient-rich waters than can be derived from coliform counts. Streptococci more accurately reflect the levels of these pathogens that are unable to multiply in such waters. Viruses are included in this category. Fecal streptococci may not provide an adequate indication of the presence of *Salmonella* under conditions that permit regrowth of this organism.

D. Occurrence on Plants

Plants and insects harbor large populations of group D streptococci which have not been shown to be of recent fecal origin. Significant numbers of enterococci have been recovered from a variety of plants and insects (Mundt et al., 1958; Mundt, 1961; Mundt, 1963; Geldreich et al., 1964; Mieth, 1961; Fitzgerald, 1947; Steinhaus,

1951; West, 1951). Characteristically, these organisms were not present on the entire test population but occurred randomly and were isolated from diverse species. These observations suggest that enterococci reside temporarily on plant and insect hosts and have the capacity for only limited reproduction in these circumstances. Such populations might originate from fecal sources and could be distributed to plants by means of insects, winds, and rains. The low incidence of fecal streptococci on plants in a wilderness region (Mundt, 1963) would tend to substantiate this theory.

However, the physiological differences between plant and fecal isolates have been the subject of numerous reports. Mundt (1975) compared the DNA composition of *S. faecium*-like and atypical heterogeneous streptococcal plant isolates with that of classified strains. Although the guanine/cytosine ratio of the atypical isolates fell within the range characteristic of streptococci, DNA/DNA homology indicated only distant relationship to *S. faecium, S. faecalis,* and *S. lactis*. Geldreich et al. (1964) observed that over one-third of the enterococci that were recovered from plants had the ability to hydrolyze starch. This characteristic was uncommon among strains isolated from water, warm-blooded animals, cold-blooded animals, insects, and soils. Mundt (1973) reported that strains of *S. faecalis* isolated from plant, animal, and insect sources could be distinguished from human isolates by their reaction in litmus milk. Ninety-eight percent of the human strains produced a reduced acid curd, whereas 82—94% of the plant, animal, and insect isolates gave a reduced rennet reaction. A subspecies that is isolated predominantly from plants has also been identified. *S. faecium* subspecies *casseliflavus* was first cultured from plant material and has since been recovered from a variety of leaves and vegetables (Mundt and Graham, 1968; Mundt et al., 1967). It has been found to occur with high frequency among insect populations as well (Martin and Mundt, 1972). These studies indicate that the enterococci occurring on plants include substantial numbers of biotypes that are specific to this environment and generally absent from fecal material.

Physiological differences characteristic of plant isolates in addition to a significantly lower incidence of fecal coliforms among plant and insect populations (Geldreich et al., 1964) suggest that enterococci recovered from these sources are not of recent fecal origin and may not have originated from fecal material at all. This conclusion is of significance because of high densities of enterococci reported in the effluent of vegetable processing plants (Geldreich and Kenner, 1969; Mundt et al., 1966). These streptococci are not generally regarded as indicative of fecal pollution. If fecal streptococci are to provide a reliable indication of health hazards associated with water, it is imperative that the origin of the enterococci isolated from plants be more fully understood. It is apparent that there is a great need for a method by which strains originating from plant and animal sources can be distinguished. Kibbey et al. (1978) have recently reported a higher incidence of multiple antibiotic resistance among recent fecal isolates than among strains recovered from soils and vegetation. They recommend drug resistance as a criterion for the recognition of fecal strains.

S. faecalis subspecies *liquefaciens* is not a plant-specific strain, but it may be distributed ubiquitously in the environment. Geldreich and Kenner (1969) have reported the recovery of high numbers of this subspecies from two unpolluted wells. High percentages of *S. faecalis* subspecies *liquefaciens* were observed among insect and soil samples (Geldreich and Kenner, 1969), and in waters that contained low densities of streptococci, this subspecies tended to predominate (Geldreich, 1970). Percent occurrence of *S. faecalis* subspecies *liquefaciens* in unpolluted or slightly polluted watersheds and wells from which streptococci were recovered ranged from 0 to 35% (Geldreich, 1976). However, this subspecies also comprises approximately 25% of the *Streptococcus* population of human feces (Geldreich and Kenner, 1969; Cooper and Ramadan, 1955). It is not absolutely clear whether high proportions of this subspecies could result from greater-than-average persistence or whether this organism is ubiquitous and does not originate from feces. In greenhouse studies,

S. faecalis subspecies *liquefaciens* was able to establish a commensal relationship with rye, corn, and cabbage plants (Mundt et al., 1962). In these experiments commensals were defined as organisms that occurred in the seed or soil, moved from the seed to emergent parts, and reproduced on the growing plant. Thus, it may be desirable to distinguish *S. faecalis* subspecies *liquefaciens* from other fecal streptococci when assessing water quality.

There is some question as to the actual effect that plant populations of enterococci have on the bacterial populations of natural waters. Aside from known high densities occurring in cannery effluents, this relationship has not been explored. Significant numbers of enterococci have not been recovered from watershed waters and soils (Lee et al., 1970; Medrek and Litsky, 1960). The frequency of enterococci on plants in these test areas was not determined. Whereas low streptococcal counts should be interpreted cautiously until this situation is resolved, high numbers of enterococci must be regarded as indicative of fecal pollution, except where vegetable processing effluents are involved.

IV. METHODOLOGY FOR RECOVERY OF FECAL STREPTOCOCCI BY MEMBRANE FILTRATION

A. Recovery Media

Selective Ingredients

Sodium azide. All the media that will be discussed derive their selective properties partially or entirely from sodium azide. Sodium azide (NaN_3) inhibits those aerobically growing organisms that obtain energy from respiration. The azide binds to the heme portion of reduced cytochrome a_3, preventing oxidation of this cytochrome (Cantarow and Schepartz, 1962). Other cytochromes are probably affected also. Electron transport is thus blocked, oxidative phosphorylation prevented, and the cell is unable to produce energy to grow. Azide also inhibits the activity of catalase by reacting with the heme group of this enzyme. Catalase is responsible for the breakdown of peroxide, a toxic product of respiration.

Under aerobic conditions, sodium azide thus inhibits the growth of most aerobes and facultative anaerobes. The streptococci and other lactic acid bacteria are unable to synthesize heme proteins and therefore lack both cytochromes and catalase (Whittenbury, 1964). Consequently, they grow well in the presence of low concentrations of sodium azide.

The toxic properties of azide are altered by pH, age, and heat and therefore certain precautions must be observed during the preparation of azide-containing media. The breakdown of azide results in the formation of hydrazoic acid, a substance that is considered more toxic than azide itself (Smuckler and Appleman, 1965). This reaction occurs particularly rapidly in acidic media, so careful adjustment of the pH of these media is critical. Azide is known to deteriorate with age. Dehydrated media containing this compound have a maximum shelf life of 2 years, after which they should be discarded (Geldreich, 1976). Although azide is also heat-sensitive (Guthof and Dammann, 1958), a variety of media have been formulated that require the addition of sodium azide prior to steam sterilization. It therefore appears that this method of preparation is acceptable provided that the concentration of azide is adjusted to compensate for some loss during autoclaving (Hartman et al., 1966).

2,3,5-Triphenyl tetrazolium chloride (TTC). TTC is included in both KF and m-Enterococcus media to produce colonies with a red pigmentation, which improves detection and facilitates counting. This coloration results from the reduction of TTC to insoluble red triphenylformazan. At pH 6.0, TTC reduction is used to differentiate *S. faecium* from *S. faecalis* (Barnes, 1956). This distinction is based on the inability of *S. faecium* to reduce TTC at this pH. At the higher pH of 7.2 recommended for KF and m-Enterococcus media, all the fecal streptococci reduce TTC. Donnelly and Hartman (1978) have observed that TTC is inhibitory to some fecal streptococci. They recommend that it should not be added to culture media but report that it can be sprayed onto agar plates subsequent to incubation. Other streptococci and lactic acid bacteria (Laxminarayana and Iya, 1953), as well as numerous other bacteria, are also capable of TTC reduction. The viridans streptococci and other lactics are most important in this regard, as they are also tolerant of sodium azide. Although these organisms generally do not survive in natural waters, neither KF nor m-Enterococcus medium is designed to inhibit their growth.

Esculin. Esculin hydrolysis coupled with bile resistance has been used to distinguish the fecal streptococci. A variety of organisms are able to hydrolyze esculin but do not ordinarily grow in the presence of bile. Among these are a number of viridans species, including *S. salivarius, S. anginosus, S. sanguis,* and *S. mutans* (MacFaddin, 1976). Only the group D streptococci are uniformly capable of both esculin hydrolysis and resistance to 40% bile (Swan, 1954; Facklam, 1973).

Esculin is a glycoside that undergoes acid hydrolysis to yield D glucose and esculitin (Harrison and van der Leck, 1909) (Fig. 4). Esculitin then reacts with the iron salts present in the medium to form a phenolic iron complex that is black to dark brown in color. Esculin-hydrolyzing colonies are distinguished by this black-brown halo, which occurs in the agar surrounding the colony. Iron salts are provided by the addition of ferric citrate to the medium.

Media Formulations

KF Streptococcus Medium. KF Streptococcus medium was formulated by Kenner et al. in 1961. This medium has since gained widespread acceptance as the medium of choice for membrane filter enumeration of fecal streptococci and is frequently employed for most probable number (MPN) determinations as well. The formulation for this medium appears in Table 4. Sodium azide is the primary selective ingredient, and differentiation results from the inclusion of 0.01% TTC. The original formulation included bromcresol purple dye as a pH indicator. This dye was used to detect positive-MPN tubes, which turned yellow as the pH was lowered by the organism. It serves no function when the medium is used for membrane enrichment, and some manufacturers (including BBL) now omit it from the

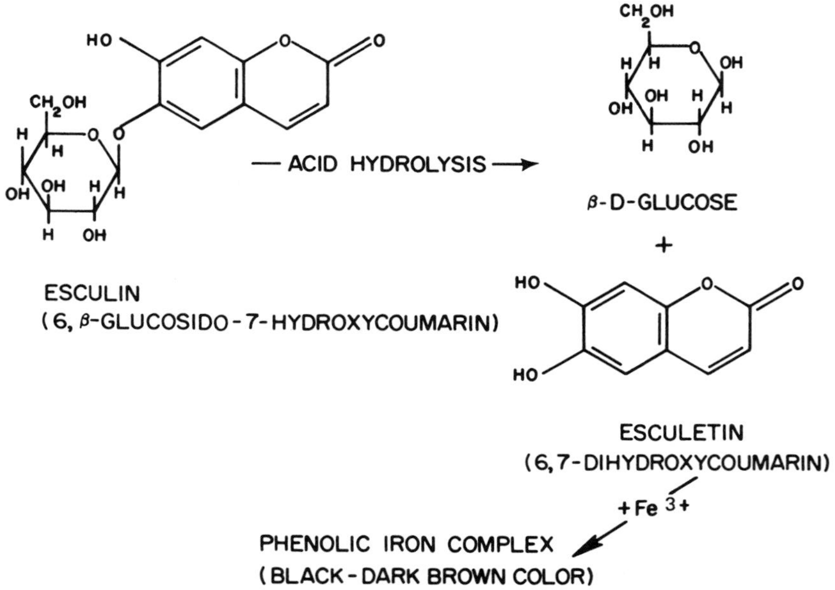

FIGURE 4 Esculin hydrolysis. (From MacFaddin, 1976.)

TABLE 4 Formulation of KF Medium

Ingredient	Grams per liter
Proteose peptone No. 3	10.0
Bacto-yeast extract	10.0
Sodium chloride	5.0
Sodium glycerophosphate	10.0
Maltose	20.0
Lactose	1.0
Sodium azide	0.4
Sodium carbonate	0.636
Bromcresol purple[a]	0.015
2,3,5-Triphenyltetrazolium chloride[b]	0.1

[a]Omitted in agar medium.
[b]Omitted in broth medium.

agar product. Original instructions called for steam sterilization of this medium for 10 min at 15 psi pressure (250°F). It has since been determined that the selectivity of the medium is enhanced if it is boiled for 5 min instead. This recommendation has been adopted by both BBL and Difco. Filter-sterilized TTC is added subsequent to heating.

Plates are counted following a 48 hr incubation at 35°C. Colonies ranging from light pink to deep red are considered fecal streptococci.

Recovery by KF Streptococcus medium significantly exceeds recovery by both m-Enterococcus medium and the azide dextrose-ethyl violet azide (Ad-EVA) MPN (Pavlova et al., 1972; Daoust and Litsky, 1975; Kenner et al., 1961; Switzer and Evans, 1974; Geldreich, 1976). Superior recovery by KF is attributed to the ability of this medium to support a greater variety of strains than can grow on either m-Enterococcus or AD-EVA media (Kenner et al., 1961). Whereas the latter media recover only enterococci, it is well documented that KF supports the growth of strains of *S. bovis* and *S. equinus* (Geldreich, 1976; Kenner et al., 1961). *S. mitus* and *S. salivarius* have also been recovered on this medium (Kenner et al., 1961). Comparisons between KF and Pfizer Selective Enterococcus medium (PSE) generally indicate equivalent recovery (Pavlova et al., 1972; Daoust and Litsky, 1975; Geldreich, 1976), although there are indications that PSE may be superior for evaluation of marine waters (Levin et al., 1975; Slanetz and Bartley, 1964). Use of PSE for membrane enrichment has been curtailed, however, by difficulties in detecting esculin hydrolysis beneath the membrane.

Although KF is currently the medium of choice for membrane recovery of fecal streptococci, this medium has a number of disadvantages. The lengthy incubation period hinders prompt action when the sanitary quality of water is questioned. Furthermore, *Staphylococcus aureus* grows well on this medium and is indistinguishable from fecal streptococci (Mossel et al., 1957; Litsky, 1977). Reports of other false positives include recovery of strains of lactobacilli (Raibaud et al., 1961; Mundt, 1976), pediococci (Mundt, 1976), aerococci (Mundt, 1976), pseudomonas, filamentous gram-negative rods, and micrococci (Slanetz and Bartley, 1964). There are also indications that this medium does not recover all of the fecal streptococci present in polluted waters. A number of recently formulated streptococci media and enrichments are reported to yield counts significantly greater than those obtained on KF (Rose and Litsky, 1965; Lin, 1974; Levin et al., 1975; Abshire, 1977; Donnelly and Hartman, 1978). Bissonnette (1974) has also demonstrated that KF and other selective media recover fecal streptococci at a much lower efficiency than does a nonselective medium (Table 5). This difference is attributed to the inability of the of the selective media to promote the growth of substantial numbers of injured cells. It is thus apparent that this formulation can be improved upon, and continued research will undoubtedly yield a superior medium.

m-Enterococcus medium. M-Enterococcus was one of the first media formulated for the enumeration of streptococci by the membrane filter technique. Proposed in 1957 (Slanetz and Bartley, 1957), it is still the only medium that has been developed specifically for the recovery of enterococci from water and wastewater by membrane filtration. The formulation of m-Enterococcus medium contains the same concentrations of sodium azide and TTC as are included in KF. Sugar and protein sources and concentrations in these media differ. The complete formulation of m-Enterococcus medium appears in Table 6. Most strains of *S. bovis* and *S. equinus* appear to require the additional enrichments contained in KF and are not able to grow on m-Enterococcus medium. Like KF, m-Enterococcus requires a 48 hr incubation period at 35°C. Pink to red colonies are regarded as enterococci.

Counts obtained using m-Enterococcus are generally lower than values obtained with either KF or PSE media (Pavlova et al., 1972;

TABLE 5 Recovery of *S. faecalis* RS 1009 Suspended in a Membrane Filter Chamber at Site EG6 with Selective and Nonselective Media by Different Enumeration Methods over a 4 Day Exposure Period

Exposure time (days)	Average TSY[b]	Percent[a] cells detected by: MPN			Pour Plate			Membrane Filtration		
		AD	EVA	PSE	KF	MEnt[c]	PSE	KF	MEnt[a]	PSE
0	100.00	78.3	78.3	108.3	91.7	91.7	91.7	91.7	83.3	83.3
1	100.0	64.8	98.6	46.5	71.8	88.7	19.7	66.2	95.8	38.0
2	100.0	7.6	11.4	1.7	3.8	51.7	0.3	1.7	55.2	1.0
3	100.0	14.9	13.4	5.5	12.8	44.7	2.1	14.9	31.9	8.5
4	100.0	33.8	41.5	2.6	7.7	40.0	0.8	11.5	29.2	0.8

[a]As a percentage of the average counts obtained by MPN, pour plate, and membrane filtration using TSY medium.
[b]Average counts on TSY medium when using MPN, pour plate, and membrane filtration.
[c]m-Enterococcus medium.
Source: Bissonnette (1974).

TABLE 6 Formulation of m-Enterococcus Agar

Ingredient	Grams per liter
Bacto-tryptose	20.0
Bacto-yeast extract	5.0
Bacto-dextrose	2.0
Sodium azide	0.4
Bacto-agar	10.0
2,3,5-Triphenyltetrazolium chloride	0.1

Daoust and Litsky, 1975; Geldreich, 1976; Kenner et al., 1961). This difference is attributed to recovery of a lesser number of strains and particularly to the inability of this medium to support growth of *S. bovis* and *S. equinus*. False positives recovered from this medium include lactobacilli (Raibaud et al., 1961; Saraswat et al., 1963) and one strain of *S. bovis* (Raibaud et al., 1961). Recovery on m-Enterococcus medium has been found to be generally equivalent to recovery by the AD-EVA MPN (Croft, 1959; Kenner et al., 1961).

Use of m-Enterococcus is advised in circumstances in which detection of animal pollution is not desired. It is obvious that the fecal coliform/fecal streptococcus ratio cannot be successfully applied to counts obtained from this medium. Samples having ratios indicative of animal pollution if cultured on KF were found to produce uncertain results when m-Enterococcus values were substituted (Geldreich, 1976).

Pfizer Selective Enterococcus medium. Although bile esculin media have long been employed for clinical identification of group D streptococci (Swan, 1954), PSE is a fairly recent formulation. Since its introduction 8 years ago, it has been used both by clinical and environmental microbiologists. The selectivity and sensitivity of this medium is generally agreed to equal that of KF medium, and PSE has the advantage of a shorter incubation time. Unfortunately, the medium is cumbersome to adapt to membrane filtration. Difficulties are encountered because esculin hydrolysis occurs in the agar beneath the membrane and is not apparent on the surface of the filter.

The selectivity of this medium is based on the ability of fecal streptococci to tolerate bile and to hydrolyze esculin. In addition to bile and esculin, the medium contains a low concentration of sodium azide, which supresses catalase-positive organisms. It must be noted that the bile concentration has been reduced from the 40% concentration standardly used for determination of bile resistance. This reduction may aid in recovery of stressed cells and less resistant strains. The complete formulation of this medium appears in Table 7.* The recommended period of incubation is 18 hr at 35°C. Colonies that develop a dark brown halo characteristic of esculin hydrolysis are regarded as fecal streptococci. Enterococcosel medium (BBL) has an identical formulation except that the agar is omitted and Bacto-Bile Esculin Azide Agar (Difco) has a comparable formula, although the concentration of sodium azide is slightly reduced. These media have not been used extensively for environmental studies, and therefore evaluation of the bile esculin media must presently be based on investigations performed using PSE.

*Pfizer has recently discontinued manufacture of PSE. We are informed, however, that this medium will soon be available from Gibco/Invenex, Chagrin Falls, Ohio 44022.

TABLE 7 Formulation of Pfizer Selective Enterococcus Medium

Ingredient	Grams per liter
Pfizer peptone C	17.0
Pfizer Peptone B	3.0
Pfizer yeast extract	5.0
Pfizer bacteriological bile	10.0
Sodium chloride	5.0
Sodium citrate	1.0
Esculin	1.0
Ferric ammonium citrate	0.5
Sodium azide	0.25
Pfizer agar	15.0

PSE was primarily developed for use with pour plates and spread plates. To utilize this medium in conjunction with membrane filtration, an overlay technique has been described (Daoust and Litsky, 1975). Membranes are overlayed with a thin layer of PSE tempered to 45°C. This overlay makes esculin hydrolysis apparent on the surface of the membrane. Recovery by this method was found to approximate recovery on KF agar.

PSE was first evaluated as a medium for primary isolation and confirmation of group D streptococci in clinical laboratories. It has proven to be satisfactory for presumptive isolation of both enterococcal and nonenterococcal group D species (Isenberg et al., 1970; Sabbaj et al., 1971; Facklam, 1973). The ability of this medium to recover fecal streptococci from environmental waters has also been assessed. Most reports indicate that PSE equals the selectivity and sensitivity of KF and exceeds the recovery rate of m-Enterococcus medium (Pavlova et al., 1972; Daoust and Litsky, 1975; Geldreich, 1976). Like KF, PSE was observed to support the growth of strains of *S. bovis* and *S. equinus* (Daoust and Litsky, 1975; Pavlova, et al., 1972; Sabbaj et al., 1971; Facklam, 1973) and *S. salivarius* (Daoust and Litsky, 1975). Recovery by PSE has been observed to exceed KF in certain studies in which marine samples (Levin et al., 1975) and bovine feces (Switzer and Evans, 1974) were tested.

Although PSE is found to be generally highly selective for group D streptococci, certain organisms that produce false-positive reactions have been recognized. For this reason, PSE is not recommended as a clinical confirmatory medium (Facklam, 1973). False positives include *Listeria monocytogenes*, which not only tolerates the bile but also hydrolyzes esculin. This organism is distinguished by its colony size and by the difference in the intensity of the blackened zone (Isenberg et al., 1970). *S. aureus* and *S. epidermidis* also grow but fail to hydrolyze esculin. More difficult to distinguish are certain strains belonging to the viridans group which produce positive reactions. *S. sanguis* and *S. mutans* both hydrolyze esculin, and certain strains are known to tolerate 40% bile (Sabbaj et al., 1971). When PSE was tested with large numbers of viridans isolates primarily of clinical origin, 9% produced false-positive reactions during the first 24 hr of incubation (Facklam, 1973). Recovery of these viridans strains may be attributed to reduction of the bile concentration to 10%. Other bile esculin media in which bile concentration is adjusted to 40% were demonstrated to recover many fewer viridans strains (Facklam, 1973). Group Q streptococci are also recovered on PSE (Facklam, 1973).

Other formulations. Several media and enrichments that have been recently developed for the enumeration of fecal streptococci are reported to surpass the recovery of the standard media.

Peptone-yeast extract-casitone enrichment (PYC). In 1965, Rose and Litsky proposed an enrichment step that yielded a twofold average increase in enterococcus recovery from sewage samples. Following sample filtration, membranes were enriched on PYC broth pads for a period of 3 hr at 35°C. These filters were subsequently transferred to m-Enterococcus agar and incubation was continued at 35°C for the remainder of the 48 hr incubation period. Recovery from sewage and river samples was compared with the standard membrane filter procedure using m-Enterococcus agar. A substantial increase in enterococcus counts was observed on the enriched membranes.

Bile broth enrichment. Lin (1974) reported a significant difference between membrane and MPN recovery of enterococci from chlorinated sewage samples. Membrane tests were performed on m-Enterococcus agar and AD-EVA broths were employed for MPN determinations. Although these recovery methods were found to be equivalent when testing unchlorinated sewage effluent, recovery from chlorinated samples was much lower when the membrane filter procedure was used. However, membranes enriched on bile broth for 2-3 hr prior to standard incubation on m-Enterococcus agar yielded counts equivalent to the MPN. Similar results were obtained if the incubation period for unenriched membranes was extended to 72 hr.

m-E media. Levin et al. (1975) have recently developed a two-step membrane filter procedure for recovery of enterococci from marine waters. The two media required for this test are referred to as mE media (Table 8). Membranes are first plated on mSD agar and incubated at 41°C for 48 hr. Selectivity is conferred not only by the elevated incubation temperature but also by the inclusion of nalidixic acid and sodium azide. Acitidione is added to inhibit fungal growth and TTC is present to discern the enterococcus colonies. MSD agar also contains esculin, which is included to induce the esculin-hydrolyzing enzyme. After 48 hr, membranes are transferred to EIA agar and incubation continues at 41°C for an additional 20-30 min. During this period esculin-hydrolyzing colonies produce a phenolic iron complex which appears as small dark spots on the medium beneath the colonies. Colonies that reduce TTC on mSD and hydrolyze esculin are considered to be enterococci. Nonenterococcal species of fecal streptococci, including *S. bovis*, *S. equinus*, *S. mitus*, and *S. salivarius*, failed to grow on these

TABLE 8 Formulation of mE Media

Ingredient	Grams per liter
mSD Medium	
Yeast extract	30.0
Actidione	0.050
Sodium azide	0.150
Peptone	10.0
Sodium chloride	15.0
Esculin	1.0
Agar	15.0
Nalidixic acid	0.24
2,3,5-Triphenyltetrazolium chloride	0.1
EIA Medium	
Yeast extract	30.0
Actidione	0.050
Sodium azide	0.150
Esculin	1.0
Ferric citrate	0.5
Agar	15.0

media. Confirmatory studies indicated that the rate of false-negative reactions was 11.7% and that the false-positive rate was 10%.

D Streptococcus-enterococcus medium (DSE). DSE medium was formulated by Abshire in 1977. When used as an MPN medium, its recovery rate was reported to exceed that of KF, m-Enterococcus, SF (Hajna and Perry, 1943), and AD broths. However, it is not suitable for membrane enrichment; used for this purpose, it produced lower counts than were obtained with m-Enterococcus agar.

The formulation for DSE medium appears in Table 9. Selection results from the inclusion of azide. A final pH of 7.6 was found to yield optimal recovery. Acid production is demonstrated by phenol red indicator, and tubes that turn yellow are assumed to contain fecal streptococci. When DSE was used with the membrane filter procedure, TTC was added.

MPN tubes were incubated for only 24 hr. An incubation temperature of 45°C was observed to yield maximal recovery at the end of this incubation period. Whereas test strains of *S. bovis* and *S. equinus* grew in this medium, an extremely low percentage of these organisms was recovered when the medium was inoculated with sewage. The overall confirmation rate was reported to be high.

Gentamicin-thallous-carbonate medium (GTC). GTC (Donnelly and Hartman, 1978) is a very recent formulation, and whether or not it can be successfully applied to membrane filtration has not yet been

TABLE 9 Formulation of DSE Medium

Ingredient	Grams per liter
Todd-Hewitt broth	30.0
Sodium chloride	5.5
Dextrose	1.0
Sodium azide	0.4
Phenol red	0.016
2,3,5-Triphenyltetrazolium chloride[a]	0.01

[a]Omitted in broth medium.

determined. GTC is an esculin medium that contains thallous acetate and gentamicin sulfate for selection of group D (Table 10). Plates are incubated for a period of 24 hr. Tests with river water and sewage indicated that GTC spread plates recovered greater numbers of group D streptococci than did either PSE or KF agars. This increase was found to be attributable to improved recovery of *S. bovis* and *S. equinus*. The percentage of false positives recovered from river samples was 13% on GTC agar, however, although no false positives were detected on KF.

B. Incubation Temperature

All the standard MPN and membrane procedures for recovery of fecal streptococci from water specify incubation at a temperature of 35°C. However, the use of elevated temperatures has been advocated from time to time, and it is appropriate to weigh the advantages of each incubation scheme. The idea of 44.5°C incubation coupled with enrichment in a selective azide medium appears to have originated with Hajna and Perry (1943), who developed SF broth for which this incubation temperature is recommended. Incubation at an elevated temperature is also required for certain recent formulations, including Abshire's DSE (1977) and Levin et al.'s mE media (1975).

Surprisingly little effort has been made to compare streptococcus recovery at 35 and 45°C, but a few studies do exist. Splittstoesser et al. (1961) compared recovery of AD and SF broths

TABLE 10 Formulation of Gentamicin-Thallous-Carbonate Medium

Ingredient	Grams per liter
Trypticase soy agar	40.0
Potassium phosphate, monobasic	5.0
Sodium carbonate	2.0
Glucose	1.0
Esculin	1.0
Thallous acetate	0.5
Ferric citrate	0.5
Tween-80	0.75 ml
Gentamicin sulfate	0.0025

incubated at 35 and 45°C. They determined that AD recovered enterococci equally well at either temperature but that there was significant reduction in SF recovery at the higher temperature. It was suggested that these results might be indicative of an interaction between temperature and azide concentration in that SF broth contains 2-1/2 times more azide than is found in AD broth. It is of interest that 35 of the 43 false positives isolated during this study were unable to grow at 45°C.

Facklam (1973) evaluated recovery of PSE and Bile Esculin medium at 35°C and 45°C. Testing was done with a total of 191 cultures that were distributed among the following groups: group D enterococci, group D and Q enterococcus-like, group D nonenterococci, viridans streptococci, and groupable streptococci. Following the first 24 hr of incubation, the elevated incubation temperature appeared to be responsible for only a small reduction in recovery among the five groups mentioned. However, after 48 hr elevated incubation temperature resulted in a 20% decrease in growth of false-positive viridans streptococci, and by 72 hr this reduction was even more dramatic.

Recent studies conducted in our laboratory (unpublished data) required speciation of streptococci isolated from sewage on KF (BBL) agar using the MF technique. Almost 20% of the 160 isolates examined were found to be false positives in that they neither produced a positive percipitin reaction with Group D antiserum nor conformed to the Sherman criteria. Nearly all of these organisms were catalase-negative, and none produced a recognizable pattern of reactions when subjected to biochemical tests for the speciation of streptococci. As many as 50% of the isolates obtained from particular sources were found to be false positives. However, 84% of these false-positive organisms failed to grow at 45°C.

Incubation at an elevated temperature thus has the advantage of reducing the number of false-positive organisms that are recovered at 35°C. A significant number of those strains that are isolated from plants and fail to conform to existing speciation schemes are also incapable of growth at 45°C. This is not a distinguishing feature of this group, however, in that a substantial number of these strains can also grow at the elevated temperature. Many false positives and some apparently plant-specific strains would be eliminated by increasing the incubation temperature to 45°C.

This advantage must be weighed against some overall reduction in recovery and, more important, against the effect that such a dramatic temperature increase might have on stressed and injured cells. Studies of fecal coliform recovery (Green et al., 1977) suggest that injured cells may not grow if immediately subjected to elevated temperatures. A preincubation period of 5 hr at 35°C prior to standard 44.5°C incubation substantially increased recovery of chlorine-injured fecal coliform bacteria. A similar two-temperature incubation scheme for recovery of fecal streptococci was reported by Burman in 1961. Membranes were incubated at 35°C for the first 4 hr and then at 45°C for the remaining 44 hr of incubation. The elevated temperature was found to increase selectivity, and most isolates were identified as recognized group D species. Standard incubation at 35°C was observed to yield high proportions of non-group D, unidentified streptococci. Effects of the elevated temperature on the sensitivity of the test were not determined. Such an incubation scheme was also suggested by Hartman et al. in 1966 but has never been widely investigated.

C. Membrane Filters

Variability among different lots and brands of membrane filters has been the subject of much discussion in recent years. Although a number of membrane comparisons have been conducted for the evaluation of fecal coliform recovery (Green et al., 1975; Lin, 1976; Dutka et al., 1974; Presswood and Brown, 1973; Schaeffer et al., 1974), the effect of filter variability on the recovery of fecal streptococci has been largely ignored. Only three studies have been published that compare recovery of fecal streptococci on a variety of membranes. Dutka et al. (1974) compared recovery of

Gelman, Sartorius, and Millipore HA membranes. The Sartorius and Millipore filters were separated into autoclave-sterilized and ethylene oxide-treated groups for further comparison. River and canal waters were used as test inocula. No conclusive difference in streptococcus recovery was observed among the membranes tested. Lin (1976) evaluated Millipore HA and HC membranes. He observed slightly superior recovery of fecal streptococci on the HC membranes when tests were performed with river and stream samples. The HC/HA ratio for these samples was 1.29. Tests conducted with sewage indicated essentially equivalent streptococcus recovery by these two membranes. Davenport et al. (1976) compared recovery of fecal streptococci from sewage and polluted river water using Millipore HA and Gelman membranes. Counts obtained with these two filters were found to be similar. Thus, no significant difference in the ability of various membranes to recover fecal streptococci has been demonstrated.

D. Membrane Filtration vs. the MPN Technique

No discussion of membrane filtration can avoid consideration of the multiple-tube dilution or MPN technique. MPN procedures for fecal streptococcus enumeration preceded development of membrane filtration methods and have provided a standard against which membrane methodology has been compared throughout its development. MPN values continue to provide the basis for evaluation of membrane recovery, owing primarily to the apparent ability of the MPN procedure to maximize resuscitation of injured cells (Bissonnette, 1974).

The MPN procedure most widely employed for recovery of fecal streptococci is still the AD-EVA MPN method proposed by Litsky et al. in 1953. The test is performed by innoculation of a minimum of three decimal dilutions of sample into three to five replicate tubes of AD broth. Tubes are incubated at 35°C and those that show turbidity after 24—48 hr are transferred to EVA broth for confirmation. The presence of fecal streptococci in EVA medium is indicated by dense turbidity or by the formation of a purple button after 24 hr of incubation at 35°C. The most probable concentration of organisms in the sample is based on the number of positive tubes in each dilution and is easily determined using an MPN table. The AD-EVA MPN recovers primarily enterococci. Most strains of *S. bovis* and *S. equinus* fail to grow in EVA (Kenner et al., 1961). Recovery by this method therefore approximates recovery by m-Enterococcus agar (Croft, 1959) but is significantly lower than that of KF and PSE media. The AD-EVA MPN is not recommended for use with marine water (American Public Health Association, 1976), possibly because of reports of frequent false positives under these circumstances (Buck, 1972). It has been suggested that PSE medium be substituted for EVA broth as the confirmatory step of this MPN. This alteration would be expected to substantially increase recovery of *S. bovis* and *S. equinus* and would eliminate the uncertainties sometimes experienced in detecting purple button formation.

An alternative MPN method was developed by Kenner et al. in 1961 and has been found satisfactory for the enumeration of fecal streptococci. This MPN method differs from AD-EVA MPN in that no confirmatory medium is required. Multiple dilutions of sample are inoculated into KF broth and tubes are incubated for 48 hr at 35°C. This medium contains bromcresol purple (a pH indicator), and tubes producing turbid growth and a bright yellow color are considered to be positive. Any tubes exhibiting marked foaming must be confirmed with a gram stain. In addition to enterococci, this MPN recovers *S. bovis, S. equinus, S. mitus,* and *S. salivarius* (Kenner et al., 1961).

Most comparisons of MF and MPN recovery of streptococci have been performed using a different selective medium for each procedure, and it is therefore difficult to ascertain the cause of observed differences in recovery. It appears that the use of a rich nonselective medium minimizes recovery differences. Bissonnette (1974) reported that injured *S. faecalis* cells were recovered with equal efficiency by MF, pour plate, and MPN techniques if TSY was used as a growth medium. When standard selective media were used for recovery of these cells, the AD-EVA MPN recovered a substantially higher percentage of organisms than any of the MF media. Data from this experiment are shown in Table 5. Such differences are frequently attributed to the fact that broth media immerse the injured cells more effectively, which is thought to enhance resuscitation. However, such differences are seldom separated from differences in the formulation of the recovery media. MPN counts with KF broth are reported to exceed recovery by membranes incubated on m-Enterococcus agar or on KF medium. Whether higher values were obtained from the KF-MPN procedure as a result of an inherent superiority of the MPN technique in resuscitating stressed cells and fragile strains or whether they are attributable to the use of bromcresol purple rather than TTC (included only in the membrane filtration medium) for detection of positives was not determined. It must also be noted that Abshire (1977) reported similar disparity between MPN and membrane streptococcus counts when an experimental medium (DSE) was used with both procedures.

In choosing between MF and MPN techniques, consideration must be given to certain factors in addition to recovery. The membrane filter procedure is more convenient and counts are available sooner than when the AD MPN is employed. Filtration is not practical, however, in situations in which turbidity, high background numbers, or the presence of metallic compounds or coagulants obscure or prohibit the development of colonies on the filter surface.

V. SUMMARY

Fecal streptococci are an important supplement to the coliform bacteria test for assessment of the sanitary quality of water. Their usefulness is derived from their inability to multiply in natural and polluted waters, their resistance to saline conditions and chlorination, and the specificity of certain streptococcal species to particular vertebrate groups. Acceptance of these organisms as fecal indicators has been limited, however, because of their occurrence on vegetation, the 48 hr period necessary for their enumeration, and because they do not occur in human feces in as high densities as do fecal coliform organisms.

KF medium is preferred for membrane enumeration of these organisms. It must be recognized, however, that currently used selective media fail to recover significant populations of injured organisms and that certain false positives occur on KF. An improved formulation would therefore be advantageous.

REFERENCES

Abshire, R. L. (1977). Evaluation of a new presumptive medium for group D streptococci. *Appl. Environ. Microbiol. 33:* 1149–1155.

Allen, L. A., Pasley, S. M., and Pierce, M. A. F. (1952). Some factors affecting the viability of fecal bacteria in water. *J. Gen. Microbiol. 7:* 36–43.

American Public Health Association (1976). *Standard Methods for the Examination of Water and Wastewater,* 14th ed. American Public Health Association, Washington, D.C.

Barnes, E. M. (1956). Tetrazoleum reduction as a means of differentiating *Streptococcus faecalis* from *Streptococcus faecium. J. Gen. Microbiol. 14:* 57–68.

Bartley, C. H., and Slanetz, L. W. (1960). Types and sanitary significance of fecal streptococci isolated from feces, sewage, and water. *Am. J. Public Health 50:* 1545–1552.

Bissonnette, G. K. (1974). Recovery characteristics of bacteria injured in the natural aquatic environment. Doctoral thesis. Montana State University, Boseman, Mont.

Bordner, R., and Winter, J., Eds. (1977). *Microbiological Methods for Monitoring the Environment: I. Water and Waste.* U.S. Environmental Protection Agency, Environmental Monitoring and Support Laboratory, Cincinnati, Ohio.

Brezenski, F. T., Russomanno, R., and DeFalco, P., Jr. (1965). The occurrence of *Salmonella* and *Shigella* in post chlorinated and non-chlorinated sewage effluents and receiving waters. *Health Lab. Sci. 2:* 40–47.

Buck, J. D. (1972). Selective detection of enterococci in marine waters. *Am. J. Public Health 62:* 419–421.

Burm, R. J., and Vaughn, R. D. (1966). Bacteriological comparison between combined and separate sewer discharges in southeastern Michigan. *J. Water Pollut. Control Fed. 38:* 400–409.

Burman, N. P. (1961). Some observations on coli-aerogenes bacteria and streptococci in water. *J. Appl. Bacteriol. 24:* 368–376.

Cantarow, A., and Schepartz, B. (1962). *Biochemistry*, 3rd ed. Saunders, Philadelphia, pp. 10–11.

Cohen, J., and Shuval, H. I. (1973). Coliforms, fecal coliforms, and fecal streptococci as indicators of water pollution. *Water Air Soil Pollut. 2:* 85–95.

Cooper, K. E., and Ramadan, F. M. (1955). Studies on the differentiation between human and animal pollution by means of fecal streptococci. *J. Gen. Microbiol. 12:* 180–190.

Croft, C. C. (1959). A comparative study of media for detection of enterococci in water. *Am. J. Public Health 49:* 1379–1387.

Daoust, R. A., and Litsky, W. (1975). Pfizer selective enterococcus agar overlay method for enumeration of fecal streptococcus by membrane filtration. *Appl. Microbiol. 29:* 584–589.

Davenport, C. V., Sparrow, E. B., and Gordon, R. C. (1976). Fecal indicator bacteria persistance under natural conditions in an ice-covered river. *Appl. Environ. Microbiol. 32:* 527–536.

Deaner, D. G., and Kerri, K. D. (1969). Regrowth of fecal coliforms. *J. Am. Water Works Assoc. 61:* 465–468.

Donnelly, L. S., and Hartman, P. A. (1978). Gentamicin-based medium for the isolation of group D streptococci and application of the medium to water analysis. *Appl. Environ. Microbiol. 35:* 576–581.

Dutka, B. J., Jackson, M. J., and Bell, J. B. (1974). Comparison of autoclave and ethylene oxide-sterilized membrane filters used in water quality studies. *Appl. Microbiol. 28:* 474–480.

Evans, F. L., III, Geldreich, E. E., Weibel, S. R., and Robeck, G. G. (1968). Treatment of urban stormwater runoff. *J. Water Pollut. Control Fed. 40:* R162–R170.

Facklam, R. R. (1972). Recognition of group D streptococcal species of human origin by biochemical and physiological tests. *Appl. Microbiol. 23:* 1131–1139.

Facklam, R. R. (1973). Comparison of several laboratory media for presumptive identification of enterococci and group D streptococci. *Appl. Microbiol. 26:* 138–145.

Faechem, R. (1975). An improved role for fecal coliform to fecal streptococci ratios in the differentiation between human and non-human pollution sources. *Water Res. 9:* 689–690.

Fitzgerald, G. A. (1947). Are frozen foods a public health problem? *Am. J. Public Health 37:* 695–701.

Gallagher, T. P., and Spino, D. F. (1968). The significance of numbers of coliform bacteria as an indicator of enteric pathogens. *Water Res. 2:* 169.

Geldreich, E. E. (1970). Applying bacteriological parameters to recreational water quality. *J. Am. Water Works Assoc. 62:* 113–120.

Geldreich, E. E. (1976). Fecal coliform and fecal streptococcus density relationships in waste discharges and receiving waters. *CRC Environ. Control 6:* 349–369.

Geldreich, E. E., and Kenner, B. A. (1969). Concepts of fecal streptococci in stream pollution. *J. Water Pollut. Control Fed. 41:* R336–R352.

Geldreich, E. E., Kenner, B. A., and Kabler, P. W. (1964). Occurrence of coliforms, fecal coliforms and streptococci in vegetation and insects. *Appl. Microbiol. 12:* 63–69.

Geldreich, E. E., Best, L. C., Kenner, B. A., and Van Donsel, D. J. (1968). The bacteriological aspects of stormwater pollution. *J. Water Pollut. Control Fed. 40:* 1861–1872.

Green, B. L., Clausen, E., and Litsky, W. (1975). Comparison of the new Millipore HC with conventional membrane filters for the enumeration of fecal coliform bacteria. *Appl. Microbiol. 30:* 697–699.

Green, B. L., Clausen, E. M., and Litsky, W. (1977). Two temperature membrane filter method for enumeration of fecal coliform bacteria from chlorinated effluents. *Appl. Environ. Microbiol. 33:* 1259–1264.

Guthof, O., and Dammann, G. (1958). Uber de Brauchbarkeit von Enterokokken-Testen zur Beurteilung von Trinkwasser und Oberflachenwasser. *Arch. Hyg. Bacteriol. 142:* 559–568.

Gyllenberg, H., Niemela, S., and Sormunen, F. (1960). Survival of bifid bacteria in water as compared with that of coliform bacteria and enterococci. *Appl. Microbiol. 8:* 20–22.

Hajna, A. A., and Perry, C. A. (1943). Comparative study of presumptive and confirmatory media for bacteria of the coliform group and fecal streptococci. *Am. J. Public Health 33:* 550–556.

Harrison, F. C., and van der Leck, J. (1909). Aesculin bile salt media for water analysis. *Zentralbl. Bakteriol. Parasitenkd. Infektionskr. Hyg. Abt. II 22:* 547.

Hartman, P. A., Reinbold, G. W., and Saraswat, D. S. (1966). Media and methods for isolation and enumeration of the enterococci. *Adv. Appl. Microbiol. 8:* 253–289.

Hendricks, C. W., and Morrison, S. M. (1967). Multiplication and growth of selected enteric bacteria in clear mountain stream water. *Water Res. 1:* 567–576.

Isenberg, H. D., Goldberg, D., and Sampson, J. (1970). Laboratory studies with a selective enterococcus medium. *Appl. Microbiol. 20:* 433–436.

Kenner, B. A., Clark, H. F., and Kabler, P. W. (1960). Fecal streptococci. II. Quantification of streptococci in feces. *Am. J. Public Health 50:* 1553–1559.

Kenner, B. A., Clark, H. F., and Kabler, P. W. (1961). Fecal streptococci. I. Cultivation and enumeration of streptococci in surface waters. *Appl. Microbiol. 9:* 15–20.

Kibbey, H. J., Hagedorn, C., and McCoy, E. L. (1978). Use of fecal streptococci as indicators of pollution in soil. *Appl. Environ. Microbiol. 35:* 711–717.

Kjellander, J. (1960). Enteric streptococci as indicators of fecal contamination of water. *Acta Pathol. Microbiol. Scand. 48, Suppl. 136:* 1–124.

Laxminarayana, H., and Iya, K. K. (1953). Studies on the reduction of tetrazolium by lactic acid bacteria. I. Dye reducing activities of different species. *Indian J. Dairy Sci. 6*(2): 75–91, in *Dairy Sci. Abstr. 16*(1): 59.

Lee, R. D., Symons, J. M., and Robeck, G. G. (1970). Watershed human use level and water quality. *J. Am. Water Works Assoc. 62:* 412–422.

Levin, M. A., Fischer, J. R., and Cabelli, V. J. (1975). Membrane filter technique for enumeration of enterococci in marine waters. *Appl. Microbiol. 30:* 66–71.

Lin, S. (1974). Evaluation of fecal streptococci tests for chlorinated secondary sewage effluents. *J. Environ. Eng. Div. Am. Soc. Civ. Eng. 100:* 253–267.

Lin, S. D. (1976). Evaluation of Millipore HA and HC membrane filters for the enumeration of indicator bacteria. *Appl. Environ. Microbiol. 32:* 300–302.

Litsky, W. (1977). Personal communication.

Litsky, W., Mallmann, W. L., and Fifield, C. W. (1953). A new medium for the detection of enterococci in water. *Am. J. Public Health 43:* 873–879.

MacFaddin, J. F. (1976). *Biochemical Tests for Identification of Medical Bacteria*. Williams & Wilkins, Baltimore, Md.

McFeters, G. A., Bissonnette, G. K., Jezeski, J. J., Thompson, C. A., and Stuart, D. A. (1974). Comparative survival of indicator bacteria and enteric pathogens in well water. *Appl. Microbiol. 27:* 823–829.

Martin, J. D., and Mundt, J. O. (1972). Enterococci in insects. *Appl. Microbiol. 24:* 575–580.

Medrek, T. F., and Litsky, W. (1960). Comparative incidence of coliform bacteria and enterococci in undisturbed soil. *Appl. Microbiol. 8:* 60–63.

Mieth, H. (1961). Untersuchungen über das Vorkommen von Enterokokken bei Jieren und Menschen. II Mitteilung: ihr Vorkommen in Stuhlproben von gesunder Menschen. *Zentralbl. Bakteriol. Parasitenkd. Infektionskr. Hyg. Abt. I. Orig. 183:* 68–69.

Morris, W., and Weaver, R. H. (1954). Streptococci as indices of pollution in well waters. *Appl. Microbiol. 2:* 282–285.

Mossel, D. A. A., von Diepen, H. M. and de Bruin, A. S. (1957). The enumeration of fecal streptococci in foods, using Packer's crystal violet sodium azide blood agar. *J. Appl. Bacteriol. 20:* 265–272.

Mundt, J. O. (1961). Occurence of enterococci: bud, blossom, and soil studies. *Appl. Microbiol. 9:* 541–544.

Mundt, J. O. (1963). Occurrence of enterococci on plants in a wild environment. *Appl. Microbiol. 11:* 141–144.

Mundt, J. O. (1973). Litmus milk reaction as a distinguishing feature between *Streptococcus faecalis* of human and nonhuman origins. *J. Milk Food Technol. 36:* 364–367.

Mundt, J. O. (1975). Unidentified streptococci from plants. *Int. J. Syst. Bacteriol. 25:* 281–285.

Mundt, J. O. (1976). Streptococi in dry and frozen foods. *J. Milk Food Technol. 39:* 413–416.

Mundt, J. O., and Graham, W. F. (1968). *Streptococcus faecium* var. *casseleflavus,* nov. var. *J. Bacteriol. 95:* 2005–2009.

Mundt, J. O., Johnson, A. H., and Khatchikian, R. (1958). Incidence and nature of enterococci on plant materials. *Food Res. 23:* 186–193.

Mundt, J. O., Coggins, J. H., Jr., and Johnson, L. F. (1962). Growth of *Streptococcus faecalis* var. *liquifaciens* on plants. *Appl. Microbiol. 10:* 552–555.

Mundt, J. O., Larson, S. A., and McCarty, I. E. (1966). Growth of lactic acid bacteria in waste waters of vegetable-processing plants. *Appl. Microbiol. 14:* 115–118.

Mundt, J. O., Graham, W. F., and McCarty, I. E. (1967). Spherical lactic acid-producing bacteria of southern grown raw and processed vegetables. *Appl. Microbiol. 15:* 1303–1308.

Nowlan, S. S., and Deibel, R. H. (1967). Group Q streptococci. I. Ecology, serology, physiology, and relationship to established enterococci. *J. Bacteriol. 94:* 291–296.

Pavlova, M. T., Brezenski, F. T., and Litsky, W. (1972). Evaluation of various media for isolation, enumeration and identification of fecal streptococci from natural sources. *Health Lab. Sci. 9:* 289–298.

Presswood, W. G., and Brown, L. R. (1973). Comparison of Gelman and Millipore membrane filters for enumerating fecal coliform bacteria. *Appl. Microbiol. 26:* 332–336.

Raibaud, P., Caulet, M., Galpin, J. V., and Mocquot, G. (1961). Studies on the bacterial flora of the alimentary tract of pigs. II. Streptococci: selective enumeration and differentiation of the dominant group. *J. Appl. Bacteriol. 24:* 285–306.

Rose, R. E., and Litsky, W. (1965). Enrichment procedure for use with the membrane filter for the isolation and enumeration of fecal streptococci in water. *Appl. Microbiol. 13:* 106–108.

Sabbaj, J., Sutter, V. L., and Finegold, S. M. (1971). Comparison of selective media for isolation of presumptive group D streptococci from human feces. *Appl. Microbiol. 22:* 1008–1011.

Saraswat, D. S., Clark, W. S., Jr., and Reinbold, G. W. (1963). Selection of a medium for the isolation and enumeration of enterococci in dairy products. *J. Milk Food Technol. 26:* 114–117.

Schaeffer, D. J., Long, M. C., and Janardan, K. G. (1974). Statistical analysis of the recovery of coliform organisms on Gelman and Millipore membrane filters. *Appl. Microbiol. 28:* 605–607.

Shuval, H. I., Cohen, J., and Kolodney, R. (1973). Regrowth of coliforms and fecal coliforms in chlorinated wastewater effluent. *Water Res. 7:* 537–546.

Silvery, J. K. G., Abshire, R. L., and Nunez, W. L., III. (1974). Bacteriology of chlorinated and unchlorinated wastewater effluents. *J. Water Pollut. Control Fed. 46:* 2153–2162.

Slanetz, L. W., and Bartley, C. H. (1957). Numbers of enterococci in water, sewage and feces determined by the membrane filter technique with an improved medium. *J. Bacteriol. 74:* 591–595.

Slanetz, L. W., and Bartley, C. H. (1964). Detection and sanitary significance of fecal streptococci in water. *Am. J. Public Health 54:* 609–614.

Slanetz, L. W., and Bartley, C. H. (1965). Survival of fecal streptococci in seawater. *Health Lab. Sci. 2:* 142–148.

Smith, D. G., and Shattock, P. M. F. (1964). The cellular location of antigens in streptococci of groups D, N, and Q. *J. Gen. Microbiol. 34:* 165–175.

Smuckler, S. A., and Appleman, M. D. (1965). Improved Staphylococcus medium No. 110: sodium azide toxicity and sources of contaminating *Bacillus* species. *Appl. Microbiol. 13:* 289.

Splittstoesser, D. F., Wright, R., and Hucker, G. J. (1961). Studies on media for enumerating enterococci in frozen vegetables. *Appl. Microbiol. 9:* 303–308.

Steinhaus, E. A. (1941). A study of the bacteria associated with thirty species of insects. *J. Bacteriol. 42:* 757–790.

Swan, A. (1954). The use of a bile-aesculin medium and of Maxted's technique of Lancefield grouping in the identification of enterococci (group D streptococci). *J. Clin. Pathol. 7:* 160–163.

Switzer, R. E., and Evans, J. B. (1974). Evaluation of selective media for enumeration of group D streptococci in bovine feces. *Appl. Microbiol. 28:* 1086–1087.

Tilton, R. C., and Litsky, W. (1967). The characterization of fecal streptococci: an attempt to differentiate between animal and human sources of contamination. *J. Milk Food Technol. 30:* 1–6.

Vasconcelos, G. J., and Swartz, R. G. (1976). Survival of bacteria in seawater using a diffusion chamber apparatus in situ. *Appl. Environ. Microbiol. 31:* 913–920.

Weibel, S. R., Anderson, R. J., and Woodward, R. L. (1964). Urban land runoff as a factor in stream pollution. *J. Water Pollut. Control Fed. 36:* 914–924.

West, L. S. (1951). *The Housefly.* Comstock, Ithaca, New York.

Whittenbury, R. (1964). Hydrogen peroxide formation and catalase activity in the lactic acid bacteria. *J. Gen. Microbiol. 35:* 13.

Wicken, A. J., and Baddiley, J. (1963). Structure of intracellular teichoic acids from group D streptococci. *Biochem. J. 87:* 54–62.

Wicken A. J., Elliot, S. D., and Baddiley, J. (1963). The identity of streptococcal group D antigens with teichoic acid. *J. Gen. Microbiol. 31:* 231–239.

5

PSEUDOMONAS AERUGINOSA: A CONTROVERSIAL INDICATOR PATHOGEN

BERNARD J. DUTKA National Water Research Institute, Canada Centre for Inland Waters, Burlington, Ontario, Canada

I. DEFINITION

Pseudomonas aeruginosa is the type species of the genus *Pseudomonas*, which consists of gram-negative, rod-shaped, motile, polar-flagellated bacteria that exhibit respiratory but never fermentative metabolism. Representatives of the genus are strict aerobes, except for those species that utilize nitrate as a terminal electron acceptor. *Pseudomonas* species require no growth factors and can multiply in mineral media containing single organic compounds that serve as the sole source of carbon and energy. *P. aeruginosa* will grow over at least a 30°C range, will tolerate relatively wide pH values, and will utilize a wide variety of organic and quaternary ammonia compounds better than most bacteria. It is proteolytic, produces cytochrome oxidase, and produces a diffusable fluorescent (fluorescein) pigment detailed under ultraviolet light.

II. INTRODUCTION

P. aeruginosa is generally considered to be a ubiquitous bacterial contaminant of surface waters and soil. It is noted for its biochemical versatility and its resistance to antibacterial agents, and it may infect a variety of plants in addition to humans and animals. Various studies have been carried out to try and establish the fecal carrier rates in humans for this organism. Sutter et al. (1967) reported carrier rates in adult populations varying from 3 to 11%. With institutionalized children, a carrier rate of about 40% was found among infants up to 6 months of age, which decreased to 5% in children between 2 and 13 years old. Sutter et al. (1968) concluded that the human intestine does not appear to be a major habitat for *P. aeruginosa,* but these bacteria seem to be a minor part of the resident flora of some individuals and are transient in others. The latter statement could explain the conclusions reached by Ringen and Drake (1952) after finding a recovery rate of 11% from human fecal samples and a rate of 90% from raw and clarified sewage, that the natural habitat of *P. aeruginosa* was human feces and sewage.

In humans, *P. aeruginosa* can cause a wide variety of infections, the importance of which has increased over the years due to the resistance of this organism to a wide range of antibiotics. *P. aeruginosa* infections are most frequent and dangerous in nurseries and among patients with cancer, burns, and tracheostomies (Bennett, 1974).

P. aeruginosa is noted for its association with outer ear infections. Although it is not generally considered a normal inhabitant of the healthy ear, where frequency of isolation from healthy ears vary from 0.5 to 1.5%, recent studies have indicated that the frequency may increase to between 10 and 20% among swimmers and among the general population in hot, humid summer weather (Wright and Alexander, 1974). Stevenson (1953) reported that ear and eye complaints accounted for more than one-half of the illnesses associated with swimming, particularly in swimming pools. The association of *P. aeruginosa* infections of the external ear with swimming is so common that the term "swimmer's ear" has been used to describe such cases (Jones, 1965), even though there was no direct evidence linking swimming and otitis media.

Reitler and Seligmann (1957) concluded from an examination of 1000 water samples collected from northern Israel that *P. aeruginosa* was not infrequently found without large numbers of *Escherichia coli*. Foster et al. (1970) agreed with these findings and stated that "in monitoring bathing water quality, supplementation of coliform tests with analysis for other organisms and with sanitary surveys would give more useful information than coliform analysis alone." Reitler and Seligmann (1957) concluded that a "Pseudomonas test should be included in the routine examination of water." Hoadley (1967) in a study on the presence of *P. aeruginosa* in surface waters concluded that *P. aeruginosa* probably does not occur in waters not affected by human activity and domestic animals, that low population levels occur in areas adjacent to human activity, and that high population levels in excess of one *P. aeruginosa* organism per milliliter may occur in waters recently contaminated by sewage. He further stated that *P. aeruginosa* appears to be a sensitive indicator of the contamination of surface waters by sewage and by municipal and barnyard runoff. Drake (1966) similarly stated that *P. aeruginosa* is rarely, if ever, found in surface water in the absence of human or animal contamination. Schiavone and Passerini (1958) expressed the opinion that *P. aeruginosa* may play an important role in waterborne epidemic outbreaks of gastrointestinal disturbances.

Hoadley (1967) found that a relationship existed between numbers of *P. aeruginosa* in water and numbers of swimmers. He found densities of 0.02 and less than 0.018 *P. aeruginosa* organism per milliliter at a beach when only an occasional swimmer was present and a population of 0.95 organisms per milliliter when the swimmer load was heavy. Hoadley (1967) and Foster et al. (1970) stated that bacteria causing eye, ear, nose and throat infections are more important than indicator bacteria in swimming waters.

Foster et al. (1970) collected 877 water samples from three freshwater bathing beaches on 57 collection dates and stated "that the extensive data collected on the densities of *P. aeruginosa* at natural bathing beaches is unique." Their data implied that there may be a connection between ear infections and the presence of *P. aeruginosa* in bathing waters.

Favero et al. (1964) and Hoadley (1968) have suggested that restrictions be placed on *P. aeruginosa* densities in bathing waters because of the association of this organism with ear infections and its presence in swimming pools and natural bathing areas.

Recently, Seyfried and Fraser (1978) conclusively demonstrated that *P. aeruginosa* colonization of the ear did occur in swimmers and that the source of the *P. aeruginosa* was the contaminated swimming pool water. In this very important study, an uncontaminated swimming pool and a pool containing *P. aeruginosa* were used to demonstrate that there is a relationship between the presence of the bacterium in swimming pool water and the colonization of the outer ear and subsequent development of infection (Seyfried and Fraser, 1978).

III. MEMBRANE FILTRATION PROCEDURES

A variety of selective media have been developed for the isolation and enumeration of *P. aeruginosa* (Drake, 1966; Goto and Bromoto, 1970); however, most of these media required large inocula and did not yield quantitative recovery of the organism. The most widely applied media was developed by Drake (1966), and its modifications were used in conjunction with the multiple-tube most probable number (MPN) technique. In 1972, Levin and Cabelli reported the development of a membrane filter procedure to isolate and enumerate *P. aeruginosa* from water samples. Their procedure was based on the use of mPA agar and an incubation temperature of 41.5°C for 48 hr. Levin and Cabelli (1972) compared the efficiency of the MF-mPA procedure to the two most widely used MPN media (Drake, 1966; American Public Health Association, 1971) in a variety of samples: river water, pond water, salt water, well water, and sewage effluents. They found that in almost every instance, the MF-mPA combination estimated the highest population, and that in this and other studies 95% or more of the typical colonies on mPA agar were verified as *P. aeruginosa*.

In 1973, Brodsky and Nixon reported on the development of a membrane filter technique using black membrane filters, MacConkey agar (Difco), 37°C incubation for 24 hr, and fluorescence under ultraviolet light to quantitatively isolate *P. aeruginosa* from swimming pools. The selectivity of this technique was improved by incubating at 42°C; however, this temperature was also found to supress the fluorescence of some *P. aeruginosa*.

Dutka and Kwan (1977) tried to apply the MF-MacConkey agar technique to the enumeration of *P. aeruginosa* from moderately contaminated natural water samples. In all dilutions where countable colonies appeared on mPA medium (Levin and Cabelli, 1972) the MacConkey agar was always overgrown with fungi and lactose and non-lactose-fermenting bacterial colonies. It was found in several water samples that there were approximately 100 coliform-type organisms for each *P. aeruginosa*. Therefore, it was impossible to obtain a *P. aeruginosa* count by the MacConkey agar procedure from any contaminated water sample, as the sample had to be diluted beyond the density of *P. aeruginosa* in the sample.

Dutka and Kwan (1977) found that by reducing the sodium thiosulfate content of mPA medium (Levin and Cabelli, 1972) to 5.0 g/liter and adding 1.5 g magnesium sulfate per liter of medium, they had a medium (mPA medium B) that showed slightly better recovery and colony definition than mPA did (Tables 1 and 2). They also found that selectivity and sensitivity were not affected by reducing the concentration of xylose to 1.25 g/liter.

In Table 1 it can be seen that both MF procedures enumerated more confirmed *P. aeruginosa* than did the MPN procedure and that mPA medium B was slightly more efficient in the more polluted sewage and effluent samples. Table 2 shows the reliability of the two MF procedures in resuscitating and enumerating confirmed *P. aeruginosa* and the percentage of atypical colonies that would have been missed by the MF procedures. There was no significant difference in the confirmation rate between the two membrane filter media except that slightly more confirmed *P. aeruginosa* colonies were recovered by

TABLE 1 Comparison of the Efficiency of MF and MPN Procedures to Enumerate *P. aeruginosa* from Sewage, Unchlorinated Effluent, and Canal and Lake Waters

Medium and procedure	Incubation period (hr)	Temp. (°C)	Sewage[a] Geometric mean (per ml)	Effluent[a] Geometric mean (per ml)	Canal[a] Geometric mean (per ml)	Lake[b] Geometric mean (per ml)
mPA, MF	48	41.5	290	83	16	12.9
mPA medium B, MF	96	41.5	1100	240	15	22
Drake medium MPN	96	41.5	830	8.3	0.85	0.31

[a]20 samples.
[b]15 samples.

TABLE 2 Percentage of Typical Colonies and Percentages of Atypical and Suspicious Colonies That Were Confirmed as *P. aeruginosa*[a]

Medium and procedure	Incubation period (hr)	Temp. (°C)	Sewage % TC	Sewage % SC	Effluent % TC	Effluent % SC	Canal % TC	Canal % SC	Lake % TC	Lake % SC	Summary n^b	Summary % Typical	Summary n^c	Summary % Suspicious
mPA, MF	48	41.5	95.0	8.3	94.0	8.3	93.0	6.0	93.3	6.7	750	93.8	415	7.5
mPA, medium B, MF	96	41.5	99.0	4.0	98.5	3.0	98.7	3.0	97.3	2.7	700	98.5	375	3.2

[a]TC, typical colonies confirmed; SC, suspicious colonies confirmed; n, number of bacteria tested.
[b]Typical colonies.
[c]Atypical colonies.

mPA medium B. This medium also produced slightly fewer atypical *P. aeruginosa* colonies.

Recently, Brodsky and Ciebin (1978) produced a modification of mPA medium B (mPA medium C) by omitting sulfapyridine and actidione. In their studies, they found that *P. aeruginosa* colonies could be counted on mPA-C medium after only 24 hr of incubation, with the same accuracy, selectivity, specificity, and precision as on mPA-B medium after 72 hr of incubation.

In preliminary evaluation tests of mPA-C medium with natural samples, we have noted poor colony definition, whose numbers and definition increased with longer incubation; unfortunately, background growth and fungi start to provide interferences with longer incubation periods. It is our belief that the mPA-C medium procedure, as proposed, still requires major field evaluation studies before it can be fully supported.

IV. METHODOLOGY

The procedures followed with mPA medium B in my laboratory to enumerate *P. aeruginosa* from natural water and sewage samples are described below.

A. Enumeration Procedure, Membrane Filtration

1. Using membrane filtration procedures and KH_2PO_4 rinse water buffer (with or without 0.1% peptone), filter sufficient sample or sample dilution to recover between 20 and 80 colonies per membrane.
2. Place membranes onto prepoured mPA medium B plates and incubate at 41.5°C for 3—4 days (4 days preferred) in a well-humidified incubator.
3. *P. aeruginosa* colonies are flat, approximately 0.8 mm in diameter, with a dark brown or greenish black center and a paler outer edge.
4. Confirmation, if required, is accomplished by streaking suspected colonies on milk medium, incubating for 24 hr, at 35—37°C, and observing for growth, hydrolysis of casein, and production of greenish pigment.

5. Confirmation should be conducted whenever a new water body is being examined or for familiarization with the technique.
6. Results should be reported as organisms per 100 ml of sample.

B. mPA Medium B Preparation

L-Lysine hydrochloride	5.0	g
NaCl	5.0	g
Yeast extract	2.0	g
Sodium thiosulfate	5.0	g
Magnesium sulfate	1.59	g
Xylose	1.25	g
Lactose	1.25	g
Sucrose	1.25	g
Phenol red	0.08	g
Ferric ammonium citrate	0.8	g
Agar	15.0	g
Distilled water	1	liter

Mix the ingredients and let stand 10–15 min; mix well and autoclave 10 min at 115°C. Cool the solution to 55°C in a water bath and add the following as dry powder.

Sulfapyridine	176.0 mg
Kanamycin sulfate	8.5 mg
Nalidixic acid	37.0 mg
Actidione	150.0 mg

A sonic sink may be used to help the dry powder dissolve. Adjust the pH to 7.1 ±0.1 Dispense in 5–7 ml quantities into glass or plastic petri dishes. Plates may be stored at least 2 weeks at 2–6°C.

C. Milk Agar Preparation (Brown and Foster, 1970)

Solution 1:

Milk powder (instant nonfat)	20.0	g
Distilled water	100	ml

Solution 2:

Nutrient broth	2.5 g
NaCl	0.5 g
Agar	3.0 g
Distilled water	100 ml

Separately sterilize solutions A and B by autoclaving at 115°C for 10 min. Cool both solutions to 55°C rapidly and combine solutions, mixing well and gently. Pour into 100 x 15 mm petri dishes approximately 30 ml per dish. Plates may be stored, inverted, at least 4 weeks at 2–6°C.

REFERENCES

American Public Health Association (1971). *Standard Methods for the Examination of Water and Wastewater,* 13th ed. American Public Health Association, Washington, D.C.

Bennett, J. V. (1974). Noscomial infections due to *Pseudomonas. J. Infect. Dis.,* Vol. 130, Suppl. No. 11, Nov., pp. 54–57.

Brodsky, M. H., and Ciebin, B. W. (1978). Improved medium for recovery and enumeration of *Pseudomonas aeruginosa* from water using membrane filters. *Appl. and Environ. Microbiol. 36:* 36-42.

Brodsky, M. H., and Nixon, M. C. (1973). Rapid method for the detection of *Pseudomonas aeruginosa* in MacConkey agar, under ultra violet light. *Appl. Microbiol. 26:* 219–220.

Brown, M. R. W., and Foster, J. H. S. (1970). A simple diagnostic milk medium for *Pseudomonas aeruginosa. J. Clin. Pathol. 23:* 172–177.

Drake, C. H. (1966). Evaluation of culture media for the isolation and enumeration of *Pseudomonas aeruginosa. Health Lab. Sci. 3:* 10–19.

Dutka, B. J., and Kwan, K. K. (1977). Confirmation of the single-step membrane filtration procedure for estimating *Pseudomonas aeruginosa* densities in water. *Appl. and Environ. Microbiol. 33:* 240-245.

Favero, M. S., Drake, C. H., and Randall, G. B. (1964). Use of staphylococci as indicators of swimming pool pollution. *Public Health Rep. 79:* 61–70.

Foster, D. H., Hanes, N. B., and Lord, S. M. (1970). A critical examination of bathing water quality standards. Presented at the 43rd Annu. Conf. Water Pollut. Control Fed., Boston, Mass. Tufts University, Dept. of Civil Engineering, Medford, Mass., p. 23.

Goto, S., and Bromoto, S. (1970). Nalidixic acid cetrimide agar. *Jap. J. Microbiol. 14:* 65–72.

Hoadley, A. W. (1967). The occurrence and behavior of *Pseudomonas aeruginosa* in surface waters. Thesis, University of Wisconsin, Madison, Wis.

Hoadley, A. W. (1968). Investigations concerning *Pseudomonas aeruginosa* in surface waters. I. Sources. *Arch. Hyg. Bakteriol. 15:* 328–332.

Jones, E. H. (1965). External otitis, diagnosis and treatment. Thomas, Springfield, Ill.

Levin, M. A., and Cabelli, V. J. (1972). Membrane filter technique for enumeration of *Pseudomonas aeruginosa*. *Appl. Microbiol. 24:* 862–870.

Reitler, R., and Seligmann, R. (1957). *Pseudomonas aeruginosa* in drinking water. *J. Appl. Bacteriol. 20:* 145–150.

Ringen, L. M., and Drake, C. H. (1952). A study of the incidence of *Pseudomonas aeruginosa* from various natural sources. *J. Bacteriol. 64:* 841–845.

Schiavone, E. L., and Passerini, L. M. D. (1958). The genus *Pseudomonas aeruginosa* in the judgment of the potability of drinking water. *Semin. Med.* (Buenos Aires) *111*(23): 1151–1157. [Abstract in *Public Health Engineering Abstracts 39*: W13 (1959).]

Seyfried, P. L., and Fraser, D. J. (1978). *Pseudomonas aeruginosa* in swimming pools related to the incidence of otitis externa infection. *Health Lab. Sci. 15:* 50–57.

Stevenson, A. H. (1953). Studies of bathing water quality and health. *Am. J. Public Health 43:* 529–538.

Sutter, U. L., Hurst, U., and Lanc, C. W. (1967). Quantification of *Pseudomonas aeruginosa* in feces of healthy human adults. *Health Lab. Sci. 4:* 245–249.

Wright, D. N., and Alexander, J. M. (1974). Effects of water on the bacterial flora of swimmers' ears. *Arch. Otolaryngol.*, Vol. 99, No. 1, Jan., pp. 15–18.

6

BIFIDOBACTERIUM

MORRIS A. LEVIN and I. GARY RESNICK† U.S. Environmental Protection Agency, Washington, D.C. and Cincinnati, Ohio*

I. INTRODUCTION

Water quality characteristics in general may be estimated by use of any of a large number of parameters. If the objective of the measurement is to provide an indication of the potential health hazard to exposed individuals, any of a variety of microbial indicators may be used to determine the presence of fecal pollution. The characteristics of ideal sewage indicator organisms have been discussed by Bonde in Sweden (1966) and more recently in 1977 by Cabelli. Both authors provide a discussion of the uses and important characteristics of indicators of pollution from this point of view. Bonde (1966) also discusses the merits of individual indicators and these are further elaborated on by Hoadley and Dutka (1977).

Although Mossel (1958), over 20 years ago, must be credited as initially proposing the use of bifidobacteria as indicators of fecal pollution, the significance of these organisms in sanitary microbiology is still unclear. Since then a number of workers (Gyllenberg et al., 1960; Evison and James, 1973; Resnick and Levin, 1977; Resnick, 1978) have examined and extended the concept of using these

Present affiliation:

*U.S. Army, Dugway Proving Ground, Dugway, Utah
†University of Texas, Austin, Texas

organisms as indicators of fecal pollution. Bifidobacteria are anaerobic bacilli which are present in the intestine in densities that exceed coliform levels by at least a factor of 10 and in some cases as much as 100-fold (Evison and James, 1974). Similarly, bifidobacteria/*Escherichia coli* ratios in sewage range from 17 to 55 (Resnick and Levin, 1977). Other characteristics of bifidobacteria which initially suggested their role as indicator organisms are (1) their potential value in permitting differentiation between human sources of fecal pollution and animal sources, (2) their inability to multiply in an extraenteral environment, and (3) survival characteristics that were considered to be similar to *E. coli*.

II. BACKGROUND

Although at present the identification of bifidobacteria to the genus level presents few problems, as recently as 1973 Poupard pointed out that in earlier literature many genera had been lumped together as *Lactobacillus bifidus*. These organisms were first described and named by Tissier in 1889 and 1900 as *Bacillus bifidus communis* (or *B. bifidus*), based on work with isolates from the stools of breast-fed infants. His work and that of Moro (1900) marked the commencement of interest in the taxonomy and significance of bifidobacteria as related to their presence in the intestine. Isolates have been obtained from intestinal specimens of humans and other animals as well as from insects. However, the major environmental source of bifidobacteria is the human being (Table 1).

TABLE 1 Environmental Niches of Bifidobacteria

Sample	Sample source				
	Human	Cattle	Pig	Pet	Insect
Intestinal material[a]	+	+	+	+	+
Feces[b]	+	-	+	-	-

[a]Based on Buchanon and Gibbons (1974).
[b]Human feces contain 10^9 organisms per gram; pig feces, 10^5 organisms per gram, using YN-6 media. Resnick and Levin (1978).

Early workers experienced difficulty because of deficiencies in culture media and techniques for cultivation of anaerobic bacteria. This was complicated by a variation in morphology reported by many workers (Norris, et al., 1950; Gyllenberg, 1955; Glick et al., 1960). In addition (or perhaps as a consequence), there is some question not only of the identity but also as to the reliability of identification criteria based on use of cultures that may have contained organisms of more than one genus (Poupard et al., 1973). The confusion that arose is reflected in the varied nomenclature applied to the genus of the isolates, beginning with *Baccillus* (1900), becoming *Bacteroides* (1919), *Bacterium* (1927), *Nocardia* (1931), and *Actinomycetes* in 1934 (Buchanon and Gibbons, 1974). Weiss and Rettger in 1934 used the designation *Lactobacillus bifidus*.

These organisms were considered variants of the same species (Weiss and Rettger, 1934), with the capability of reversion from the branched bifid form to the unbranched *L. acidophilus* form. This confusion may have been due to the fact that *L. acidophilus* exists in an intimate relationship with bifidobacteria in the intestine and may have been carried as a contaminant in what were thought to be pure cultures of bifidobacteria. As a facultative anaerobe, the lactobacillus would be capable of growth and survival in an anaerobic environment. Orla-Jensen in 1924 proposed that *L. bifidus* be placed in a separate genus (*Bifidobacterium*), but this reclassification waited years to appear in the eighth edition of *Bergey's Manual for Determinative Bacteriology* (Buchanon and Gibbons, 1974). Although an exhaustive review of the classification of these organisms is beyond the scope of the chapter, a summary is necessary to permit an understanding of the definition of the speciation in this genus. The literature prior to 1935 has been reviewed by Weiss (1933) and Weiss and Rettger (1934). Poupard et al. reviewed the literature in 1973, with emphasis on classification and physiology.

Eggerth (1935) provided the first indication of species differences among the bifidobacteria using fermentation patterns and divided the genus into two sections, group I and group II. The work of Reuter in 1963 provided eight species and several variants based on morphology (colonial and cellular), carbohydrate fermentations, and types of acids produced from glucose fermentation. His conclusions, based on isolates from adults and children, were that the organisms should be classified within the tribe Lactobacilleae, family Lactobacillaceae, and genus *Bifidobacterium*. Further data defining and describing the genus have been produced by numerous workers. De Vries et al. (1967) and Scardovi and Trovatelli (1965) discovered the fructose-6-phosphate pathway in bifidobacteria, a carbohydrate fermentation pathway not found in any of the other genera formerly associated with bifidobacteria. Comparison of DNA base composition gave further evidence that these organisms were not closely related to the lactobaccilli (Sebald et al., 1965; Gasser and Mandel, 1968). In addition, bifidobacteria have been shown to differ from lactobaccilli in phospholipid composition (Exterkate and Veerkamp, 1969) and to serologically differ from other gram-positive filamentous bacteria. Mitsouka in 1969 confirmed and extended the classification, examining isolates from both human and nonhuman sources. Using carbohydrate fermentation patterns as a basis, he retained the species designation of Reuter (1963) and added two new ones.

The current classification of bifidobacteria is heavily dependent on the work of Scardovi et al. (1971a), Scardovi and Crociani (1974), Scardovi and Trovatelli (1974, Scardovi and Zani (1974). These workers applied the techniques of DNA hybridization, as well as conventional taxonomic techniques, to produce the basis for the classification scheme presented in the eighth edition of *Bergey's Manual*.

A. Definition of Bifidobacteria

Bifidobacteria can be defined as strictly anaerobic, non-spore-forming, nonmotile, gram-positive, morphologically thick pleomorphic rods. They may exhibit branching, bulbous clubs, coryneforms, buds, spheroids, and bifurcated Y and V forms when freshly isolated from fecal sources (the morphology is influenced by nutritional conditions and may appear difference when isolated from an aquatic environment). They are catalase-negative, with acetic and lactic acids as the major fermentation products from glucose. Although previous workers have stated that lesser amounts of formic and succinic acids and ethanol are produced and that a 3:2 ratio of acetic acid to lactic acid was critical in confirming the identity of an isolate, recent work by Lauer and Kondler (1976) has shown that ratios in excess of 4:1 may be obtained from most species. These authors demonstrated that bifidobacteria produce acetic acid via phosphoroclastic splitting of pyruvate, particularly while in the log phase of growth. Therefore, the ratio must simply be in excess of 1:1. Thus, tests to identify and quantitate the fermentation products by gas-liquid chromotography are needed for identification to the generic level.

Eleven species are listed in the current (eighth) edition of *Bergey's Manual*. As could be anticipated, as more strains are isolated and characterized, proposals for combining species and the elevation of biotypes to species level have been forthcoming. For example, Scardovi et al. (1971) suggested consolidation of *B. infantis, B. lactentis,* and *B. liberorum* on the basis of genetic homogeneity. He also found no basis for species differentiation between *B. longum* Reuter and *B. longum* subspecies animalis, and pointed out that although *B. breve* and *B. parvulorum* differ phenotypically, they are genetically similar.

In 1974, Scardovi and Trotavelli proposed that *B. longum* Reuter subspecies *animalis* Mitsouka biotype a be elevated to species rank, as *B. animalis* Mitsouka, and at the same time proposed two new subgroups isolated from sewage. In the same year *B. magnum* was proposed as a new species isolated from rabbit feces (Scardovi and Zani, 1974), and Scardovi and Crociani (1974) present three new species based on DNA homology and colonial morphology. The most recent classification scheme appears in the *Anaerobe Laboratory*

TABLE 2 Species of the Genus *Bifidobacterium*

Species	Arabinose	Xylose	Ribose	Gluconate	Cellobiose	Lactose	Mannitol	Melezitose	Salicin	Starch	Trehalose
Bifidum	-	-	-	-	+	+	-	-	-	-	-
Adolescentis	+	+	+	+	+	+	V	V	+	V	V
Catenulatum	+	+	+	+	+	+	-	-	-	-	-
Angulatum	+	+	+	-	-	+	-	-	+	+	-
Dentium	+	+	+	+	+	+	+	+	+	+	+
Infantis	-	-	+	-	V	+	-	-	V	V	V
Liberorum[a]	-	+	+	-	V	+	-	V	V	V	V
Lacentis[a]	-	+	+	-	V	+	+	-	-	V	V
Breve	-	-	+	-	+	+	+	V	V	V	V
Parvulorum[a]	-	-	+	-	V	+	+	-	-	+	-
Longum	+	+	+	-	-	+	-	+	-	V	V
Pseudolongum	+	+	+	-	V	V	-	V	V	+	-
Suis	+	+	-	-	-	+	-	-	-	-	-
Asteroides	+	+	+	+	+	-	-	-	+	-	-
Indicum	-	-	+	+	+	-	-	-	+	-	-
Coryneforme	+	+	+	+	+	-	-	-	+	-	-
Magnum	+	+	+	-	-	+	-	-	-	-	-

[a]Uncertain of species designations.

Manual (second edition, 1973) and is based on work (primarily by Scardovi et al.) prior to 1971. Table 2 presents a summary of species to date.

III. MATERIALS AND METHODS

In essence, assignment of an isolate to the genus *Bifidobacterium* is based on gram stain (morphology and gram-positive reaction), strictly anaerobic growth, absence of spores, and the release of acetic and lactic acids in a 1:1 or greater ratio from the metabolism of glucose. The procedure for identifying and quantitating volatile and nonvolatile fatty acids is given in Fig. 1. Samples for fatty acid analysis are prepared according to instruction given by Holdeman and Moore (1972). Speciation requires the examination of biochemical reactions under strict anaerobic conditions and measurements of DNA homology.

Techniques and methodology for growth and manipulation of anaerobes have been detailed by a number of workers (Holdeman and Moore, 1972; Sutter et al., 1975). Fortunately, members of the genus *Bifidobacterium*, while requiring reduced oxygen tension for replication, are capable of surviving in air for long periods relative to other anaerobic genera. Hence, the term aeroduric has been used to describe these organisms. It is for this reason that the enumeration of bifids is feasible in routine water quality laboratories. As described later, the methodology can be considered a relaxed anaerobic technique, suitable for aeroduric organisms.

There are several methods available for obtaining anaerobic culture conditions, the two most convenient being glove boxes and anaerobic jars. Two types of glove boxes are available commercially. The one used in this laboratory is a tent arrangement supported by a frame of metal tubing and positive pressure maintained at 37°C. One can also purchase a solid type constructed of stainless steel or fiberglass. Both boxes maintain an anaerobic atmosphere by employing a palladium-coated catalyst to scavenge oxygen, which is removed from the atmosphere as water. This catalyst is rendered inactive by

Volatile fatty acids and alcohols[a]

including: acetic, propionic, isobutyric, n-butyric, isovaleric, n-valeric, isocaproic, n-hexanoic, and ethanol.

Column length: 6 ft X 1/8 in.
Liquid phase: FFAP 6%
Support: Poropak Q, mesh size 80-100
Carrier gas: helium,
Flow rate: 30 ml/min
Sample size: 0.5 μl
Detector: flame ionization detector
Temperature
Column: 200°C
Injector: 250°C
Manifold: 250°C
Attenuation: X1

Nonvolatile fatty acids[b] (as methyl esters)

including: lactic, pyruvic, and succinic.

Column length: 6 ft X 1/8 in.
Liquid phase: 20% Degs
Support: Poropak Q, mesh size 80-100
Carrier gas: helium
Flow rate: 30 ml/min
Sample size: 0.2 μl of chloroform extract
Detector: flame ionization detector
Temperature
Column: 150°C
Injector: 200°C
Manifold: 200°C
Attenuation: X32

[a]Sutter, et al. (1975).
[b]Holdeman and Moore (1972).

FIGURE 1 Parameters for the isothermal separation of short-chain fatty acid by-products of glucose fermentation by gas chromatography.

moisture and hydrogen sulfide but can be regenerated by heating at 160° for several hours. It is recommended that the catalyst be regenerated once a month to avoid problems. Removal of the water (formed by the combination of hydrogen and oxygen) is most conveniently achieved by keeping Drierite in flat trays in the glove box. Drierite was chosen as a desiccant because it provides a colorimetric indication (blue changing to colorless) of saturation with water and can also be regenerated by heating at 160°C for several hours. The hydrogen required for removal of oxygen in the tent is supplied in a gas mixture of 85% nitrogen, 10% hydrogen, and 5% carbon dioxide. At this concentration of hydrogen, with moderate care there is no danger of explosion. If materials are passed in and out of the tent, the hydrogen concentration can often be lowered to 5%. Carbon dioxide is included for stimulation of growth of bacteria, and nitrogen is used as an inert carrier. Since the exact ratio of gases in the mixture is not critical, an unanalyzed gas mixture in size 200 cylinders, which can be obtained for approximately $45 per cylinder, can be used.

Entrance and exit to both the rigid and soft glove boxes is achieved through a gas evacuation replacement chamber. Before material in the entrance chamber can be taken into the hood, the chamber must either be evacuated (0.969 atm) or flushed twice with gas and then filled with the hydrogen/carbon dioxide/nitrogen mixture. Flushing consists of evacuation of the chamber to 0.668 atm and allowing gas to replace the air. The evacuation technique can be used when no liquids are involved, and flushing is required when there is a possibility of boiling over of liquids due to reduced pressure. To reduce the cost of operation, prepurified nitrogen can be used to flush the chamber. Gas must be added to the hood on an intermittent basis, as required, to maintain positive pressure.

Manipulations inside the chambers are conducted with rubber gloves, which unfortunately can be easily damaged. If care is taken, however, rubber gloves can be kept intact for a long period. Precautions include using vinyl-covered test tube racks to eliminate sharp edges, use of a microscope slide dispenser to eliminate handling slides, and use of equipment with rounded edges. If a leak occurs, it can usually be detected by a decrease in the positive pressure of the chamber and can be repaired with adhesive tape. The gloves are usually the culprits, but leaks may occur at seams or around the gaskets in the entrance of the interchange. When the leak is small, it may be difficult to detect, and in these instances an agent such as mercaptan, which can be detected by olfaction, can be released inside the hood (Malligo, 1977). The opening can then be located by smelling around the exterior of the tent. It should be noted that a very small leak may be ignored. The solid type of glove box eliminates a good deal of problem surface area because all walls are made of stainless steel or fiberglass. Nevertheless, the front panel for the gloves is removable and there are numerous gaskets.

The proper functioning of the system can be monitored by keeping agar plates containing Eh indicators in the tent. Use of an indicator such as methylene blue or resazurin is sufficient, but a panel of indicators can be employed to estimate a value for the Eh. Thus, use of methylene blue (E_0' +0.011 V), resazurin (E_0' -0.05 V), indigo disulfonate (E_0' -0.125 V), janus green (E_0' -0.252 V), neutral red (E_0' -0.325 V), and methyl viologen (E_0' -0.440 V) would permit continual visual monitoring of the performance of the anaerobic system. Sterilization of culture loops is accomplished with an incandescent heating device. Our work is carried out with good results in a soft glove box purchased from COY Laboratory Products, Ann Arbor, Michigan.

In appreciation of budget considerations, the use of the BBL Gas-Pak system for anaerobic culture was compared with the glove box. The Gas-Pak system is relatively inexpensive and simple to use. It consists of a plastic cylinder with a gas-tight lid and an indicator strip coated with a methylene blue solution. A palladium-coated catalyst is used in the jar. The hydrogen required is generated by a chemical reaction that takes place in a foil envelope placed in the jar. The reaction is initiated by adding 10 ml of water to the envelope immediately before sealing the jar. Anaerobic conditions are reached slowly in the Gas-Pak, compared to the anaerobic hood, which is always maintained at an Eh below -0.160 V. However, no significant differences, either in plate-to-plate variability or in overall recovery, could be detected when comparing Gas-Pak containers with use of an anaerobic hood.

Prereduced media are recommended for the culture and biochemical testing of fastidious anaerobic organisms, but this added expenditure of time and money does not appear to be necessary in handling bifids. Purple broth base has been used for fermentation tests, and it has been shown that the bromcresol blue indicator yields comparable results to that derived with PY base (Holdeman and Moore, 1972) and a pH electrode for reading final pH of test medium. The method of preparation of the fermentation tubes depends on the type of carbohydrate employed (Fig. 2). Heat-labile compounds (arabinose, xylose, and ribose) must be filter-sterilized and added to the autoclaved basal media. Heat-stabile compounds may be added prior to autoclaving . Six milliliters of broth is dispensed into screw-cap tubes, and the tubes are stored at 4°C in darkness. They are placed in the hood 24 hr before use. Inoculation of the tubes is accomplished by using 1 drop of a turbid culture (24-48 hr of incubation at 37°C) grown in peptone yeast glucose broth (Holdeman and Moore, 1972).

Reinforced clostridial agar (RCA, BBL) was used as a control medium and to produce isolates for storage. The problem of maintaining a stock culture collection is complicated by the fact that these organisms produce copious amounts of acid and can reduce the pH of the medium to a toxic level (4.5) very rapidly. Passage once

Carbohydrate	Percent
Arabinose	0.5
Xylose	1.0
Ribose	0.5

Carbohydrate added aseptically to purple broth base after autoclaving.

Carbohydrate	Percent
Gluconate	1.0
Cellobiose	1.0
Lactose	1.0
Mannitol	1.0
Salicin	1.0
Trehalose	0.5

Purple broth base and carbohydrate sterilized by autoclaving.

FIGURE 2 Method of preparation of fermentation tubes.

a week in chopped meat medium is required to maintain viability. For long-term storage, lypohilization in skim milk is suggested, with storage at -40°C.

If it is the intent of the researchers to determine the species present, the fermentation patterns in Table 2 must be consulted. If these tests alone are used for speciation, some species will be indistinguishable from others. This "lumping" of phenotypically similar species, although disturbing from a strictly taxonomic point of view, will not affect the conclusions as to the sanitary significance of the observation. Although separation of the isolates is possible by DNA homology testing (Scardovi et al., 1971a), the cost and logistical problems involved preclude the handling of a large number of isolates over a short time period.

There are presently two membrane filter techniques in use for isolation and quantitation of bifidobacteria. The medium of Evison and James (1973, 1974) is chemically defined and is differential and selective; the YN-6 medium (Resnick and Levin, 1977) developed in our laboratory is a complex medium that is also differential and selective. The choice was made to pursue development of a complex medium for several reasons. Defined media for fastidious organisms are expensive in formulation and difficult to prepare. In addition, when the nutritional requirements for the various species of bifidobacteria are considered, the total of the required nutrients is so close to a complex media that only a very slight gain in selectivity can be expected relative to a medium containing undefined ingredients.

A. YN-6 Medium

YN-6 medium was developed for the selective cultivation of bifidobacteria on membrane filters. The formulation and directions for preparation are presented in Table 3. The ingredients indicated were chosen for either their growth promoting capacity or their ability to confer a selective or differential advantage. The concentration of each component was varied individually to determine the optimum level for recovery of bifidobacteria. RCA spread plates

TABLE 3 YN-6 Medium Preparation

Add	
Yeast extract	20 g
Peptone	10 g
Lactose	10 g
Casamino acids	8 g
Sodium chloride	3.2 g
Bromcresol green	0.3 g
H_2O	1 liter
Boil	
10 min and cool	
Add	
Cysteine hydrochloride	0.4 g
Naladixic acid	80 mg
pH	
Adjust to pH 6.9 with 10 N sodium hydroxide	
Add	
Agar	15.0 g
Autoclave	
15 min at 15 psi	
Cool	
to 60°C	
Add	
Neomycin, 1 ml of stock solution (containing 2.5 mg/ml)	
Dispense	
4 ml volumes to tight-lid 50 mm petri dishes	
Store	
4°C in dark	

were used to determine 100% recovery. Results typical of those observed appear in Table 4. As can be seen, the optimum concentration of bromcresol green was selected by comparing the effects of varying concentrations of the dye in the final medium to recovery on YN-6 basal medium. The optimum concentrations of other ingredients were determined in a similar manner. Casamino acids (Difco) and Polypeptone peptone (BBL) were chosen as a source of amino acids. Yeast

TABLE 4 Evaluation of Medium Ingredients

Bromcresol green[a]	Organisms/ml[b] (x 10^7)	% Recovery[c]
0.00	4.2 ±0.32	100
0.02	4.0 ±0.44	95
0.03	5.0 ±0.66	100
0.04	1.7 ±0.46	40

[a]Concentrations in YN-6 medium base (g/100 ml).
[b]Based on three replicate plates per dilution; 95% confidence limits.
[c]TN-6 basal medium recovery designated as 100%.

extract (Oxoid) is included to satisfy vitamin requirements. The fermentable carbohydrate energy source, lactose, was selected over arabinose (used in Evison's medium) to permit recovery of more species of human origin. If arabinose were used as the sole carbohydrate, *B. bifidum*, *B. breve*, and *B. infantis* would not be recovered. Neomycin sulfate (ICN Pharmaceuticals, Inc.) and naladixic acid (Calbiochem) are included as selective agents to inhibit growth of gram-positive and gram-negative rods, respectively. Bromcresol green is present as a selective agent as well as an indicator of pH. Thus, fermentation of lactose results in a dark green colony. Cysteine hydrochloride acts as a reducing agent. The medium once prepared is dispensed into 50 mm petri dishes with tight lids. The medium is stored in the dark at 4°C and can be kept for 2 weeks. The medium is poised at pH 6.9 prior to autoclaving. This yields maximum recovery of bifidobacteria and imparts a dark blue color on the medium.

Standard membrane filtration procedures [*Standard Methods*, Part 906 (American Public Health Association, 1975)] are used for preparing organisms for growth on YN-6. The glass filtering apparatus is sterilized between samples by exposure to ultraviolet light for 5 min. A minimum sample volume of 10 ml is filtered. Phosphate-buffered saline (NaCl, 0.85 g; Na_2HPO_4, 0.25 g; NaH_2PO_4, 0.056 g; 100 ml H_2O) is used as dilutent. To facilitate the exchange of gases over the filters for removal of oxygen, a hot probe should be used to pierce the covers of the petri dishes. The inverted plates are incubated 48 hr at 37°C; the plates should be placed either in an anaerobic tent or a Gas-Pak system within 1 hr after filtering.

Colonies of bifidobacteria on YN-6 are 1-2 mm in diameter, green (light-dark), circular, entire, convex or pulvinate, smooth, butyrous, and opaque. A dissecting microscope (20X magnification) must be used to aid identification.

Confirmation is accomplished by establishing the requirement for anaerobic conditions, gram stain, and other pertinent tests (see Fig. 3). In practice, when using YN-6 medium, gram staining of the

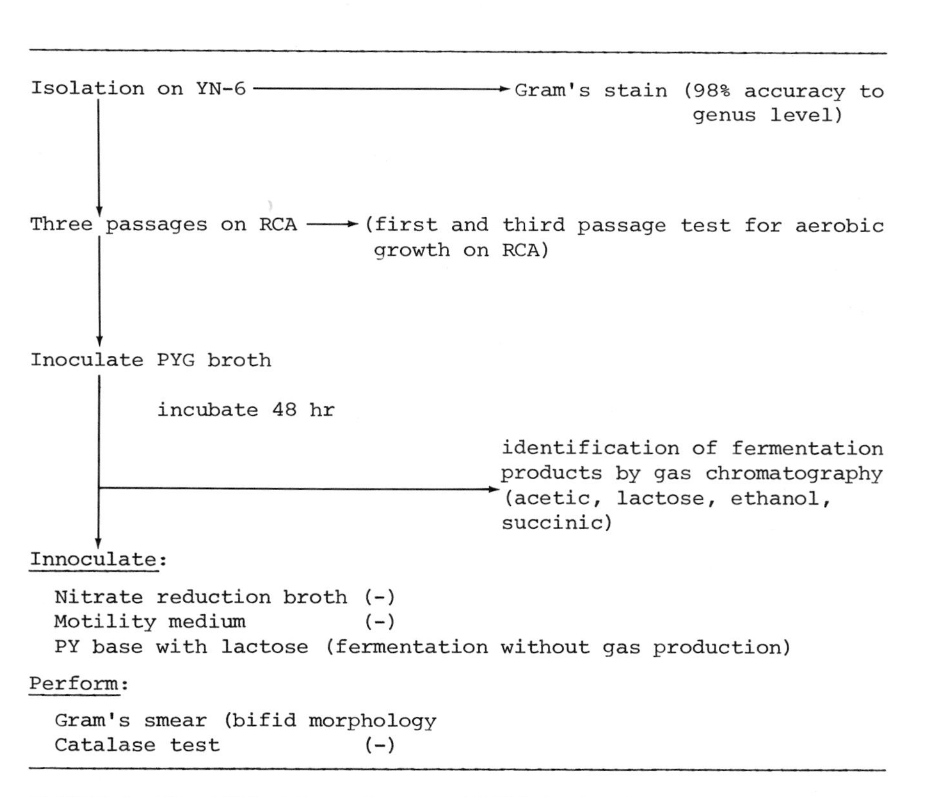

FIGURE 3 Identification of genus *Bifidobacterium*.

isolate provides identification to the genus level with 98% accuracy. This confirmation procedure can be eliminated after sufficient experience in observing colonial morphology has been acquired.

The effect of physical and chemical differences between membrane filters on the recovery and colonial morphology of bacterial isolates has been described (Bordner et al., 1977). All isolation techniques that employ membrane filters should be evaluated on the various types and brands available to determine which membranes are best suited for the procedure. In appreciation of the variations between different types of membrane filters, a comparison of several brands was conducted. We compared six different types of filters from five companies. The results of the comparison appear in Table 5. The dramatic effect of the filters on percent recovery is obvious. Unfortunately, the ability of the filter to permit the maximum recovery of the bacterial type being selected for cannot be the only consideration. The use of YN-6 medium and selection for bifidobacteria provides an excellent example of the interaction between several significant parameters. The colonial morphology of isolates on this medium is a differential characteristic of the technique and must be maintained. Thus, although it would be easy to arbitrarily establish an acceptable level and apply statistical analysis to select a filter, this is not possible. Nevertheless, the Nuclepore and Schliecher and Schuell filters are not acceptable from a percent recovery point of view. Discrimination between the remaining filters is more difficult. Gelman and Sartorius filters exhibited similar recovery and colonial characteristics, which rendered them acceptable for use when attempting to recover bifidobacteria on YN-6 medium. The colonies were approximately 2 mm in diameter and conformed in all respects to the description given above.

The colonies observed on Millipore HC filters exhibited unusual morphology. The 64% recovery was achieved by selecting the most likely colonies. No clear differentiation between positive and

TABLE 5 Evaluation of Membrane Filters[a]

Filter type	Lot number	% Recovery[b]	False positive[c]	False negative[d]	Background[e]
Gelman GN-6	81891	100 ±5	5	0	8.4
Millipore HA (certified)	C8J622777	76 ±10	0	0	32
Millipore HC	C7M20059	64 ±15[f]	-[f]	100[f]	35
Sartorius	13756	111 ±5	5	0	ND[g]
Nuclepore	N040CPR	0	-	-	-
Schliecher and Schuell	70/4	18 ±1	15	0	0

[a]YN-6 medium was used to assay septic tank samples. All filters were 47 mm in diameter. The Millipore HC filter had a designated pore size of 0.7 μm; all others were designated 0.45 μm pore size.
[b]Two experiments were conducted with five replicate plates used at each dilution. Gelman GN-6 was arbitrarily selected as a standard.
[c]Percent of typical colonies that could not be confirmed as bifidobacteria.
[d]Percent of atypical colonies that were confirmed as bifidobacteria.
[e]Expressed as percent of typical colonies.
[f]Plates were extremely difficult to count, owing to unusual colonial morphology. No clear differentiation between typical and atypical colonies was possible.
[g]No data.

negative colonies was possible. The Millipore HA filter (certified), on the other hand, yielded colonies that were one-third larger (3 mm) and much easier to distinguish from the background colonies than any of the other filter types examined. This enhances the discriminatory ability of the operator and reduces the incidence of false-positive colonies. Another filter-medium interaction evident from the data in Table 5 is the variation in background level. As can be seen, Gelman filters reduced the background level by approximately 66%. Although this is not significant when using this medium and septic tank samples, it may be an important factor in another environment.

Based on all of the foregoing considerations, it is suggested that either Gelman GN-6, Sartorius (47 mm diameter membrane, 0.45 μm pore size, white, gridded), or Millipore HA (certified) filters be used with the YN-6 medium. The small difference in recovery between the HA filters and the Gelman or Sartorius filters is overcome by the ease of counting and the reduced frequency of false-positive colonies.

B. Evaluation of YN-6 Medium

An overall recovery of 91.7% of pure cultures of various bifidobacteria has been achieved on YN-6 medium (Table 6). All the cultures were derived from recent isolates obtained from RCA agar plates.

TABLE 6 Recovery of Bifidobacteria Species on YN-6 Using Fresh Cultures

Environmental isolate	% Recovery[a]
B. longum (B29)	88
B. longum (B30)	88
B. adolescentis (23)	66
B. adolescentis (B306)	100
B. infantis (30-14)	100
B. infantis (W-2)	100
B. longum (B22)	100
Average	91.7

[a] $\frac{\text{Organisms/ml on YN-6 filter}}{\text{Organisms/ml on RCA spread plate}}$.

TABLE 7 Recovery of Stressed Cultures on YN-6 Medium[a]

Time (hr)	Buffer (%)	% Recovery[b]	
		Marine water	Fresh water
0	100	100	100
3	100	70	91
9	62	34	9

[a]Stressed by exposure to buffer or fresh or salt menstrum at 6°C.
[b] $\frac{\text{Organisms/ml on YN-6 filter}}{\text{Organisms/ml on RCA spread plate}}$.

When the cultures were stressed by exposure to fresh or salt water for up to 6 hr, as great as 60% recovery was observed (Table 7).

Other data (obtained in this laboratory with cells stressed in buffered saline) indicate that environmental samples cannot be stored for a prolonged period even though the organisms survive well (3-9 hr). This would suggest that samples should be assayed within 3 hr of collection, if possible.

Sewage and river samples were used to compare recovery of bifidobacteria on YN-6 medium to recoveries obtained on the medium developed by Evison (1973). The data indicate that the Evison medium is less selective and achieves a lower overall recovery of bifidobacteria. The Evison medium was overgrown with gram-positive facultative anaerobic bacilli when used with sewage samples from which YN-6 medium recovered 30×10^4 bifidobacteria per milliliter. Similar results were obtained with natural samples. The Evison medium recovered at least 90% less bifidobacteria while allowing a large number of background colonies to develop.

C. Distribution of Bifidobacteria

The interpretation of quantitative data concerning levels of bifidobacteria in environmental samples requires a body of information relating the data obtained to fecal pollution or potential hazards. In an attempt to begin to amass the required data base, quantitative data have been collected from the examination of sewage, fecal specimens, and environmental samples. Initially, fecal samples were

TABLE 8 Quantitation of Bifidobacteria in Feces of Healthy Adults[a]

Species	Range (organisms/g)	Arithmetic Mean (organisms X 10^8)
B. longum	2×10^6-2×10^9	28.83
B. adolescentis	1×10^8-6×10^9	38.67
B. breve	3×10^6-3×10^8	1.54
B. bifidum	1.3×10^8	0.59
B. thermophilum	3×10^6	0.01
B. infantis	3×10^6	0.01

[a]Based on 22 healthy adults.

obtained from healthy persons and from a variety of animals. All samples were assayed within 3-5 hr of collection, and data are based on three replicate plates obtained from serial 10-fold dilutions of a 1 g sample. All colonies that were present at the appropriate dilution were identified. As can be seen (Table 8), *B. longum* and *B. adolescentis* comprise 96.5% of the population, with *B. breve* providing an additional 2.2%. The contribution of other species is negligible; one isolate that was unclassifiable by use of biochemical patterns was obtained. No organisms were isolated from similar samples obtained from dogs (14), horses (6), cats (4), chickens (23), cows (15), goats (2), sheep (4), turkeys (2), and beavers (4). The distribution of the component species in a sewage treatment plant is compared with the human data in Table 9. It is clear that the

TABLE 9 Frequency of Isolation of Component Species of Bifidobacteria

Species	Sample source	
	Human feces[a]	Sewage (raw)
B. longum	41.4[b]	39.6
B. adolescentis	55.5	35.4
B. breve	2.2	2.5
B. infantis	0.01	Not found
B. thermophilum	0.01	6.1
B. bifidum	0.59	10.0

[a]Based on 22 healthy adults.
[b]Percent of total isolates.

distribution within the plant is similar to the distribution of species in human feces in that *B. longum* and *B. adolescentis* constitute the major portion of the population. The discrepencies in the levels of *B. bifidum* and *B. thermophilum* between sewage and feces may be due to the presence of fecal material from children in sewage, whereas the data for fecal isolates were obtained with feces from adults.

D. Fructose-6-phosphoketolase

It has been reported (Vries et al., 1967) that members of the genus *Bifidobacterium* utilize a pathway for glucose catabolism uncommon in other organisms. The key enzyme in this pathway is fructose-6-phosphoketolase (F6PPK), which acts as a catalyst in the cleavage of fructose-6-phosphate to acetyl phosphate and erythrose-4-phosphate. The progress of this reaction can be followed by quantitation of the acetyl phosphate produced (Lipmann and Tuttle, 1945). Scardovi et al. (1971b) found that the F6PPK from different species varied in electrophoretic mobility in starch-gel electrophoresis. Although the distance migrated was not constant within species, it seemed to be related to the habitat from which the particular bacterium was isolated. Using *B. globosum* isolated from the rumen of a cow as a reference strain, these workers observed a migration distance for FPPK of 80 mm and assigned that distance a value of 10. The values from *L. bifidum* isolated from human feces and *B. asteroides* isolated from bee intestine were 14.25 and 15.75, respectively. Values for other isolates from animal, human, and insect sources were consistent with these values. Although the extraction of enzymes and determination of their electrophoretic mobility is not feasible for the routine water quality laboratory, it may serve as a useful took for identifying the source of fecal pollution in particular cases.

E. Survival

Survival characteristics of "bifid bacteria" were first described in 1960 by Gyllenberg et al. They compared the survival of bifidobacteria, coliforms, and enterococci by suspending human fecal samples in a variety of environmental waters. "Bifid bacteria" were enumerated on the medium of Gyllenberg and Niemela (1959). No confirmatory tests were conducted of presumptive positive colonies. They concluded that "bifid bacteria" have survival characteristics similar to coliform bacteria. They also observed 51% survival after 48 hr at room temperature and 32% survival after 12 days of exposure to 4°C. Evison and James (1973, 1974), measuring anaerobic lactobacilli, conclude that there were no significant differences between the survival of these organisms and coliforms, *E. coli*, and fecal streptococci. The major influence on survival appeared to be temperature.

Resnick (1978) employed pure cultures as well as fecal samples in a controlled study in 1977. The survival of the inoculum in autoclaved buffer, filter-sterilized marine waters, and fresh water was determined at 4°C, 12°C, and 20°C over a 24 hr period. When pure cultures were used, RCA spread plates were employed for enumerating the viable organisms present, and YN-6 medium was employed for sewage inoculum.

The data obtained (Figs. 4, 5, and Table 7) differ markedly from Gyllenberg et al. (1960). It is possible that this difference was due to either false-positive colonies of lactobacilli, which are known to survive well at refrigerator temperatures, or to a difference in the source of the organisms, in that Gyllenberg used fresh fecal material as an inoculum whereas Resnick used raw sewage. The use of raw sewage was dictated by the fact that organisms entering the treatment plant have been stressed by a lengthy exposure to the sewerage system. The data obtained upon exposure to fresh water supported the observations of Evison and James (1973, 1974) that temperature has a major effect on the survival of bifidobacteria. However, in a marine environment (32°/oo), no effect of temperature was observed. In both cases, however, the decrease in viability was >26 times faster than that of *E. coli* (Vaccaro et al., 1950). It is of interest to note that survival of strains of pure cultures exhibited marked variability in survival characteristics. Thus, the

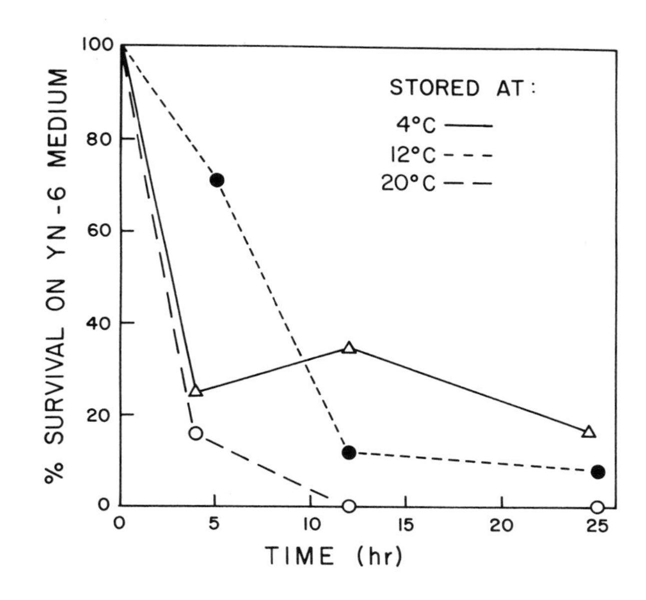

FIGURE 4 Survival of *Bifidobacterium* in fresh water.

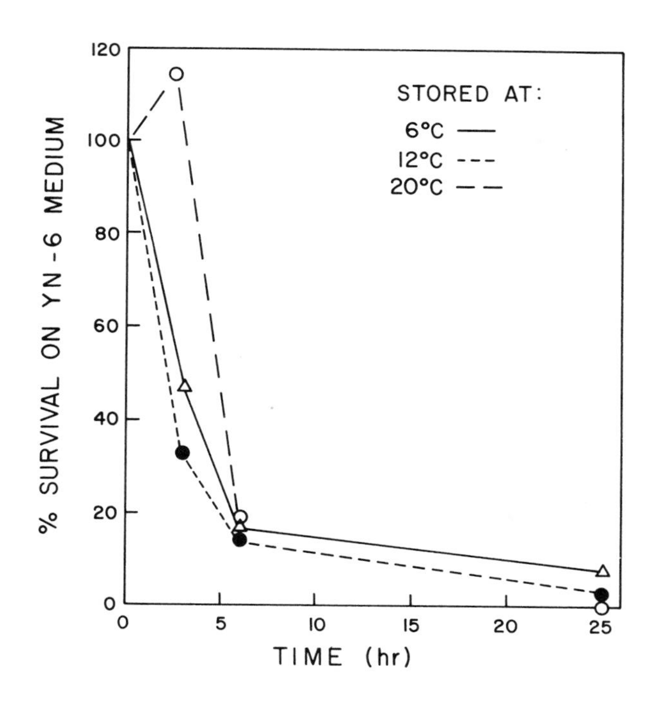

FIGURE 5 Survival of *Bifidobacterium* in marine water (32°/oo).

T90 values ranged from 9.8 hr to 22 hr for strains of *B. longum* and 23 hr to 82 hr for strains of *B. adolescentis*.

The survival of bifidobacteria as affected by exposure to chlorine was studied in a similar manner to determine the effect of postsecondary chlorination in a treatment plant.

It was found that a residual chlorine concentration as low as 0.2 ppm after 15 min contact time reduced the density of bifidobacteria from 7.9 X 10^4/100 ml to <10/100 ml. These data correlate well with observations in treatment plants.

F. Environmental Surveys

Few data are available describing the distribution of bifidobacteria under environmental conditions. The work of Evison and James (1973, 1974), with data from Africa and England, and the work of Resnick (1978), providing data from the New England area, are the only two sources of information. Although these authors used different techniques, some of their data appear to be comparable (Table 10).

The differences between these observations at the upstream and downstream sampling locations may be attributable to water temperature, geographical location, or to the utilization of different techniques. The Evison procedure may have included the densities of anaerobic lactobacilli with bifidobacteria, and these may have predominated above and below the outfall. However, it is clear that bifidobacteria were present in high densities at the outfall and at the time the samples were taken. These observations were extended to other locations and the ratio between bifidobacteria and *E. coli* has been calculated in several rivers (Table 11) and in sewage (Table 12).

TABLE 10 Environmental Levels of Bifidobacteria

Location	Bifidobacteria/100 ml X 10^3		
	Qued Sebou River[a]	Pawtuxet River[b]	Thames River[b]
Upstream (ca. 5 km)	5.0	0.00	0.060
Outfall	3.5	220.00	1.300
Downstream (ca. 22 km)	3.5	0.14	0.007

[a]Evison and James (1974).
[b]Resnick (1978).

TABLE 11 Relationship Between Bifidobacteria and *E. coli* in River Water

Location	Distance from outfall (km)	Organisms/100 ml X 10^3		
		Bifidobacteria[a]	*E. coli*[b]	B/E Ratio
Pawtuxet River	0	22.0	5.2	4.82
	1.2	6.0	34.0	0.17
	8.8	0.0	10.0	0.00
	0	220,000	100,000	2.20
	1.6	14.0	35.0	0.46
Thames River	0.8	6.4	4.7	14.68
	0	130	9.0	14.41
	0.8	0.76	1.1	0.71
Seekonk River		13.3	90	0.14
Non point source		62.6	90	0.68

[a]Bifidobacteria were enumerated using YN-6 medium.
[b]*E. coli* estimates were obtained by the mTEC procedure of Dufour et al. (1975).
Source: Resnick (1978).

TABLE 12 Relationship Between Bifidobacteria and *E. coli* in sewage

Location	Organisms/100 ml X 10^3		
	Bifidobacteria	*E. coli*	B/E Ratio
University STP			
Raw	83.3	1.5	55.5
Primary	36.5	2.1	17.1
Secondary	9.8	8.7	11.2
Warwick STP			
Raw	2.3	2.9	0.79
Narragansett STP			
Raw	4.4	0.5	88
Septic Tank	1140	260	4.38

Source: Resnick (1978).

The river data clearly indicate that bifidobacteria can be detected in samples taken at stations located in close proximity to an outfall. The dramatic decrease in the bifidobacteria/*E. coli* (B/E) ratio at downstream locations is an expression of the fact that bifidobacteria are markedly more sensitive to environmental stress than are *E. coli*. This conclusion is substantiated by the survival studies described above and would indicate that the mere presence of these organisms in a river should be a cause for alarm, because their presence reflects a nearby source of sewage. Thus, even when the *E. coli* density is as low as 52 organisms per milliliter, the presence of 230 bifidobacteria per milliliter (B/E = 7.84) in the vicinity of the outfall indicates that there has been a malfunction of the treatment plant or the presence of a small source in the general area.

The data obtained from the treatment plants also provided an indication of the relative survival rates of bifidobacteria and *E. coli*. The University STP serves a very small system, with a short flow time to the plant along the sewer lines. Thus, the B/E ratio is high at the time of entry to the plant (55.5) and lower at the exit (11.2). The other plants reflect the effect of a much larger sewer system and a longer travel time.

IV. CONCLUSIONS

There have been numerous articles raising doubts about the efficacy of the various indicators currently in use and about the standards that are linked to particular organisms. Thus, bathing beach standards are considered far too strict by some (Moore, 1977), too lenient by others (Senn et al., 1963), and based on the wrong indicator organisms by still other workers (Cabelli et al., 1977). In addition, the paucity of epidemiological evidence relating incidence of disease to indicator density makes answering these questions difficult in the extreme.

However, there is little doubt that the presence of sewage and/or fecal pollution must be determined and quantitated in some manner in order to permit officials to establish safeguards for the health of the general public. There is also little doubt that an indicator (or perhaps more than one indicator) of some type must be measured, because it is logistically impossible to continually monitor all the potential pathogens. As indicated above, numerous authors have elaborated on the characteristics of the ideal indicator organisms. Bifidobacteria passes many of these characteristics. They are present in the feces of normal adults at levels in excess of coliforms and enterococci. They are not able to multiply in extraenteral environmentals and can survive for moderate periods of time. They can be quantitated with only a moderate amount of specialized equipment and training. They are highly specific, in that they have not been isolated from other than human feces using the YN-6 medium and associated techniques with one exception (swine feces).

The chief drawback is extreme sensitivity to chlorination and only moderate survival times, 1/26 that of coliforms (Resnick, 1978). These factors can, however, be construed to be advantages. The presence of bifidobacteria can only mean nearness of a source of unchlorinated human fecal pollution, and the level would provide an indication of how close to the source the sample was taken.

The work of Scardovi et al. (1971b) would indicate that these organisms can be used to differentiate clearly and quantitatively between human and animal sources of pollution. Although examination of fecal samples from seven species of common animals were negative, bifidobacteria were isolated from swine feces. Determination of the electrophoretic type of FPPK present should permit identification of the source of each isolate obtained from the membrane filter.

REFERENCES

APHA, AWWA, WPCF (1975). *Standard Methods for the Examination of Water and Wastewater*. 14th ed. American Public Health Association, Washington, D.C.

Bonde, G. J. (1966). Bacterial methods for estimation of water pollution. *Health Lab. Sci. 3:* 112.

Bordner, R. H., Firth, C. F., and Winter O. A., Eds. (1977). *Symposium on the Recovery of Indicator Organisms Employing Membrane Filters,* September. EPA-600/9-77-024 and ASTM.

Buchanon, R. E., and Gibbons, N. E., Eds. (1974). *Bergey's Manual of Determinative Microbiology.* 8th ed. Williams & Wilkins, Baltimore, Md.

Cabelli, V. J. (1977). Indicators of recreational water quality in bacteria, pp. 222–238. In A. W. Hoadley and B. J. Dutka (Eds.), *Bacterial Indicators/Health Hazards Associated with Water.* American Society for Testing and Materials, Philadelphia.

Cabelli, V. J., Levin, M. A., Dufour, A. P., and McCabe, L. J. (1975). In *Discharge of Sewage from Sea Outfalls.* Pergamon Press, London.

Dufour, A. P., Strickland, E. R., and Cabelli, V. J. (1975). A membrane filter procedure for enumerating thermotolerant *E. coli. Proc., 9th Natl Shellfish Sanit. Workshop,* D. S. Wilt (Ed.), Charleston, S.C., U.S. Food and Drug Administration.

Eggerth, A. H. (1935). The gram-positive non-spore-bearing anaerobic bacilli of human feces. *J. Bacteriol. 30:* 277–299.

Evison, L. M. (1973). Personal communication.

Evison, L. M., and James, A. (1973). A comparison of the distribution of intestinal bacteria in British and East African water sources. *J. Appl. Bacteriol. 36:* 109.

Evison, L. M., and James, A. (1974). Bifidobacterium as an indicator of faecal pollution in water. *Proc. 7th Int. Conf. Water Pollut. Res., Paris.*

Exterkate, F. A., Veerkamp, J. H. (1969). Biochemical changes in *Bifidobacterium bifidum* var. pennsylvanicus after cell wall inhibition. I. Composition of lipids. *Biochem. Biophys. Acta 176:* 65–77.

Gasser, F., and Mandel, M. (1968). Deoxyribonucleic acid base composition of the genus *Lactobacillus. J. Bacteriol. 96:* 580–588.

Glick, M. C. T., Sall, T., Zilliken, F., and Mudd, F. (1960). Morphological changes of *Lactobacillus bifidus* var. pennsylvanicus produced by a cell wall precursor. *Biochem. Biophys. Acta 37:* 361–363.

Gyllenberg, H. G. (1955). The development of the "straight rod type" of *Lactobacillus bifidus. J. Gen. Microbiol. 13:* 394–397.

Gyllenberg, H., and Niemela, S. (1959). A selective method for the demonstration of bifid bacteria (*L. bifidus*) in materials tested for faecal contamination. *Maataloustiet. Aikak. 31:* 94.

Gyllenberg, H., Niemela, S., and Sormunen, T. (1960). Survival of bifid bacteria in water as compared with that of coliform bacteria and enterococci. *Appl. Microbiol. 8:* 20.

Hoadley, S. W., and Dutka, B. J., Eds. (1977). *Bacterial Indicators/Health Hazards Associated with Water.* American Society for Testing and Materials, Philadelphia.

Holdeman, L. V., and Moore, W. E. C. (1972). *Anaerobe Laboratory Manual.* Virginia Polytechnic Institute and State University, Blacksburg, Va.

Lauer, E., and Kandler, O. (1976). Mechanism of the variation of the acetate/lactate ratio during glucose fermentation by bifidobacteria. *Arch. Microbiol. 110* (2/3): 271–278.

Levin, M. L. (1977). Bifidobacteria as water quality indicators, pp. 131–138. In A. W. Hoadley and B. J. Dutka (Eds.), *Bacterial Indicators/Health Hazards Associated with Water.* American Society for Testing and Materials, Philadelphia.

Lipmann, F., and Tuttle, L. C. (1945). A specific micromethod for determination of acyl-phosphates. *J. Biol. Chem. 159:* 21–28.

Malligo, J. E. (1977). Use of organoleptic tracer (mercaptan) for testing for leaks in safety equipment. *Appl. Environ. Microbiol. 34:* 861–862.

Mitsuoka, T. (1969). Vergleichinde Untersuchiengen über die Bifidobakterien aus dem Verdauungstrakt von Menschen and Tieren. *Zentralb. Bakteriol. Parasitenkd. Infektionskr. Hyg. Abt. Orig. 210:* 52–64.

Moore, B. (1977). The EED bathing water directive. *Marine Pollut. Bull. 8* (12): 268.

Moro, E. (1900). Über die nach Gram färbbaren Bacillen des Sauglingsstuhles. *Wien. Klin. Wochenschr. 13:* 114–115.

Mossel, D. A. A. (1958). The suitability of bifidobacteria as part of a more extended bacterial association, indicating faecal contamination of foods. *7th Int. Congr. Mikcrobiol. Stockholm, Sweden.* Abst. Papers, p. 440.

Norris, R. F., Flanders, T. Tomarelli, R. M., and Gyorgy, P. (1950). The isolation and cultivation of *Lactobacillus bifidus.* A comparison of branched and unbranched strains. *J. Bacteriol. 60:* 681–696.

Orla-Jensen, S. (1924). La classification des bactéries lactiques. *Lait 4:* 468–474.

Poupard, J. A., Husain, I., and Norris, R. F. (1973). Biology of the *Bifidobacterium. Bacteriol. Rev. 37:* 136–165.

Resnick, I. G. (1978). *Bifidobacterium,* indicator of human fecal pollution. Dissertation, University of Rhode Island, West Kingston, R.I.

Resnick, I. G., and Levin, M. A. (1978). Unpublished data.

Resnick, I. G., and Levin, M. A. (1977). Enumeration of *Bifidobacterium* in aquatic and fecal samples. *Abst. Annu. Meet. Amer. Soc. Microbiol.*, p. 263.

Reuter, G. (1963). Vergleichende Untersuchungen über die Bifidus-Flora im Sauglingsand Erwacksenstuhl. *Zentralbl. Bakteriol. Parasitenkd. Infektionskr. Hyg. Abt. I Orig. 191:* 486—507.

Scardovi, V., and Crociani, F. (1974). *Bifidobacterium catenulatum, Bifidobacterium dentium,* and their deoxyribonucleic acid homology relationships. *Int. J. Syst. Bacteriol. 24:* 6—20.

Scardovi, V., Trovatelli, L. D., Zani, G., Crociani, F., and Matteuzzi, D. (1971a). Deoxyribonucleic acid homology relationships among species of the genus *Bifidobacterium. Int. J. Syst. Bacteriol. 21:* 276—294.

Scardovi, V., Sgorbati, B., and Zani, G. (1971b). Starch gel electrophoresis of fructose-6-phosphate phosphoketolase in the genus *Bifidobacterium. J. Bacteriol. 106:* 1036—1039.

Scardovi, V., and Trovatelli, L. D. C. (1965). The fructose-6-phosphate shunt as a pecular pattern of hexose degradation in the genus *Bifidobacterium. Ann. Microbiol. Enzymol. 15:* 19—29.

Scardovi, V., and Trovatelli, L. D. (1974). *Bifidobacterium animalis* (Mitsouka) comb. nov. and the "minimum" and "subtile" groups of new *Bifidobacterium* found in sewage. *Int. J. Syst. Bacteriol. 24:* 21—28.

Scardovi, V., and Zani, G. (1974). *Bifidobacterium magnum* sp. Nov., a large, acidophizic *Bifidobacterium* isolated from rabbit feces. *Int. J. Syst. Bacteriol. 24:* 29—34.

Sebald, M., Gasser, F., and Weiner, H. (1965). Teneur 60% et classification. Application au groupe des bifidobactéries et à quelques génies voisins. *Ann. Inst. Pasteur 109:* 251—269.

Senn, C. L., Berger, B. B., Jensen, E. C., Ludwig, H., Romes, H., and Shapiro, M. A. (1963). Coliform standards for recreational water. *J. Sanit. Eng. Div. 89:* 57.

Sutter, V. L., Vargo, V. L., and Finegold, S. M. (1975). *Wadsworth Anaerobic Bacteriology Manual.* Wadsworth Hospital Center, Los Angeles, Calif.

Tissier, M. H. (1889). La réaction chromophile d'eschérichel et le bactérium coil. *C.R. Acad. Sci. 51:* 943—945.

Tissier, H. (1900). Recherches sur la flore intestinale des nourrissons (état normal et pathologique). Thesis, Paris.

Vaccaro, R. F., Briggs, M. P., Corey, C. L., and Ketchum, B. H. (1950). Viability of *Escherichia coli* in seawater. *Am. J. Public Health 40:* 1257—1266.

Weiss, J. D. (1933). *Lactobacillus bifidus* Tissier and its biological position in the group of aciduric organisms. Dissertation, Yale University, New Haven, Conn.

Weiss, J. E., and Rettger, L. F. (1934). *Lactobacillus bifidus*. *J. Bacteriol*. *28*: 501–521.

7

VIBRIO AND AEROMONAS

---

*R. R. COLWELL and JAMES B. KAPER** University of Maryland, College Park, Maryland

I. VIBRIOS

Vibrios are common in estuarine, coastal, and deep ocean water. In freshwater lakes and streams, *Vibrio* species can also be found, but it is less clear how significant they are in ecosystem processes. Both the salt-requiring and freshwater vibrios include species pathogenic for humans, the most important species being *V. parahaemolyticus* and *V. cholerae,* respectively. In the case of the latter, cholera has been recognized as a human disease since the turn of the century, with *Vibrio cholerae* believed to enter water directly via discharge from cholera victims or indirectly, in wastewater and runoff from areas endemic for cholera (i.e., in which overt cases of cholera, persons in the incubation stage, or healthy carriers are present), or to exist in brackish water areas (Colwell, 1980). Transmission of *V. parahaemolyticus*, on the other hand, is assumed to occur via contaminated seafoods, with estuarine crustaceans the suspect reservoir of *V. parahaemolyticus* (Kaneko and Colwell, 1975). Massive epidemics of cholera are a historical phenomenon, with records including a recent pandemic of the El Tor biotype of *V.*

**Present affiliation:* University of Washington, Seattle, Washington

TABLE 1 Differentiation of Related Genera Frequently Isolated from the Same Source in Nature

Characteristic	*Vibrio*	*Aeromonas*	*Plesiomonas*	*Photobacterium*	*Lucibacterium*	*Pseudomonas*	*Spirillum*	*Campylobacter*
Morphology	Straight or curved rod	Straight rod	Straight rod	Straight rod	Straight or curved rod	Straight rod	Helical	Spirally curved rod
Diffusible pigment	None	None[a]	None	None	None	None or green-fluorescent	None or green-fluorescent	None
Motility	+[b]	+	+	+	+	+	+	+
Flagella	Polar	Polar	Lophotrichous	Polar	Peritrichous (usually)	Polar	Lophotrichous	Polar
Carbohydrate metabolism	Fermentative	Fermentative	Fermentative	Fermenative	Fermentative	Respiratory or not metabolized	Respiratory	Not metabolized

Gas production from carbohydrates	-	v	-	+	-	-	-	-
Luminescence	v	-	-	+	+	-	-	-
Oxidase	+	+	+	v	+	+	+	-
0/129 sensitivity	+	-	+	+	-	-	-	-
"Round bodies" or "cysts" produced	+	-	-	-	-	-	+	+
DNA base composition (% G + C)	40-50	57-63	51	39-42	45-46	58-70	38-65	30-35

[a]Species of *Aeromonas* may produce a brown pigment.

[b]Symbol code: +, positive or present; -, negative or absent; v, variable.

Source: Reprinted with permission from *Handbook of Microbiology*, 1973, *Vibrios* and *Spirilla*, by R. R. Colwell, p. 102. Copyright The Chemical Rubber Co., CRC Press, Inc., Boca Raton, Fla.

cholerae. A number of serious, but more geographically restricted, outbreaks of *V. parahaemolyticus* have occurred because this organism was first identified about 25 years ago. Pathogenic species of vibrios have, therefore, been known for many years as agents of disease and their presence recognized as a health threat.

Until recently, *Vibrio* species have not been considered by clinicians on a routine basis as a causative agent of enteric disease in countries where cholera was absent. Recently documented cases of *Vibrio* species associated with food-poisoning outbreaks, wound infections, and otitis media have alerted microbiologists and public health workers to consideration of vibrios as potential agents of disease. The potential health hazard of *Vibrio* species and their role in waterborne disease have now been recognized.

A. Taxonomy

The genus *Vibrio* comprises short, curved or straight, gram-negative asporogenous rods. The organism is usually motile by means of a single polar flagellum and is facultatively anaerobic, with both a respiratory and a fermentative metabolism. In addition, vibrios are oxidase-positive, form nitrites from nitrates, produce acid but no gas from glucose, and grow in media of pH 6-10. The overall base composition of the deoxyribonucleic acid (DNA) of vibrios ranges from 40 to 50 mol % guanosine + cytosine (G + C). They are usually sensitive to novobiocin and 2,4-diamino-6,7-diisopropyl pteridine (0/129).

Vibrios can be readily isolated from saltwater and freshwater samples but are often confused, on initial isolation with other genera, such as *Aeromonas* and *Spirillum*. Several distinguishing characteristics that must be determined to identify vibrios properly are given in Table 1.

Five species of *Vibrio* are listed in the eighth edition of *Bergey's Manual of Determinative Bacteriology*, and these include *V. cholerae*, *V. parahaemolyticus*, *V. anguillarum*, *V. marinus* (synonym, *V. fischerii*), and *V. costicola*. Many workers have assigned separate species status to *V. alginolyticus*, but it is designated biotype 2 of *V. parahaemolyticus* in *Bergey's Manual* (Shewan and Véron, 1974). *V. alginolyticus* can be distinguished from *V. parahaemolyticus* by a number of characteristics, including the fact that *V. parahaemolyticus* has been associated unequivocably with acute gastroenteritis in humans, whereas *V. alginolyticus* rarely, if ever, has been implicated in outbreaks of gastroenteritis, its pathogenic manifestation usually being limited to wounds and burn infections exposed to seawater (Rubin and Tilton, 1975; English and Lindberg, 1977; Pien et al., 1977). Differentiating characteristics of *V. alginolyticus, V. parahaemolyticus*, and other *Vibrio* species are listed in Table 2.

B. Significance

The significance of *Vibrio* species as indicators of potential health hazards associated with water is increasingly evident. *V. parahaemolyticus* and *V. alginolyticus* have been isolated from artesian waters in Florida (Koburger and Lazarus, 1974), as well as from estuarine waters located within 500 miles of the Arctic circle (Vasconcelos et al., 1975). There is a great deal of interest in *V. parahaemolyticus, V. alginolyticus,* and related vibrios, not only in the United States, but also in the Netherlands, Great Britain, India, Japan, and other countries, because the organism can be found in significant numbers in nutrient-rich waters. Since the ecology of *Vibrio* species, with the exception of *V. parahaemolyticus*, is poorly understood, the significance of vibrios as indicators of water quality is not yet fully clarified. The primary indicator organism for water quality estimation has, for decades, been *Escherichia coli,* with the presence of *E. coli* being interpreted as indicative of fecal contamination and implying the presence of pathogenic bacteria. *E. coli*, well-established procedurally as an indicator species, is not clearly established as a pathogen in the same way as *V. cholerae, V. parahaemolyticus, V. alginolyticus,* and *V. anguillarum* have been, although many cases of disease due to enteropathogenic *E. coli* have been reported (Sack, 1975). The significance of large numbers of *V. parahaemolyticus*, for example, in

TABLE 2 Features Useful in Differentiating and Characterizing Species of the genus *Vibrio*

Characteristic	*V. cholerae*	*V. parahaemolyticus*	*V. alginolyticus*	*V. anguillarum*	*V. marinus*	*V. costicolus*
Rod shape	+[a]	+	+	+	+	+
Motility	+	+	+	+	+	+
Single polar flagellum	+	+	+	+	v	+
Lophotrichous flagella	-	-	-	-	v	-
Gram reaction	-	-	-	-	-	-
Diffusible pigment	-	-	-	-	-	-
Luminescence	-	-	-	-	-	-
Pathogenicity for humans or animals	+	+	+	+	-	-
DNA base composition (% G + C)	46-49	44-46	44-46	44-45	40-44	50
Indole reaction	+	+	+	+	-	-
Methyl red reaction	v	+	-	+	+	-
Voges-Proskauer reaction	v	-	+	+	-	+
Citrate utilization	+	+	+	-	-	-
Citrulline utilization	-	-	nt	+	-	nt
Sensitivity						
0/129	+	+	+	+	+	v
Novobiocin	+	+	+	+	+	+
Penicillin, 10 units	+	-	-	-	-	nt
Polymyxin, 300 units	v	-	-	v	+	nt
Streptomycin, 10 μg	+	-	-	-	+	nt
Growth at:						
0% NaCl	+	-	-	+	+	-
1% NaCl	+	+	+	+	+	-
7% NaCl	v	+	+	v	+	+
10% NaCl	-	-	+	-	-	+

5°C	-	-	-	+	+	+
20°C	+	+	+	+	+	+
37°C	+	+	+	v	-	v
42°C	+	+	+	-	-	-
Acid production from:						
Arabinose	-	v	-	-	-	-
Inositol	-	-	-	-	v	-
Mannitol	+	+	+	+	+	v
Mannose	+	+	+	+	+	+
Salicin	-	-	-	-	v	-
Sucrose	+	-	+	+	v	+
Gelatin liquefaction	+	+	+	+	+	v
Hydrolysis						
Casein	+	+	+	+	+	-
Starch	+	+	+	+	v	-
Tween 80	+	+	+	+	+	+
H_2S production (on lead acetate agar)	-	-	-	-	-	-
Lecithinase (egg yolk)	+	+	+	nt	nt	v
Arginine dihydrolase	-	-	-	+	-	+
Lysine decarboxylase	+	+	+	-	+	-
Ornithine decarboxylase	+	+	+	-	-	-
Hemolysis	+	+	+	v	-	-
Swarming on TSA with 3% NaCl	-	-	+	-	-	-

[a]+, positive; -, negative; v, variable; nt, not tested.

Source: Reprinted with permission from *Handbook of Microbiology*, 1973, *Vibrios* and *Spirilla*, by R. R. Colwell, p. 99. Copyright The Chemical Rubber Co., CRC Press, Inc., Boca Raton, Fla.

an embayment of an estuary, is not, however, known to be a public health hazard in every circumstance, whereas the occurrence of large numbers of *V. parahaemolyticus* in shellfish destined for human consumption is unequivocably recognized as a public health threat (Food and Drug Administration, 1976). Yet to be proven is a correlation between standard pollution indexes, such as coliform counts, and number of *Vibrio* species in water or food. On the contrary, it has been found that no significant relationship can be detected between incidence of *V. parahaemolyticus*, either as simply present or in large numbers, and high counts of *E. coli* in estuaries (Thompson et al., 1976). In fact, *V. alginolyticus* may prove to be a more useful standard than *V. parahaemolyticus* for purposes of water quality evaluation, since it has been found to occur in greater numbers than *V. parahaemolyticus* in natural waters. A proportional relationship between the numbers of *V. alginolyticus* and *V. parahaemolyticus* has, indeed, been suggested (Baross, 1980).

V. cholerae has recently achieved notoriety in the United States since it was found to occur in Chesapeake Bay (Colwell and Kaper, 1977; Colwell et al., 1977, 1980) and in other estuaries of the United States (Jonas, 1979). The ecology and enterotoxin production of the estuarine isolates have been extensively studied, showing that the distribution of *V. cholerae* in Chesapeake Bay is limited by salinity; that is, *V. cholerae* is isolated only from water of salinity in the range 4-17% (Kaper et al., 1979). No station in the Chesapeake Bay where the water salinities were <4% or >17% harbored *V. cholerae*. In addition, results of statistical analysis showed no correlation between incidence of fecal coliforms and *V. cholerae*, whereas incidence of *Salmonella* spp., measured concurrently, was found to correlate with fecal coliforms, the salmonellae being isolated only in areas of high fecal coliform counts. A seasonal cycle could not be determined for *V. cholerae* because the organism was presently only at very low levels (ca. 1-10 cells/liter) throughout the year. Although none of the Chesapeake Bay isolates were found to be agglutinable in *V. cholerae* O-I antiserum, the majority of the strains tested were toxigenic, as evidenced by Y-1 adrenal cell and rabbit ileal loop tests. Strains not agglutinating in O-I antiserum, formerly referred to in the older literature as NAG (non-agglutinable) or NCV (noncholera vibrios) are now classified as *V. cholerae* (Shewan and Véron, 1974). Pathogenicity of the non-O-I-agglutinable strains has only recently been recognized. Several reports of outbreaks of cholera-like disease associated with such strains include symptoms ranging from mild diarrhea to severe cholera (Hughes et al., 1978; McIntyre et al., 1965). Not only have the non-O-I-agglutinable strains been repeatedly isolated from cases of diarrhea, but they have also been implicated in septicemia and meningoencephalitis (Hughes et al., 1978; Fearington et al., 1974). The status of infections caused by the non-O-I-agglutinable strains of *V. cholerae* in the United States was recently reviewed (Hughes et al., 1978).

Evidence therefore exists that a pathogenic vibrio rarely isolated from local cases in the United States has been found to be ubiquitous in Chesapeake Bay. Indeed, it has been concluded that *V. cholerae* is an autochthonous estuarine bacterial species resident in Chesapeake Bay (Kaper et al., 1979; Colwell, 1980). Currently, other investigators are attempting to isolate *V. cholerae* from estuarine environments elsewhere in the United States (R. Sizemore, 1980; W. Fraser, 1980). Thus, the geographic distribution of this species in the natural environment may be more accurately assessed in the future.

Methods for isolation that have proved successful include an MPN series, using alkaline peptone broth, a loopful of which is subsequently streaked onto TCBS agar. Because water samples must be filtered to concentrate the vibrios, a membrane filter (MF) procedure would be extremely useful. Presently, membrane filters are employed only to concentrate samples, with the filters subsequently placed into tubes of broth. The very small numbers of vibrios present in Chesapeake Bay (1-10 cells/liter) create problems in filtering large volumes of turbid estuarine waters, even if satisfactory selective and differential media were available, the major obstacle being that unpolluted estuarine water samples contain 10^6-10^7 bacteria/liter, but only 1-10 cells/liter of *V. cholerae*. Shellfish, which filter feed and have the potential to concentrate *V. cholerae*, are a likely source of infection, more so than estuarine water, because ingestion of ca. 10^8 cells is considered to be the infectious dose for *V. cholerae* (Cash et al., 1974). Therefore, routine monitoring of shellfish for *V. cholerae* and *V. parahaemolyticus* is recommended.

C. Isolation Methods and Media for Vibrio spp.

A variety of media have been devised for the isolation and characterization of *Vibrio* species. Few media have been designed specifically for membrane filtration application. The most commonly employed medium for isolation of *V. cholerae* and *V. parahaemolyticus* is the thiosulfate citrate bile salts sucrose (TCBS) agar, which acts to inhibit most of the usual fecal flora, by inclusion of bile salts and sodium citrate and adjustment of the pH to 8.6 and provides differentiation of species by fermentation of sucrose. *V. cholerae* and *V. alginolyticus* ferment sucrose, producing yellow colonies indicated by the bromthymol blue and thymol blue included in the medium. *V. parahaemolyticus* is not a sucrose-fermenting organism and therefore produces a bluish or blue-green colony on TCBS agar. A simple membrane filtration technique has been suggested, in which TCBS agar may be employed without modification (Millipore Corp., 1970). However, while TCBS is a medium selective for *V. cholerae* and *V. parahaemolyticus*, the selectivity of the medium is often insufficient to suppress growth of other organisms, such as *Proteus* or *Aeromonas*. Furthermore, differentiation among pathogenic vibrios and a variety of other, as yet poorly characterized, species of *Vibrio* is not always successful, necessitating a variety of follow-up screening tests. Various modifications of TCBS have been proposed, such as addition of high concentrations of NaCl (Morris et al., 1976) and use of various bile salts derivatives

(Golten and Scheffers, 1975). It is interesting that significant variation in selectivity has been reported among brands of TCBS agar (McCormack et al., 1974; Nicholls et al., 1976).

A variety of media have been devised that purport to be selective. The *Vibrio* agar of Tamura et al. (1971) contains sodium citrate, sodium deoxycholate, ox bile, and sodium lauryl sulfate, adjusted to pH 8.5. These constituents comprise the selective agents. Differentiation is achieved by adding sucrose, the fermentation of which is indicated by water blue, which imparts a bluish gray tint to colonies of *V. cholerae*. The gelatin-taurocholate-tellurite agar of Monsur (1961) is selective for *V. cholerae* (i.e., the taurocholate and tellurite are selective and gelatin hydrolysis differentiates the *V. cholerae* from other organisms growing on the agar). However, this medium may require up to 48 hr for development of typical colonies. The medium of Simidu et al. (1977), suggested for enumerating *Vibrio* spp. in the environment, employs sodium deoxycholate as the selective agent and glucose fermentation at pH 8.0 as the differentiating feature. Barua (1974) recently reviewed the clinical literature concerning media for isolation and identification of *V. cholerae*.

Although more attention has been paid to enumeration, isolation, and characterization of *V. parahaemolyticus* than *V. cholerae* and other *Vibrio* spp., the first concerted effort spent in developing methods for *V. cholerae* was by Japanese investigators, who used agents such as Teepol (a neutral detergent), bile salts, dyes, high salt concentrations, and alkaline pH to select for *V. cholerae*. Glucose-salt-Teepol broth contains methyl violet and Teepol and is adjusted to pH 9.4 (Akiyama et al., 1963). The arabinose-ethyl violet broth of Horie et al. (1964) contains ethyl violet and is adjusted to pH 8.6.

A membrane filtration technique, the first devised for vibrios, was described by Horie et al. (1967). Enumeration of *V. parahaemolyticus* was effected using an arabinose-ammonium sulfate-cholate (AAC) medium, containing sodium cholate and adjusted to pH 8.6.

After filtration of the sample, the membrane was placed on AAC medium and incubated at 42°C. Growth at 42°C is an important differential characteristic of *V. parahaemolyticus*. Yellow colonies growing on the filter on AAC medium, being arabinose fermenters, were *V. parahaemolyticus*. Unfortunately, the medium tends to underestimate the true incidence of *V. parahaemolyticus* since arabinose fermentation is a character that is variable among strains of this species (see Table 2).

The water blue-alizarin yellow agar (WA) (Eiken Chem. Co.) incorporates water blue and alizarin yellow in a medium for distinguishing between *V. parahaemolyticus* and *V. alginolyticus*. In addition to these dyes, WA also contains beef extract, peptone, sodium chloride, Teepol, and sucrose. Fermentation of sucrose by *V. alginolyticus* results in a reduction in the pH of the medium, thereby causing the medium to assume a blue hue, due to the dye, water blue. Alkalinization of the medium by sucrose negative strains of *V. parahaemolyticus* results in an orange-yellow color, arising from the alizarin yellow. The WA medium can be used to detect *V. cholerae*, because *V. cholerae* produces blue-colored colonies, being sucrose-fermenting.

A semiselective medium was devised by Baross and Liston (1970), employing starch hydrolysis as a differential characteristic. The Baross and Liston medium contains no selective agents and is adjusted to pH 7.5. Unfortunately, overgrowth by organisms other than vibrios will often be encountered with this medium.

In 1971, Twedt and Novelli conducted a systematic study of media and media constituents used for isolation of *V. parahaemolyticus*. Penicillin incorporated into an alkaline pH medium proved to be more useful than a variety of selective agents, including potassium tellurite, sodium deoxycholate, Teepol, and 6% NaCl, for isolation of *V. parahaemolyticus*. Starch hydrolysis was employed as a differential indicator system, with the final formulation including peptone (2%), yeast extract (0.2%), cornstarch (0.5%), NaCl (3.0%), penicillin (2 units/ml), and agar (1.5%), pH 8.0. Use of the medium with membrane filtration was suggested by the authors, but no data were provided.

Vanderzant and Nickelson (1972) modified the medium of Twedt and Novelli (1971) slightly, using 7% NaCl and 1% cornstarch, finding superior results, compared with other media recommended for recovery of *V. parahaemolyticus*. The method of Vanderzant and Nickelson (1972) employed enrichment in trypticase soy broth (TSB) to which 7% NaCl had been added. However, the medium was not used in a membrane filtration procedure.

In 1976, Watkins et al. reported on a method using membrane filtration and specifically designed for enumeration of *V. parahaemolyticus*. The primary isolation medium, as described, was based on the ability of *V. parahaemolyticus* to grow in 3% NaCl, at high pH (i.e., 8.6) and at 41°C. Sodium cholate was used to inhibit growth of gram-positive organisms. Galactose was included in the medium as a carbohydrate source for *V. parahaemolyticus*. The closely related species, *V. alginolyticus*, was inhibited by inclusion of copper sulfate. A series of tests were employed for rapid biochemical identification of *V. parahaemolyticus* colonies, without transfer of each individual colony onto test media (Dufour and Cabelli, 1975). Instead, the membrane filters themselves, on which the colonies were growing, were transferred successively to galactose and sucrose fermentation media, followed by the oxidase test. The procedure can be carried out in a 30 hr period, with a reported 95% accuracy. Compared with other methods for enumeration of *V. parahaemolyticus*, this MF procedure yielded consistently higher recoveries, according to Watkins et al. (1976). Unfortunately, the method proved unsatisfactory for recovery of *V. parahaemolyticus* from Chesapeake Bay (Kaper et al., 1979). The MF technique was found to be too selective, with higher recoveries obtained when an MPN procedure employing a galactose-ethyl violet broth for enrichment was used, followed by streaking onto TCBS, with the individual colonies appearing on TCBS thereupon being screened for a "minimum plexus" of biochemical characteristics (Kaper et al., 1980). If proven reliable and accurate, the MF technique of Watkins et al. (1976) represents a significant step forward in development of a successful isolation and enumeration method for *Vibrio* spp.

II. AEROMONAS

Species of the genus *Aeromonas* resemble *Vibrio* spp. in many biochemical characteristics and also in their natural habitat. Aeromonads have been found repeatedly to occur in unpolluted fresh and salt water as well as in sewage and human feces (Schubert, 1974; von Graevenitz and Mensch, 1968). *Aeromonas* spp. are known to cause disease in humans and in a variety of animals, including frogs, snakes, pet turtles, freshwater fish, and salmonid fish (Bullock, 1961; Boulanger et al., 1977; Page, 1961; Caselitz, 1966; McCoy and Seidler, 1973; Simidu and Kaneko, 1973). In seawater, *Aeromonas* and *Vibrio* may constitute up to ca. 45% of the total aerobic, heterotrophic bacterial flora and more than 70% of the heterotrophic bacterial flora associated with phyto- and zooplankton (Simidu et al., 1971). Thus, *Aeromonas* spp. are ubiquitous in the environment as well as being frequently isolated in the clinical environs.

A. Taxonomy

The taxonomy of the genus *Aeromonas* has undergone many changes in recent years, with universal acceptance of any one classification system for the genus yet to be achieved. The definition of the genus given in *Bergey's Manual of Determinative Bacteriology* (1974) is as follows: gram-negative rods; facultative anaerobes; carbohydrates broken down to acid or acid and gas; oxidase-positive; nitrates reduced to nitrites; insensitive to the vibriostatic compound 0/129; and G + C content of 57-63 mol%.

Differentiation of *Aeromonas* from other closely related organisms is given in Table 1.

The genus *Aeromonas* is, in general, divided into three species, the exact delineation of these species being dependent upon the authority cited. A species universally agreed upon is *A. salmonicida*, which differs from the other aeromonads in being nonmotile, indole negative, unable to grow at 37°C, and producing a brown diffusible pigment on nutrient agar. In the eighth edition of *Bergey's Manual* (Schubert, 1974), the description provided follows the classification of Schubert (1969), who recognizes two additional species: (1) *A. hydrophila,* with subspecies *hydrophila, anaerogenes,* and *proteolytica*; and (2) *A. punctata,* with subspecies *punctata* and *caviae*. Ewing and co-workers (1961, 1974) presented an identification scheme widely followed in clinical microbiology laboratories, which includes all of the above within a single species, *A. hydrophila* (i.e., including *A. liquefaciens, A. formicans,* and *A. punctata*). In a recent taxonomic study, Popoff and Véron (1976) also combined *A. hydrophila* and *A. punctata* into a single species, *A. hydrophila,* and described a new species, *A. sobria*. The biochemical characteristics of *A. hydrophila,* according to the various classification schemes cited, are listed in Table 3.

TABLE 3 Biochemical Characteristics of *A. hydrophila*

Test or substrate	Ewing	*Bergey's Manual* (8th ed.)	Popoff and Véron
Hydrogen sulfide (TSI or KI agar)	-[a]	-	-
Urease	-	-	-
Indole	+	+	+
Methyl red	+		v
Voges-Proskauer	v		
Citrate (Simmons)	v		v
KCN (growth)	v	+	+
Motility	+	+	v
Gelatin	+	+	+
Lysine decarboxylase	-	v	-
Arginine dihydrolase	v	+	+
Ornithine decarboxylase	-	-	-
Phenylalanine deaminase	v		-
Glucose			
Acid	+	+	+
Gas	v	v	v
Lactose	-		v
Sucrose	v	v	+
Arabinose	v	v	+
Mannose	+		+

TABLE 3 (Continued)

Test or substrate	Ewing	*Bergey's Manual* (8th ed.)	Popoff and Véron
Mannitol	+	+	+
Dulcitol	-	-	-
Salicin	v		+
Adonitol	-	-	-
Inositol	-	-	-
Sorbitol	v		v
Raffinose	-		-
Rhamnose	-		-
Maltose	+	+	
Xylose	-	-	-
Trehalose	+	+	+
Glycerol	v		+
Esculin	v	v	+
Melezitose	-	-	
Melibiose	-		-
Lipase (corn oil)	+		+
Nitrate to nitrite	+		+
Oxidation-fermentation	F	F	F
Oxidase	+	+	+
Catalase	+	+	+
Growth on MacConkey agar	+	+	
ONPG			+
Casein		+	
DNase		+	+
Butanedial dehydrogenase		+	v
Growth on 0% NaCl	+	+	+
0/129 resistance	+	+	

[a]+, positive or present; -, negative or absent; v, variable.
Source: Data from Ewing and Hugh (1974), Schubert (1974), and Popoff and Véron (1976). Popoff, M., and M. Véron (1976). A taxonomic study of the *Aeromonas hydrophila-aeromonas punctata* group. *J. Gen. Microbiol. 94:* 11-22. Copyright Cambridge University Press, London, England.

Deficiencies in the taxonomy of *Aeromonas* were revealed by DNA homology studies in which strains of *A. salmonicida* were found to be more closely related to the neotype strain of *A. hydrophila* ATCC 7966 than were half of the *A. hydrophila* strains (Beach et al., 1973). Homologies among the strains of *A. hydrophila* ranged from 33 to 100%. *A. hydrophila* ATCC 7966 was used as reference and conditions of hybridization were not stringent. Clearly, a great deal of work needs to be done on the classification of *Aeromonas*.

B. Significance of Aeromonas spp. as a Human Pathogen

The role of *Aeromonas* spp., particularly *A. hydrophila*, as a human pathogen has become increasingly evident in recent years. Fatal and nonfatal infections have been reported, with a variety of clinical manifestations, including septicemia (Tapper et al., 1975; Dean and Post, 1967; Ketover et al., 1973; von Graevenitz and Mensch, 1968; Zajc-Satler, 1972), meningitis (Qadri et al., 1976), corneal ulcers (Feaster et al., 1978), wound infections (Hanson et al., 1977; Fraire, 1978; Shackelford et al., 1973), peritonitis (Salto and Schick, 1973), and acute diarrheal disease (Ljungh et al., 1977; Zajc-Satler, 1972; Wadström et al., 1976; Annapurna and Sanyal, 1977; von Graevenitz and Mensch, 1968). Infections have been reported for both healthy and immunologically compromised hosts. Several reviews of the role of aeromonads in human infections have been published recently (von Graevenitz and Mensch, 1968; Zajc-Satler, 1972; Washington, 1972). Laboratory studies have demonstrated that aeromonads are capable of producing enterotoxin, hemolysin, and cytotoxin (Bernheimer and Avigad, 1974; Annapurna and Sanyal, 1977; Wadström et al., 1976; Ljungh et al., 1977; Ljungh et al., 1978; Sanyal et al., 1975). Some strains of *Aeromonas* can produce fluid accumulation in ligated ileal loops of adult rabbits comparable to that of toxigenic strains of *Vibrio cholerae* (Annapurna and Sanyal, 1977).

Waterborne infections of *Aeromonas* are well documented. In the reviews of von Graevenitz and Mensch (1968) and Washington (1972) are cited a large number of cases showing *Aeromonas* infections occurring as a result of contact with water and soil. Hanson et al. (1977) described a case whereby a scalp laceration was sustained by a patient while diving in shallow fresh water, the laceration subsequently becoming infected with *A. hydrophila*. Fraire (1978) cites details of a wound infection traced to exposure to water in a swimming pool. Rather more bizarre is a case of bacteremia following infection of a mosquito bite via stagnant water (Tapper et al., 1975). Feaster et al. (1978) reported the occurrence of a corneal ulcer in a man struck in the eye with a piece of seashell, the infectious agent proving to be *A. hydrophila*.

C. Isolation Methods and Media for Aeromonas spp.

Very few media have been developed specifically for the isolation, identification, and culture of *Aeromonas* spp. Since the aeromonads share many characteristics with the *Enterobacteriaceae*, differentiation is often difficult. Indeed, it was as late as 1960 that the cytochrome oxidase test was recommended for use in distinguishing *Aeromonas*, an oxidase-positive organism, from the *Enterobacteriaceae*, the members of which are oxidase-negative (Ewing and Johnson, 1960). Recent studies have shown that the oxidase reaction can be media-dependent; a strain of *A. hydrophila* can be oxidase-negative when grown on selective media, but oxidase-positive subsequent to growth on nonselective media (McGrath et al., 1977). In fact, it has been observed that strains of *Aeromonas* isolated from activated sludge gave only weak or negative oxidase activity upon initial isolation, and IMViC reactions identical to the pattern for *E. coli*. However, additional tests were done at a later date, showing the strains to be oxidase-positive aeromonads (Neilson, 1978).

A medium developed for identification of *A. salmonicida* was reported by Gutsell and Snieszko (1946), which contained a simple nutrient formula of tryptic digest of casein, yeast extract, and sodium chloride, providing for optimal growth of the organism but was bereft of selective or differential agents. Eddy (1960) used a peptone water agar, containing 5% oxalated horse blood and 30% dialyzed skimmed milk, to isolate and culture aeromonads. After incubation at 30°C for 48 hr, colonies that were both hemolytic and caseinolytic were picked and screened by passage through an expanded set of tests.

The taxonomy and distribution of *Aeromonas* spp., employing DSF agar, which contains sodium sulfite and dextrin, have been reported by Schubert (1967), who used membrane filtration with the DSF medium (1976). In 1973, Shotts and Rimler described a medium for isolation of *A. hydrophila*, subsequently named the Rimler Shotts (RS) medium, which employs sodium deoxycholate and novobiocin as selective agents, and is adjusted to neutral pH (i.e., pH 7.0). Differentiation is provided by inclusion of a H_2S indicator system, together with maltose, ornithine, lysine, and a pH indicator (bromthymol blue). Yellow colonies appearing on the agar are maltose fermenters, whereas yellow-green or green colonies are capable of decarboxylating lysine or ornithine, or both. *A. hydrophila* produces yellow colonies on the RS agar, this species being maltose-positive and ornithine- and lysine-negative. The lysine reaction should be considered to be questionable, because not all classification schemes are in agreement as to the lysine reaction (see Table 3). A modification of this medium that is used in our laboratory does not contain lysine.

The RS medium has been used to isolate and enumerate *Aeromonas* spp. in several studies carried out in the Potomac River (Chen et al., 1975), Chesapeake Bay (Kaper et al., 1979), and a thermally polluted lake in South Carolina. In the South Carolina study, the numbers of *A. hydrophila* found when a nearby nuclear reactor was operating were significantly higher than when it was not in operation (Fliermans et al., 1977), suggesting that elevated water temperatures caused by heated effluent from the reactor may plan an important role in initiating *Aeromonas* sp.-induced fish epizootics. Fliermans et al. (1977) employed a membrane filter technique, whereby membrane filters were placed directly on the RS medium after filtration of the water sample. Typical yellow colonies appearing on the RS agar that were also oxidase-positive were identified as *A. hydrophila* without any subsequent testing.

A filtration method for enumeration of *A. hydrophila* was reported by Cabelli (1973), who carried out a study of recreational waters. Like the RS medium, the medium devised by Cabelli (1973) also included sodium deoxycholate and novobiocin as selective agents. Yellow colonies appearing on the filter placed on the agar medium were dextrose fermenters. These colonies were marked on the plate and urease and oxidase tests were performed on the colonies picked directly from the plate to the test substrate. The colonies were also confirmed for oxidation-fermentation, indole, and hemolysis reactions. From the study, it was concluded that *A. hydrophila* demonstrates a seasonal distribution, with the numbers of *A. hydrophila* being lowest in the winter. Greatest numbers of *A. hydrophila* were found in polluted waters.

A survey of the composition of media recommended for *Aeromonas* was reported by McCoy and Pilcher (1974), who compared nutrients, selective agents, and indicators. Sodium lauryl sulfate (SLS) was found to be a good selective agent for all strains of *Aeromonas* examined. Of the other inhibitors tested, sodium deoxycholate, crystal violet, methylene blue, brilliant green, and selenite were found to be inhibitory to one or more of the strains of *Aeromonas*. Bromthymol blue appeared to be a more sensitive pH indicator than bromphenol blue, bromcresol purple, bromcresol green, chlorphenol red, and neutral red. Glycogen was selected as the fermentable carbohydrate of value in differentiation among the species of *Aeromonas* and between *Aeromonas* and other genera. The formula of the medium, peptone beef extract glycogen agar (PBG), comprised the following (grams per liter): peptone, 10; beef extract, 10; glycogen, 4; NaCl, 5; SLS, 0.1; bromthymol blue, 0.1; agar 15, and pH adjusted to 6.9-7.1.

The PBG medium is used as a pour plate, which is subsequently overlayered with 2% agar. After incubation at 37°C for 18-24 hr, colonies of *A. hydrophila* growing on PBG medium were larger and more intensely yellow than any of the glycogen-fermenting strains of *Enterobacteriaceae*. *A. salmonicida* could also be detected on PBG medium if the inoculated medium was incubated at 25°C for 3 days. The PBG medium was employed to study intestinal bacteria of small green pet turtles (McCoy and Seidler, 1974), but has not been used in membrane filtration applications.

Except where indicated otherwise, the studies cited here employed conventional media, such as MacConkey, EMB, SS, and deoxycholate agar for isolation, identification, and culturing of aeromonads. Thus, methods and media specifically for isolation of *Aeromonas* spp. are relatively new. Because the aeromonads are widely distributed in the aquatic environment, a standard membrane filtration technique would be of great value in gaining an understanding of the biology and ecology of the genus *Aeromonas*.

III. FUTURE WORK

Membrane filtration techniques offer good potential for recovery of *Vibrio* and *Aeromonas* species from natural waters, both freshwater and marine. For *Vibrio*, the commonly employed media contain bile salts or its derivatives, high salt concentration, and alkaline pH for selection and starch hydrolysis, and carbohydrate fermentation reactions for differentiation. However, alkaline pH cannot be relied upon for selection of marine *Vibrio* species (Gilmour et al., 1976). Decarboxylation reactions, gelatin hydrolysis, and the hydrolysis of compounds such as Tween 80 and casein offer promise in indicator systems. A variety of selective agents remain to be tested for MF methods and application.

The focus of attention has been on *V. parahaemolyticus*, with little attention paid to problems of enumerating *V. cholerae*, *V. alginolyticus*, *V. anguillarum*, and related estuarine and marine *Vibrio* species. Clearly, there is much yet to be done in this area of diagnostic environmental microbiology, with respect to the vibrios.

Aeromonas is a genus that is even less explored than *Vibrio*, with respect to development of indicator systems. The ecology, pathogenicity, and public health significance of the aeromonads are incompletely understood. Few media and only a smattering of methodologies have been devised for enumeration, isolation, and differentiation of *Aeromonas* sp. Thus, improved methods for selection and enumeration are badly needed. Ampicillin has been proposed as a selective agent for *Aeromonas* in broth culture (Chen et al., 1975), but a filtration method employing a medium incorporating ampicillin has not yet been reported. Elevated temperatures, comparable to those used for isolation and differentiation of *V. parahaemolyticus*, or fecal coliforms cannot be applied in the case of *Aeromonas*, because so many *Aeromonas* spp. are psychrotrophic (Rouf and Rigney, 1971). Selected carbohydrate fermentation reactions (e.g., mannitol, inositol, trehalose, and xylose) offer promise for differentiation among species, as do the decarboxylase, gelatinase, lipase, and DNase reactions (see Table 3). Differential characteristics employed in the future in diagnostic media require an improved classification of this genus.

REFERENCES

Akiyama, S., Takizawa, K., and Ohara, Y. (1963). Application of teepol to isolation media for *Vibrio parahaemolyticus*. *Jap. J. Bacteriol. 18:* 255—256.

Annapurna, E., and Sanyal, S. C. (1977). Enterotoxicity of *Aeromonas hydrophila*. *J. Med. Microbiol. 10:* 317—323.

Baross, J. (1980). Personal communication.

Baross, J., and Liston, J. (1970). Occurrence of *Vibrio parahaemolyticus* and related hemolytic vibrios in marine environments of Washington State. *Appl. Microbiol. 20:* 179—186.

Barua, D. (1974). Laboratory diagnosis of cholera. In *Cholera*, D. Barua and W. Burrows (Eds.). Saunders, Philadelphia, pp. 85—126.

Beach, P., Seidler, R. J., and Mandel, M. (1973). The aeromonads: DNA base composition and polynucleotide homologies. *Am. Soc. Microbiol. Annu. Meet., Abstr. Annu. Meet.*, p. 42.

Bernheimer, A. W., and Avigad, L. S. (1974). Partial characterization of aerolysin, a lytic exotoxin from *Aeromonas hydrophila*. *Infect. Immun. 9:* 1016—1021.

Boulanger, Y., Lallier, R., and Cousineau, G. (1977). Isolation of enterotoxigenic *Aeromonas* from fish. *Can. J. Microbiol. 23:* 1161—1164.

Bullock, G. L. (1961). The identification and separation of *Aeromonas liquefaciens* from *Pseudomonas fluorescens* and related organisms occurring in diseased fish. *Appl. Microbiol. 9:* 587–590.

Cabelli, V. J. (1973). The occurrence of aeromonads in recreational waters. *Am. Soc. Microbiol. Annu. Meet., Abstr. Annu. Meet.,* p. 32.

Caselitz, F. H. (1966). *Pseudomonas-Aeromonas* und ihre humanmedizinische Bedeutung. *Gustav Fisher, Jena,* pp. 1–23.

Cash, R. A., Music, S. I., Libonati, J. P., Snyder, M. J., Wenzel, R. P., and Hornick, R. B. (1974). Response of man to infection with *Vibrio cholerae*. I. Clinical, serologic and bacteriologic responses to a known inoculum. *J. Infect. Dis. 129:* 45–52.

Chen, P. K., Ellenberger, R. S., and White, M. (1975). Selective enrichment of *Aeromonas liquefaciens* by using ampicillin. *Am. Soc. Microbiol. Annu. Meet., Abstr. Ann. Meet.,* p. 187.

Colwell, R. R. (1973). *Vibrios* and *Spirilla*. In *Handbook of Microbiology,* A. I. Laskin and H. A. Lechevalier (Eds.). CRC Press, Inc., Boca Raton, Fla., pp. 229–236.

Colwell, R. R. (1980). Human pathogens in the aquatic environment. In *Proc. Conf. Aquatic Microbial. Ecology,* R. R. Colwell and J. Foster (Eds.). Univ. MD Sea Grant Publication, College Park, MD, pp. 337–344.

Colwell, R. R., and Kaper, J. (1977). *Vibrio* species as bacterial indicators of potential health hazards associated with water. In *Bacterial Indicators/Health Hazards Associated with Water,* ASTM STP 635, A. W. Hoadley and B. J. Dutka (Eds.). American Society for Testing and Materials, Philadelphia, pp. 115–125.

Colwell, R. R., Kaper, J., and Joseph, S. W. (1977). *Vibrio cholerae, Vibrio parahaemolyticus,* and other vibrios: occurrence and distribution in Chesapeake Bay. *Science 198:* 394–396.

Dean, H. M., and Post, R. M. (1967). Fatal infection with *Aeromonas hydrophila* in a patient with acute myelogenous leukemia. *Ann. Intern. Med. 66:* 1177–1179.

Dufour, A. P., and Cabelli, V. J. (1975). Membrane filter procedure for enumerating the component genera of the coliform group in seawater. *Appl. Microbiol. 29:* 826–833.

Eddy, B. P. (1960). Cephalotrichous, fermentative gram-negative bacteria: the genus *Aeromonas*. *J. Appl. Bacteriol. 23:* 216–249.

English, V. L., and Lindberg, R. B. (1977). Isolation of *Vibrio alginolyticus* from wounds and blood of a burn patient. *Am. J. Med. Technol. 43:* 989–993.

Ewing, W. H., and Hugh, R. (1974). *Aeromonas*. In *Manual of Clinical Microbiology*, E. H. Lennette, E. H. Spaulding, and J. P. Truant (Eds.). American Society for Microbiology, Washington, D.C., pp. 230–237.

Ewing, W. H., and Johnson, J. G. (1960). The differentiation of aeromonads and C27 cultures from Enterobacteriaceae. *Int. Bull. Bacteriol. Nomencl. Taxon. 10:* 223–230.

Ewing, W. H., Hugh, R., and Johnson, J. G. (1961). *Studies on the Aeromonas Group*. Center for Disease Control, Atlanta, Ga.

Fearington, E. L., Rand, C. H., Jr., Mewborn, A., and Wilkerson, J. (1974). Noncholera *Vibrio* septicemia and meningoencephalitis. *Ann. Intern. Med. 81:* 401.

Feaster, F. T., Nisbet, R. M., and Barber, J. C. (1978). *Aeromonas hydrophila* corneal ulcer. *Am. J. Ophthal. 85:* 114–117.

Fliermans, C. B., Gorden, R. W., Hazen, T. C., and Esch, G. W. (1977). *Aeromonas* distribution and survival in a thermally altered lake. *Appl. Environ. Microbiol. 33:* 114–122.

Food and Drug Administration. (1976). *Bacteriological Analysis Manual*, 4th ed. Food and Drug Administration, Washington, D.C.

Fraire, A. E. (1978). *Aeromonas hydrophila* infection. *J. Am. Med. Assoc. 239:* 192.

Fraser, W. (1980). Personal communication.

Gilmour, A., Allan, M. C., and McCallum, M. F. (1976). The unsuitability of high pH media for the selection of marine *Vibrio* species. *Aquaculture 7:* 81–87.

Golten, C., and Scheffers, W. A. (1975). Marine vibrios isolated from water along the Dutch coast. *Neth. J. Sea Res. 9:* 351–364.

Gutsell, J. S., and Snieszko, S. F. (1946). Dosage of sulfamerazine in the treatment of furunculosis in brook trout, *Salvelinus fontinalis*. *Trans. Am. Fish. Soc. 76:* 82–96.

Hanson, P. G., Standridge, J., Jarrett, F., and Maki, D. G. (1977). Freshwater wound infection due to *Aeromonas hydrophila*. *J. Am. Med. Assoc. 238:* 1053–1054.

Horie, S., Saheki, K. Kozima, T., Nara, M., and Sekine, Y. (1964). Distribution of *Vibrio parahaemolyticus* in plankton and fish in the open sea. *Bull. Jap. Soc. Sci. Fish. 30:* 786–791.

Horie, S., Saheki, K., and Okuzumi, M. (1967). Quantitative enumeration of *Vibrio parahaemolyticus* in sea and estuarine waters. *Bull. Jap. Soc. Sci. Fish. 33:* 126–130.

Hughes, J. M., Hollis, D. G., Gangarosa, E. J., and Weaver, R. E. (1978). Noncholera vibrio infections in the United States.

Clinical, epidemiologic, and laboratory features. *Ann. Intern. Med. 88:* 602–606.

Jonas, R. (1979). Personal communication.

Kaneko, T., and Colwell, R. R. (1975). Adsorption of *Vibrio parahaemolyticus* onto chitin and copepods. *Appl. Microbiol. 29:* 269–274.

Kaper, J., Lockman, H., Colwell, R. R., and Joseph S. W. (1979). Ecology, serology, and enterotoxin production of *Vibrio cholerae* in Chesapeake Bay. *Appl. Environ. Microbiol. 37:* 91–103.

Kaper, J., Lockman, H., Remmers, E., and Colwell, R. R. (1980). Unpublished data.

Kaper, J. Remmers, E., and Colwell, R. R. (in press). A medium for the presumptive identification of *Vibrio parahaemolyticus*. *J. Food Protection*.

Ketover, B. P., Young, L. S., and Armstrong, D. (1973). Septicemia due to *Aeromonas hydrophila:* clinical and immunologic aspects. *J. Infect. Dis. 127:* 284–290.

Koburger, J. A., and Lazarus, C. R. (1974). Isolation of *Vibrio parahaemolyticus* from salt springs in Florida. *Appl. Microbiol. 27:* 435–436.

Ljungh, A., Popoff, M., and Wadström, T. (1977). *Aeromonas hydrophila* in acute diarrheal disease: detection of enterotoxin and biotyping of strains. *J. Clin. Microbiol. 6:* 96–100.

Ljungh, A., Wretlind, B., and Wadström, T. (1978). Evidence for enterotoxin and two cytolytic toxins in human isolates of *Aeromonas hydrophila*. In *Toxins: Animal, Plant and Microbial*. Proc. 5th Int. Symp., P. Rosenberg (Ed.). Pergamon Press, New York, pp. 947–960.

McCormack, W. M., DeWitt, W. E., Bailey, P. E., Morris, G. K., Soeharjono, P., and Gangarosa, E. J. (1974). Evaluation of thiosulfate-citrate-bile salts-sucrose agar, a selective medium for the isolation of *Vibrio cholerae* and other pathogenic vibrios. *J. Infect. Dis. 129:* 497–500.

McCoy, R. H., and Pilcher, K. S. (1974). Peptone beef extract glycogen agar, a selective and differential *Aeromonas* medium. *J. Fish. Res. Board Can. 31:* 1553–1555.

McCoy, R. H., and Seidler, R. J. (1973). Potential pathogens in the environment: isolation, enumeration, and identification of seven genera of intestinal bacteria associated with small green pet turtles. *Appl. Microbiol. 25:* 534–538.

McGrath, V. A., Overman, S. B., and Overman, T. L. (1977). Media-dependent oxidase reaction in a strain of *Aeromonas hydrophila*. *J. Clin. Microbiol. 5:* 112–113.

McIntyre, O. R., Feeley, J. E., Greenough, W. B., III, Benenson, A.S., Hassan, S.I., and Saad, A. (1965). Diarrhea caused by non-cholera vibrios. *Am. J. Trop. Med. Hyg. 14:* 412–418.

Millipore Corporation. (1970). A Membrane Filter Technique for Detecting *Vibrio parahaemolyticus* and *Vibrio comma*. Application Note 136. Millipore Corp., Bedford, Mass.

Monsur, K. A. (1961). A highly selective gelatin-taurocholate-tellurite medium for the isolation of *Vibrio cholerae*. *Trans. R. Soc. Trop. Med. Hyg. 55:* 440–442.

Morris, G. K. DeWitt, W. E., Gangarosa, E. J., and McCormack, W. M. (1976). Enhancement by sodium chloride of the selectivity of thiosulfate citrate bile salts sucrose agar for isolating *Vibrio cholerae* Biotype El Tor. *J. Clin. Microbiol. 4:* 133–136.

Neilson, A. H. (1978). The occurrence of aeromonads in activated sludge: isolation of *Aeromonas sobria* and its possible confusion with *Escherichia coli*. *J. Appl. Bacteriol. 44:* 259–264.

Nicholls, K. M., Lee, J. V., and Donovan, T. J. (1976). An evaluation of commercial thiosulphate citrate bile salt sucrose agar (TCBS). *J. Appl. Bacteriol. 41:* 265–269.

Page, L. A. (1961). Experimental ulcerative stomatitis in king snakes. *Cornell Vet. 51:* 258–266.

Pien, F., Lee, K., and Higa, H. (1977). *Vibrio alginolyticus* infections in Hawaii. *J. Clin. Microbiol. 5:* 670–672.

Popoff, M., and M. Véron (1976). A taxonomic study of the *Aeromonas hydrophila-Aeromonas punctata* group. *J. Gen. Microbiol. 94:* 11–22.

Qadri, S. M. H., Gordon, L. P., Wende, R. D., and Williams, R. P. (1976). Meningitis due to *Aeromonas hydrophila*. *J. Clin. Microbiol. 3:* 102–104.

Rouf, M. A., and Rigney, M. M. (1971). Growth temperatures and temperature characteristics of *Aeromonas*. *Appl. Microbiol. 22:* 503–506.

Rubin, S. J., and Tilton, R. C. (1975). Isolation of *Vibrio alginolyticus* from wound infections. *J. Clin. Microbiol. 2:* 556–558.

Sack, R. B. (1975). Human diarrheal disease caused by enterotoxigenic *Escherichia coli*. *Annu. Rev. Microbiol. 29:* 333–353.

Salto, R., and Schick, S. (1973). *Aeromonas hydrophila* peritonitis. *Cancer Chemother. Rep. 57:* 489–491.

Sanyal, S. C., Singh, S. J., and Sen, P. C. (1975). Enteropathogenicity of *Aeromonas hydrophila* and *Plesiomonas shigelloides*. *J. Med. Microbiol. 8:* 195–198.

Schubert, R. H. W. (1967). Das Vorkommen der Aeromonaden in oberirdischen Gewässern. *Arch. Hyg. 150:* 688–708.

Schubert, R. H. W. (1969). Infrasubspecific taxonomy of *Aeromonas hydrophila* (Chester 1901) Stanier 1943. *Zentrabl. Bakteriol. Parasitenkd. Infektionskr. Hyg. Abt. I Orig. 211:* 406–408.

Schubert, R. H. W. (1974). Genus *Aeromonas* Kluyver and van Niel. In *Bergey's Manual of Determinative Bacteriology,* 8th ed., R. E. Buchanan and N. E. Gibbons (Eds.). Williams & Wilkins, Baltimore, Md., pp. 345–349.

Schubert, R. (1976). The detection of aeromonads of the "hydrophila-punctata-group" within the hygienic control of drinking water. *Zentrabl. Bakteriol. Parasitenkd. Infectionskr. Hyg. I Abt. Orig. B 161:* 482–497.

Shackelford, P. G., Ratzan, S. A., and Shearer, W. T. (1973). Ecthyma gangrenosum produced by *Aeromonas hydrophila. J. Pediatr. 83:* 100–101.

Shewan, J. M., and Véron, M. (1974). Genus *Vibrio* Pacini. In *Bergey's Manual of Determinative Bacteriology,* 8th ed., R. E. Buchanan and N. E. Gibbons (Eds.). Williams & Wilkins, Baltimore, Md., pp. 340–345.

Shotts, E. G., and Rimler, R. (1973). Medium for the isolation of *Aeromonas hydrophila. Appl. Microbiol. 26:* 550–553.

Simidu, U., and Kaneko, E. (1973). A numerical taxonomy of *Vibrio* and *Aeromonas* from normal and diseased marine fish. *Bull. Jap. Soc. Sci. Fish. 39:* 689–703.

Simidu, U., Ashino, K., and Kaneko, E. (1971). Bacterial flora of phyto- and zoo-plankton in the inshore water of Japan. *Can. J. Microbiol. 17:* 1157–1160.

Simidu, U., Kaneko, E., and Taga, N. (1977). Microbiological studies of Tokyo Bay. *Microb. Ecol. 3:* 173–191.

Sizemore, R. (1980). Personal communication.

Tamura, K., Shimada, S., and Prescott, L. M. (1971). Vibrio agar: a new plating medium for isolation of *Vibrio cholerae. Jap. J. Med. Sci. Biol. 24:* 125–127.

Tapper, M. L., McCarthy, L. R., Mayo, J. B., and Armstrong, D. (1975). Recurrent *Aeromonas* sepsis in a patient with leukemia. *Am. J. Clin. Pathol. 64:* 525–530.

Thompson, C. A., Vanderzant, C., and Ray, S.M. (1976). Relationship of *Vibrio parahaemolyticus* in oyster water and sediment and bacteriological and environmental indices. *J. Food Sci. 41:* 117–122.

Twedt, R. M., and Novelli, R. M. E. (1971). Modified selective and differential isolation medium for *Vibrio parahaemolyticus*. *App. Microbiol. 22:* 593–599.

Vanderzant, C., and Nickelson, R. (1972). Procedure for isolation and enumeration of *Vibrio parahaemolyticus*. *Appl. Microbiol. 23:* 26–33.

Vasconcelos, G. J., Stang, W. J., and Laidlaw, R. H. (1975). Isolation of *Vibrio parahaemolyticus* and *Vibrio alginolyticus* from estuarine areas of southeastern Alaska. *Appl. Microbiol. 29:* 557–559.

von Graevenitz, A., and Mensch, A. H. (1968). The genus *Aeromonas* in human bacteriology. *N. Engl. J. Med. 278:* 245–249.

Wadström, T., Ljungh, A., and Wretlind, B. (1976). Enterotoxin, haemolysin and cytotoxin protein in *Aeromonas hydrophila* from human infections. *Acta Pathol. Microbiol. Scand. Sect. B. 84:* 112–114.

Washington, J. A. (1972). *Aeromonas hydrophila* in clinical bacteriologic specimens. *Ann. Intern. Med. 76:* 611–614.

Watkins, W. D., Thomas, C. D., and Cabelli, V. J. (1976). Membrane filter procedure for enumeration of *Vibrio parahaemolyticus*. *Appl. Environ. Microbiol. 32:* 679–684.

Zajc-Satler, J. (1972). Morphological and biochemical studies of 27 strains belonging to the genus *Aeromonas* isolated from clinical sources. *J. Med. Microbiol. 5:* 263–265.

8

ISOLATION AND ENUMERATION OF FUNGI USING MEMBRANE FILTRATION

JAMES P. SHERRY National Water Research Institute, Canada Centre for Inland Waters, Burlington, Ontario, Canada

ANSAR A. QURESHI *Ministry of the Environment, Rexdale, Ontario, Canada*

I. INTRODUCTION

The fungi are a large group of heterotrophic microorganisms, ranging in habit from the morphologically simple unicellular yeasts and chytrids to the more complex members of the basidiomycetes. Characteristically, fungi are achlorophyllus, eucaryotic heterotrophs (Hawker, 1966) that have shown an ability to adapt to an extensive range of habitats and to flourish in often quite extreme environments. The virtually ubiquitous occurrence of fungi is a result of their versatile physiologies, prolific reproductive systems, and of the ability of some fungi to survive and even grow in low- and high-temperature environments (Cooney and Emerson, 1964; Cochrane, 1965; Deverall, 1968; Smith, 1969). The majority of fungi are saprophytes and may be found wherever organic matter occurs, whereas many others are either parasitic or pathogenic.

Whether a fungus is an indigenous member of a particular ecosystem or an active alien, its continued growth and reproduction is dependent on the ability to degrade and utilize previously synthesized organic materials. For this purpose fungi are equipped with broad enzymatic complements (Cochrane, 1965) and with the bacteria

---

**Present affiliation:* Alberta Environmental Centre, Vegreville, Alberta, Canada

may rightly be considered as nature's scavengers. Both molds and yeasts have been reported from diverse aquatic systems (Sparrow, 1968; Cooke, 1970; Jones, 1971). In the aquatic environment fungi may be present in both the water column and sediment (Hedrick and Soyugenc, 1967; Dick, 1971).

Those fungi found in natural waters may be divided into various groups, each showing a different degree of adaptation to life in the aquatic environment. The aquatic phycomycetes, whose members include genera of the Chytridiomycetes and the more morphologically complex Oomycetes, are probably the most highly adapted group, and occur in virtually all types of aquatic systems. They are characterized by the production of motile asexual reproductive spores, which is a distinct advantage to an organism that must propagate, locate, and colonize fresh substrates in water. The successful adaptation of the aquatic phycomycetes to the aquatic habitat is illustrated by their importance as pathogens of other aquatic organisms such as algae (Masters, 1976), fish (Scott and O'Bier, 1962; Willoughby, 1968), and crayfish (Weston, 1941). As a consequence of their motile reproductive phase, fast growth rates, and coarse hyphae, some Oomycetes may be important as early colonizers of organic substrates in waters (Park, 1972a).

The characteristic and often multiradiate conidia of the aquatic hyphomycetes also represent an advantageous adaption to life in water. Such elaborate conidial modifications may serve to anchor the conidia onto suitable substrata (Webster, 1959), as flotation aids, and consequently, as the aquatic hyphomycetes rely on water currents and wave action for dissemination, dispersal aids (Ingold, 1976). The aquatic hyphomycetes participate in the degradation of plant litter, and their importance in the aquatic food chain has been demonstrated (Barlocher and Kendrick, 1976).

A variety of yeasts and molds more commonly associated with the soil microflora have been consistently recovered from polluted and nonpolluted waters and sewage (Cooke, 1957; Park, 1972a). Stormwater runoff, land drainage, sewage treatment plant discharges, and air deposition may each contribute in various degrees to the "geofungal" content of natural aquatic systems. The origins, ecological classification, and possible roles of these immigrant heterotrophs have been considered in some detail by Cooke (1961), Dick (1971), Park (1972a, b), and Qureshi and Dutka (1974). Although the exact functional role of "geofungi" within the aquatic environment has yet to be elucidated, Cooke (1976) has suggested a model life-cycle for an active geofungal immigrant within an aquatic habitat.

The fact that members of the foregoing fungal groups are consistently present in many and varied aquatic habitats is strongly suggestive that they may have a functional role therein. It is postulated that fungi, as heterotrophs, participate in the degradation and recycling of organic materials in the aquatic environment. However, in the absence of detailed quantitative studies, this proposition has yet to be reliably confirmed or even demonstrated for most aquatic systems.

Quantitatively, as pointed out by Sparrow (1968), we know comparatively little about fungi in water. Their significance, distributions, and relationships to other ecological parameters are poorly understood and will only be established through future studies. In this chapter, we consider how the membrane filtration (MF) technique can facilitate the isolation and enumeration of fungi in water, thus contributing to our understanding of the significance of these organisms in the aquatic environment.

II. TECHNIQUES USED TO STUDY FUNGI IN AQUATIC ENVIRONMENTS

A. Qualitative Techniques

Traditionally, most of the information on the occurrence, distribution, and ecological roles of fungi in aquatic environments has been obtained using qualitative methods. Such methods include both in situ and in vivo baiting of water with suitable substrates and have been used to isolate Chytridiomycetes (Sparrow, 1968), Oomycetes (Dick, 1976), and lignicolous fungi (Jones, 1971, 1976) from water. In an effort to overcome the inherent shortcomings of these methods,

Ulken and Sparrow (1968) adopted a most probable number (MPN) approach that involved baiting with pollen grains to estimate numbers of chytrid propagules in lake water. Gaertner (1968) and Bremer (1976) also described a quantitative pollen-baiting technique for use in the isolation of lower marine fungi.

Much useful information relating to the structure of aquatic fungal communities and fungal colonization patterns on various substrates may be acquired through the collection and in vitro incubation of naturally occurring particulate organic matter, a procedure that induces some, but not necessarily all, of the fungi active in the organic material to sporulate. Park (1972a), while studying the mycofloras of a series of Irish lotic waters, found this technique advantageous because it provides an indication of those fungi that are actually active within the substrate, whereas baiting procedures may induce otherwise dormant species to become active.

B. Quantitative Techniques

In addition to qualitative information, there is also a requirement for quantified data concerning the distribution of fungi in aquatic environments. Such data could help to determine: (1) the origins and significance of fungi present in aquatic environments; (2) the seasonal distribution patterns of aquatic fungi in relation to various physical, chemical, environmental, and microbial parameters; (3) the effects of pollutants on the composition of aquatic fungal populations; (4) the relationship among fungal distribution patterns, the occurrence of indicator fungi, and pollutant inputs from extraneous sources; and (5) the public health hazards posed by the presence of potentially pathogenic fungi in fresh or marine waters.

However, the interpretation of quantitative data obtained using mycological techniques presents some problems. Unlike enumerations of unicellular bacteria, it cannot be assumed that the fungal colonies that grow on a spread plate or membrane filter arise from equivalent reproductive units. Colonies may form through germination and growth of both asexual and sexual reproductive spores, mycelial fragments, chlamydospores, or aggregations of any of these. Thus, a 1:1 relationship may not exist between the number of colonies formed and the number of fungal spores in a water sample. To circumvent this problem, each colony is considered to originate from a colony-forming unit (CFU). In addition, since all isolation media are to some extent selective, their use permits enumeration of only that component of the fungal flora capable of growth on the medium in use.

A further difficulty is that a single fungal reproductive structure may produce from few to many reproductive units. Therefore, enumeration techniques may favor heavily sporulating fungi. However, the majority of viable fungal spores present in a habitat are genetically and biochemically complete entities, possessing the capacity to germinate and colonize fresh substrates under favorable conditions. Thus, the more numerous spores of the heavily sporulating fungi may reflect an increased ability to produce offspring that will encounter a micro environment favorable for their germination and growth. In other words, for certain fungi, increased spore production and associated numerical dominance may represent an ecologically significant advance.

A number of quantitative techniques that are essentially adaptations of techniques used by soil microbiologists have been used to study fungi in the aquatic environment. The spread plate technique (SPT) is essentially a modification of the dilution (Barron, 1971) and soil plate (Warcup, 1950) techniques routinely used by mycologists in studies of soil fungi. When used to enumerate fungi in water and sediments, this technique is largely selective for the geofungal segment of the aquatic mycoflora. For the successful application of SPT, the sample's geofungal content should be at least 50 CFU/ml. Such fungal levels are common in polluted rivers, inshore lake waters, sewage, and industrial effluents.

Cooke (1954a, 1963, 1971) originally developed the basic methodology used as a dilution plate technique. Water samples at appropriate dilutions were incorporated into plates of aureomycin-rose bengal-glucose-peptone agar (ARGPA) and were then incubated at room temperature for 1 week. Harvey (1952) also used a standard dilution technique on cornmeal and potato dextrose agars to isolate Fungi Imperfecti from water. Using ARGPA agar to study fungi in deritus from nonpolluted freshwater, Park (1972a) observed that spreading of the dilution sample over the surface of a prepoured agar plate, was more convenient and allowed uniform colony development. Beech and Davenport (1971) described a spread plate technique for the isolation of yeasts from suspension. The SPT has also been used to enumerate and isolate filamentous fungi from lake (Qureshi and Dutka, 1974) and estuarine (Cooke and la Course, 1975) water.

Both the pour and spread plate techniques have been used in the quantification of aquatic Oomycetes, particularly saprolegniaceous fungi, in freshwater bodies, as reviewed by Dick (1976). Willoughby and Pickering (1977) monitored Saprolegniaceae in a hatchery pond by spreading water samples (0.5 ml portions) onto plates of glucose-yeast extract-penicillin-streptomycin agar. After 48 hr incubation at 15°C the saprolegniaceous colonies, which had reached a diameter of 5-13 mm, were enumerated and subsequently cultured on hemp seeds in water to facilitate sporulation and identification. Similarly, Ho (1975) used SPT on a selective agar medium to enumerate saprolegniaceous propagules in fresh water.

An inherent weakness in the SPT is that low fungal populations are not detectable at statistically analyzable levels. Often in mycological analyses of offshore and nonpolluted waters, enumeration plates may contain from 1 to 10 colonies and thus do not provide accurate estimates of fungal populations. Use of the MF technique can help overcome this problem by concentrating the fungi in a water sample into detectable and analyzable numbers.

III. USE OF MEMBRANE FILTRATION FOR THE ISOLATION AND ENUMERATION OF FUNGI

As applied to studies of the various aquatic fungal groups, the MF technique involves filtration of a known volume of water through a membrane filter with subsequent direct microscopic examination and/or incubation of the membrane on a suitable agar medium. The choice of membrane filter, type of growth medium, type and concentration of microbial inhibitors incorporated in the medium, and incubation conditions are dependent on both the nature of the fungi under investigation and the sample source, as discussed in the following sections.

A. Aquatic Phycomycetes

Fuller and Poyton (1964) described a continuous flow centrifugation technique for the concentration of sparsely distributed phycomycetes from large volumes of water. The concentrate so obtained was plated and incubated on an agar medium suitable for chytrids. An alternative approach was developed by Miller (1967), who used MF to concentrate and isolate phycomycete propagules from fresh and marine water by filtration through 0.8, 1.2, or 3.0 μm pore size membrane filters. The concentrate was resuspended in 0.5 ml of filtrate and was then streaked onto the surface of a low nutrient agar medium that contained antibiotics. After incubation, the developing chytrid colonies were transferred to a suitable chytrid medium. Each of the foregoing methods could, with suitable modifications, be used to provide data on the spatial and seasonal distribution of phycomycetes within freshwater bodies. However, the ability of the relatively fragile asexual reproductive spores (zoospores) of the phycomycetes to withstand the stresses imposed by membrane filtration has yet to be determined and would have a direct bearing on the reliability of data obtained using this technique. Ulken and Sparrow (1968), however, found both centrifugation and membrane filtration procedures unsatisfactory in the enumeration of chytrid zoospores in a freshwater lake.

B. Aquatic Hyphomycetes

The MF technique can also be used to quantify the aquatic hyphomycete spora of natural waters (Iqbal and Webster, 1973). Water samples (250-500 ml) are filtered through 8 μm membrane filters. A few drops of 0.1% cotton blue in lactic acid are then poured onto the filter to kill and stain the spores. The filters are then transferred to glass petri dishes, flooded with fresh 0.1% cotton blue in lactic acid, and heated at 50-60°C for 45-55 min so as to render the filters sufficiently transparent for low-power microscopic examination. With this method, it is possible to determine both the number of spores and also, because of the readily identifiable nature of aquatic hyphomycete spores, the concentration of individual species in water samples. Membrane filtration has been successfully used to estimate numbers of aquatic hyphomycete conidia in freshwater streams (Barlocher et al., 1978; Iqbal and Webster, 1973). Further such studies could provide additional information on the possible relationships between aquatic hyphomycetes and other ecological parameters.

C. Geoaquatic Fungi

For the purpose of this chapter, the term "geoaquatic fungi" is used to describe those fungi (including both molds and yeasts) that are mainly of soil origin but, following transportation by various means and pathways, are present in aquatic environments. Geoaquatic fungi are not true aquatic fungi in the strict sense of the term and, in general, are poorly adapted to an aquatic existence.

Molds

Little use has been made of MF to enumerate geoaquatic molds. Hedrick and Soyugenc (1967) and Hedrick et al. (1968) used 1.2 μm membrane filters to isolate yeasts and molds from Lake Ontario and Lake Superior sediment and water samples. After filtration of samples, the membranes were placed on agar media and incubated at 25°C for 48-72 hr. The number and type of fungal colonies appearing on the membranes were then recorded, but the results were not expressed quantitatively. In these investigations, the following geoaquatic molds were predominantly isolated: *Penicillium, Aspergillus, Alternaria, Cladosporium, Fusarium,* and *Trichoderma*.

Fungi responsible for the production of earthy, moldy, or musty tastes in tap water have been isolated using the MF technique

(Burman, 1965; Burman et al., 1969). After filtration of water samples the membranes were placed on a selective agar medium (Appendix, formula 12) and incubated for 7 days at 22°C; fungal colonies were then enumerated, isolated, and identified.

Qureshi and Dutka (1976) compared five commercial brands of autoclave- and ethylene oxide-sterilized membrane filters for their ability to recover fungi from natural waters. Samples were filtered through 0.45 μm membranes, which were then placed on modified aureomycin-rose bengal-glucose-peptone agar (MARGPA) and modified streptomycin-terramycin-malt extract agar (MSTMEA) (Appendix, formulas 13 and 14) and were incubated at 15°C for 5 days. The total number of geoaquatic fungal colonies developing on each membrane were then determined. Figure 1 illustrates the appearance of both mold and yeast colonies on some of these membranes. The results of this study indicated that various brands of membrane filters differ in their ability to recover viable fungi from natural waters, with the Gelman GN-6 filters being superior to other membranes tested.

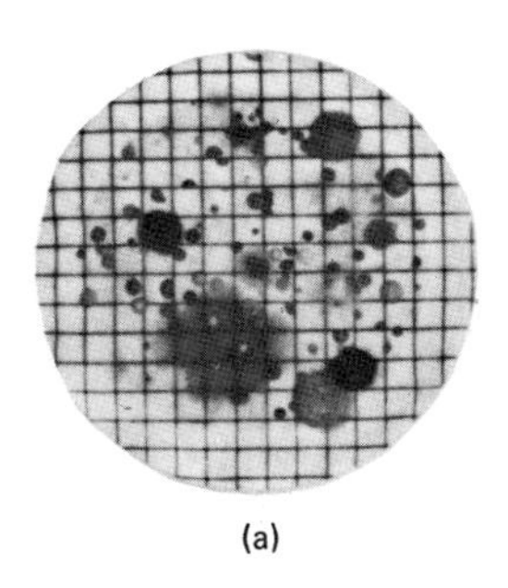

(a)

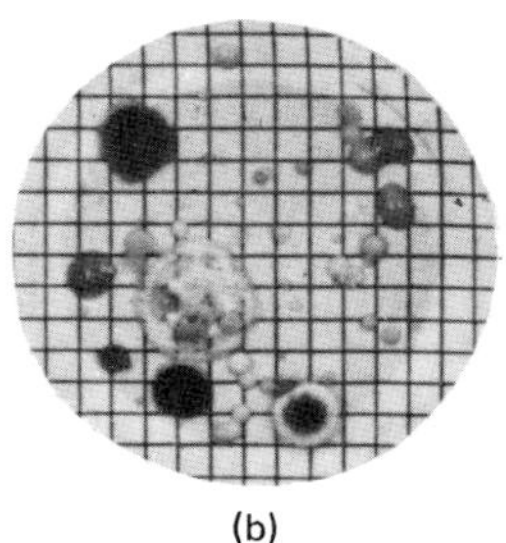

(b)

FIGURE 1 Recovery of geoaquatic fungi on membrane filters plated on MARGPA (a) and MSTMEA (b) after filtration of 30 ml aliquots of water samples: (a) Oxoid N47/45 (autoclave-sterilized); (b) Johns-Manville 045M (ethylene oxide-sterilized).

TABLE 1 Summary of Studies in Which Membrane Filtration Was Used to Recover Yeasts from the Aquatic Environment

Sample type	Membrane pore size (μm)	Isolation medium (Appendix formula number)	Incubation		Population densities	Predominant genera	Investigator(s)
			Temp. (°C)	Time (days)			
Marine water	-	Seawater agar (1)	-	-	Generally <100/liter	*Rhodotorula*, *Candida*, *Trichosporon*, *Debaryomyces*, *Cryptococcus*	Roth et al. (1962)
Marine water	0.45	Yeast isolation medium (8)	12	-	Range 1-200/liter; Average 1-14/liter	26 species belonging to Ascomycetes, Basidiomycetes, Deuteromycetes	Fell (1974)
Marine water	0.6	Yeast enumeration agar (11)	10	-	Range 10-20,000/liter	-	Schneider (1978)
Fresh, estuarine, and marine water	0.45	Variety (4)	22-24 ↓ 8-15	Until colonies visible	Range <10-500/100 ml	*Candida*, *Cryptococcus*, *Rhodotorula*, *Debaryomyces*	Ahearn et al. (1968)
Fresh, estuarine, and marine water	0.45	Y agar (9)	20 and 37	5-7 and 2-3	-	*Torulopsis* and *Candida* more dominant than *Cryptococcus* and *Trichosporon*	Buck (1975)

Great Lakes water	1.2	YM (2)	28	2	1-120/100 ml	*Cryptococcus*, *Torulopsis*, *Rhodotorula*	Hedrick et al. (1964)
Lake water	1.2	YM (5)	17 or 25	3-5	Range 1->430/100 ml; generally <20/100 ml	*Cryptococcus*, *Rhodotorula*, *Candida*	Cook (1970)
Lake water	0.45	m-12 and YM (6 and 5)	15-17; occasionally 25	3	Range 1->400/100 ml; generally <20/100 ml	*Cryptococcus*, *Rhodotorula*, *Torulopsis*	Meyers et al. (1970)
Fresh water (polluted)	1.2	Yeast isolation medium (7)	25	3-7	Maximum of 27,000/100 ml; average 3,000/100 ml	*Candida*, *Cryptococcus*, *Rhodotorula*	Woollett and Hedrick (1970)
Fresh water (polluted)	-	Acidified malt extract agar	Room temp.	-	0-30,000/liter	*Candida*, *Trichosporon*, *Rhodotorula*, *Torulopsis*, *Cryptococcus*	Spencer et al. (1970)
Fresh water (polluted)	0.45	Yeast extract agar (15)	12-14	Up to 14 days	0-9,500/100 ml	*Rhodotorula*, *Candida*, *Cryptococcus*	Simard and Blackwood (1971)
Sediment core samples	1.2	Yeast isolation medium (3)	24-30	2	1-250/6 ml or 0.6 g dry weight	*Cryptococcus*, *Rhodotorula* more frequent than *Candida*, *Torulopsis*, *Sporobolomyces*, and *Hansenula*	Hedrick et al. (1966)

Furthermore, ethylene oxide-sterilized membranes generally yielded lower counts than autoclave-sterilized filters.

Yeasts

The MF technique has been used frequently to isolate and enumerate yeasts in both marine and fresh water. From the synopsis of such studies presented in Table 1, it is apparent that the MF procedures used to recover yeasts from water differ in several aspects, including filter pore size, enumeration medium, and incubation conditions.

Membrane filters in a range of pore sizes (0.45-1.2 μm) have been used to recover yeasts from water. For the filtration of small volumes ($\leq$100 ml) of nonturbid water, 0.45 μm pore size membrane are usually adequate. However, when yeast levels are low and large sample volumes ($\geq$100 ml) must be filtered, or when samples contain excessive suspended matter, it is preferable to use larger pore size membranes (e.g., 1.2 μm).

Also, a variety of enumeration media have been used to resuscitate and grow those yeasts entrapped on the membrane surface (see the Appendix). In general, media prepared with filter-sterilized seawater and sterile distilled water have been used to examine marine and fresh waters, respectively. Ahearn et al. (1968) observed that media of high organic content gave maximal yeasts recoveries from fresh, estuarine, and marine waters. Two problems must be overcome when using agar media to isolate yeasts from aquatic environments. First, the growth of bacteria, which are plentiful in natural waters, must be prevented; this is achieved by either medium acidification and/or the inclusion of antibiotics in the agar medium (see the Appendix). Although numerous antibiotics have been employed to prevent bacterial growth, in most instances optimal combinations or concentrations of antibiotics have not been established. Buck (1975), however, observed that yeast counts decreased with levels of chloramphenicol in excess of 400 mg/liter, whereas media without chloramphenicol and with 100 mg/liter always allowed bacterial overgrowth of the membranes. Second, to prevent mold overgrowth of the membrane surface, mold colonies should be suppressed. Media incorporation of rose bengal and low-temperature (10-17°C) incubation have been used to facilitate yeast development while retarding mold growth. Enumeration of yeast colonies is usually feasible after 3-5 days of incubation.

It is apparent that in the absence of a standardized procedure, the selection of filter pore size, enumeration medium, and incubation conditions should be made in relation to the nature of water samples and the segment of the yeast flora under examination.

D. Pathogenic Fungi

Many fungi, including yeasts, are pathogenic to humans (Emmons et al., 1977). Spores or other propagules of such fungi, when present in recreational water, particularly bathing water, may be a potential public health hazard to water recreationists. Such a situation would be most likely to occur in waters that are subject to inputs of extraneous contamination, such as inadequately treated sewage or untreated stormwater runoff. Overloading of beaches with bathers may also result in accumulations of pathogenic fungi in both the sand and water.

Since the MF technique permits the analysis of large volumes of water, it should prove useful in examining the occurrence of pathogenic fungi in water. Kraus and Tiefenbrunner (1975) used membrane filtration followed by plating onto Sabouraud agar to isolate *Trichophyton spp.*, casual agents of dermatophytic infections in humans, from 1 liter samples of swimming pool water. Attempts (Sherry, 1977) to isolate dermatophytes from freshwater bathing beaches using the MF technique have proved unsuccessful, possibly because of the lack of a sufficiently selective isolation medium.

The yeast *Candida* is associated with various "thrush"-type infections in humans. *Candida albicans* is both the most common and most pathogenic species of *Candida* associated with human diseases. This organism (Gentles and La Touche, 1969) is the causal agent of a variety of superficial mycotic infections, such as oral and vaginal thrush; skin infections, especially in interdigital locations; and ocular infections (Beneke and Rogers, 1970; Emmons et al., 1977). *C. albicans* is a member of the body's normal saprophytic microflora (Winner and Hurley, 1964). Under certain conditions the yeast's role is altered from that of a commensal to a pathogen. Trauma caused by surgical operations, or environmental conditions that cause skin maceration, and debilitating disease are important factors contributing to infection. With the widespread use of broad-spectrum antibiotics and oral contraceptives, *C. albicans* has become increasingly important as a cause of mycoses in humans in recent years.

Whereas the normal mode of infection is probably by person-to-person contact or through a change in the status of the host's native *C. albicans* flora, the possibility of infection from a contaminated environment also exists, particularly if dermal maceration occurs either through occupational (housewives, fruit canners) or recreational (bathers) hazards. Although the occurrence of *C. albicans* in recreational waters has yet to be directly and conclusively associated with outbreaks of candidiasis among bathers, Brisou (1975) commented on instances of vaginal infections caused by *Candida* among females frequenting marine beaches contaminated with intestinal fungi. Thus, the occurrence of *C. albicans* in recreational waters may have public health significance.

Using the MF technique, Kraus and Tiefenbrunner (1975) isolated *C. albicans* from swimming pool waters that had a free chlorine content of <0.35 mg/liter. Monitoring *C. albicans* in natural waters has been greatly facilitated by the development of a membrane filter procedure for its enumeration (Buck and Bubucis, 1978). The enumeration procedure takes advantage of the resistance of *C. albicans* to high concentrations of cycloheximide and its ability to grow at 37°C. The selective medium (mCA agar) incorporates a characteristic indicator system and is highly selective for *C. albicans*. Details of the methodology, which involves filtration of water samples through 1.2 μm membrane filters, and confirmatory procedures are described by Buck and Bubucis (1978), who examined the occurrence of *C. albicans* in rivers, estuaries, and marine bathing beaches. Sherry et al. (1979a) also successfully used this procedure to determine the incidence of *C. albicans* in selected Lake Ontario bathing beaches.

TABLE 2 Recovery of *C. albicans* Cells, Freshwater Stressed at 4°C and 20°C, on Various 1.2 μm Membrane Filters

Manufacturer	Stress duration at 4°C (days)				Stress duration at 20°C (days)			
	0	1	4	11	0	1	4	11
Sartorius	89.3[a]	98.1	83.9	77.5	81.0	85.8	86.5	10.3
Gelman	92.9	92.0	96.1	92.0	86.4	86.3	87.3	52.6
Millipore	86.7	95.5	80.8	78.8	79.1	69.4	84.4	5.7
Schleicher and Schuell	101.6	101.9	99.7	111.3	87.3	78.9	95.3	31.5
Johns-Manville	99.5	96.2	89.0	94.6	86.2	77.5	89.2	18.8
Mean recovery	94.0	96.7	89.9	90.8	84.0	79.6	88.5	23.8
$\frac{R(max) - R(min)}{R(max)} \times 100$	14.7	9.7	19.0	30.4	9.4	19.6	11.4	89.2

[a]Expressed as percent of viable cells recovered using SPT on Sabouraud dextrose agar.

However, the use of MF introduces a number of variables into any microbial enumeration methodology. Different membrane filters vary in physical and chemical characteristics (Qureshi and Dutka, 1976; Standridge, 1976; Tobin and Dutka, 1977) that may affect their relative performances. In addition, the actual filtration process may further stress fungi in water samples.

Little difference has been found in the comparative abilities of membrane filters in the 0.45-1.2 μm pore size range to recover "nonstressed" *C. albicans* cells from standard cell suspensions in sterile buffer, using a noninhibitory nutrient agar (Buck and Bubucis, 1978) and the highly selective mCA agar (Sherry et al., 1979b). However, preliminary results of experiments using mCA agar suggest that if *C. albicans* cells are subjected to a prefiltration "stress" treatment in filter-sterilized water, differences in the abilities of membrane filters to recover these cells in a viable condition become apparent. For example, as shown in Table 2, when washed cells of an environmental isolate of *C. albicans* were stressed by suspension in filter-sterilized (0.2 μm) offshore Lake Ontario water, the mean recoveries for different stress intervals were higher at 4°C than at 20°C. In general, Schleicher and Schuell membranes gave the highest mean recoveries; Gelman membranes, however, apparently recovered the largest number of *C. albicans* cells that had been stressed for 11 days at 22°C. Furthermore, the differential between the best and poorest membranes increased, during the stress period of 0-11 days, from 14.7 to 30.4% at 4°C and from 9.4 to 89.2% at 20°C.

IV. FUNGI AS POLLUTION INDICATOR ORGANISMS

As the great majority of fungi are saprophytic in nature, they should respond to inputs of organic materials in a given environment. Therefore, it is generally assumed that fewer fungi are present in relatively unpolluted waters than in moderately to heavily polluted waters (Cooke, 1965a). Although less attention has been given to the presence of fungi in polluted waters than to other microorganisms, such as bacteria and algae, it has become apparent in recent years that fungi can be used as indicators of eutrophication and fecal contamination in water.

Contamination of water with fecal material is both aesthetically offensive and potentially dangerous, because the majority of agents of waterborne diseases are discharged from the intestines of infected persons (Wolf, 1972). As pathogenic microorganisms are also known to occur in the feces of both domestic and wild animals (Geldreich, 1972), fecal contamination from wastes of warm-blooded animals may also have public health significance. Inadequately treated and untreated sewage treatment plant discharges, untreated domestic and industrial effluents, storm sewer discharges, and urban and agricultural land drainage are all possible sources of feacal contamination in receiving waters. Fecal contaminants, through a process of nutrient enrichment, can also contribute to the accelerated eutrophication of natural aquatic ecosystems. Thus, monitoring the presence of fecal contamination in water has a dual aspect, that of detecting the presence of human pathogens and associated health hazards and of determining long-term enrichment effects in the aquatic environment.

In addition to bacteria, the microflora of feces from warm-blooded animals contains a variety of molds, yeasts, viruses, and protozoans (Geldreich, 1972, 1976; Cooke, 1976). It has been known since the 1950s, mainly through the work of Cooke (1954a-c, 1957, 1959, 1965b, 1970), that fungi are present in large numbers at every stage of the sewage treatment process, and occur in elevated numbers in waters that have been contaminated with sewage. Cooke (1965a) states that waters with low organic content generally contain large numbers of true aquatic fungi and few aquatic hyphomycetes and geofungi. However, as organic pollutants are added and the water becomes moderately to heavily polluted, the number of pollutant-tolerant species and the number of individuals of these species increases. In heavily polluted waters, geofungi become predominant and the aquatic fungal population decreases drastically. Harvey (1952) examined the relationship of true aquatic fungi to water pollution and found that species of *Saprolegnia, Achlya,* and *Dictyuchus* were persistently absent from heavily polluted waters, whereas only members of *Aphanomyces* showed affinity for polluted waters, thereby implying that these genera could be useful in determining water quality.

Using the MF technique to enumerate yeasts, Ahearn et al. (1968) noted that the occurrence and distribution patterns of certain yeasts could be employed as indicators in characterizing waters with various types of organic content. Meyers et al. (1970), who used the MF technique to study the yeast flora of Lake Champlain, observed that yeast populations generally exceeded 300 CFU/100 ml in areas directly affected by industrial or urban effluents, as opposed to base-line levels of approximately 20 CFU/100 ml. In these areas of the lake, physiologically distinct species predominated, which suggested to the authors the possibility of using yeasts as indicators of water quality. Similar conclusions were drawn by Woollett and Hedrick (1970), who used the MF technique in a survey of 14 polluted freshwater habitats and observed that the generic composition of the predominant segment of the yeast population appeared to be dependent on the nature of the pollutants. Species of *Rhodotorula* were present in low to heavily polluted waters at all locations. However, the genus *Cryptococcus* was a major component of the yeast population only in low and nonpolluted waters, and populations of *Candida* were predominant in organically rich waters that were contaminated with industrial and domestic wastes.

Cook (1970) also used MF to record elevated yeast populations at a polluted site on Lake Champlain and, on the basis of the consistent isolation of *Candida albicans* near effluents from college dormitories, confirmed the previous suggestion of Ahearn et al. (1968) that this opportunistic pathogen could serve as an indicator of recent fecal contamination. Simard (1971), who enumerated yeast populations in the St. Lawrence River by MF, suggested the possibility of using "pink yeasts" (especially *Rhodotorula glutinis*) as indicators of fecal pollution. However, as Meyer (1974) pointed out, the ubiquitous, oxidative yeasts such as *Rhodotorula* may not be suitable indicators of enrichment with organic materials, although local increases in numbers of these species may reflect an imbalance in the normal environment. Strongly fermentative yeasts, such as members of the genus *Candida,* which are predominant in sewage-affected waters, are potentially more suitable as water quality indicators. Of those species of *Candida* that have been detected in water, most attention is currently being directed toward *C. albicans* as a specific indicator of fecal contamination. Sherry et al. (1979a), using an MF procedure for the enumeration of *C. albicans*, observed an apparent association between the occurrence of *C. albicans* in lake water and the presence of elevated levels of bacterial fecal pollution indicators. This observation suggests that the inclusion of *C. albicans* in sanitary surveys of bathing beaches could usefully complement data obtained using conventional fecal pollution indicators.

Qureshi and Dutka (1974) and El-Shaarawi et al. (1977) studied the occurrence, distribution, and relationship of fungi with certain physical, chemical, and bacteriological parameters in Lake Ontario adjacent to the Niagara River. Their results suggested that the numbers (as determined by the spread plate and MF techniques) and distribution patterns of geoaquatic fungi could be used as indicators of water pollution, the degree of eutrophication, and to ascertain the zone of influence of point source pollution in inshore waters.

From the foregoing, it is apparent that estimates of fungal populations, by means of MF and other techniques, may prove to be a valuable aid in determining water quality and evaluating the extent and significance of pollution.

V. CONCLUSIONS

Membrane filtration is but one of a number of techniques used to study fungi in water; it has been widely used for the isolation and quantification of fungi, providing a growing body of data on the occurrence, distribution, and ecological significance of fungi in aquatic environments.

As an analytical tool, the MF technique has proved useful, particularly in evaluating water quality, because it is simple, rapid, reproducible, and permits both quantitative and qualitative analyses of relatively large sample volumes. This facilitates the enumeration of low-density populations of bacteria and fungi that might otherwise escape detection in water. However, it has been established in recent years that various commercial brand membranes differ in their ability to recover microorganisms from the aquatic environment. Realizing the significance of this factor, we would stress the need for establishing standardized procedures for testing the efficiency of membrane filters; this topic is considered in further detail in Chap. 2.

Future development of improved selective media and procedures will probably lead to the more frequent and widespread use of the MF technique in the enumeration and isolation of water-borne fungi, resulting in an improved understanding of the occurrence and significance of this important microbial group in the aquatic environment.

APPENDIX: Formulas for Media

1. Seawater agar (Roth et al., 1962)

Glucose	10.0 g
Peptone	5.0 g
Yeast extract	1.0 g
Agar	*
Aureomycin	100 mg†
Chloramphenicol	20 mg†
Streptomycin sulfate	20 mg†
Seawater	1 liter

2. YM agar (Hedrick et al., 1964)

Glucose	10.0 g
Yeast extract	5.0 g
Malt extract	5.0 g
Peptone	10.0 g
Agar	15.0 g
Streptomycin	100 mg
Penicillin	100 mg
Chloromycetin	50 mg
Distilled water	1 liter

3. Yeast isolation medium (Hedrick et al., 1966)

Basal salts and vitamins (Wickerham, 1946)	‡
Glucose	10.0 g/liter
Ammonium sulfate	5.0 g/liter
Agar	*
Streptomycin	100 mg/liter
Penicillin	10^5 units/liter
Chloromycetin	50 mg/liter
Lactic acid (85%)	1.5 mg/liter

Prior to addition of the agar, the pH of medium was adjusted to 4.0-4.5.

4. Yeast isolation media (Ahearn et al., 1968)

A.	Dextrose	10.0 g
	Malt extract	3.0 g
	Peptone	5.0 g
	Yeast extract	3.0 g
	Agar	*
	Chloramphenicol	0.5 g
	and/or Lactic acid (85%)	10 ml

---

*Agar concentration not specified.

†Originally expressed as mg%.

‡Quantity not specified.

B.	Dextrose	10.0 g
	Yeast extract	1.0 g
	Nutrient agar	23.0 g
	Peptone	4.0 g
	Chloramphenicol	0.5 g
C.	Dextrose	20.0 g
	Neopeptone	10.0 g
	Yeast extract	1.0 g
	Agar	*
	Rose bengal	0.2 g
	Chloramphenicol	0.5 g
D.	Malt extract	0.1 g
	Yeast extract	0.1 g
	Cornmeal agar	8.0 g
	Agar	5.0 g
	Chloramphenicol	0.5 g
E.	Inositol	10.0 g
	Yeast nitrogen base	7.0 g
	Agar	*
	Chloramphenicol	0.5 g
	Actidione	0.4 g

These media are calculated for preparation in 1 liter volumes. In estuarine waters, both distilled and filter-sterilized gulf stream water were used. Medium A was employed routinely for isolation of yeasts from fresh waters; medium B, with seawater, was used for salt waters.

5. YM agar (Cook, 1970)

Glucose	10.0 g
Peptone	10.0 g
Yeast extract	3.0 g
Malt extract	3.0 g
Agar	20.0 g
Declomycin	150 mg
Penicillin	100 mg
Chloramphenicol	100 mg
Distilled water	1 liter

6. m-12 agar (Meyers et al., 1970)

Dimalt-20	20.0 g
Bactopeptone	5.0 g
Bacto-yeast extract	3.0 g
Agar	*
Distilled water	1 liter

*Agar concentration not specified.

Following sterilization, the cooled medium was acidified to pH 4.3-4.5 with sterile lactic acid (10%) to retard bacterial growth. Antibiotics such as chloramphenicol were added to selected media.

7. Yeast isolation medium (Woollett and Hedrick, 1970)

Glucose	10.0 g/liter
Ammonium sulfate	5.0 g/liter
Agar	20.0 g/liter

Foregoing constituents were dissolved in basal salts Wickerham, 1951).

Vitamin solution	1.0 ml
Lactic acid	0.15 ml
Penicillin	1.0 mg
Aureomycin	0.5 mg
Chloromycetin	0.5 mg

Foregoing constituents were added to each 100 ml of agar medium.

8. Yeast isolation medium (Fell, 1974)

Difco nutrient agar	23.0 g
Agar	2.0 g
Glucose	20.0 g
Yeast extract	1.0 g
Water	1 liter

pH adjusted to 4.5 with HCl.

9. Y agar (Buck, 1975)

Glucose	20.0 g
Difco nutrient agar, pH 6	23.0 g
Malt extract	2.0 g
Difco yeast extract	1.0 g
Distilled water	1 liter

Various concentrations of chloramphenicol were added after autoclaving.

10. mCA agar (Buck and Bubucis, 1978)

Glycine	10.0 g
Maltose	30.0 g
Bismuth ammonium citrate	5.0 g
Na_2SO_3	3.0 g
Chloramphenicol	0.5 g
Cycloheximide	1.5 g
Distilled water	900 ml

Add the foregoing constituents to distilled water, heat to 50°C, and adjust pH to 7.1 with 1.0 NHCl. Add 15 g of agar and bring mixture slowly to the boiling point; a milky precipitate will form. Continue boiling gently for 1 min. Cool to 45-50°C in a water bath, add 100 ml of membrane-filtered (0.45 μm) yeast nitrogen base (Difco) prepared at a 10X concentration. Check the pH, which should be 6.7-6.8. Dispense the medium to a depth of 4-5 mm in sterile plastic petri dishes (60 x 15 mm) and store in the dark at 4°C.

11. Yeast enumeration agar (Schneider, 1978)

Difco peptone	5.0 g
Meat extract	2.4 g
Glucose	20.0 g
Yeast extract	1.0 g
$FePO_4 4H_2O$	0.01 g
Agar	15.0 g
Aged seawater	500 ml
Distilled water	500 ml
Streptomycin sulfate	0.25 g
Binotal	0.25 g

After sterilization, pH was adjusted with lactic acid (10%) to 4-4.5 and the antibiotics were added to the solution.

12. Acid rose bengal-kanamycin agar (Burman et al., 1969)

Mycological peptone (Oxoid)	5.0 g
Dextrose	10.0 g
KH_2PO_4	1.0 g
$MgSO_4 \cdot 7H_2O$	0.5 g
Rose bengal	0.7 g
Agar (Oxoid)	10.0 g
Distilled water	1 liter

The pH was adjusted to 5.4, and the medium was sterilized at 115°C for 10 min. Kanamycin was added to a final concentration of 100 μg/ml before pouring plates.

13. Modified aureomycin-rose bengal-glucose-peptone agar (MARGPA) (Dutka, 1978)

Glucose	10.0 g
Peptone	5.0 g
KH_2PO_4	1.0 g

$MgSO_4 \cdot 7H_2O$	0.5 g
Agar	20.0 g
Distilled water	1 liter
Rose bengal	0.035 g
Aureomycin HCl	200 mg

The pH was adjusted to 5.4, and the medium was sterilized at 121°C for 15 min. Membrane filter (0.2 μm) sterilized aureomycin was added before pouring plates.

14. Modified streptomycin-terramycin-malt extract agar (MSTMEA) (Dutka, 1978)

Malt extract	30.0 g
Peptone	5.0 g
Agar	15.0 g
Distilled water	1 liter
Streptomycin	200 mg
Terramycin	200 mg

The pH was adjusted to 5.4, and the medium was sterilized at 121°C for 15 min. Membrane filter (0.2 μm) sterilized streptomycin and terramycin were added before pouring plates.

15. Yeast extract agar (Simard and Blackwood, 1971)

Yeast extract	5.0 g
Peptone	10.0 g
Glucose	20.0 g
Agar	20.0 g
Distilled water	1 liter

After sterilization, the medium was acidified with sterile 10% lactic acid to a final pH of 4.5.

REFERENCES

Ahearn, D. G., Roth, F. J., Jr., and Meyers, S. P. (1968). Ecology and characterization of yeasts from aquatic regions of South Florida. *Mar. Biol. 1:* 291-308.

Barlocher, F., and Kendrick, B. (1976). Hyphomycetes as intermediaries of energy flow in streams. In *Recent Advances in Aquatic Mycology,* E. B. G. Jones (Ed.). Elk Science, London, pp. 435–446.

Barlocher, F., Kendrick, B., and Michaelides, J. (1978). Colonization and conditioning of *Pinus resinosa* needles by aquatic hyphomycetes. *Arch. Hydrobiol. 81:* 462–474.

Barron, G. L. (1971). Soil fungi. In *Methods in Microbiology*, vol. 4, C. Booth (Ed.). Academic Press, New York, pp. 405–427.

Beech, F. W., and Davenport, R. R. (1971). Isolation, purification, and maintenance of yeasts. In *Methods in Microbiology*, vol. 4, C. Booth (Ed.). Academic Press, New York, pp. 155–181.

Beneke, E. S., and Rogers, A. L. (1970). *Medical Mycology Manual*. Burgess, Minneapolis, Minn., p. 226.

Bremer, G. B. (1976). The ecology of marine lower fungi. In *Recent Advances in Aquatic Mycology*, E. G. B. Jones (Ed.). Elk Science, London, pp. 313–333.

Brisou, J. (1975). Yeasts and fungi in marine environments. *Bull. Soc. Fr. Mycol. Med. 4:* 159–162.

Buck, J. D. (1975). Distribution of aquatic yeasts--effect of incubation temperature and chloramphenicol concentration on isolation. *Mycopathologia 56:* 73–79.

Buck, J. D., and Bubucis, P. M. (1978). Membrane filter procedure for enumeration of *Candida albicans* in natural waters. *Appl. Environ. Microbiol. 35:* 237–242.

Burman, N. P. (1965). Taste and odour due to stagnation and local warming in long lengths of piping. *Proc. Soc. Water Treat. Exam. 14:* 125–131.

Burman, N. P., Oliver, C. W., and Stevens, J. K. (1969). Membrane filtration technique for the isolation from water of coli-aerogenes, *Escherichia coli*, faecal streptococci, *Clostridium perfringens*, actinomycetes and microfungi. In *Isolation Methods for Microbiologists*, D. A. Shapton and G. W. Gould (Eds.), Academic Press, London. Soc. Appl. Bacteriol. Tech. Ser. No. 3, pp. 127–134.

Cochrane, V. W. (1965). *Physiology of Fungi*. Wiley, New York, p. 524.

Cook, W. L. (1970). Effects of pollution on the seasonal population of yeasts in Lake Champlain. In *Recent Trends in Yeast Research*, D. Ahearn (Ed.), Georgia State, Atlanta. *Spectrum, Monogr. Ser. Arts and Sci. 1*: 107–112.

Cooke, J. C., and la Course, J. R. (1975). A preliminary study of microfungi from the Connecticut River estuary. *Bull. Torrey Bot. Club 102:* 1–6.

Cooke, W. B. (1954a). Fungi in polluted water and sewage. II. Isolation technique. *Sewage Ind. Wastes 26:* 661–674.

Cooke, W. B. (1954b). Fungi in polluted water and sewage. I. Literature review. *Sewage Ind. Wastes 26:* 539–549.

Cooke, W. B. (1954c). Fungi in polluted water and sewage. III. Fungi in a small polluted stream. *Sewage Ind. Wastes 26:* 790–794.

Cooke, W. B. (1957). Checklist of fungi isolated from polluted water and sewage. *Sydowia 1:* 146—175.

Cooke, W. B. (1959). Trickling filter ecology. *Ecology 40:* 273—291.

Cooke, W. B. (1961). Pollution effects on the fungus population of a stream. *Ecology 42:* 1—18.

Cooke, W. B. (1963). *A Laboratory Guide to Fungi in Polluted Waters, Sewage and Sewage Treatment Systems*. U.S. Dept. of Health, Education and Welfare, Cincinnati, Ohio, p. 132.

Cooke, W. B. (1965a). The enumeration of yeast populations in a sewage treatment plant. *Mycologia 57:* 696—703.

Cooke, W. B. (1965b). Fungi in sludge digesters. *Proc. 12th Ind. Waste Conf.,* Purdue University, pp. 6—17.

Cooke, W. B. (1970). *Our Mouldy Earth--A Study of the Fungi in Our Environment with Emphasis on Water*. U.S. Dept. of the Interior, Federal Water Pollution Control Administration, R. A. Taft Research Centre, Advanced Waste Treatment Research Laboratory, Cincinnati, Ohio.

Cooke, W. B. (1971). The role of fungi in waste treatment. *CRC Crit. Rev. Environ. Control 1:* 581—619.

Cooke, W. B. (1976). Fungi in sewage. In *Recent Advances in Aquatic Mycology,* E. B. G. Jones (Ed.). Elk Science, London, pp. 389—434.

Cooney, D. G., and Emerson, R. (1964). *Thermophilic Fungi: An Account of Their Biology, Activities and Classification*. W. H. Freeman, San Francisco, p. 188.

Deverall, B. J. (1968). Psycrophiles. In *The Fungi: An Advanced Treatise, vol. 3, The Fungal Population,* G. C. Ainsworth and A. S. Sussman (Eds.). Academic Press, New York, pp. 129—135.

Dick, M. W. (1971). The ecology of Saprolegniaceae in littoral muds with a general theory of fungi in the lake ecosystem. *J. Gen. Microbiol. 65:* 325—337.

Dick, M. W. (1976). The ecology of aquatic phycomycetes. In *Recent Advances in Aquatic Mycology,* E. B. G. Jones (Ed.). Elk Science, London, pp. 513—542.

Dutka, B. J., Ed. (1978). *Methods for Microbiological Analysis of Waters, Wastewaters and Sediments*. Inland Waters Directorate, Canada Centre for Inland Waters, Burlington, Ontario.

Emmons, C. W., Binford, C. H., Utz, J. P., and Kwan-Chung, K. J. (1977). *Medical Mycology*. Lea & Febiger, Philadelphia, p. 592.

Fell, J. W. (1974). Distribution of yeasts in the water masses of the southern oceans. In *Effects of the Ocean Environment on*

Microbial Activities, R. R. Colwell and R. Y. Morita (Eds.). University Park Press, Baltimore, Md., pp. 510–523.

Fuller, M.S., and Poyton, R. O. (1964). A new technique for the isolation of aquatic fungi. *Bioscience 14:* 45–46.

Gaertner, A. (1968). Eine methode des quantitativen Nachweises niederer, mit Pollen koederbarer Pilze im Meerwasser und in Sediment. *Veroeff. Inst. Meeresforsch., Bremerhaven 3:* 75–92.

Geldreich, E. E. (1972). Water borne pathogens. In *Water Pollution Microbiology,* R. Mitchell (Ed.). Wiley-Interscience, New York, pp. 207–241.

Geldreich, E. E. (1976). Fecal coliform and fecal streptococcus density relationships in waste discharges and receiving waters. *CRC Crit. Rev. Environ. Control 6:* 349–369.

Gentles, J. C., and La Touche, C. J. (1969). Yeasts as human and animal pathogens. In *The Yeasts,* A. H. Rose and J. S. Harrison (Eds.). Academic Press, New York, pp. 107–182.

Harvey, J. V. (1952). Relationship of aquatic fungi to water pollution. *J. Water Pollution Control Fed. 24:* 1159–1164.

Hawker, L. E. (1966). *Fungi.* Hutchinson, London, p. 216.

Hedrick, L. R., and Soyugenc, M. (1967). Yeasts and moulds in water and sediments of Lake Ontario, *Proc. 10th Conf. Great Lakes Res.,* pp. 20–30. Int. Assoc. Great Lakes Res., Ann Arbor, Mich.

Hedrick, L. R., Soyugenc, M., and Larsen, L. (1966). Yeasts in sediment core samples from Lake Michigan. *Proc. 9th Conf. Great Lakes Res.,* pp. 27–37. Publ. No. 15, Great Lakes Res. Div., Univ. Mich., Ann Arbor, Mich.

Hedrick, L. R., Cook, W., and Woollett, L. (1968). Yeasts and molds in Lake Superior water and some of its tributaries. *Proc. 11th Conf. Great Lakes Res.,* pp. 538–543. Int. Assoc. Great Lakes Res., Ann Arbor, Mich.

Hedrick, L. T., Soyugenc, M., DuPont, P., and Ambrosini, R. (1964). Yeasts in Lake Michigan and Lake Erie. *Proc. 7th Conf. Great Lakes Res.,* pp. 77–83. Publ. No. 11, Great Lakes Res. Div., Univ. Mich., Ann Arbor, Mich.

Ho, H. H. (1975). A selective medium for the isolation of *Saprolegnia* spp. from fresh water. *Can. J. Microbiol. 21:* 1126–1128.

Ingold, C. T. (1976). The morphology and biology of freshwater fungi excluding phycomycetes. In *Recent Advances in Aquatic Mycology.* E. B. G. Jones (Ed.). Elk Science, London, pp. 335–357.

Iqbal, S. H., and Webster, J. (1973). Aquatic hyphomycete spora of the River Exe and its tributaries. *Trans. Br. Mycol. Soc. 61:* 331–346.

Jones, E. B. G. (1971). Aquatic fungi. In *Methods in Microbiology*, vol. 4, C. Booth (Ed.). Academic Press, New York, pp. 335–365.

Jones, E. B. G. (1976). Lignicolous and algicolous fungi. In *Recent Advances in Aquatic Mycology*, E. B. G. Jones (Ed.). Elk Science, London, pp. 1–49.

Kraus, H., and Tiefenbrunner, F. (1975). Randomized testing of a number of swimming pools in the Tyrol for the presence of *Trichomonas vaginalis* and pathogenic fungi. *Zentrabl. Bakteriol. Parasitenkd. Infektionskr. Hyg. I Abt. Orig. B. 160:* 286–291.

Masters, M. J. (1976). Freshwater phycomycetes on algae. In *Recent Advances in Aquatic Mycology*, E. B. G. Jones (Ed.). Elk Science, London, pp. 489–512.

Meyer, S. A. (1974). Yeasts: species and distribution as indicators of water quality. Presented at Annu. Meet. (Sess. 190) Am. Soc. Microbiol., Chicago.

Meyers, S. P., Ahearn, D. G., and Cook, W. L. (1970). Mycological studies of Lake Champlain. *Mycologia 62:* 504–515.

Miller, C. E. (1967). Isolation and pure culture of aquatic Phycomycetes by membrane filtration. *Mycologia 59:* 524–527.

Park, D. (1972a). Methods of detecting fungi in organic detritus in waters. *Trans. Br. Mycol. Soc. 58:* 281–290.

Park, D. (1972b). On the ecology of heterotrophic microorganisms in fresh water. *Trans. Br. Mycol. Sco. 58:* 291–299.

Qureshi, A. A., and Dutka, B. J. (1974). A preliminary study on the occurrence and distribution of geo-fungi in Lake Ontario near the Niagara River. *Proc. 17th Conf. Great Lakes Res.*, pp. 653–662. Intern. Assoc. Great Lakes Res., Ann Arbor, Mich.

Qureshi, A. A., and Dutka, B. J. (1976). Comparison of various brands of membrane filters for their ability to recover fungi from water. *Appl. Environ. Microbiol. 32:* 445–447.

Roth, F. J., Ahearn, D. G., Fell, J. W., Meyers, S. P., and Meyer, S. A. (1962). Ecology and taxonomy of yeasts isolated from various marine substrates. *Limnol. Oceanogr. 7:* 178–185.

Schneider, J. (1978). Fungi. In *Microbial Ecology of a Brackish Water Environment*, G. Rheinheimer (Ed.). Springer-Verlag, New York, pp. 90–102.

Scott, W. H., and O'Bier, A. H. (1962). Aquatic fungi associated with diseased fish and fish eggs. *Prog. Fish Cult. 24:* 3–15.

El-Shaarawi, A., Qureshi, A. A., and Dutka, B. J. (1977). Study of microbiological and physical parameters in Lake Ontario adjacent to the Niagara River. *J. Great Lakes Res. 3:* 196–203.

Sherry, J. (1978). Unpublished data.

Sherry, J. P., Kuchma, S. R., and Dutka, B. J. (1979a). The occurrence of *Sandida albicans* in Lake Ontario bathing beaches. *Can. J. Microbiol. 25*: 1036–1044.

Sherry, J. P., Kuchma, S. R., Zarzour, J., and Dutka, B. J. (1979b). Occurrence and significance of *Candida albicans* in Lake Ontario bathing beaches. *Sci. Ser. No. 98*. Inland Waters Directorate, National Water Research Institute, CCIW, Burlington, Ontario, p. 31.

Simard, R. W. (1971). Yeasts as an indicator of pollution. *Mar. Pollut. Bull. 2:* 123–125.

Simard, R. W., and Blackwood, A. C. (1971). Yeasts from the St. Lawrence River. *Can. J. Microbiol. 17:* 197–203.

Smith, G. (1969). *An Introduction to Industrial Mycology*. Edward Arnold, London, p. 390.

Sparrow, F. K., Jr. (1968). Ecology of freshwater fungi. In *The Fungi: An Advanced Treatise,* vol. 3, *The Fungal Population,* G. C. Ainsworth and A. S. Sussman (Eds.). Academic Press, New York, pp. 41–93.

Spencer, J. F. T., Gorin, P. A. J., and Gardner, N. R. (1970). Yeasts isolated from the South Saskatchewan, a polluted river. *Can. J. Microbiol. 16:* 1051–1057.

Standridge, J. H. (1976). Comparison of surface pore morphology of two brands of membrane filters. *Appl. Environ. Microbiol. 31:* 316–319.

Tobin, R. S., and Dutka, B. J. (1977). Comparison of the surface structure, metal binding, and fecal coliform recoveries of nine membrane filters. *Appl. Enviorn. Microbiol. 34:* 69–79.

Ulken, A., and Sparrow, F. K. (1968). Estimation of chytrid propagules in Douglas Lake by the MPN pollen grain method. *Veroeff. Inst. Meeresforsch., Bremerhaven 11:* 83–88.

Warcup, J. H. (1950). The soil plate method for isolation of fungi from soil. *Nature, Lond. 166:* 117–118.

Webster, J. (1959). Experiments with spores of aquatic hyphomycetes. 1. Sedimentation and impaction on smooth surfaces. *Ann. Bot. 12:* 595–611.

Weston, W. H. (1941). The role of the aquatic fungi in hydrobiology. In *A Symposium on Hydrobiology*. University of Wisconsin Press, Madison, Wis., pp. 129–151.

Wickerham, L. J. (1946). A critical evaluation of the nitrogen assimilation tests commonly used in the classification of yeasts. *J. Bacteriol. 52*: 293–301.

Wickerham, L. J. (1951). Taxonomy of yeasts. *U.S. Dept. Agric. Tech. Bull. 1029:* 1–56.

Willoughby, L. G. (1968). Atlantic salmon disease fungus. *Nature 217:* 872–873.

Willoughby, L. G., and Pickering, A. D. (1977). Viable Saprolegniaceae spores on the epidermis of the salmonid fish *Salmo trutta* and *Salvelinus alpinus*. *Trans. Br. Mycol. Soc. 68:* 91–95.

Winner, H. I., and Hurley, R. (1964). *Candida albicans*. Little, Brown, Boston.

Wolf, H. W. (1972). The coliform count as a measure of water quality. In *Water Pollution Microbiology*, R. Mitchell (Ed.). Wiley-Interscience, New York, pp. 333–345.

Woollett, L. L., and Hedrick, L. R. (1970). Ecology of yeasts in polluted waters. *Antonie van Leeuwenhoek J. Microbiol. Serol. 36:* 427–435.

9

MEMBRANE FILTERS IN VIROLOGY

SAGAR M. GOYAL and CHARLES P. GERBA Baylor College of Medicine, Houston, Texas

I. VIRUS CONCENTRATION FROM WATER

A. Introduction

Membrane filters are used in virology primarily for the estimation of virus size, removal of aggregates from virus suspensions, removal of extraneous material, and concentration of viruses from dilute suspensions. The first three depend on the physical entrapment of the virus by a limiting pore size; the last is dependent on adsorption. In this chapter, we are concerned mainly with the last application, as it has received the greatest amount of study.

The use of membrane filters in water virology is a unique one. In this procedure, viruses are removed from aqueous systems by adsorption to filters whose nominal pore size may be many times the diameter of the viruses. Membrane filters are used primarily for this purpose, although depth filters have also been used (Gerba et al., 1978a). To enhance virus adsorption, acid and salts are usually added to the water being processed before passage through the membrane. Adsorbed viruses are then eluted with a small volume of eluent, usually a high-pH buffer or a soluble organic. Using these techniques, viruses can be concentrated from 3780 liters (1000 gal) of tap water to a volume of 20-40 ml. This technique was largely developed around the need for the detection of enteric viruses (i.e., viruses excreted in the feces) in sewage and sewage-contaminated waters.

The development of the membrane filter adsorption-elution methodology for virus concentration from drinking waters, natural waters, and wastewaters during the late 1960s and early 1970s significantly advanced the field of environmental virology. Prior to that time only qualitative detection methods were available, or sample volumes were limited to 1 gal or less.

B. Early Development of Concentration Methods

A number of methods are available to detect and quantitate viruses in water, but most of them suffer from one or more limitations, and hence no single method can be used to detect viruses in all types of waters (i.e., waste, natural, and treated waters). Most of these methods have exploited various physicochemical properties of the virions [e.g., physical adsorption, precipitation, phase partitioning, sedimentation, filtration, migration under the influence of an electric field, and the immunochemical reactions of the viruses (Sobsey, 1976)]. Virus recovery depends greatly on water quality and is influenced by such factors as particulates, organics, and salts.

It is obvious that the ideal method for virus concentration from water should be (1) capable of processing large amounts of a variety of waters in the least possible time, (2) sensitive enough to concentrate most types of viruses known to be present in water and wastewater, (3) easy to perform and economical to use, and (4) able to detect viral aggregates and viruses adsorbed to suspended solids.

An extensive body of literature describing virus interactions with filter surfaces has been developed in recent years. However, this interaction was reported as early as 1931, when Elford (1931) recommended that bacterial viruses should be suspended in protein diluents to avoid loss of virus during filtration through collodion membranes made of cellulose nitrate. During the late 1940s and early 1950s, several investigators described the use of gauze pads to detect animal viruses in sewage.

Metcalf (1961) first reported the use of membrane filters to recover viruses from aqueous suspensions. He discovered that influenza virus could be retained on Millipore membranes (types HA, VM, and VF) despite the fact that the membrane porosities exceeded the diameter of the virus. Viruses were recovered by grinding the membranes in a mortar and pestle and resuspending the pulp.

Cliver (1965) reported that poliovirus type 1 and coxsackievirus B2 were adsorbed to Gelman (cellulose acetate) and Millipore (cellulose nitrate) membranes. Cliver was interested primarily in filtering debris and bacteria from food extracts so that the extract might be assayed for virus. Adsorption of viruses to the filter was thus not desirable and it was found that treatment of the membrane with serum or gelatin effectively prevented viral adsorption.

Cliver later reported that enteroviruses (polio, coxsackie, echo) could be adsorbed to Millipore membranes in the course of filtration of experimentally contaminated deionized water, tap water, phosphate-buffered saline, urine, or throat washings. The adsorbed viruses could be eluted from the filter surfaces by treatment with phosphate-buffered saline plus 30% chicken serum (Cliver, 1967a).

Wallis and Melnick (1967a) reported the first application of virus-adsorbing filters when they described a technique for concentrating enteroviruses from crude virus harvests by adsorption onto Millipore membranes. They found that viruses could be adsorbed efficiently to nitrocellulose membranes in the presence of salts (Earle's salt solution, NaCl, or $MgCl_2$) if the crude virus harvests were first treated with resins to remove the organic components that interfered with virus adsorption. The viruses could be eluted from the membrane by washing with small volumes of fetal calf serum, and in most cases, 80- to 100-fold concentrations were achieved.

In a subsequent publication, Wallis and Melnick (1967b) reported that viruses could be concentrated from sewage by adsorption onto Millipore cellulose nitrate membranes. Interfering organics were removed by treatment of the sewage with anionic resins prior to filtration. Salts ($MgCl_2$, 0.05 M) were then added to the resin-treated sewage to enhance virus adsorption to the filter surface. The viruses were eluted by homogenization of the filter in Melnick's medium B containing 10% fetal bovine serum. Using this technique, 2795 enteroviruses were isolated from 1 gal samples of sewage during a 7 month period in 1966. During the same period, only four isolates were recovered from unconcentrated sewage. This report represented a significant advance in environmental virology in that it was the first application of the virus-adsorbing potential of microporous membranes to detect human enteroviruses in polluted waters.

Rao and Labzoffsky (1969) suggested that low concentrations of viruses could be detected in large volumes of clean water by the membrane filter technique. In this laboratory study, a fiberglass prefilter (Millipore AP20) and a 0.45 μm pore size cellulose ester filter (Millipore HA) were used in series as the viral adsorbents. Calcium salts (200 mg/liter) were added to the river water samples (500 ml) to enhance virus adsorption. Adsorbed viruses were eluted with 3% beef extract and recoveries averaged 92%.

Berg et al. (1971) reported a technique for recovering small quantities of enteroviruses from clean waters on Millipore cellulose nitrate membranes. Their method consisted of adding Na_2HPO_4 (final concentration, 0.05 M) to the water and sufficient citric acid to produce a pH of 7. Viruses were eluted from the filters by sonication of the membranes in 3% beef extract. Good recoveries of enteroviruses and reovirus type 1 were reported, but the method was only applicable to finished waters of high quality because of clogging problems.

Wallis et al. (1972a) studied the effects of various salts on adsorption of enteroviruses to Millipore membranes. They discovered that trivalent salts ($AlCl_3$) could be used effectively to enhance viral adsorption at only 1% of the concentration required for divalent salts ($MgCl_2$). In that report, they also described a reconcentration procedure whereby viruses recovered from filters were readsorbed to smaller-diameter filters and eluted with smaller volumes of eluent, thus greatly reducing the amount of eluent to be assayed. The eluent used in those experiments was 0.05 M glycine adjusted to pH 11.5 with NaOH. The basic glycine was an extremely effective eluent, enabling recovery of up to 100% of adsorbed viruses by passing small volumes of the pH 11.5 glycine through the filters. The significance of this report lies in the fact that it described methodology that greatly facilitated the processing of large volumes of water for the determination of enterovirus contamination.

C. Wallis-Melnick Virus Concentrator

Simultaneously with the development of efficient methods for adsorption and elution of enteroviruses from membrane filters, Wallis and Melnick described the development of a portable virus concentrator for concentrating viruses in the field based on this technique. The original model (Wallis et al., 1972b, c) was a continuous-flow-through apparatus in which incoming water was passed through a series of five non-virus-adsorbing clarifying textile filters followed by treatment with an anion-exchange resin to remove organics before adsorption of the virus onto 293-mm-diameter cellulose nitrate disk filters. Salts ($MgCl_2$ added to a final concentration of 0.05 M) were injected prior to the viral adsorption step to enhance adsorption. Viruses were eluted from the filter in situ with 1 liter volumes of glycine (0.05 M) adjusted to pH 11.5. Eluates from the virus adsorbent were reconcentrated on smaller-diameter cellulose nitrate membranes, which could be eluted with small volumes (approximately 20 ml) of pH 11.5 glycine. Using the concentration/reconcentration procedure, the virus concentrator was capable of recovering 80% of exogenously added viruses during laboratory processing of 300 gal (1134 liters) of tap water.

Further refinement of the portable virus concentrator resulted in a unit in which the water was first clarified through a series of orlon or polyester depth cartridge filters (5 and 1 μm porosity) and a Tween 80-treated, 1 μm porosity cotton depth cartridge filter; the viruses were then adsorbed to a fiberglass or cellulose acetate cartridge depth filter 1 μm in porosity (Homma et al., 1973). In addition to these changes, $AlCl_3$ (0.005 M) was used instead of $MgCl_2$ to enhance virus adsorption, since a 100-fold less concentration was needed.

Based on the prototype virus concentrator developed by Wallis and Melnick, the Carborundum Company (Niagara Falls, N.Y.) made commercially available a self-contained virus concentrator, the Aquella Virus Concentrator (Fig. 1).

These early systems were found to efficiently concentrate viruses from large volumes of tap water and smaller volumes of sewage and seawater, but several problems soon became evident. Suspended matter was so great in wastewater and other turbid waters that it tended to clog the filters and/or greatly reduce the flow

FIGURE 1 Apparatus (Aquella Virus Concentrator) for concentrating viruses from large volumes of water. (Courtesy of the Carborundum Company, Niagara Falls, N.Y.)

rate. The use of the clarifying filters in front of the adsorbing filters reduced the magnitude of this problem, but often resulted in the decreased efficiency of virus recovery because of the loss of solid-associated virus on the clarifying filters (Homma et al., 1973; Metcalf et al., 1974a). This problem became greater as solids accumulated on the prefilters (Homma et al., 1973; Sobsey et al., 1977).

Further study by Sobsey et al. (1973) resulted in a modified version of the portable virus concentrator. In this system tap water was acidified to pH 3.5 with 1 N HCl and then passed through a virus adsorber consisting of a fiberglass cartridge depth filter (K-27) and a 142 mm diameter, 0.65 μm pore size Cox membrane filter in series. The adsorbed viruses were eluted with 1 liter of pH 11.5 glycine-NaOH and reconcentrated to a volume of 10 ml by adsorption to and elution from 47 mm diameter, 5, 1, and 0.45 μm pore size Cox filters in series. With this method small quantities of poliovirus type 1 in 100 gal (378 liter) volumes of tap water were concentrated nearly 40,000-fold, with average recoveries of 77%. This methodology was also found useful for concentration of viruses from 50-100 gal amounts of seawater (Sobsey et al., 1977) and well water beneath a wastewater land treatment site (Gilbert et al., 1976).

Unfortunately, this system could not be effectively scaled up to process larger volumes at reasonable flow rates. Because of the limited surface area of the flat-disk adsorbent filters, maximum flow rates of only 3 gal (ca. 11 liters) per minute could be achieved with finished tapwater. Under field conditions using turbid seawater, average flow rates ranged from 0.5-1 gal (ca. 2-4 liters) per minute when 50 gal (189 liter) samples were processed (Sobsey et al., 1977).

A final problem was that humic acid and other organic compounds were also concentrated from water onto the filters during operation of the virus concentrator (Farrah et al., 1976a). These compounds were eluted from the filters along with the virus and formed an insoluble precipitate when the eluate was neutralized. They seriously

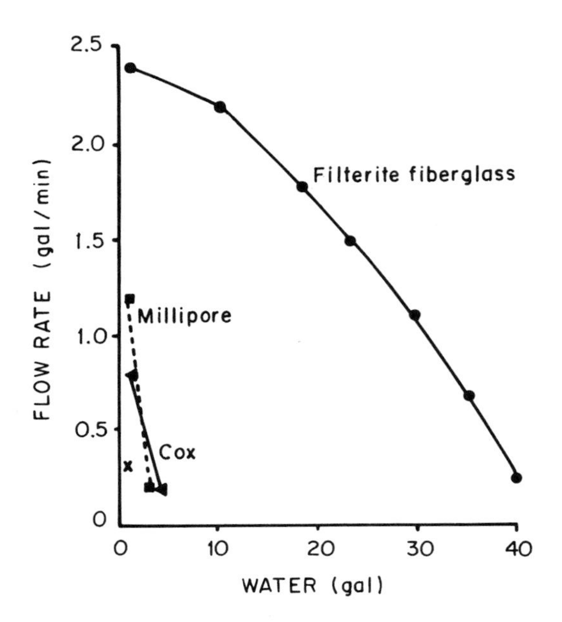

FIGURE 2 Flow rate versus filter type. X, Acropor.

interfered with any attempted reconcentration of the initial eluate using membrane filters (Sobsey et al., 1977) when processing greater than 100 gal of tap water or smaller volumes of seawater and sewage.

To overcome these limitations, Farrah et al. (1976b) tested a variety of membrane filters and found that Filterite fiberglass membrane filters (Duo-Fine series) were far less easily clogged than Cox (series AA), acrylonitrile polyvinyl chloride copolymer filters (Acropor series) and nitrocellulose (Millipore) of approximately the same rated pore size (Fig. 2). All filters adsorbed greater than 90% of poliovirus added to tap water at pH 3.5. However, the 47 mm diameter Acropor, Cox, and Millipore filters clogged after processing less than 20 liters of water, whereas the Filterite processed 150 liters of water before clogging. Filterite Duo-Fine filters are manufactured as 10 in. (ca. 25.4 cm) long, pleated cartridges (Fig. 3), which fit easily into the plastic see-through housing described

FIGURE 3 Partially unfolded Filterite filter, showing the pleated nature of the filter.

in earlier models of the Wallis-Melnick virus concentrator that were used to house depth filters (Wallis et al., 1972b). Since the surface area of a 10 in. Filterite filter is 280 times that of a 47 mm diameter filter, a pleated cartridge filter would be able to process 40,000 liters of similar-quality tap water before becoming clogged. Flow rates of up to 37.8 liters/min (10 gal/min) were obtained with the pleated membrane filter cartridges. Large volumes of tap water could be processed without the need of a prefilter, but a prefilter (a 3 μm pore size Filterite or a K-27 spun fiberglass depth filter) was used to spare the 0.25 or 0.45 μm final adsorbing filter and allow reuse of the pleated filter. The 0.45 μm pore size filter was used with sewage and seawater to reduce clogging problems. Using these filters, seeded poliovirus could be recovered from 472-1900 liters (125-500 gal) of tap water, 378 liters (100 gal) of seawater, and 19-190 liters (5-50 gal) of secondarily treated sewage with an average efficiency of 52%, 53%, and 50%, respectively. Because of clogging problems with humic acid and other organics eluted from the filters in the initial concentration step, reconcentration was accomplished by a combination of aluminum flocculation followed by hydroextraction (Farrah et al., 1977a).

Another advantage of this system is that the filters could be reused several times without loss of concentration efficiency after autoclaving or soaking overnight in a concentrated solution of sodium hydroxide, thus reducing operational costs of the virus concentrator (Farrah et al., 1977b).

Wallis and Melnick (1967c) demonstrated that viruses could be concentrated on aluminum hydroxide flocs. In a recent study, Farrah et al. (1978a) made use of this observation to concentrate poliovirus from tap water. Low concentrations of aluminum chloride in tap water were found to form flocs that adsorbed viruses and were subsequently retained by membrane filters at pH values near neutrality. Tap water treated with 2×10^{-5} M aluminum chloride showed a slight decrease in pH (<0.5), a slight increase in turbidity, and enhanced removal of poliovirus by membrane filters. Pleated Filterite filters (which are more resistant to clogging than nitrocellulose or epoxy-asbestos-fiberglass filters) were used to entrap the aluminum floc containing the adsorbed virus. Virus was quantitatively recovered by treating the filters with 0.05 M glycine (pH 11.0) for 3 min. After neutralization, the eluate was reconcentrated to smaller volumes by adsorption to aluminum hydroxide followed by hydroextraction (Farrah et al., 1977a). Using these procedures, virus from 1000 liters of water was recovered in a final eluate of 20-80 ml with a mean recovery of 70%.

This procedure resulted in (1) elimination of the need to acidify the water sample to pH 3.5 before processing and (2) reduction of the amount of aluminum chloride to 1/25 of that previously used (Farrah et al., 1976b; Wallis et al., 1972a).

D. Other Membrane Filter Adsorption-Elution Systems

Several other systems similar to the Wallis-Melnick virus concentrator have been studied, but they have seen only limited use to date. Hill et al. (1972) reported on the development of a technique for concentrating viruses (low multiplicities) from large volumes (100 gal) of tap water and estuarine water. Millipore MF tube cartridge filters (cellulose nitrate) were used to adsorb viruses; $MgCl_2$ (final concentration, 0.05 M) was added to enhance virus adsorption; and the pH of the water sample was adjusted to 4.0. Filters containing adsorbed viruses were eluted with 0.05 M carbonate-bicarbonate buffer at pH 9.0. Eluates were reconcentrated by the aqueous polymer two-phase separation technique. Using this procedure, 67% of the exogenously added viruses could be recovered.

This group (Jakubowski et al., 1974) later described the use of epoxy-fiberglass tube filters (Balston) for the adsorption of viruses from tap water after adjustment to pH 3.5 and addition of 0.0005 M $AlCl_3$. Elution was accomplished with pH 11.5 glycine buffer. The recovery of seeded poliovirus from 100 gal (378 liter) volumes of tap water ranged from 42 to 57% using 8 μm pore size tube filters.

Hill et al. (1976) compared four microporous virus-adsorbent media as to their sensitivity for reliably detecting low levels of poliovirus in large volumes of tap water. The virus adsorbent media tested were (1) Millipore MF cellulose-nitrate membrane filters, (2) Cox AA epoxy-fiberglass-asbestos filters, (3) Fulflo K-27 yarn-wound fiberglass depth filters, and (4) Balston epoxy-fiberglass filters. Viruses were adsorbed to the media at pH 3.5 (no salts added) and eluted with glycine buffer at pH 11.5. At mean virus input levels of 1-2 plaque-forming units per 380 liters of tap water, the detection reliability was 66% in 76 samples of 1900 liter volumes. No significant difference in virus detection sensitivity was observed among the various virus-adsorbent media tested. A major innovation of this system was the use of a proportioner pump for the injection of acid and salts. Previous concentrator systems had depended upon injection from a pressurized tank and, as the flow rate through the system decreased (due to clogging of the filters), it was necessary to periodically adjust the amount of acid and salts being injected. The amount of acid and salts being injected by the proportioner pump changes automatically with changes in flow rate (Jakubowski et al., 1974). This method is not suitable for the high flow rates possible with pleated membrane filters (Gerba et al., 1978a) because the orifice in the proportioner limits the maximum flow rate, but it is certainly the method of choice for use with lower flow rates.

Unfortunately, tube filters suffer from many of the drawbacks inherent with flat-disk membrane filters. The Balston tube filters clog more readily than pleated membrane filters and cannot be used even with moderately turbid water. They cannot be used above an in-line pressure of 25 psi, and thus high flow rates cannot be achieved. This limits their practical use to only finished tap water of good quality. In contrast, pleated filters can operate with an in-line pressure of up to 100 psi. A list of filters commonly used for virus concentration is given in Table 1.

TABLE 1 Filters Commonly Used for Virus Concentration

Composition	Manufacturer	Code
Nitrocellulose	Millipore Corp., Bedford, Mass.	HA
Epoxy-fiberglass-asbestos	Cox Instrument Corp., Detroit, Mich.	Series AA
Epoxy-fiberglass	Filterite Corp., Timonium, Md.	Duo-Fine series
Cellulose, modified anion-exchange resin, and inorganic filter aid	AMF, CUNO Division, Meriden, Conn.	Zeta-plus
Borosilicate glass microfiber-epoxy resin	Balston, Inc., Lexington, Mass.	Filter tubes
Fiberglass	Commercial Filter Division, Carborundum Co., Lebanon, Ind.	Wound fiber depth filter (K-27)

E. Virus Concentration from Seawater

Methods for the concentration of viruses from large volumes of estuarine water have been hampered by the often-high turbidities characteristic of these waters (Hill et al., 1972; Metcalf et al., 1974a). Hill et al. (1974a) described a method using Celite as a filtering aid in processing 15-100 gal amounts of estuarine water with turbidities of 8.5-80 Jackson turbidity units (JTU) through flat membrane filters. However, virus recoveries were very poor, ranging from 0.4 to 2.2% when low numbers of viruses (49-692 plaque-forming units) were concentrated. In another modification of the membrane adsorption method, Metcalf et al. (1974b) first clarified seawater through three clarifying filters with final adsorption occurring on two sets, each containing a fiberglass depth filter and a flat Cox membrane filter. Sample sizes as large as 155 gal were processed using this system, but again only low recoveries were achieved. Also, in this system viruses tended to adsorb to suspended matter trapped on the clarifying filters when large volumes were processed (Metcalf et al., 1974a).

Payment et al. (1976) reported on the use of pleated filters to concentrate viruses from estuarine waters. They were able to recover seeded poliovirus with an average efficiency of 53% in 378 liter (100 gal) volumes of turbid estuarine water. It is important to note that optimal adsorption of virus in seawater to the Filterite pleated membranes requires a minimum $AlCl_3$ concentration of 0.0015 M.

From our experience, 1000 gal (3780 liters) of tap water and 25-75 gal (95-284 liters) of sewage or seawater can be processed in this manner without difficulty. Because of the lower cost, the fiberglass depth filter (K-27) may be used instead of the 3 μm porosity filter in the processing of tap water and 50-100 gal amounts of sewage and seawater. These methods have been used successfully to isolate naturally occurring enteric viruses from seawater along the upper Texas coast (Gerba et al., 1977a; Goyal et al., 1978). A diagrammatic representation of the concentration scheme is shown in Fig. 4.

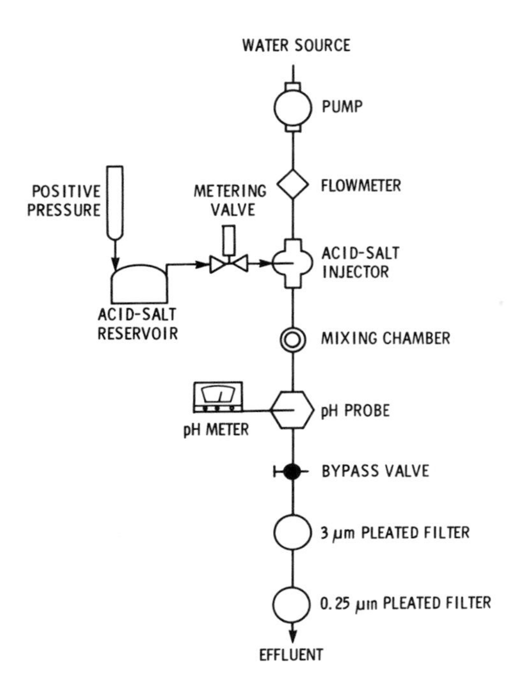

FIGURE 4 Diagrammatic representation of the virus concentrator for use with large volumes of water. (Reproduced by permission from American Society for Microbiology.)

F. Virus Concentration from Sewage

The use of filters for the concentration of virus from raw sewage was first described by Homma et al. (1973). In that study raw sewage was first passed through a series of clarifying filters and adjusted to a pH of 3.5 and $AlCl_3$ was added to enhance virus adsorption to a spun fiberglass depth filter (model K-27). No membrane filters were used in this procedure. Elution of seed virus from the fiberglass filter yielded 81% of the total input virus in a 1 liter eluate. Because of the relatively high concentration of virus normally present in raw sewage, reconcentration was not attempted.

Recent studies by Gerba et al. (1978b) on solid-associated viruses in discharges from activated sludge and trickling filter sewage treatment plants indicated that the percent of solid-associated enteroviruses ranged from 3 to 100% of the total virus in the discharge. The association of viruses with solids (e.g., clays, organic matter, etc.) is known to protect them from inactivation in natural waters (Gerba and Schaiberger, 1975a; Smith et al., 1978) and from the action of disinfectants. Thus, there is a need to have methods available for their detection. Gerba et al. (1978a) demonstrated that poliovirus adsorbed to suspended matter in secondarily treated sewage was eluted easily from the surface of solids by high-pH glycine buffer. This indicated that surface solid-associated viruses in sewage effluents that became entrapped by adsorbent filters would be eluted along with the freely suspended viruses that became adsorbed to the filter and thus would be present in the eluate from the filter.

G. Reconcentration

Eluates obtained from the elution of membrane filters after processing large volumes of water (400-2000 liters) must be reduced to a smaller volume before assay of the sample for viruses by standard tissue culture techniques. The volume of such eluates usually ranges from 1 to 2 liters and must be reconcentrated to a 20-50 ml volume before assay. Techniques that have been used for reconcentration include two-phase separation (Shuval et al., 1969; Hill et al., 1974b), hydroextraction (Cliver, 1967b), precipitation with inorganic salts (Wallis et al., 1972a), continuous-flow ultracentrifugation (Anderson et al., 1969; Scutt, 1971), electroosmosis and forced-flow electrophoresis (Sweet et al., 1971), and readsorption to and elution from smaller-diameter membrane filters (Wallis et al., 1972b, c).

Recently, Katzenelson et al. (1976) described organic flocculation as a second-step concentration (reconcentration) of viruses from water. They used 300 ml of 3% beef extract (pH 9.0) for elution of viruses from the adsorbent filters (Cox AA, M780, 293 mm diameter). Reconcentration of this eluate was carried out by lowering the pH to 3.5 with dropwise addition of 2 N HCl, which resulted in the flocculation of proteins. The sediment obtained after light centrifugation was dissolved in 15 ml of 0.15 M Na_2HPO_4 (pH 9), thus achieving a concentration factor of 20. Table 2 lists different reconcentration methods that have been used by various investigators.

H. Factors Influencing Viral Adsorption and Elution

Membrane filter concentration methodology depends upon the adsorption of viruses onto the filter surface, followed by elution. The adsorption step is brought about by surface-chemical forces, as determined by the surface properties of the virus and membrane and the chemical composition of the suspending solution. Although few studies have been done on the mechanisms of virus adsorption to filters, Mix (1974) has suggested that several mechanisms may be involved. These include hydrophobic bonding between nonpolar aliphatic and aromatic groups on the surfaces of the virus and filter, hydrogen bonding between polar groups on the surfaces of the virus and filter, and salt bridging between negatively charged groups on the surfaces of the virus and filter by adsorbed cations obtained from the solution to be filtered. Factors known to influence adsorption of viruses to filter surfaces are listed in Table 3. Retention or passage of virus through filters is a physicochemical phenomenon that depends upon many variables and uncontrollable factors, such as size of the virus, pore size of the filter, presence of salts, pH, soluble organics, flow rate, and amount of liquid.

Composition of Filters and Their Pore Size

Cliver (1968) compared the virus affinities of a number of filters and found that fluorinated vinyl, acrylic polyvinyl chloride-type polycarbonate, and cellulose triacetate membranes could adsorb viruses, whereas nylon adsorbed significantly less virus. Membranes made of cellulose nitrate, fiberglass, asbestos, aluminum alginate, and cellulose acetate have been shown repeatedly to adsorb viruses

TABLE 2 Methods Used for Reconcentration

Type of water sampled	Initial volume (liters)	Eluent	Reconcentration method	Final volume (ml)	Overall efficiency (%)	Reference
Tap and estuarine water	390	5X nutrient broth, pH 9.0	Aqueous polymer two-phase separation	2-3	67	Hill et al. (1972)
Tap water	378.5	0.05 M glycine, pH 11.5	Adsorption-elution from smaller-diameter disk filters	10	77	Sobsey et al. (1973)
Sewage effluent	390	0.05 M glycine, pH 11.5	Pretreatment of eluate with activated carbon and ion-exchange resin before adsorption to and elution from smaller filters	40	40	Farrah et al. (1976c)
Tap water	1900	0.05 M glycine, pH 10.5	Aluminum hydroxide flocculation	20	40-50	Farrah et al. (1976b)
Estuarine water of low turbidity	390	0.05 M glycine, pH 11.5	Ferric chloride precipitation	10-20	∿100	Payment et al. (1976)
Estuarine water of high turbidity	390	0.05 M glycine, pH 11.5	Aluminum hydroxide flocculation	10-100	50-70	Payment et al. (1976)
Tap water	500	3% beef extract, pH 9	Organic flocculation	15	75	Katzenelson et al. (1976)
Highly turbid estuarine water	190	0.05 M glycine, pH 11.5	Ferric chloride precipitation	15-20	41	Sobsey et al. (1977)
Estuarine water	400	0.05 M glycine, pH 11.5	Aluminum hydroxide flocculation followed by hydroextraction	10-40	50	Farrah et al. (1977a)

TABLE 3 Factors Influencing Virus Adsorption to Filter Surfaces

1. Composition of filter
2. Flow rate
3. Ratio of pore diameter to virus diameter
4. pH
5. Cations
6. Presence of substances competing with virus for adsorption (i.e., proteins)

under the proper conditions (Berg et al., 1971; Wallis et al., 1972a; Sobsey et al., 1977; Gerba et al., 1978a). Recently, positively charged membranes (Zeta-plus, AMF, Cuno) made of cellulose, modified anion-exchange resins, and inorganic filter aids have been shown to have excellent virus adsorption properties at neutral pH (Sobsey and Jones, 1979).

The lack of virus adsorption on the polycarbonate membranes was considered by Mix (1974) to be due to the polar nature of the carbonate bond and the considerably reduced membrane contact area for the Nucleopore-type filter as compared with the Millipore-type filter. The lack of adsorption on cellulose triacetate membranes was due to the lack of any unesterified nucleophilic hydroxyl groups in the cellulose surface and hence little capacity in the cellulose surface for binding cations (Mix, 1974).

The area of the membrane through which a solution is filtered greatly affects the amount of virus adsorbed to a filter. Since there are a finite number of adsorption sites per unit membrane area, virus adsorption decreases with decrease in membrane area (Mix, 1974). The adsorption capacity of a filter surface is so greatly increased by the addition of acid and salts that the adsorption capacity of a filter is never exceeded in practice. In addition to adsorbing to the filter, viruses adsorb to particulate matter and organics (Metcalf et al., 1973) entrapped by the filter during processing of natural waters, thereby actually increasing the number of adsorption sites.

Salts

Viruses are colloidal particles that are negatively charged at pH values near neutrality (Bitton, 1975). The filters are also negatively charged at this pH and hence little adsorption occurs. However, virus does adsorb to filters at neutral pH if salts are present (Wallis et al., 1972a), and adsorption can be enhanced if the pH is acidic and salts are present (Homma et al., 1973; Farrah et al., 1976b). It has also been found that 0.5 mM aluminum chloride is as effective as 50 mM magnesium chloride to enhance virus adsorption (Wallis et al., 1972a). The superiority of aluminum chloride over magnesium chloride was considered to be the result of greater amplification of electrostatic forces developed at adsorbent surfaces at acid pH levels (Metcalf et al., 1974a). Aluminum can actually reverse the charge of a filter's surface from negative to positive at pH values of both 3.5 and 7.0 (Nordin et al., 1967; Carlson et al., 1968). Kessick and Wagner (1978) demonstrated that aluminum reversed the charge on fiberglass-epoxy material at pH 3.5 and at 7.0. They also demonstrated that at an aluminum chloride concentration of 0.005 M, the filter material was positively charged in tap water but negatively charged in secondarily treated sewage effluent. This was thought to be due to the fact that aluminum might have preferentially interacted with soluble organics in the sewage effluent (Kessick and Wagner, 1978).

High concentrations of cations can result in a decrease in virus adsorption to filters. Thus, concentrations of NaCl above 0.4 M have been observed to reduce virus adsorption to cellulose nitrate filters (Millipore) (LaBelle, 1978). Cliver (1965) reported virus elution from filters with 1 M $MgCl_2$.

pH

Control of pH is thought to be of considerable importance in virus adsorption to filters (Mix, 1974). At high pH levels, viruses are strongly negatively charged, whereas at low pH values they are positively charged. Like proteins, viruses have no net charge at their isoelectric point and are susceptible to aggregation and adsorption (Mix, 1974). Poliovirus has two isoelectric points (i.e., at pH 7.1 and 4.5). Below pH 4.5, poliovirus is positively charged, and a marked enhancement of virus adsorption to filters occurs in solutions below this pH if competing organics are not present.

In the absence of organics, poliovirus begins eluting from membrane filters above pH 9.0 and increases to pH 11.5. Above pH 11.5, enteroviruses are rapidly inactivated and elution efficiency decreases. Adenoviruses, rotaviruses, and reoviruses are rapidly inactivated above pH 10.0, and organic eluents should be used when attempts are made to isolate these viruses.

Using microelectrophoresis, Kessick and Wagner (1978) studied the surface charge characteristics of the filter surfaces and compared them to those of viruses under a variety of conditions. They measured the electrophoretic mobilities of filter material particles in aqueous suspensions as a function of pH, ionic strength, salt type, and protein concentrations and found out that in the presence of KCl/HCl, the filter materials exhibited a net negative charge at pH 2-7 and that the least negative charge was found to occur at low pH values. This was thought to be due to the presence of ionized carboxyl groups on the cellulosic surface. Such carboxyl groups are known to be present on purified natural cellulose (Sookne and Harris, 1954; Kessick and Wagner, 1978). Alternatively, it was thought that many nonesterified hydroxyl groups contained within the cellulosic material may serve as good hydrogen bonders and perhaps form hydrogen bonds with hydroxyl ions from the solution (Kessick and Wagner, 1978).

In fiberglass-epoxy material, however, the net negative charge was greatest at pH 5, which may be due to ionization of -SiOH groups on the surface of fiberglass. Because viruses could be efficiently adsorbed to filters under acidic conditions [when the virus has a net positive charge as demonstrated by Mandel (1971) and Mix (1974)], it was suggested that the negative charge on the filters attracts them (Kessick and Wagner, 1978).

Effect of Organics

The presence of organic and proteinaceous substances in the virus suspension medium interferes with virus adsorption to membranes, presumably by competing for membrane adsorption sites. These viral adsorption interfering substances have been referred to as MCC (membrane coating components) (Wallis and Melnick, 1967a) and can be effectively removed by passing the liquid sample through an anion-resin column [Dowex 1-X8 (Cl^-), 100-200 mesh] (Wallis and Melnick, 1967a).

Virus is unable to adsorb to membrane surfaces that have been precoated with competitive adsorbents such as serum or whey proteins (Mix, 1974) or polyvinylpyrrolidone (Yoshino and Morishima, 1971). Similarly, viruses already adsorbed to the membrane surface can be eluted with protein at high pH (Berg, 1967; Wallis and Melnick, 1967a, b). In addition to proteins, chelating and wetting agents also can exchange for the virus and elute it from the filter surface. These high-molecular-weight adsorbents inhibit virus adsorption to membranes by forming bonds with the filter surface and making it difficult for a virus to displace them even if the virus is capable of adsorbing strongly to the membrane surface (Mix, 1974). Hence, elution following adsorption requires significant alterations in the chemical environment to enable simultaneous disruption of the multiple virus adsorption bonds (Mix, 1974).

Since organics can compete with viruses for adsorption onto membrane filters, they have been used to elute viruses from filter surfaces (Table 4). Basic amino acids and casein (Bitton et al., 1978), nutrient broth, yeast extract, beef extract, fetal bovine serum (Konowalchuk and Speirs, 1971; Hill et al., 1974a; Katzenelson et al., 1976), Tween 80 (Gerba, 1978), and tryptose phosphate broth (Farrah et al., 1978b) have been used to elute viruses from filter surfaces. Recommended minimum eluent volumes for different size filters are given in Table 5. Some membrane-bound viruses will elute at neutral pH with most of these organics, but best results are obtained at a pH between 9.0 and 10.0. At this pH, elution is

TABLE 4 Substances Used for Elution of Viruses from Membrane Filters

Eluent	Elution pH	Remarks	Reference
Glycine	11.5	pH must be adjusted to 9.0 or below within 5 min of collection to prevent viral inactivation	Farrah et al. (1976b)
Glycine	10.5	To obtain maximum elution at this pH, the eluent must be passed through the filter at least five times	Farrah et al. (1976b)
Beef extract	9.0	Volume reduction upon reconcentration is not as great as with glycine	Katzenelson et al. (1976)
Nutrient broth, 5X	9.0		Hill et al. (1974a)
Tryptose phosphate broth	9.0		Farrah et al. (1978b)
Arginine, lysine	9.0		Farrah and Bitton (1978)

TABLE 5 Recommended Minimum Eluent Volumes

Filter diameter (mm)	Filter area (cm^2)	Eluent volume[a] (ml)
13	0.8	0.4
25	3.9	2.0
47	13.8	8.0
90	45.5	25.0
142	97.0	50.0
293	468.0	240.0

[a]This is the volume of pH 11.5 glycine-NaOH buffer required to efficiently elute virus off the membrane surface indicated. The amount of pH 1.0 glycine-HCl buffer needed to neutralize the eluate increases the final eluate volume by approximately 20%.

similar to that obtained with pH 11.5 glycine buffer. A handicap with the use of organic eluents is that volume reduction during reconcentration is not as great as with high-pH glycine eluents.

Flow Rate

Virus adsorption to filters decreases with increase in flow rate through a filter (Scutt, 1971), because a minimum contact time is required for an electrostatic bond to form between the virus and the adsorption site. Also, forces tangential to internal membrane surfaces arising from fluid velocity and shear stress should not be greater than adsorptive forces holding the virus particle to the membrane surface. The flow rate at which virus breakthrough would occur is dependent on factors influencing the strength of the electrostatic forces interacting between the virus and membrane. Thus, Homma et al. (1973) observed that at a flow rate of up to 0.76 gal/min, 96% of the virus present in sewage was adsorbed by the glass filter. As the flow rate increased to 1.66 and 2.5 gal/min, the rate of virus adsorption decreased to 76 and 66%, respectively.

I. Generalized Methodology for Concentration of Viruses from Water

It is difficult to suggest any standard membrane filter methodology for the concentration of viruses from water because of the rapid development of new techniques in this area. A tentative standard method for virus concentration from finished tap waters appeared in the 15th edition of *Standard Methods for the Examination of Water and Wastewater* (American Public Health Association, 1978).

A generalized methodology for virus concentration from tap water, seawater, and sewage similar to that which will appear in the next edition of that book, is given in Fig. 5. This figure incorporates several of the commonly used approaches for concentration from large volumes (100-1900 liters). A listing of the different animal viruses for which membrane filter concentration has been reported is given in Table 6.

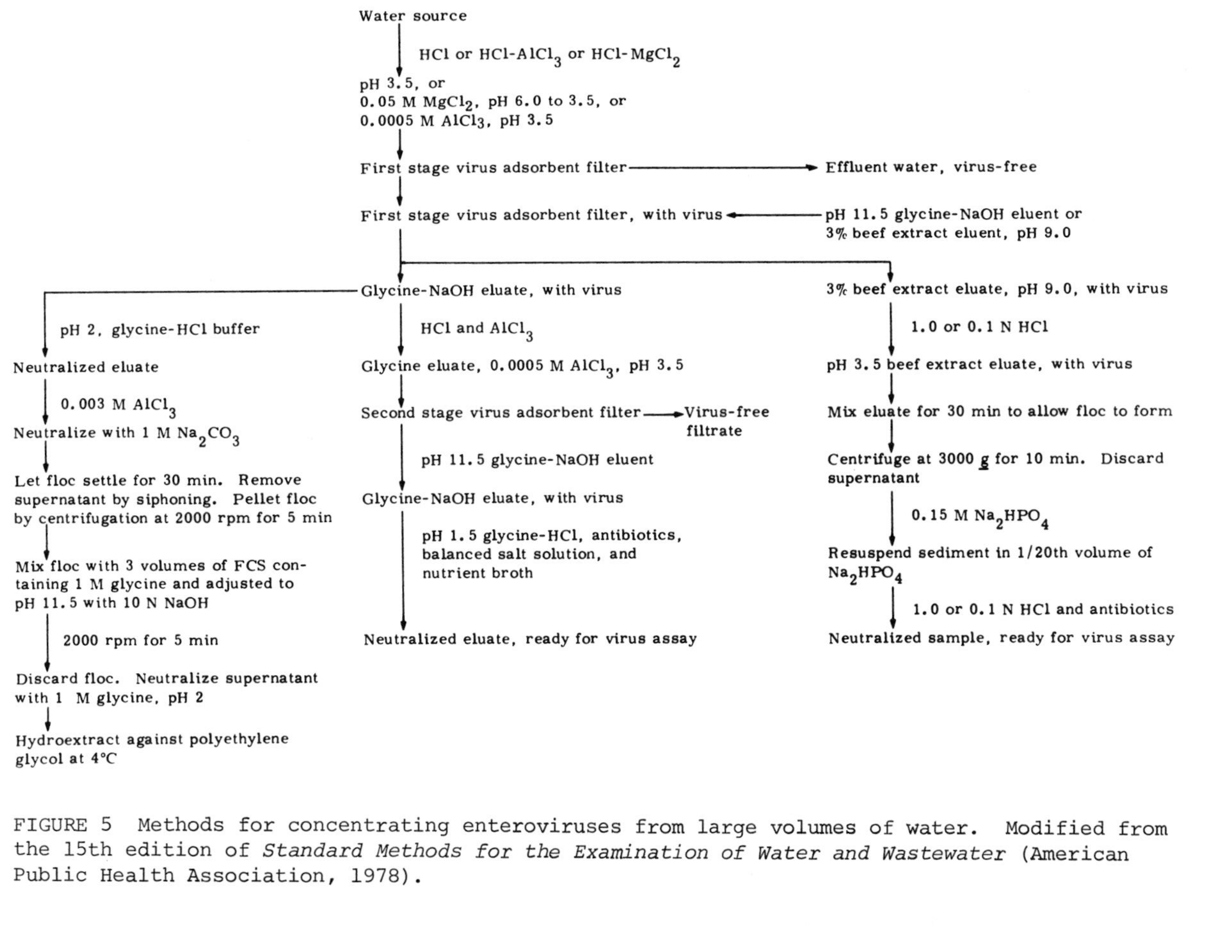

FIGURE 5 Methods for concentrating enteroviruses from large volumes of water. Modified from the 15th edition of *Standard Methods for the Examination of Water and Wastewater* (American Public Health Association, 1978).

TABLE 6 Methods for Concentration of Viruses by Membrane Filters

Virus	Reference, remarks
Enteroviruses (polio, coxsackie, echo)	Gerba et al. (1978a) Katzenelson et al. (1976)
Adenovirus	Fields and Metcalf (1975)
Reovirus	The membrane filter method has not been tested for efficiency of recovery, but isolation of reovirus from sewage has been reported by Gilbert et al. (1976) using the method of Sobsey et al. (1973)
Rotavirus (simian virus SA11)	Farrah et al. (1978b)
Herpesvirus	Tschider et al. (1974)
Hepatitis A virus	This virus can be concentrated using the methods described by Gerba et al. (1978a) for enteroviruses (Gerba et al., 1980)
Parvovirus	Minute virus of mice (MVM) was adsorbed to Zeta-plus filters at pH 7.5 and eluted at pH 10 with beef extract with an average recovery of 80% (Putala and Sobsey, 1978)

Because of the complexity of the methodology for virus concentration from large volumes of water, an evaluation of the efficiency of the method using a vaccine strain of poliovirus should be attempted by anyone not familiar with the technique, before attempts are made to isolate naturally occurring viruses. In addition, because virus concentration methodology is so dependent on the nature of the water being tested, it is beneficial to first evaluate recovery efficiency with artificially seeded virus to obtain an idea of the efficiency of virus recovery from the water being examined.

II. OTHER USES

A. Detection of Viruses in Landfill Leachates, Marine Sediments, and Soil

In addition to the direct concentration of viruses from water, membrane filter adsorption-elution methodology can be used to concentrate viruses from leachates of solid waste landfill sites and from solutions used to elute viruses adsorbed to soil and sediment. The methodology for concentrating enteroviruses from leachates and soil-sediment eluates is essentially identical to that for tap water and involves adjustment to pH 3.5 and addition of $AlCl_3$ (Sobsey et al., 1974).

In the elution of enteroviruses from soil and marine sediment, 0.05 M edetic acid (EDTA) in high-pH glycine buffer is used (Gerba et al., 1977b; Hurst et al., 1979). To obtain good adsorption of the virus onto membrane filters from eluent, it is necessary to add 0.06 M $AlCl_3$ at pH 3.5 to overcome the chelating effect of EDTA.

B. Concentration of Viruses from Cell Culture Harvests

Viruses in cell culture harvests often need to be purified and concentrated. In 1967, Wallis and Melnick (1967a) reported a method for concentration of enteroviruses from tissue culture harvests by adsorption to and elution from membrane filters. Ordinarily, viruses present in cell culture fluids do not adsorb to membrane filters because of the presence of membrane-coating components (MCC) which interfere with virus adsorption. These authors were able to remove MCC from tissue culture fluids by using a series of treatments, including extraction with chloroform, precipitation with protamine sulfate, and ion-exchange chromatography. Viruses in such treated cell culture harvests were found to be readily adsorbable to nitrocellulose membranes. Because this method was limited to rather small volumes of virus harvests, Henderson et al. (1976) reported a modification that used a combination of precipitation with cationic detergent and ion-exchange chromatography to remove the MCC from cell culture harvests. Virus was then adsorbed to epoxy-fiberglass-asbestos membrane filters and recovered by elution with a small volume of high-pH buffer.

A simpler method for removing MCC from cell culture harvests (Farrah et al., 1978b) consists of blending the harvests with fluorocarbon (trichlorotrifluoroethane; Freon 113) and adsorbing the virus in acidified harvests to epoxy-fiberglass filters. Using this procedure, enteroviruses and a simian rotavirus (SA11) were concentrated 100-fold from 400 ml of cell culture harvest with an efficiency of 99%.

C. Filtration, Sizing of Viruses, Removal of Aggregates, and Collection of Solid-Associated Viruses

Filtration of viruses through membrane filters can be used to remove undesired contaminants (i.e., bacteria, cell debris, particulates, etc.) and viral aggregates and for sizing of new viral agents. Unlike the use of membrane filters for virus concentration, it is desirable to prevent virus adsorption to the membrane surface for the foregoing purposes. Because of the affinity of most membrane filters for viruses, it is first necessary to treat them to minimize virus adsorption. Virus adsorption is most effectively prevented by treatment of the filter with a solution of proteinaceous material or a detergent. Animal sera are most commonly used, although solutions of gelatin (Cliver, 1965), polyvinylpyrrolidone (a plasma expander used in blood transfusion) (Yoshino and Morishima, 1971), sodium lauryl sulfate (Ver et al., 1968), and Tween 80 (Wallis et al., 1972a) have also been used. Tween 80 and sodium lauryl sulfate cannot be used with viruses containing a lipid membrane, such as herpesviruses, because of their sensitivity to inactivation by detergents. Viral adsorption can be minimized by incorporation of these membrane-coating substances (MCS) into the solution being filtered or by pretreatment of the filter by passage of the MCS through the filters or soaking them in solutions containing the MCS. Of these methods, adsorption is best minimized by passage of a solution of MCS through a filter immediately before viral filtration. Virus adsorption to some types of membrane filters is negligible when the virus is suspended in distilled water, and this technique has also been used for virus filtration (Cliver, 1965, 1968).

The same procedure as described above can be used to remove viral aggregates from suspension. Removal of aggregates is useful in the preparation of monodispersed virus suspensions or in determining the extent of viral aggregation in a virus suspension (Ver et al., 1968; Gerba and Schaiberger, 1975b).

Treated membrane filters have also been used to demonstrate the adsorption of viruses onto suspended solids in seawater (Gerba et al., 1978b) and in the recovery of solid-associated viruses in wastewater (Gerba et al., 1978a). Gerba et al. (1978b) collected sewage solids on treated membrane filters and recovered the solid-associated viruses trapped on the filters by elution of the viruses from the solids with 0.05 M, pH 11.5 glycine buffer. In an alternative procedure, solids collected on the filter were scraped off with a spatula into trypticase soy broth before direct assay. Using this procedure, the authors were able to determine the size of the solids viruses were associated with and the percentage of the total virus population associated with solids in secondarily treated wastewater

REFERENCES

American Public Health Association (1978). *Standard Methods for the Examination of Water and Wastewater,* 15th ed. American Public Health Association, Washington, D.C.

Anderson, N. G., Waters, D. A., Nunley, C. E., Gibson, R. F., Schilling, R. M., Denny, E. C., Cline, G. B., Babelay, E. F., and Perardi, T. E. (1969). K-series centrifuges. I. Development of the K-11 continuous-sample-flow-with-banding centrifuge system for vaccine purification. *Anal. Biochem. 32:* 460–494.

Berg, G. (1967). Introduction. In *Transmission of Viruses by the Water Route,* G. Berg (Ed.). Interscience, New York, pp. 1–2.

Berg, G., Dahling, D. R., and Berman, D. (1971). Recovery of small quantities of viruses from clean waters on cellulose nitrate membrane filters. *Appl. Microbiol. 22:* 608–614.

Bitton, G. (1975). Adsorption of viruses onto surfaces in soil and water. *Water Res. 9:* 473–484.

Bitton, G., Feldberg, B. N., and Farrah, S. R. (1978). Personal communication.

Carlson, G., Woodard, F., Wentworth, D., and Sproul, O. (1968). Virus inactivation on clay particles in natural waters. *J. Water Pollut. Control Fed. 40:* R89–R106.

Cliver, D. O. (1965). Factors in the membrane filtration of enteroviruses. *Appl. Microbiol. 13:* 417–425.

Cliver, D. O. (1967a). Enterovirus detection by membrane chromatography. In *Transmission of Viruses by the Water Route,* G. Berg (Ed.). Interscience, New York, pp. 139–149.

Cliver, D. O. (1967b). Detection of enteric viruses by concentration with polyethylene glycol. In *Transmission of Viruses by the Water Route,* G. Berg (Ed.). Interscience, New York, pp. 109–121.

Cliver, D. O. (1968). Virus interactions with membrane filters. *Biotech. Bioeng. 10:* 877–889.

Elford, W. J. (1931). A new series of graded collodion membranes suitable for general bacteriological use, especially in filterable virus studies. *J. Pathol. Bacteriol. 34:* 505–535.

Farrah, S. R., and Bitton, G. (1978). Personal communication.

Farrah, S. R., Goyal, S. M., Gerba, C. P., Wallis, C., and Shaffer, P. T. B. (1976a). Characteristics of humic acid and organic compounds concentrated from tapwater using the Aquella Virus Concentrator. *Water Res. 10:* 897–901.

Farrah, S. R., Gerba, C. P., Wallis, C., and Melnick, J. L. (1976b). Concentration of viruses from large volumes of tapwater using pleated membrane filters. *Appl. Environ. Microbiol. 31:* 221–226.

Farrah, S. R., Wallis, C., Shaffer, P. T. B., and Melnick, J. L. (1976c). Reconcentration of poliovirus from sewage. *Appl. Environ. Microbiol. 32:* 653–658.

Farrah, S. R., Goyal, S. M., Gerba, C. P., Wallis, C., and Melnick, J. L. (1977a). Concentration of enteroviruses from estuarine water. *Appl. Environ. Microbiol. 33:* 1192–1196.

Farrah, S. R., Gerba, C. P., Goyal, S. M., Wallis, C., and Melnick, J. L. (1977b). Regeneration of pleated filters used to concentrate enteroviruses from large volumes of tapwater. *Appl. Environ. Microbiol. 33:* 308–311.

Farrah, S. R., Goyal, S. M., Gerba, C. P., Wallis, C., and Melnick, J. L. (1978a). Concentration of poliovirus from tap water onto membrane filters with aluminum chloride at ambient pH levels. *Appl. Environ. Microbiol. 35:* 624–626.

Farrah, S. R., Goyal, S. M., Gerba, C. P., Conklin, R. H., Wallis, C., Melnick, J. L., and DuPont, H. L. (1978b). A simple method for concentration of enteroviruses and rotaviruses from cell culture harvests using membrane filters. *Intervirology 9:* 56–59.

Fields, H. A., and Metcalf, T. G. (1975). Concentration of adenovirus from seawater. *Water Res. 9:* 357–364.

Gerba, C. P. (1978). Personal observations.

Gerba, C. P., and Schaiberger, G. E. (1975a). Effect of particulates on the survival of virus in seawater. *J. Water Pollut. Control Fed. 47:* 93–103.

Gerba, C. P., and Schaiberger, G. E. (1975b). Aggregation as a factor in loss of viral titer in seawater. *Water Res. 9:* 567–571.

Gerba, C. P., Goyal, S. M., Smith, E. M., and Melnick, J. L. (1977a). Distribution of viral and bacterial pathogens in a coastal canal community. *Mar. Pollut. Bull. 8:* 279–282.

Gerba, C. P., Smith, E. M., and Melnick, J. L. (1977b). Development of a quantitative method for the detection of enteroviruses in estuarine sediments. *Appl. Environ. Microbiol. 34:* 158–163.

Gerba, C. P., Farrah, S. R., Goyal, S. M., Wallis, C., and Melnick, J. L. (1978a). Concentration of enteroviruses from large volumes of tap water, treated sewage, and seawater. *Appl. Environ. Microbiol. 35:* 540–548.

Gerba, C. P., Stagg, C. H., and Abadie, M. G. (1978b). Characterization of sewage solid-associated viruses and behavior in natural waters. *Water Res. 12:* 805–812.

Gerba, C. P., Hollinger, F. B., Farrah, S. R., Goyal, S. M., and Melnick, J. L. (1980). Concentration and detection of hepatitis A in sewage. (In preparation.)

Gilbert, R. G., Gerba, C. P., Rice, R. C., Bouwer, H., Wallis, C., and Melnick, J. L. (1976). Virus and bacteria removal from wastewater by land treatment. *Appl. Environ. Microbiol. 32:* 333–338.

Goyal, S. M., Gerba, C. P., and Melnick, J. L. (1978). Prevalence of human enteroviruses in coastal canal communities. *J. Water Pollut. Control Fed. 50:* 2247–2256.

Henderson, M., Wallis, C., and Melnick, J. L. (1976). Concentration and purification of enteroviruses by membrane chromatography. *Appl. Environ. Microbiol. 32:* 689–693.

Hill, W. F., Jr., Akin, E. W., Benton, W. H., and Metcalf, T. G. (1972). Virus in water. II. Evaluation of membrane cartridge filters for recovering multiplicities of poliovirus from water. *Appl. Microbiol. 23: 880-888.*

Hill, W. F., Jr., Akin, E. W., Benton, W. H., Mayhew, C. J., and Metcalf, T. G. (1974a). Recovery of poliovirus from turbid estuarine water on microporous filters by the use of Celite. *Appl. Microbiol. 27:* 506–512.

Hill, W. F., Jr., Akin, E. W., Benton, W. H., Mayhew, C. J., and Jakubowski, W. (1974b). Apparatus for conditioning unlimited quantities of finished waters for enteric virus detection. *Appl. Microbiol. 27:* 1177–1178.

Hill, W. F., Jr., Jakubowski, W., Akin, E. W., and Clarke, N.A. (1976). Detection of virus in water: sensitivity of the tentative standard method for drinking water. *Appl. Environ. Microbiol. 31:* 254–261.

Homma, A., Sobsey, M. D., Wallis, C., and Melnick, J. L. (1973). Virus concentration from sewage. *Water Res. 7:* 945–950.

Hurst, C. J., and Gerba, C. P. (1979). Development of a quantitative method for the detection of enteroviruses in soil. *Appl. Environ. Microbiol. 37*: 626–632.

Jakubowski, W., Hoff, J. C., Anthony, N. C., and Hill, W. F., Jr. (1974). Epoxy-fiberglass adsorbent for concentrating viruses from large volumes of potable water. *Appl. Microbiol. 28:* 501–502.

Katzenelson, E., Fattal, B., and Hostovesky, T. (1976). Organic flocculation: an efficient second-step concentration method for the detection of viruses in tap water. *Appl. Environ. Microbiol. 32:* 638–639.

Kessick, M. A., and Wagner, R. A. (1978). Electrophoretic mobilities of virus adsorbing filter materials. *Water Res. 12:* 263–268.

Konowalchuk, J., and Speirs, J. I. (1971). An evaluation of three agents for eluting adsorbed enterovirus from Millipore filters. *Can. J. Microbiol. 17:* 1351–1355.

LaBelle, R. (1978). Personal communication.

Mandel, B. (1971). Characterization of type 1 poliovirus by electrophoretic analysis. *Virology 44:* 554–568.

Metcalf, T. G. (1961). Use of membrane filters to facilitate the recovery of virus from aqueous suspensions. *Appl. Microbiol. 9:* 376–379.

Metcalf, T. G., Wallis, C., and Melnick, J. L. (1973). Concentration of viruses from seawater. In *Advances in Water Pollution Research* (Proc. 6th Int. Conf. Water Pollut. Res., Jerusalem, June 1972), S. H. Jenkins (Ed.). Pergamon Press, Oxford, pp. 109–115.

Metcalf, T. G., Wallis, C., and Melnick, J. L. (1974a). Environmental factors influencing isolation of enteroviruses from polluted surface waters. *Appl. Microbiol. 27:* 920–926.

Metcalf, T. G., Wallis, C., and Melnick, J. L. (1974b). Virus enumeration and public health assessments in polluted surface water contributing to transmission of viruses in nature. In *Virus Survival in Water and Wastewater Systems,* J. F. Malina and B. P. Sagik (Eds.). Center for Research in Water Resources, Austin, Tex., pp. 57–70.

Mix, T. W. (1974). The physical chemistry of membrane-virus interaction. *Dev. Ind. Microbiol. 15:* 136–142.

Nordin, J. S., Tsuchiya, H. M., and Fredrickson, A. G. (1967). Interfacial phenomena governing adhesion of chlorella to glass surfaces. *Biotech. Bioeng. 9:* 545–558.

Payment, P., Gerba, C. P., Wallis, C., and Melnick, J. L. (1976). Methods for concentrating viruses from large volumes of estuarine water on pleated membranes. *Water Res. 10:* 893–896.

Putala, W. A., and Sobsey, M. D. (1978). Evaluation of the tentative standard method for detecting parvoviruses in tapwater. *Abstr. Annu. Meet. Am. Soc. Microbiol.,* p. 199.

Rao, N. U., and Labzoffsky, N. A. (1969). A simple method for the detection of low concentration of viruses in large volumes of water by the membrane filter technique. *Can. J. Microbiol. 15:* 399–403.

Scutt, J. E. (1971). Virus retention by membrane filters. *Water Res. 5:* 183–185.

Shuval, H. I., Fattal, B., Cymbalista, S., and Goldblum, N. (1969). The phase-separation method for the concentration and detection of viruses in water. *Water Res. 3:* 225–240.

Smith, E. M., Gerba, C. P., and Melnick, J. L. (1978). Role of sediments in the persistence of enteroviruses in the estuarine environment. *Appl. Environ. Microbiol. 35:* 685–689.

Sobsey, M. D. (1976). Methods for detecting enteric viruses in water and wastewater. In *Viruses in Water,* G. Berg, H. L. Bodily, E. H. Lennette, J. L. Melnick, and T. G. Metcalf (Eds.). American Public Health Association, Washington, D.C., pp. 89–127.

Sobsey, M. D., and Jones, B. L. (1979). Concentration of poliovirus from tapwater using positively charged microporous filters. *Appl. Environ. Microbiol. 37:* 588–595.

Sobsey, M. D., Wallis, C., Henderson, M., and Melnick, J. L. (1973). Concentration of enteroviruses from large volumes of water. *Appl. Microbiol. 26:* 529–534.

Sobsey, M. D., Wallis, C., and Melnick, J. L. (1974). Development of methods for detecting viruses in solid waste landfill leachates. *Appl. Microbiol. 28:* 232–238.

Sobsey, M. D., Gerba, C. P., Wallis, C., and Melnick, J. L. (1977). Concentration of enteroviruses from large volumes of turbid estuary water. *Can. J. Microbiol. 23:* 770–778.

Sookne, A. M., and Harris, M. (1954). Chemical nature of cellulose and its derivatives--base exchange properties. In *Cellulose and Cellulose Derivatives,* vol. 1, E. Ott, H. M. Supurlin, and M. W. Grafflin (Eds.). Interscience, New York, pp. 208–215.

Sweet, B. H., McHale, J. S., Hardy, K. J., and Klein, E. (1971). Concentration of virus from water by electro-osmosis and forced flow electrophoresis. *Prep. Biochem. 1:* 77–89.

Tschider, S. R., Berryhill, D. L., and Schipper, I. A. (1974). Membrane concentration of infectious bovine rhinotracheitis virus from water. *Appl. Microbiol. 28:* 1030–1032.

Ver, B. A., Melnick, J. L., and Wallis, C. (1968). Efficient filtration and sizing of viruses with membrane filters. *J. Virol. 2:* 21–25.

Wallis, C., and Melnick, J. L. (1967a). Concentration of enteroviruses on membrane filters. *J. Virol. 1:* 472–477.

Wallis, C., and Melnick, J. L. (1967b). Concentration of viruses from sewage by adsorption on Millipore membranes. *Bull. World Health Organ. 36:* 219–225.

Wallis, C., and Melnick, J. L. (1967c). Concentration of viruses on aluminum phosphate and aluminum hydroxide precipitates. In *Transmission of Viruses by the Water Route,* G. Berg (Ed.). Interscience, New York, pp. 120–138.

Wallis, C., Henderson, M., and Melnick, J. L. (1972a). Enterovirus concentration on cellulose membranes. *Appl. Microbiol. 23:* 476–480.

Wallis, C., Homma, A., and Melnick, J. L. (1972b). Apparatus for concentrating viruses from large volumes. *J. Am. Water Works Assoc. 64:* 189–196.

Wallis, C., Homma, A., and Melnick, J. L. (1972c). A portable virus concentrator for testing water in the field. *Water Res. 6:* 1249–1256.

Yoshino, K., and Morishima, T. (1971). Filtration of viruses through Millipore membranes coated with polyvinyl-pyrrolidone. *Arch. Ges. Virusforsch. 35:* 399–401.

10

APPLICATION OF EPIFLUORESCENCE MICROSCOPY TO THE ENUMERATION OF AQUATIC BACTERIA CONCENTRATED ON MEMBRANE FILTERS

GILL G. GEESEY California State University, Long Beach, California

J. WILLIAM COSTERTON University of Calgary, Calgary, Alberta, Canada

I. INTRODUCTION

The estimation of bacterial numbers and biomass has always been an important step in evaluating the microbiology of natural waters from both a public health and an ecological point of view. In the past, microbiological determinations have been based on direct counting, culturing, or chemical techniques. These methods, however, have certain limitations when applied to some natural aquatic systems. Classical direct enumeration of stained bacteria is often difficult in waters containing large amounts of particulate material due to nonspecific staining of the detritus (Jannasch and Jones, 1959). It is also sometimes difficult to detect those bacteria attached to particles. Cultural methods have always produced underestimations of the total numbers of bacteria present due to the selective nature of the growth medium. Plate count methods also fail to reveal the numbers of bacteria that occur in aggregates or microcolonies (Rodina, 1967). Chemical methods, such as the determination of cellular ATP, are often an estimation of algal rather than bacterial biomass in productive waters (Jassby, 1975). However, the recent application of an assay for lipopolysaccharide, a cell wall component of gram-negative bacteria, holds much promise as an accurate chemical means of estimating bacterial biomass in natural waters (Watson et al., 1977). This method alone, however, provides little information on the size and morphology of microorganisms.

Recent technological advances in fluorescent illumination systems (Kraft, 1975) has led to the successful application of a direct microscopic method for routine enumeration of bacteria in a wide variety of aquatic environments. By combining epiilumination with fluorescence microscopy, many limitations associated with the classical direct enumeration methods are overcome. In view of its recent introduction to the field of aquatic microbiology, we feel justified in acquainting the reader with the design of the epifluorescent microscope before proceeding with considerations of sample preparation.

In spite of the recent application of epifluorescence microscopy to aquatic microbiology, a number of modifications to the technique have appeared in the literature (Jones and Simon, 1975; Hobbie et al., 1977). Unfortunately, rigorous comparisons between some of the methods have not yet been carried out. Thus, it is not possible to advocate one protocol over another at this time. Many modifications, though, address the more important considerations associated with the epifluorescence method, and in fact, arose in response to some of the problems that have been encountered when employed for bacterial enumeration studies in natural waters. Hopefully, the following discussion will expedite more widespread use of the technique.

II. PRINCIPLES OF EPIFLUORESCENCE MICROSCOPY

The principle of this new microscopic method is based on an optical system that transmits excitation energy (short-wave radiation) to a specimen containing a fluorochrome and collects the fluorescent light emitted by the specimen (long-wave radiation) for observation. Excitation energy is generated by a light source housed between the ocular and objective lenses of the microscope (Fig. 1). The light

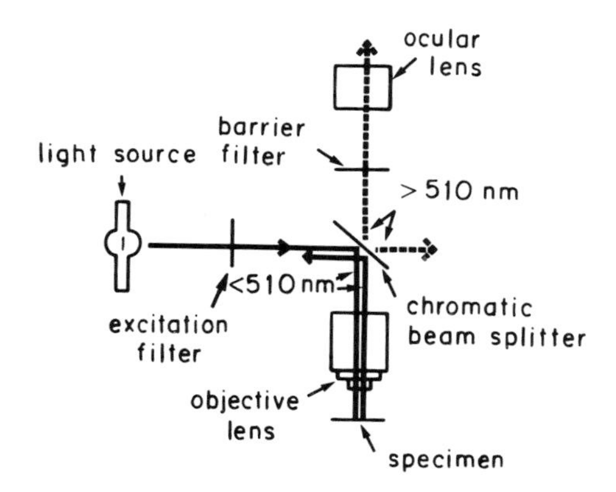

FIGURE 1 Schematic diagram of light pathway in epifluorescence microscopy.

source must be compatible with the fluorochrome used to "stain" the bacterial cells. Acridine orange (AO), a commonly employed stain, is excited by light of wavelengths between 450 and 490 nm. Since xenon and mercury vapor lamps are particularly efficient emitters of light in this range of wavelengths, they provide an excellent source of excitation energy for AO fluorescence. A 100 W halogen lamp has also been used successfully with the Zeiss epifluorescent microscope (Hobbie et al., 1977).

The light is conditioned by an excitation filter which is transparent to wavelengths that stimulate fluorescence but is suppressive to other wavelengths (Fig. 1). The light is further conditioned by a chromatic beam splitter (Fig. 1). Excitation wavelengths intercepted from the excitation filter are reflected toward the specimen while longer wavelengths are allowed to pass through the beam splitter into a collector or a window for lamp centering control. The excitation light is then focused on the specimen by the objective lens, which takes the place of a condenser lens.

Fluorescent light emitted by the specimen is collected by the objective lens and because of its longer wavelength, passes through the beam splitter and into the ocular tube. The beam splitter thus promotes the efficient separation of excitation and fluorescent light needed for successful viewing of the stained specimen. For AO-stained specimens, optimal separation is achieved by using a beam splitter capable of deflecting wavelengths less than 510 nm while transmitting longer wavelengths (acridine orange fluoresces in the green and red region of the spectrum). Before reaching the ocular lens, the fluorescent light is intercepted by a barrier filter, which further suppresses background illumination. For observation of fluorochrome-stained bacteria, a 100X oil-immersion objective lens with high numerical aperature and a 10X ocular lens are recommended for enumeration studies. Since reflected light is used in this system, problems of specimen thickness are minimized. Samples of AO-stained bacteria retained on membrane filter surfaces can therefore be seen as brightly fluorescing cells against a dark background.

III. STAINS

A number of fluorochromes have been tested for their effectiveness in staining bacteria for epifluorescence microscopic enumeration. AO has received the widest application, primarily due to the intense and prolonged fluorescence that it produces (Strugger, 1948). A final concentration of 10 mg/ml AO has been recommended for optimum staining of bacteria recovered from water (Bell and Dutka, 1972; Zimmerman and Meyer-Reil, 1974). The fluorochrome solution should be filtered just prior to use, as the stain may become contaminated with bacteria during storage.

Acridine-based dyes have been shown to react strongly with nucleic acids in living cells (Yamabe, 1973). Studies have revealed, however, that chemically fixed and heat-fixed cells can also exhibit AO fluorescence, so that the technique cannot necessarily be used as a measure of viable cells (Jones and Simon, 1975; Hobbie et al., 1977).

As mentioned above, acridine orange can produce either a green or a red fluorescence, depending on the nature of the nucleic acids with which it reacts. This has created a controversy in the past, with some investigators suggesting that only green cells be counted whereas others recommended that red cells be counted. Hobbie et al. (1977) postulated that the color differences were related to the DNA/RNA ratio that exists in a bacterium. Because the physiological significance of the color of fluorescence is not yet completely understood, both red and green cells are enumerated for an estimate of total bacterial numbers (Meyer-Reil, 1977; Geesey et al., 1978). Clearly, more studies need to be conducted before this problem is resolved.

Jones (1974) and Jones and Simon (1975) found that staining bacteria with Euchrycine-2GNX produced higher cell counts than did subsamples stained with AO. Daley and Hobbie (1975) suggested, however, that differences arising from the use of various acridine-based dyes were less important than considerations concerning the choice of light source and filter combination for a particular fluorochrome. For AO-stained specimens, they recommended using a Zeiss Standard 18 microscope fitted with an IV FL epifluorescence condenser, a 100 W halogen lamp, a filter combination that passes 450-490 nm blue light, a 510 beam splitter, and an LP 528 barrier filter. An HBO 50 mercury burner has also provided satisfactory results when combined with a KP 490 exciter filter, 510 beam splitter, and an LP 520 barrier filter (Geesey et al., 1978).

IV. SAMPLE PREPARATION

The ability to detect individual fluorochrome-labeled cells on a membrane filter provides an excellent opportunity to estimate the number of bacteria concentrated from water samples by filtration techniques. Several variations in sample preparation have evolved for the direct enumeration of bacteria by epifluorescence microscopy. Bell and Dutka (1972) and Zimmerman and Meyer-Reil (1974) concentrated the bacteria on the membrane before staining. A fixation step was included to stabilize the cells after they had been collected on the membrane. Bell and Dutka (1972) used 2 N hydrochloric acid, whereas Zimmerman and Meyer-Reil (1974) employed a

2% formaldehyde solution. A freshly prepared 5% glutaraldehyde solution has also been used for fixation (Geesey et al., 1978). In the latter, fixation was carried out by transferring the membrane containing the bacteria to a sterile plastic petri dish containing a cotton ball saturated with the fixative. Samples prepared in this manner were rapidly fixed and easily transported and stored for later staining and counting. This method is particularly useful when samples are collected in remote waters where there is a delay between sample collection and examination. The samples could be stored for at least 2 weeks without noticeable reduction in counts. Staining was performed by placing the membrane in a filter apparatus and overlaying with fluorochrome solution. After an appropriate period (usually 2 min), the stain was removed by applying negative pressure. When prepared by this method, the majority of bacteria fluoresced orange.

Francisco et al. (1974), Jones and Simon (1975), and Daley and Hobbie (1975) added the fluorochrome directly to the water sample before concentrating the bacteria on membrane filters. Darkened Millipore and Sartorius membrane filters were found to be most suitable for this purpose. Daley and Hobbie (1975) also found that by adding fixative directly to the water samples (2% formaldehyde final concentration), bacteria were stable on slide mounts for 2 weeks. Samples prepared in this manner generally exhibited green fluorescence.

V. EXAMINATION

Whereas Zimmerman and Meyer-Reil (1974) dried the membranes before mounting on glass slides, several investigators have stressed the importance of examining "wet" membrane filters (Jones, 1974; Hobbie et al., 1977). However, no comprehensive studies have yet been conducted to determine what effect drying has on cell counts. It is necessary to saturate the membrane with either a mounting medium (Zimmerman and Meyer-Reil, 1974) or a nonfluorescing, low-viscosity immersion oil (Hobbie et al., 1977). The membranes are mounted on a microscope slide, saturated with immersion oil, overlaid with a glass cover slide, and viewed under oil. The number of fluorescent bacteria observed in a number of randomly chosen microscope fields are used to estimate the total number of cells in the water sample according to the formula

$$N = \frac{n \times S}{s \times V}$$

where N = number of bacteria per milliliter of sample

n = average number of bacteria per area of membrane delineated by net micrometer

S = working surface area of filter, mm^2

s = area enclosed by net micrometer as measured by a stage micrometer at the same magnification, mm^2

V = volume of sample filtered, ml [as described by Sorokin and Overbeck (1972)]

Jones (1974) recommended that n should be determined from counts obtained from 50 different areas of the membrane.

VI. MEMBRANES

The choice of membrane filters used to trap and concentrate the bacteria from water samples has received considerable attention in recent years. Zimmerman and Meyer-Reil (1974) were the first to recommend polycarbonate membranes (Nuclepore). Jones and Simon (1975) and Daley and Hobbie (1975), however, obtained higher bacterial counts using cellulose nitrate filters produced by Millipore and Sartorius. This was due partially to the variable distribution of bacterial cells on the polycarbonate membranes, a phenomenon resulting from the uneven wettability of the latter. Polycarbonate membranes are superior to other membranes for enumeration studies because they trap the bacteria on a single plane, which makes counting easier. Hobbie et al. (1977) have subsequently recommended treatment of polycarbonate membranes (Nuclepore) with several drops of a 0.5% solution of a surfactant (Wayfos, Phillip A. Hunt Chemical Corp., East Providence, R.I.) to achieve a more even cell distribution.

Attention has also been directed to the problem of background fluorescence produced by the various membranes. Zimmerman and Meyer-Reil (1974) included a destaining step using isopropyl alcohol to decrease nonspecific adsorption of AO by the polycarbonate membranes as well as by trapped detrital particles. This was carried out on the filtration apparatus in a manner similar to that of the staining step but required an additional filtration step. Our experience has been that the destaining time depends on the amount of detritus present in each sample. Unfortunately, the numerous filtrations associated with this technique are time consuming and, in addition, there is accumulating evidence which suggests that the physical force exerted on the bacteria from vacuum filtration promotes cell lysis (Jones and Simon, 1977). It is therefore recommended that a vacuum no greater than 125 mm Hg be applied during any filtration step.

Francisco et al. (1974) recommended the use of blackened Millipore membranes to increase the contrast between stained bacteria and the background. Methods of darkening membranes not commercially available in black have been reported (Jones and Simon, 1975; Hobbie et al., 1977). This has led to the use of membranes with pore sizes smaller than 0.45 μm.

Membrane filter pore size is an important consideration when concentrating bacteria from natural waters. Recent studies have shown that many bacteria in low-nutrient waters are smaller than 0.45 μm in diameter (Geesey et al., 1977). It is therefore desirable to use membranes with a pore size of 0.2 μm or smaller.

VII. SAMPLE VOLUME

Bacterial distribution on membranes has been shown to be affected by the sample volume filtered. Jones and Simon (1975) found that volumes less than 6 ml tended to produce higher cell densities at the periphery than near the center of the membrane. It is therefore necessary to dilute samples containing high bacterial densities prior to filtration. Samples collected from oligotrophic systems, on the other hand, may require filtration of 20-40 ml volumes of undiluted water to obtain high-enough cell densities on the membrane for accurate count estimates. In any case, it is advisable to count 20-25 microscope fields along a transect of the membrane.

VIII. CLUMPING OF BACTERIA

The application of epifluorescence microscopy to the estimation of bacterial biomass provides a long-awaited means of studying sessile bacterial populations in aquatic environments, particularly those associated with sediment and other submerged surfaces. It is well known that sessile bacteria in these habitats often exist in aggregates or clumps (Rodina, 1967; Khiyama and Makemson, 1973). This tends to lead to an uneven distribution of bacteria on the membrane following filtration (Geesey, 1977). A significant number of bacterial clumps can be dispersed by vortexing and blending of samples prior to filtration (Geesey et al., 1978). We recommend, however, that the bacteria be fixed first to reduce any cellular damage that may occur during the dispersion steps. Individual cells in the smaller remaining clumps are usually distinguishable when stained and examined by epifluorescent microscopy (Meyer-Reil, 1977).

The adoption of the destaining step recommended by Zimmerman and Meyer-Reil (1974) makes it possible to distinguish cells from detrital material even when moderate amounts of the latter are present. There is, however, an annoying amount of background fluorescence associated with sediment samples prepared in this manner, which makes routine examination and enumeration difficult. The development of a more specific staining technique could, therefore, improve conditions tremendously. Estimates of sessile bacterial biomass based on the direct enumeration technique described above compared favorably with other biomass estimation techniques. Colonization studies of glass slides implanted in the sediment of the Fraser River in British Columbia, Canada, yielded 25 pg of bacterial lipopolysaccharide (LPS) per square millimeter of surface based on the Lymulus lysate endotoxin assay (Watson et al., 1977).

Direct counts obtained by epifluorescence microscopy yielded 4.4 X 10^3 bacteria per square millimeter, which when converted to bacterial LPS using the factor 2.78 fg of LPS per bacterium (Watson et al., 1977) was equivalent to 12 pg of bacterial LPS per square millimeter. Even better agreement between biomass estimates based on these techniques may be obtained when bacterial cell volumes are taken into consideration during the direct counting procedure.

IX. CONCLUSION

Clearly, epifluorescence microscopy provides an effective means of estimating bacterial concentrations as well as an understanding of the size and morphology of microorganisms in natural waters. Studies conducted in a variety of freshwater systems in Alberta, Canada, have yielded bacterial concentrations that are 1-3 orders of magnitude greater than those obtained by traditional colony count methods (Costerton, 1979). The technique has also led to the discovery that a large fraction of bacteria in marine waters are much smaller in size than was previously realized (Ferguson and Rublee, 1976; Watson, 1978). Epifluorescence microscopy thus offers the potential of providing very accurate bacterial biomass estimates in a wide range of aquatic environments.

Future studies on the fluorescence properties of stained bacteria will probably promote a better understanding of the physiological state of the organisms in their natural environment. When combined with heterotrophic activity and autoradiography studies, estimations of total bacterial concentration and biomass obtained by epifluorescence microscopy should provide a means of evaluating the activities of individual cells and populations of cells.

REFERENCES

Bell, J. B., and Dutka, B. J. (1972). Bacterial densities by fluorescent microscopy. *Proc. Conf. Great Lakes Res. 15:* 15–20.

Costerton, J. W. (1979). Unpublished data.

Daley, R., and Hobbie, J. E. (1975). Direct counts of aquatic bacteria by a modified epifluorescence technique. *Limnol. Oceanogr. 20:* 875–882.

Ferguson, R. L., and Rublee, P. (1976). Contribution of bacteria to standing crop of coastal plankton. *Limnol. Oceanogr. 21:* 141–145.

Francisco, D. E., Mah, R. A., and Rabin, A. C. (1974). Acridine orange-epifluorescence technique for counting bacteria in natural waters. *Trans. Am. Microsc. Soc. 92:* 416–412.

Geesey, G. G. (1977). Unpublished data.

Geesey, G. G., Richardson, W. T., Yeomans, H. G., Irvin, R. T., and Costerton, J. W. (1977). Microscopic examination of natural sessile bacterial populations from an alpine stream. *Can. J. Microbiol. 23:* 1733–1736.

Geesey, G. G., Mutch, R., Costerton, J. W., and Green, R. B. (1978). Sessile bacteria: an important component of the microbial population in small mountain streams. *Limnol. Oceanogr. 23:* 1214–1223.

Hobbie, J. E., Daley, R. J., and Jasper, S. (1977). Use of Nuclepore filters for counting bacteria by fluorescence microscopy. *Appl. Environ. Microbiol. 33:* 1225–1228.

Jannasch, H. W., and Jones, G. E. (1959). Bacterial populations in seawater as determined by different methods of enumeration. *Limnol. Oceanogr. 4:* 128–139.

Jassby, A. (1975). An evaluation of ATP estimations of bacterial biomass in the presence of phytoplankton. *Limnol. Oceanogr. 20:* 646–648.

Jones, J. G. (1974). Some observations on direct counts of freshwater bacteria with a fluorescence microscope. *Limnol. Oceanogr. 19:* 540–543.

Jones, J. G., and Simon, B. M. (1975). An investigation of errors in direct counts of aquatic bacteria by epifluorescence microscopy, with reference to a new method for dyeing membrane filters. *J. Appl. Bacteriol. 39:* 317–329.

Jones, J. G., and Simon, B. M. (1977). Increased sensitivity in the measurement of ATP in freshwater samples with a comment on the adverse effect of membrane filtration. *Freshwater Biol. 7:* 253–260.

Khiyama, H. M., and Makemson, J. C. (1973). Sand beach bacteria; enumeration and characterization. *Appl. Microbiol. 26:* 293–297.

Kraft, W. (1975). The technology of new fluorescence illumination systems. *Mikroscopie 31:* 129–146.

Meyer-Reil, L. A. (1977). Bacterial growth rates and biomass production, pp. 223–236. In G. Rheinheimer (Ed.), *Microbial Ecology of a Brackish Water Environment*. Ecol. Stud. 25. Springer-Verlag, Berlin.

Rodina, A. G. (1967). On the forms of existence of bacteria in water bodies. *Arch Hydrobiol. 63:* 238–242.

Sorokin, Y. I., and Overbeck, J. (1972). Direct microscopic counting of microorganisms, pp. 44–47. In Y. I. Sorokin and H. Kadota (Eds.), *Techniques for the Assessment of Microbial Production and Decomposition in Fresh Waters*. IBP Handbook No. 23, Blackwell Scientific, Oxford.

Strugger, S. (1948). Fluorescence microscope examination of bacteria in soil. *Can. J. Res. Ser. C 26:* 188–193.

Watson, S. W. (1978). Personal communication.

Watson, S. W., Novitsky, T. J., Quinby, H. L., and Valois, F. W. (1977). Determination of bacterial number and biomass in the marine environment. *Appl. Environ. Microbiol. 33:* 940–946.

Yamabe, S. (1973). Further fluorospectrophotometric studies on the binding of acridine orange with DNA. Effects of thermal denaturation of DNA and additions of spermine, kanamycin, dihydrostreptomycin, methylene blue and chlorpromazine. *Arch. Biochem. Biophys. 154:* 19–27.

Zimmerman, R., and Meyer-Reil, L. A. (1974). A new method for fluorescence staining of bacterial populations on membrane filters. *Kiel. Meeresforsch. 30:* 24–27.

11

MEMBRANE FILTRATION: A EUROPEAN VIEW

G. I. BARROW *Centre for Applied Microbiology and Research, Porton Down, Salisbury, Wiltshire, England*

*NORMAN P. BURMAN** *Thames Water Authority, London, England*

*VERA G. COLLINS** *Freshwater Biological Association, Ambleside, Cumbria, England*

I. INTRODUCTION

In this chapter, we review briefly the current use of membrane filtration procedures and their application to bacteriological studies on water quality in some European countries. It is therefore limited essentially to indicator bacteria of sanitary and ecological significance. It will inevitably be incomplete, not only because of the large geographical area concerned, but also because of problems of language and communication. At the time of writing, most European countries are engaged in revising their methods for the bacteriological examination of waters, with particular emphasis on recommended standard or reference methods--including membrane filtration techniques--for potable supplies and environmental protection, often with eventual international agreement in view.

To meet the requirements of the European Economic Community (EEC) for harmonization, microbiological methods are being extended for use in monitoring surface waters for abstraction for potable supplies, bathing and recreational waters, and environmental quality

---

**Present affiliation:* Dr. Burman and Dr. Collins are retired.

standards for rivers, lakes, and estuaries. There are, at present, nine member states within the EEC--Belgium, Denmark, Eire (Ireland), France, Germany, the Netherlands, Italy, Luxembourg, and the United Kingdom. The work of the bacteriological experts of each of the member states is coordinated by the attendance of representatives at meetings convened by the Health Protection Directorate of the Commission of the European Communities. Some EEC directives concerned with the quality of fresh and saline waters have already been notified to member states, other directives exist as published drafts under discussion, and some exist as proposals only.

For European countries that are not members of the EEC, there are other international coordinating bodies, such as the World Health Organization (WHO), the International Organization for Standardization (ISO), and the Council for Mutual Economic Assistance (CMEA). In connection with its long-term program on environmental pollution control in Europe, WHO is sponsoring the production of a comprehensive *Manual on Analysis for Water Pollution Control*. The stated purpose of the manual is to provide governmental and other agencies, particularly health and environmental administrations, with information and recommendations to assist them in setting up water pollution control programs and the establishment of a unified system for freshwater and wastewater analysis and the recording of results. Moreover, the manual is designed to serve as a technical guide for scientists and engineers active in the field of water quality management and environmental protection and to provide, if possible, methods of analysis applicable to an average European water pollution control laboratory.

The International Organization for Standardization (ISO) established Technical Committee 147 on Water Quality (ISO TC 147) with a view to standardizing methodology wherever possible. At present this Technical Committee has six subcommittees covering the main subject areas of methodology for water quality, including microbiology. Each subcommittee has working groups of experts dealing with selected or specialized topics within each subject area. In microbiology there are working groups for each of the main indicator organisms of sanitary significance, as well as for some pathogens of humans and animals. Many European countries, represented by nominated experts affiliated with their respective national "standards" organizations, participate actively in this work. The working group meetings of ISO offer an open forum for free and full exchange of information on methods. Indeed, the "primary task of each ISO technical committee and sub-committee is to promote, by efficient decision-making, the preparation of documents which can be adopted as International Standards with the least possible delay and which can be implemented as widely as possible throughout the world without substantial change." It is probable that draft international proposals for methods for the isolation and enumeration of some of the indicator organisms of sanitary importance will soon be available.

The Council for Mutual Economic Assistance first produced its own *Standard Methods* for water quality control during 1965—1966. A revised second edition consisting of four parts, on chemical, radiochemical, biological, and microbiological methods, respectively, was published during 1973—1975. A new edition of the CMEA *Standard Methods* is now in preparation, and its contents will be based on the results of the experience and use of the methods contained in the second edition, with the addition of any developments in methodology.

This information on some of the main international coordinating bodies active in Europe in the field of water quality and environmental pollution control is given in an attempt to provide an impartial background and thus avoid undue emphasis on the membrane filter methodology of any one country. Much of the information has been abstracted or summarized from documentary material from national and international agencies as well as from other published work.

II. MEMBRANE FILTRATION: PROCEDURES AND APPLICATIONS

A. Choice of Membranes

The examination of water was one of the first microbiological applications of membrane filters (Mulvany, 1969). The choice of membrane filters used in laboratories concerned with water quality is governed by cost, the nature of the water to be analyzed, and the purpose of the analysis, as well as by the type of laboratory. Four main types of commercial membranes are currently available in Europe--of British, Czechoslovakian, German, and American manufacture, respectively. Exceptionally, a few laboratories make their own membranes, usually for particular purposes, and some methods and uses for these procedures are given by Daubner and Peter (1974). These authors also give a comprehensive review (in German) of the many varieties of membrane filters available throughout Europe. The importance of adequate quality control of membranes and the resulting comparability of filtration characteristics in relation to microbiological use cannot be too strongly emphasized.

The choice of plain, grid-marked, or colored filters seems to be determined mainly by the purpose of the examination and the individual worker. Thus, for estimating total viable counts of bacterial populations, Jones (1972) used 0.22 μm plain membranes for freshwater studies, whereas Chaina (1968) used 0.45 μm plain membranes for seawater. In work concerned with the application of epifluorescence microscopy to direct counting of aquatic bacteria on membranes, Jones and Simon (1975) used a variety of commercially available filters: their results illustrate many of the difficulties that may be encountered. Of the commercially prepared black filters they used, Sartorius and Selectron membranes gave the most satisfactory fluorescent image, as judged by low background fluorescence, smooth surface, and good contrast. The color and surface texture of Gelman membranes was uneven; and with Millipore membranes the black dye used gave unacceptably high background fluorescence. Subsequent batches of Sartorius membranes gave higher background counts, owing to the presence of fluorescing particles of bacterial size, thus necessitating careful batch checking. Filterite Micro-Sep membranes, at one time the only black membranes produced with 0.22 μm pores, are no longer available; although they gave good fluorescent images, the membranes did not filter evenly, giving many blank patches. Not only did this reduce the effective filtration area, but it affected the actual distribution of microorganisms. A similar problem was encountered with Selectron membranes with wide grid lines, which apparently interfered with filtration, and because of this Jones and Simon used ungridded membranes for all their quantitative work. When white or green membranes were dyed, they found that the highest counts were obtained with Nucleopore polycarbonate membranes, partly because of their thickness (10 μm) and partly because their structure retained bacteria on a single plane. However, difficulties with uneven wettability subsequently occurred.

The influence of test conditions on counts was also studied by Jones and Simon (1975). It is well known that membrane filters with a pore size of >0.5 μm do not retain all bacteria, and in an attempt to improve the accuracy of direct counts on membranes, Jones and Simon tried several staining procedures, the most successful of which used the commercial dye Dylon. Counts on different membranes were compared as a percentage of those obtained on Sartorius 0.45 μm membranes, which they considered to be the most satisfactory commercial product at that time. In general, counts obtained with commercially prepared and Dylon-stained 0.45 μm membranes were, however, consistently higher, the differences depending on the proportions of small bacteria in the samples. Of the Dylon-stained cellulose ester membranes, Millipore products gave the best fluorescent image and higher counts. Dylon-stained Nucleopore 0.2 μm membranes often yielded even higher counts, but there was much more background fluorescence. Although accessory light filters eliminated this, some batches of Nucleopore membranes gave uneven bacterial distribution, so that it was impossible to determine the actual filtration area and thus the numbers of bacteria. The apparently high counts on Filterite (0.22 μm) membranes were probably due in part to the same problem. On average, the counts they obtained on 0.45 μm membranes were about 75% of those obtained on 0.22 μm cellulose ester membranes and 65% of those on 0.2 μm polycarbonate membranes, suggesting that bacteria either passed through or became embedded in membranes with the larger pore size. Counts of bacteria in filtrates from Sartorius membranes (0.45 μm), as judged by filtration with 0.22 μm Dylon-stained Millipore filters, showed that 98-99.8% of the initial bacterial content had been retained and that any further reduction in numbers in the filtrate, due to membrane blockage, occurred only after filtration of 25 ml or more. Jones and Simon therefore considered that direct bacterial counts on membranes of pore size 0.45 μm could be lower than the true value because organisms may become embedded in the body of the membrane and thus remain undetectable by epifluorescence microscopy.

For routine procedures, especially when a large number of water samples are processed for bacteriological analysis, grid marks as an aid to counting are often preferred. Indeed, for this reason, ruled membranes are recommended in the United Kingdom for the bacteriological examination of drinking water (Report, 1969). However, an apparent inhibitory effect of printed grid lines on membranes was observed by some workers. The discussion following the paper given by Burman (1967a) illustrates this well. This phenomenon seems to have been considerably reduced now, but depending on the nature of the work and the organisms sought as well as the degree of accuracy required, plain membranes may be used for critical analyses with a grid projected from a suitable light source to facilitate the counting of colonies. The effect of grid markings on membrane filters was also studied intensively by Niemela (1965) in relation to the overlap of colonies as an inherent source of count error. This work is discussed later.

B. Methods for Handling Membrane Filters

Most of the main commercial firms that supply membrane filters and apparatus in Europe also provide excellent "technical literature" on the general methodology and uses of membrane filtration. There is good evidence that such advice is followed closely by many laboratories in Europe.

For ease of manipulation in many colony counting methods, the most popular size of membrane appears to be 47-50 mm in diameter with a pore size of 0.45 μm, particularly where the detection and enumeration of waterborne pathogens and indicator organisms is concerned. Most organisms of "sanitary significance" in water are retained on or within the surface of membrane filters with a pore size of 0.45 ±0.2 μm; however, certain groups of organisms (e.g., vibrios, spirochetes, and pseudomonads) may not be completely retained. For such organisms, filters with a pore size of 0.22 ±0.02 μm are to be preferred for more precise work, especially if these organisms are actually sought.

Many commercial manufacturers now supply their membrane filters in presterilized packs ready for immediate use, but although convenient for small numbers of samples, such packaging significantly increases the cost of laboratory procedures. Such membranes must also be handled with great care, as they are usually very friable and tear easily. Even without prepackaging, some laboratories recycle membranes after use, for reasons of economy. Because some commercially available membranes may be more suitable for recycling than others, this may become an additional factor in the choice of membrane by individual laboratories. It is important to note that the possible reuse of membranes is strictly limited by the nature of the medium previously used as well as the organisms then sought. However, the procedure for recycling membranes after use for the enumeration of coliform organisms, described in Report (1969), may prove useful in emergency situations.

The choice of filter holder and apparatus are also determined partly by cost, and the choice of suction or pressure filtration essentially by the nature of the sample and the objective of the work. Laboratories processing large numbers of samples invariably prefer simple but efficient filter holders, either as single units or as a multiple system incorporated in a manifold, usually with suction. Although metal filter holders and funnels can also be sterilized directly by flame if necessary, for practical purposes, transparent apparatus that will withstand repeated exposure to moist heat, such as immersion in boiling water, has the additional advantage that the volume of water for filtration can be readily seen. We have found that plastic holders and funnels of the magnetic type are simple and easy to use and can be resterilized rapidly in boiling water between samples (Fig. 1). It is convenient to collect a large volume of filtrates for disposal in a central reservoir or tank fitted with a draining tap.

FIGURE 1 Membrane filter holder and funnel of the magnetic type.

For special purposes, a single-unit apparatus designed to filter a number of samples simultaneously through one membrane was designed by Jannasch (1954). The collection of representative samples of water from great depths, as at sea, poses special problems both of sampling and of subsequent handling in the laboratory. The latter include the decrease in pressure, resulting in disturbance of the gaseous equilibrium, and increase in temperature, which may affect obligate barophiles and psychrophiles, respectively. To overcome some of these problems, Williams (1969) devised a membrane filtration apparatus for use at great depths. When at the required depth, the water under its own pressure is allowed to enter the cylinder through the membrane filter, a nonreturn valve preventing backflow when raised. This apparatus eliminates the need for much handling work; it is necessary only to transfer the membrane to a suitable culture medium and to measure the volume of filtered water in the cylinder. It has the disadvantage that only one microbiological operation can usually be performed on any one sample (Collins et al., 1973).

C. Isolation of Indicator Organisms

The following information on membrane methods, including media, time, and temperature of incubation, currently used for indicator organisms of sanitary significance in Europe has been compiled as a brief summary from published work, in particular from the survey report produced by Ormerod (1974) for use in the preparatory stages of the section on bacteriological examination for the forthcoming WHO publication, *Manual on Analysis for Water Pollution Control*. In this manual, approved "alternative" rather than "reference" cultural methods are given, each with recommended media for different organisms, because both membrane filter and multiple-tube techniques are regarded as equivalent procedures. Where other media are found to give similar results after adequate evaluation, they may also be regarded as being equivalent (Suess, 1977). In addition, the report of a working group of experts on bacteriological analyses of drinking water, published by the Commission of the European Communities (CEC, 1977), contains much detailed information, including media and technical procedures. A useful summary of the constituents of media used in Europe is also given by Ormerod (1974). The membrane filter methods employed in Britain and described in Report No. 71 are given separately in the Appendix to this chapter.

Indicator organisms in water may be in a stressed condition either as a result of a disinfection process or from long sojourn in an unfavorable environment. Any comparison between bacteriological techniques for the eumeration of such organisms should therefore preferably be made on water containing stressed organisms. In Great Britain this has been achieved by initial treatment of water with chloramine. This may be done by cooling the water to near 0°C, adding excess ammonia (2 ml of a sterile 0.38% solution of ammonium chloride to 1 liter) followed by sufficient hypochlorite solution to give a final concentration of available chlorine in the water of 1 mg/liter. After standing for 15-60 min at near 0°C, the chlorine is neutralized with 0.5 ml of a sterile 3% solution of sodium thiosulfate to a 1 liter sample of water. After this treatment, coliform organisms can be recovered more readily by some techniques than by others. This method was very useful in comparative studies on different selective media for coliform organisms (Public Health Laboratory Service, 1968).

Coliform Organisms and Presumptive Escherichia coli

Many European countries seem to favor the use of one or other of the various Endo media for coliform organisms. In Germany, Endo agar is recommended for total coliforms (20-44 hr/37°C) and presumptive *E. coli* 24 hr/44°C) with subsequent biochemical confirmation. As an alternative, absorbent nutrient pads saturated with Endo broth may be used with airtight containers incubated for 24 hr at 37°C for total coliforms and at 42-44°C for presumptive *E. coli*. The standard methods for Eastern Europe (CMEA, 1968) also recommends Endo agar at 37°C for 24 ±2 hr for total coliforms with subculture of colonies at

43°C and further identification when necessary. The Norwegian Institute for Water Research uses Endo broth with membranes at 37°C for 18 hr for total coliforms and membrane fecal coliform (mFC) broth at 44.5°C for 24 hr for thermostable coliforms. In France, the lactose TTC tergitol agar of Chapman, modified by Buttiaux, is used at 30-37°C for total coliforms and at 44°C for fecal coliforms. In Britain and Denmark, 0.4% enriched Teepol broth is used, with a short period of lower-temperature incubation for resuscitation of stressed organisms. In the forthcoming WHO *Manual,* lactose bile rosolic-acid medium (M-FC broth) will be recommended for fecal coliforms (Suess, 1977). With Endo media, differentiation of coliform colonies is not always easy, and many regard these preparations as generally unstable, variable in performance, and sometimes inhibitory to *E. coli*. Both Endo and membrane fecal coliform media gave poor results in the technical seminar for member countries of the EEC held at Lyons in 1975 (CEC, 1977). In addition, Endo media gave poor results in the recent feasibility study within the EEC during 1976 on the distribution and use of simulated water samples for comparative bacteriological analysis carried out by Barrow et al. (1978).

As can be appreciated from this limited surevy, there is some variation in the media used and the conditions of primary incubation, as well as in the extent of subsequent confirmation. The majority of countries incubate membranes for total coliforms either at 35 or 37°C, with a tolerance of ±0.5°C, so that, in fact, slightly differing coliform populations may be measured. In addition, many laboratories in practice examine and record the results of their water tests as the first task each morning, often irrespective of the actual time the tests were set up during the previous day(s). Thus, although many procedures, for example, recommend incubation for 24 hr, the actual incubation times may in fact vary from 14 to 24+ hr, a range that may itself produce considerable variations in results. Too little attention seems to have been given to this practicality. As far as is known, only in Britain, and to a lesser extent in

Denmark, has attention been drawn to the importance of a lower initial temperature during incubation as a simple way to resuscitate some of the bacterial cells--whatever the nature of the organism sought--damaged by chemical treatment or stressed in other ways. If water samples received throughout the day are processed in batches and placed in a dual temperature incubator timed for the complete cycle to start toward the end of the working day, then in practice all the samples will receive *not less* than 4 hr of lower-temperature preincubation and most will receive considerably more. The same rationale applies also to multiple-tube tests; indeed, there is much to be said for not warming tubes to incubation temperature before inoculation, a practice that theoretically would not aid recovery of stressed organisms. It must be remembered also that differing climatic and geographical conditions are likely to influence the incidence and distribution of competing organisms. As far as the safety of drinking water in relation to health is concerned, the detection of small numbers of coliform organisms, if present, is far more important than great accuracy and reproducibility with large numbers. As with glutamate medium for multiple-tube tests (Gray, 1964), the development of chemically defined membrane media for coliform organisms would eliminate much of the variation in current methodology and go a long way toward standardization.

The concept of "fecal coliforms" also requires care in interpretation; as currently defined in water bacteriology, this term applies to all coliform organisms, irrespective of genus, which are able to ferment lactose at 44°C with the production of both acid and gas. It can thus include many coliforms that are not, in fact, fecal in origin and which can multiply in the aquatic environment under suitable conditions. In contrast, *E. coli* is an obligate intestinal parasite of warm-blooded animals, including humans; it fails to multiply in water, in which its survival time is limited; and in sewage, it greatly outnumbers any pathogens. For these reasons, in Britain and in Germany (Müller, 1977) the use of presumptive *E. coli* is considered preferable to "fecal coliforms" as the indicator of fecal pollution (Barrow, 1977). The term "fecal coliform," however, is used by the World Health Organization as well as in EEC water directives.

Fecal Streptococci

In Europe almost all countries that employ membrane filtration for fecal streptococci use the glucose azide medium of Slanetz and Bartley (1957) with general satisfaction. This itself is a strong reason for its continued use. This medium will also be recommended in the WHO *Manual* (Suess, 1977). The time and temperature of incubation is usually 48 hr at 37°C; Sartorius, however, suggests 30-48 hr, and France incubates at 35°C. As far as is known, Britain is the only country that recommends incubation at 44-45°C for 44 hr following a preliminary period of 4 hr at 37°C for resuscitation of stressed cells. This dual incubation procedure yields counts of fecal streptococci, as defined in U.K. Report No. 71, which closely parallel those of confirmed results given by the British 72 hr multiple-tube method with the glucose azide broth of Hannay and Norton (1947). Although Slanetz and Bartley claimed that following incubation at 36°C for 48 hr, all the streptococci isolated could be classified as enterococci, in Britain it was found that many unclassified strains were isolated, especially from surface waters (Burman, 1961). For routine purposes, however, it is impracticable to confirm that all colonies of streptococci which grow on membranes belong to Lancefield's group D. Indeed, it is rather doubtful if either cultural or serological confirmatory tests are often used in practice in Europe. The choice of methods thus appears to lie between those which give higher counts but more false-positive results and those with greater specificity but lower counts. The method recommended for the United Kingdom was chosen as a compromise between these extremes (Burman et al., 1969), and contrary to some schools of thought, it is both simple and practicable for routine use. Relevant extracts on methodology for fecal streptococci from the British Report are given separately in Sec. III. In selecting methods, careful attention must be paid to the nature of the water and the objective of the test in relation to the degree of accuracy and precision needed. Despite the recent development of other media to yield greater numbers of fecal streptococci, such as solid azide media containing esculin and kannamycin, little, if any, comparative work on dual-temperature incubation procedures seems to have been done outside Britain.

Sulfite-Reducing Clostridia

There are very few published accounts available in Europe on the use of membrane filtration for the enumeration of sulfite-reducing clostridia, in particular *Clostridium perfringens (welchii)*, possibly because it is not usually necessary to use this organism in routine practice as an indicator of fecal contamination. A simple adaptation of the multiple-tube technique for spore counts, using Wilson and Blair's glucose sulfite medium, was described by Taylor and Burman (1964) and updated by Burman et al. (1969). Briefly, this consisted of heating the water sample to 75-80°C for 10 min before filtration. Following this, the membrane is carefully placed face (grid side) down on the surface of well-dried plates of Wilson and Blair's medium, and another layer of the same medium is poured on top. After incubation anaerobically for 1-5 days at 37-45°C, depending on the purpose of the examination, *Cl. perfringens (welchii)* produces colonies with large black halos due to the reduction of the sulfite. If the extent of pollution is completely unknown, a series of 2-fold rather than 10-fold dilutions should be filtered, as the black halos are so large that they may easily become confluent, even with relatively small numbers of colonies. A very similar membrane technique and medium for spores of *Cl. perfringens (welchii)*, differing only in minor detail, is recommended by CMEA *Standard Methods* (1968) and will be recommended in the WHO *Manual* (Suess, 1977).

Discussions in ISO have revealed that a number of variations in technique are used in Europe, and these are being collated in an attempt to produce an international standard membrane method. SPS agar seems to be preferred to Wilson and Blair's medium; but there are wide variations in the concentration of sulfite used; the iron may be added as ferrous sulfate, ferrous citrate, or ferrous ammonium citrate; and there are many variations of layering techniques. Some place the membrane on the bottom of the dish; some pour a layer of medium first; all pour a layer of medium on top. Some place the membrane face down and some face up; some use the same medium for both layers; some use a noninhibitory nutrient agar for one layer either above or below the membrane. Incubation temperatures vary from 37°C to 45°C.

For clostridial spores, the instructions for heating the test samples are sometimes not entirely clear. In practice, the time for adequate pasteurization will vary with the volume of water and the nature of the container, and it is important that all the volume of water to be examined be brought to the specified temperature and held at it for the required time.

Pseudomonads and Pseudomonas aeruginosa

The use of membrane filter techniques is not mentioned in the definitive works on the taxonomy of pseudomonads by Jessen (1965) and Stanier et al. (1966). Indeed, few studies have been reported in Europe on the enumeration of pseudomonads by this method for water quality assessment. In Britain, extensive work on the occurrence of fluorescent pseudomonads, especially *Pseudomonas aeruginosa,* in potable water supplies and surface waters has been carried out by the Thames Water Authority. The ability of fluorescent pseudomonads to utilize a wide variety of organic compounds in solution and the ability of *P. aeruginosa,* in particular, to multiply on materials commonly used in contact with potable water (Burman and Colbourne, 1977) has led to the development of a number of enumeration techniques. Membrane filtration, together with a modification of King's A broth for the enumeration of fluorescent pseudomonads, has now been superseded by a mutliple-tube procedure with repli dishes. These are sterile square plastic petri dishes containing 5 X 5 separate compartments, each about 8 ml in volume, so that sample dilutions may be made directly in them and culture medium then added. King's B broth containing fungizone and erythromycin incubated at 22°C for

4 days has been found to give good recovery of *P. fluorescens*, *P. putida*, and *P. aeruginosa* from both raw and chlorinated waters. Membrane filtration has, however, proved successful as a method for the enumeration of *P. aeruginosa* in chlorinated waters. The membranes are placed onto absorbent pads saturated with Drake's modification of King's A broth and incubated at 37°C for 48 hr. Typical colonies usually produce the blue-green pigment pyocyanine and fluorescein, which can be detected by illumination under ultraviolet light. This method, coupled with subsequent confirmation by subculture onto milk agar (Brown and Scott Foster, 1970) incubated at 41.5°C for 24 hr, has been found to give consistently reliable results (Colbourne, 1978).

In an in-depth comparative study on different methods and media for the enumeration of *P. aeruginosa* from different waters, Jawad (1976) concluded that with the waters used and under the test conditions specified, the addition of 0.1% cetrimide to the membrane Pseudomonas agar of Levin and Cabelli (1972) resulted in a greatly improved medium. In this modified medium, not only were fecal streptococci inhibited, but all the presumptive colonies examined were subsequently confirmed as *P. aeruginosa* by the method of Brown and Scott Foster (1970).

III. REPORT NO. 71: MEMBRANE METHODS IN THE UNITED KINGDOM FOR COLIFORM ORGANISMS AND FECAL STREPTOCOCCI

The World Health Organization, in recommending membrane methods for organisms indicative of fecal pollution (WHO, 1971), suggested that either those employed in America, as described in *Standard Methods for the Examination of Water and Wastewater*, or those used in Britain, as given in Report No. 71, *The Bacteriological Examination of Water Supplies*, should be used. Relevant extracts on membrane methods recommended for coliform organisms and fecal streptococci from the current British Report are quoted verbatim in the Appendix to this chapter, and for context and clarity, the sections on multiple-tube methods and confirmatory tests are also included.

Readers who wish to have details of the standards of bacterial quality recommended as well as information regarding the appendixes or references mentioned should consult the report itself. Although this report is at present being revised, the principles of the methods described are unlikely to be changed significantly. The current position has been summarized by Barrow (1977): comparative work with both membrane and multiple-tube media is in progress; a suitable substitute for Teepol 610, manufacture of which has now ceased, will be recommended; and it is likely that the use of single-tube confirmatory tests at 44°C for both indole production and the fermentation of lactose (or other suitable carbohydrate) for presumptive *E. coli* will be recommended. It is also probable that as a laboratory routine, the temperature of initial incubation of membranes for maximum recovery of both total coliforms and presumptive *E. coli*, including stressed organisms, will be 5 ±1 hr at 30°C.

IV. MEMBRANE FILTRATION: SOME GENERAL APPLICATIONS

In considering the published work on membrane filter methods for water in Europe, it is perhaps not surprising that studies carried out in Britain should have received much attention, and the following papers, listed in chronological order, should be consulted: Taylor (1955); Taylor et al. (1953); Taylor (1957); Taylor et al. (1955); Burman (1955); Taylor (1959, 1961, 1963); Burman (1960, 1961); Taylor and Burman (1963, 1964); Burman (1965); Burman (1967a, b); Burman et al. (1969); and Report (1969). Most of these papers, in particular the detailed review by Burman et al. (1969), are concerned with the enumeration of indicator organisms of sanitary importance and their role in water quality assessment. The membrane filter technique is, however, widely applicable to the isolation of many other organisms, including pathogens, not only from water but also from other sources.

In discussing the isolation of actinomycetes, fungi, and yeasts, Burman et al. (1969) were particularly concerned with their role in the production of earthy, moldy, or musty tastes or odors in potable water due to their growth in long lengths of piping, especially under warm conditions. For isolation of fungi and yeasts, they recommend acid rose bengal kannamycin agar prepared as follows: mycological peptone (Oxoid), 5 g; dextrose, 10 g; KH_2PO_4, 1 g; $MgSO_4 \cdot 7H_2O$, 0.5 g; rose bengal, 0.07 g; agar (Oxoid), 10 g; water to 1 liter; pH 5.4, sterilized at 115 psi for 10 min. Kanamycin is added before pouring plates to give a final concentration of 100 μg/ml in the final medium. After filtration, membranes are placed face up in the normal manner and incubated at 22°C for 7 days. The medium is highly selective for fungi and yeasts. Aerobic sporing bacilli occasionally produce a few small colonies which are easily recognizable. The rose bengal suppresses common rapidly growing spreading fungi so that slow-growing species can be readily isolated; if necessary, the concentration of rose bengal may be increased up to 0.7 g/liter. Those which have been readily isolated in association with moldy-tasting water include strains of *Cephalosporium* sp., *Verticillium* sp., *Trichoderma sporulosum, Nectria viridescens, Phoma* sp. similar to *P. eupyrena,* and *Phialophora* sp. similar to *P. fastigiata*.

For actinomycetes, most media and methods have usually been designed for their isolation from soil. Water presents quite different problems, as the ratio of bacteria to actinomycetes is usually much higher in water than in soil, thus necessitating a more selective method. The medium of Lingappa and Lockwood (1962), containing chitin as the sole source of C and N, is recommended by Burman et al. (1969) for use with the membrane filtration technique for actinomycetes in water. In their paper, precise instructions are given for the preparation of colloidal chitin. As chitin is insoluable in water, it cannot diffuse through membranes and would thus not be available to organisms on the surface. The membrane must therefore be placed face down on the surface of a well-dried medium and incubated for 4 hr at 22-30°C. The membrane is then carefully removed and discarded; the culture plate is returned to the incubator for

1-8 weeks. The organisms thus imprinted on the medium grow as normal surface colonies. The addition of antibiotics, such as pencillin and polymixin B, sometimes recommended for the isolation of actinomycetes from soil, were found to be ineffective for water samples and indeed inhibitory to some strains. Colonial appearances are not typical on chitin agar, so that subcultures on more nutrient media are essential for isolation and identification.

Pretreatment of the sample is often practiced, such as by carefully controlled chlorination (Burman et al., 1969), but heating is even more effective. Heating for 2 hr at 44°C before filtration is one method, but drying and heating the membrane after filtration or heating it after placing it on the medium for a similar time and temperature are equally effective. Higher temperatures for shorter times, with or without drying, and drying without heating have also been used. To obtain optimum isolation of actinomycetes, higher temperatures and/or longer times may be used, but for enumeration 2 hr at 44°C (an incubation temperature commonly used in waterworks laboratories) gives the best results on samples with large numbers of interfering organisms. Samples containing few interfering organisms are more likely to give higher counts without heating (Burman, 1973).

Other workers have described the application of membrane filters to a variety of different organisms for different purposes. Thus, Miller (1963) reported on a simple and apparently efficient method for removing, from the surface of membranes, microorganisms derived from air and water for plate counts on nutrient agar for industrial quality control purposes. The trapped organisms were removed from membranes by a simple washing process using either glass beads or a magnetic stirrer. The efficiency of removal of organisms was confirmed experimentally by incorporating the membranes in nutrient agar.

Attenborough and Scarr (1957) described an adaptation of the membrane filter technique for quality control work on canner's sugar. Their detailed paper presents some practical applications for the enumeration of thermophilic spores in sugar. These workers adapted the membrane filter technique as used in water examination and described by Goetz and Tsuneishi (1951) and Taylor et al. (1955). Aerobic spore counts on black membranes were made using absorbent pads saturated with dextrose tryptone broth without indicator. In particular, counts of *Clostridium thermosaccharolyticum* were made by a deep culture method for membranes in which the relevant colonies were visually identifiable, and *Cl. nigrificans* was estimated by culture on rolled membranes inserted into tubes of agar medium.

Another application of membrane filtration was described by Chiori et al. (1965). These workers tested the effect of the retention properties of the filters when a bactericidal solution was filtered through membranes. Their investigations showed that after treatment with a 0.1% (wt/vol) aqueous solution of cetrimide and washing with sterile water, cellulose acetate membranes retained sufficient cetrimide to inhibit the growth of many Gram-positive organisms and spores. Such differential retention could in part explain the success obtained by Jawad (1967) when she reported the inhibition of fecal streptococci on the addition of 0.1% cetrimide to the membrane Pseudomonas agar of Levin and Cabelli (1972) for the enumeration of *P. aeruginosa*.

The application of the membrane filter technique to the enumeration of viable bacterial populations in both the marine and freshwater environment is described by Collins et al. (1973). After filtration of a suitable volume of the sample, the membrane is placed on a plate of agar medium appropriate for the bacterial population sought. Thus, for lake water, 0.22 μm membranes (Millipore U.K. Ltd.) are used and then placed on casein-peptone-starch (Collins and Willoughby, 1962) medium. After appropriate incubation, such as 2 days at 10°C or 36 hr at 15°C, the membranes are removed, dried, and stained with 1% aqueous erythrosin for 30 min. After further drying, the membranes are washed twice in distilled water, redried, cleared with immersion oil or cedarwood oil, and examined microscopically. The colonies remain pink when the excess stain is washed out of the membrane. All staining and washing procedures are done by allowing the solutions to soak through membranes from below, using only membrane-filtered stains and water. Full details of this method with examples of its use are given by Jones (1972). For seawater, Chaina (1968) used 2.0 ml test volumes filtered through 0.45 μm membranes, which were incubated for 5 days at 20°C. The resulting colonies on the membrane were counted directly; three counts were made on each sample with each of the media used. By counting directly without staining, colonies can be selected for subsequent subculture and qualitative examination of the flora as described by Collins et al. (1973). Membrane filtration was one of the methods used in the study of the distribution of *Vibrio parahaemolyticus* in British coastal waters in the survey described by Ayres and Barrow (1978).

Pioneering work on microscopic examination and direct counting of plankton and bacterial populations in the aquatic environment was done by Cholodny (1928) with bacteria in river water. Rasumov (1932) used the same method, but observed the microorganisms on the surface of membranes by the simple expedient of clearing the filters with immersion oil. This technique, after further development by Jannasch (1953), led to the commercial production of an ingenious membrane filter holder. The papers by Jannasch (1958) and Jannasch and Jones (1959) summarize well the value and limitations of the membrane filter technique as applied to heterotrophic bacterial populations in natural waters. For the cultivation and identification of fungi, Funder and Johannessen (1957) adopted Jannasch's method so that they were able to eliminate the need for culture plates and microcultures on slides. Jones (1975) described the direct quantitative estimation of selected organisms, including the iron bacterium *Leptothrix ochracea,* by membrane filtration.

A further development of the membrane filter technique for the direct enumeration of aquatic bacteria has been the application of epifluorescence microscopy to give a better assessment of total numbers within natural bacterial populations. Some of the developmental work originated in the food industry, with studies on the use of membrane filters and fluorescent microscopy for estimating the numbers of yeast cells in beverages (Cranston and Calver, 1974). Following this, Paton and Jones (1974) devised a quicker and simpler method using incident fluorescence microscopy and fluorogenic esters. These techniques were incorporated and further developed for the direct counting of bacteria in fresh water by Jones (1974). The methodology involved in the treatment of membrane filters for use in epifluorescence microscopy is discussed fully by Jones and Simon (1975), and the methods derived from their work have since been applied to the study of some groups of planktonic bacteria in fresh water (Jones, 1978).

The definitive monograph, *Membranfilter in der Mikrobiologie des Wassers,* by Daubner and Peter (1974) is in urgent need of translation. However, the clear style of the text should enable most readers to understand it in general. This monograph gives extensive coverage of European views on membrane filter techniques and, for all practical purposes, also provides a complete summary of the European literature on the use of membrane filtration for direct counts for the years 1958–1974. Mulvany's *Membrane Filter Techniques in Microbiology* (1969) also gives a useful survey of membrane filter techniques and their applications. In addition, the excellent recent treatise by Bonde (1977), *Bacterial Indication of Water Pollution,* should be studied closely. Finally, no survey of the European literature on membrane filter techniques would be complete without reference to *Methods in Aquatic Microbiology* by the late A. G. Rodina, revised and translated from the Russian by Colwell and Zambruski (1972). In this book, the chapter "Quantitative Determination of Micro-organisms in Water and Sediment" gives a detailed description of Rodina's method of treating membranes.

V. MEMBRANE FILTRATION: SOME GENERAL CONSIDERATIONS

A. Basic Concepts: A Cautionary Tale

An important European publication relevant to the basic concept of membrane filtration for water bacteriology is that of Niemela (1965). In this paper, the author has surveyed and critically reviewed, *inter alia,* sources of error in membrane filtration work, including the distribution of colonies on membranes, the physical effects of grid lines, the problems of colony overlap, and the principles to be applied in the correction of overlap error. Other subject areas extensively covered, both theoretically and experimentally, include the influence of incubation time and depth of agar on colony diameter and of the filtered volume on the distribution of cells, surface-tension error, staining of colonies, initial points of growth of bacterial colonies and their enumeration, density of growth points in relation to sample volume, measurement of colony diameter and area, visual colony counts, and parallel samples on different media. In his discussion, Niemela challenges one of the basic concepts of enumeration procedures with membrane filters:

> The statement frequently appears in microbiological literature that the colony count is *statistically* reliable when the number of colonies per plate is between 30 and 300. The origin of this misconception has not been traced, and what the term statistically reliable implies remains obscure. The statement is obviously not based on any theoretically valid statistical concepts, and its practical basis must necessarily also be rather weak, because the size of the colonies is never mentioned in the same connection.
>
> It has become evident from the experiments reported that in any colony-counting procedure, where the overlap error is taken into account, the colony size is almost equally as important as their number. If numerical limits, in terms of colony numbers alone, are wanted, one thing is clear. The closer the mean colony number per plate is to zero, the less biased the estimate of the mean will be. The farther the colony number is from zero, the more important the colony size becomes. Owing to the fact that the colony size is inversely proportional to the colony number, the colony size becomes

important at a very low density of colony centres. In the normal membrane colony counts, a 5% overlap error will be expected before the number of colonies reaches 500.

There are several examples of misinterpretations of enumerational data due to failure to recognize the importance of colony size. The question of an *optimum incubation time* is one of these. It has been claimed (Beling and Maier, 1954; Emmenegger, 1956) that an incubation exceeding a certain time limit will cause a diminution of the colony count due to the overlap. This is undoubtedly true, but the time (16-24 hours) after which the overlap was believed to start exerting its effect has been misjudged. The preliminary tests in this paper already showed that the growth of *E. coli* colonies has normally practically stopped by 20 hours, and most of the overlap has taken place long before that.

Another typical mistake is connected with the choice of the optimal nutrient medium. The comparison between several alternatives is always based on the colony number. Several nutrient media are inoculated with samples from the same bacterial suspension. The one that gives the highest yield of colonies is termed the best. In such situations a medium which is *the most favourable* for growth, giving the largest colonies and, consequently, the greatest overlap error, will be judged *the worst*, unless even greater chemical inhibitory effects are operative.

B. Some Advantages and Disadvantages of Membrane Filtration

One of the main disadvantages of the membrane filtration technique is that of cost. This was considered carefully by Taylor and Burman (1964), who pointed out that the most expensive item in the method is the membrane itself. To reduce the costs involved for large numbers of samples with their routine tests for potable supplies, they devised a method for recycling membranes of British manufacture and found that it was possible to reuse them as often as 10 times. This should not, however, be interpreted as encouragement for the general reuse of membranes: the many limitations and inherent sources of error in so doing are obvious.

With coliform populations in hard alkaline rivers and waters from wells in chalk deposits, Burman (1967b) showed that inability to differentiate anaerogenic strains immediately was an important disadvantage of the membrane method. Based on the usual definition of coliform organisms--the production of both acid and gas from lactose--a proportion of false-positive results will therefore almost always occur because gas formation is not detected by the membrane method. This, however, errs on the side of safety insofar as pollution and health risks are concerned. Where such results may be significant, as in distribution systems--samples from which should normally be negative--any positive results that do occur can be confirmed by subculture to lactose peptone water, nutrient-agar, and MacConkey agar for further identification of the colonies.

Membranes cannot generally be used for turbid waters with low bacterial counts. Turbidity due to particulate matter can interfere with filtration itself, and the deposit retained on the membrane surface may interfere with bacterial growth. Membranes are also unsatisfactory for samples of water containing small numbers of coliforms among many non-lactose-fermenting organisms growing on the membrane (Burman, 1967a). Indeed, apart from the difficulties caused by overcrowding of colonies and the growth of interfering organisms closely resembling those of the indicator organisms sought, Bonde (1977) concluded that membrane filtration required more technical skill and personal judgment than the multiple-tube method. Nor is prefiltration with coarse filters usually suitable for quantitative work because it is well known that bacteria often become attached to particulate matter.

The positive, apparently advantageous characteristic which, as stated by Mulvany (1969), "makes a membrane filter such a valuable tool is that all particles both biological and non-biological which are larger than the pore size will be positively retained on the filter surface"; however, this may sometimes be a disadvantage.

Thus, when applied to surface and other waters used for abstraction for potable supplies, the efficiency of the membrane technique may be impaired by the turbidity of the water due to naturally occurring particulate matter composed mainly of phytoplankton and organic detritus. The association of bacteria in water with particulate matter is well known and is recognized, for example, by microbial ecologists working with natural populations in the environment (Collins, 1963, 1977) and by epidemiologists in relation to the occurrence of waterborne infectious disease (Moore, 1971).

The retention properties of membrane filters have occasionally posed problems for certain purposes, as for example in factory quality control situations (Miller, 1963), but these have generally been overcome by suitable washing techniques. Work on the bactericidal properties of certain solutions showed that membranes may retain some substances and thus produce an inhibitory effect on bacterial growth (Chiori et al., 1965). However, as mentioned earlier, the same inhibitory effect may sometimes be advantageous for the selective examination of certain groups of bacteria--such as pseudomonads--in mixed populations (Jawad, 1976). For other purposes, as in the quality control of potable supplies where large numbers of routine samples must be examined or where large volumes of water are to be tested, the outstanding advantage of the membrane filtration technique is the speed with which direct bacterial counts, including *E. coli,* are obtained. This enables prompt corrective action to be taken when required and allows a water treatment plant or supply to be put back into service more quickly when negative results are obtained. In the laboratory, there is a considerable saving in technical labor as well as in the quantities of media and glassware required. There is also a significant saving in the cost of materials (Report, 1969). Another advantage of membranes is the ability to filter samples on site in field work and thus avoid undue delays in samples reaching the laboratory. Such membranes may be posted to the laboratory or delivered when convenient. The use of membrane transport or storage media for this purpose was described by Panezai et al. (1965). They showed experimentally that there was virtually little loss of recovery of coliform organisms on membranes stored for 2 days. It must be remembered that the physicochemical characteristics of membranes may vary not only with different manufacturing processes but also with the methods of sterilization, so that quality control assurance and careful selection for the purpose required is essential.

VI. CONCLUSIONS

Bearing in mind the limitations of geography and language, any conclusions must of necessity take the form of a summary of the current state of the art of the subject areas reviewed.

The brief survey of the available literature indicates that membrane filtration techniques are used throughout Europe for a variety of purposes, especially for the control of water quality by means of bacteriological analysis, with general satisfaction either instead of or as an alternative to the multiple tube method. Many countries in Europe have developed recommended or standard methods of their own. For example, those at present used in the United Kingdom are described in *The Bacteriological Examination of Water Supplies* (Report, 1969); although this report is at present under revision, the essential principles of the methods will remain unchanged. Similarly, Eastern European countries have the CMEA manual of standard methods, also currently being revised. Most European countries now participate in the work of the International Organization for Standardization, which is likely in the future to be the principal coordinating body for the formulation of agreed standard methods for adoption, or otherwise, by appropriate national organizations. European and other countries will also have access to the *Manual of Analysis for Water Pollution Control* to be published by WHO as an international reference source for recommended methods for unified systems of water quality assessment and environmental pollution control. It must be recognized, however, that a number of WHO as well as ISO recommended methods are likely to contain compromise agreements on points of technical detail.

In addition to these organizations concerned with the production of standard, recommended, or reference methods, countries that belong to the European Economic Community must also take into account the mandatory requirements contained in directives on water quality and environmental protection. One of the main areas of discussion within the Community, particularly in the physicochemical field, has been the concept of environmental "water quality objectives"--preferred by the United Kingdom--in contrast to that of uniform or "fixed emmission" standards favored by other countries.

The controversial subject of recommended or standard bacteriological methods, in contrast to reference or even mandatory procedures, is also an important issue, but it is not as simple and clear cut as in the physicochemical field. The final choice of methods, including media, for the bacteriological examination of water--whether for drinking, recreational use, catchment control, or environmental protection--will inevitably be influenced by the various conditions and circumstances existing in individual countries and laboratories. Approved and agreed reference methods will ultimately be needed before bacterial standards for different waters can be applied meaningfully in relation to their use at international levels. It must be recognized, however, that recommended but alternative methods--which may vary from one country to another, depending on climate, water sources, economic considerations, and laboratory practice--are likely to remain in use for some considerable time. Although ideal, it may be impracticable and perhaps unjustifiable to expect countries to adopt international reference methods--which probably include some technical compromises--without essential comparison in field trials with their own best methods which have stood the test of time. Universal standards for water quality, in contrast to objectives, are indeed still a long way off, and it is important meanwhile that ideals be reconciled with practicalities (Barrow, 1977).

APPENDIX: THE BACTERIOLOGICAL EXAMINATION OF WATER SUPPLIES*

Part III. Technical

Technique of Bacteriological Examination of Water

In the following pages the various procedures recommended for the bacteriological examination of water are set out in full. Each test is described separately so that it may be used either as part of a complete bacteriological examination, or alone as a routine daily or weekly test. The purpose of this detailed description is not to prescribe a rigid set of rules from which any deviation is to be deprecated, but rather to specify techniques which are likely to be accompanied by little experimental error and to give reproducible results. It is left for the individual worker to decide whether any changes in technique will result in greater convenience or economy without appreciable loss in accuracy or in comparability with the results obtained in other laboratories.

General laboratory hygiene. As the water bacteriologist attaches significance to very small numbers of coliform organisms and *E. coli*, and as these organisms are very common in man and his environment, and are freely cultivated in the laboratory, special precautions should be taken to avoid contamination on reception and during examination. The risk is especially great in laboratories where samples of faeces, sewage or other heavily contaminated materials are examined.

Wherever practicable a separate part of the laboratory should be set aside for examining the water samples; it should be secluded as far as possible from the passage to-and-fro of other workers. Bench surfaces should be swabbed down with disinfectant daily, and

---

*Note that the references cited in this report are not included in the reference list at the end of this chapter. Similarly, the Appendexes and page numbers mentioned refer to the report itself. (Source: Reports on Public Health and Medical Subjects, No. 71. Her Majesty's Stationery Office, London, 1969. Reprinted with the permission of the Controller of Her Brittanic Majesty's Stationery Office, London, England.)

immediately after accidental spillages. Windows should be kept closed. All technicians should wash their hands before commencing examination of samples, and should preferably wear a clean laboratory coat that is reserved for this purpose. Boxes for the transport of samples should preferably be made of metal or plastic, so that they can be disinfected with a chlorine solution at weekly intervals. They should not be used for carrying anything other than samples of water for bacteriological examination and should not be stood on benches.

The necessary apparatus and its preparation are described in detail in Appendix A and methods for the preparation of media in Appendix B. These methods should be strictly followed if uniform results are to be obtained. For the same reason the use of certain standard commercial preparations is occasionally recommended.

The examination should be commenced immediately after receipt of the sample at the laboratory. If more than six hours elapse between the collection of the sample and its examination this should be expressly noted on the report.

Section VII. The Count of Presumptive Coliform Organisms

As the number of coliform organisms in water may be small, large volumes may have to be examined in order to detect a single organism. Direct inoculation techniques on solid agar media are therefore not practicable and other methods must be used. Two procedures available are the multiple tube method, sometimes known as the dilution method or the most probable number (MPN) method, and membrane filtration.

1. *Multiple Tube Method*

In the multiple tube method of counting bacteria, measured volumes of the water to be tested, or of one or more dilutions of it, are added to tubes containing a liquid differential medium. It is assumed that, on incubation, each tube which received one or more viable organisms in the inoculum will show growth and the differential reaction appropriate to the organisms sought and the medium used. Provided negative results occur in some tubes, the most probable number of organisms in the original sample may be estimated from the number of tubes giving a positive reaction; statistical tables of probability are normally used for this purpose.

Since coliform organisms ferment lactose with the production of acid and gas, the media used for the presumptive coliform count contain lactose and an indicator of acidity. The tubes of media also contain an inverted inner (Durham) tube for the detection of gas production. In the United Kingdom inhibitory substances are usually added to suppress the growth of other organisms which may be present in water, and thus make the medium more selective for coliform organisms. Inhibitors commonly used include bile salts, as in MacConkey broth, and Teepol 610 as in Jameson and Emberley's (1956) medium. Selective media can also be made by choosing chemically defined nutrients which can be utilized by only a limited number of bacteria as in the various modifications of Folpmers' (1948) glutamic acid medium. In all these media the production of acid and gas is presumed to be due to the growth of coliform organisms.

Choice of medium. MacConkey broth has the disadvantage that the two main constituents, bile salts and peptone, vary considerably in their inhibitory and nutrient properties. The variability of the bile salts has been overcome by substituting Teepol 610 in Jameson and Emberley's medium, and the variability of all ingredients has been overcome in chemically-defined glutamate media. These media also have the advantage that Teepol 610 is much cheaper than bile salts and the ingredients of glutamate media are the cheapest of all.

A comparison of these media (PHLS Standing Committee on the Bacteriological Examination of Water Supplies, 1968a, 1969) showed that Jameson and Emberley's Teepol broth was a good alternative to MacConkey broth but that Gray's (1964) improved formate lactose glutamate medium gave results superior in most respects to MacConkey broth and Teepol broth and was superior to any of the other glutamate media which have been described. This work also extended previous work with MacConkey broth (PHLS Water Sub-committee 1953a) and confirmed that, with very few exceptions, the production of acid and gas in these three media when inoculated with unchlorinated water and incubated at 37°C for 48 hours indicates the presence of coliform organisms. Since the presumption that the production of acid and gas is due to the presence of coliform organisms requires further tests to confirm its correctness, this reaction is generally referred to as a "presumptive positive coliform reaction". The presumption is, however, in the United Kingdom, almost always correct for unchlorinated waters incubated at 37°C for 48 hours. Even though some aerobic spore-bearing organisms may cause false presumptive positive reactions in glutamate media, Gray's improved formate lactose glutamate is regarded as the most satisfactory medium for general use.

The sampling error of the coliform count. Various workers, of whom the first was McCrady (1915), have put forward formulae, based on the laws of probability, with which to estimate the number of organisms that are present in 100 ml of water when any given proportion of the tubes inoculated show growth, or acid and gas production, or other characteristic change. The various mathematical approaches have been reviewed by Eisenhart and Wilson (1943); Cochran (1950) has given an excellent introduction to the principles involved in the estimation of bacterial densities by dilution methods; and McCrady (1918), Hoskins (1934) and Swaroop (1938, 1951) have published tables that are particularly suitable for use in water analysis (see Appendix C). By means of these tables it is possible to report the most probable number (MPN) of coliform organisms in 100 ml of water.

The multiple tube method has a very large sampling error. Confidence limits for the MPN are given in tables in International Standards for Drinking Water (WHO, 1963) which should be consulted if required. For the 11-tube (1×50 ml, 5×10 ml, 5×1 ml) method and for the 15-tube (5×10 ml, 5×1 ml, 5×0·1 ml) method the upper limit generally lies between twice and three times the MPN and the lower limit generally lies between a third and a quarter of the MPN. For any given estimation it is possible that the true result lies beyond these limits but this will occur only in an average of 5 per cent of all such estimations and therefore the upper limit can, for practical purposes, be regarded as the maximum number of bacteria the sample might contain.

Technique of the multiple tube coliform count. The sample bottle should be inverted rapidly several times in order to distribute any deposit uniformly throughout the water. After the mouth of the bottle has been flamed, some of its contents should be poured off, the stopper or cap replaced, and the bottle well shaken. Any dilutions required, either for this test or for the plate count, should next be prepared as described below.

(a) Choice of volumes for inoculation. With waters expected to be of good quality one 50 ml volume and five 10 ml volumes of water should be inoculated into the medium chosen; with waters of doubtful quality one 50 ml, five 10 ml, and five 1 ml volumes should be used; with more polluted waters five 1 ml volumes of a 1 in 10 dilution should also be included and the single 50 ml volume may be omitted. With heavily polluted waters dilutions of 1/100, 1/1,000 or higher may be required to give some negative reactions and thus obtain a finite figure for the most probable number.

(b) Diluent. One-quarter-strength Ringer's solution should be used for making the dilutions, but sterilized tap water may be employed if it has been shown to be free from germicidal activity. Water from copper pipes is best avoided.

(c) Filling the dilution bottles with diluent. The sterile diluent should be measured out in 90 ml volumes with aseptic precautions into sterile dilution bottles. Alternatively, volumes of 9 ml may be used although this results in a slight loss of accuracy of dilution. A simpler but less accurate method is to sterilize the tubes or bottles in the autoclave after the diluent has been measured into them. This takes no account of the loss by evaporation during sterilization if cotton-wool plugs are used. Provided care is taken in the use of the autoclave, e.g. allowance is made for expansion and the pressure is raised and lowered gradually, screw-capped containers can safely be sterilized after the diluent has been measured out, without loss of volume.

(d) Preparation of dilutions. After the sample bottle has been thoroughly shaken (as described above), one or more tenfold dilutions, according to the nature of the sample, should be made in dilution bottles, test-tubes or screw-capped containers, by carrying over a suitable quantity of water to nine times its volume of diluent, mixing thoroughly, carrying over a similar volume of the diluted water to a further nine volumes of diluent, and so on. A separate single mark pipette should be used for making each dilution.

During the process of pipetting, care must be taken that the whole of the inner surface of the pipette is thoroughly wet. This can be ensured by sucking up the water several times. Excess water should be removed from the outer surface of the point of the pipette by touching the inside of the sample bottle before withdrawing.

(e) Inoculation of the culture medium. The same careful technique should be used as in making the dilutions. The 1 ml volumes are pipetted into tubes containing 5 ml of single-strength culture medium (Appendix B); the larger volumes (10 ml and 50 ml) are similarly added to their own volumes of double-strength culture medium (Appendix B). For volumes of 1 ml and 10 ml, sterile straight-sided pipettes should be used but for the 50 ml volume a bulb pipette is required. Bulb pipettes of this size are awkward to use, and an alternative method for the 50 ml volume is to use large screw-capped containers or bottles marked at the 50 ml and 100 ml levels. These have been previously filled to the 50 ml mark with double-strength culture medium, and sterilized. The necks of both containers should be flamed, and the water to be tested poured into the medium up to the 100 ml mark. With good technique the error in volume is negligible and there is at least no greater danger of contamination than in using a bulb pipette.

(f) Incubation and examination of the cultures. The inoculated tubes should be incubated at 37°C and examined after 13-24 hours. All those tubes showing acid and a bubble of gas in the inverted inner (Durham) tube, and those in which gas appears on tapping, should be regarded as "presumptive positive" tubes. Each of these should then be subcultured to a tube of confirmatory medium and to a tube of peptone water both for incubation at 44°C in order to give a rapid indication of the presence of *E. coli* (see also p. 28). At the same time a tube of confirmatory medium should also be inoculated for incubation at 37°C to confirm that the "presumptive positive tubes" do contain coliform organisms (see also Section VIII). The remaining tubes should be re-incubated and examined after another 24 hours. Any more tubes in which acid and gas formation become apparent should similarly be regarded as "presumptive positives" and examined for the presence of *E. coli* and for the confirmation of the presence of coliform organisms. With unchlorinated waters it may be sufficient to examine for the presence of *E. coli* without confirming the presence of other coliform organisms in all the presumptive positive tubes. With chlorinated waters, since false presumptive results may be due to the presence of aerobic or anaerobic spore-bearing bacilli, confirmation that coliform organisms are present is essential.

It has been recommended that acid production together with any detectable amount of gas should be regarded as a presumptive positive reaction. This is true of all tubes examined after 24 hours' incubation but when gas is first seen at the 48 hours' examination the acceptance of small amounts of gas may produce, with some water sources, too high a proportion of presumptive positive results which cannot be confirmed as being due to coliform organisms. In such circumstances it may be necessary to use the older rule that sufficient gas to fill the concavity of the Durham tube is the minimum needed before a presumptive positive result after 48 hours' incubation can be recorded. Some true positive results, however, may be missed if small amounts of gas are ignored (PHLS Standing Committee on the Bacteriological Examination of Water Supplies, 1968a).

(g) Reporting the results. Appendix C gives three sets of figures (Tables I, II and III), taken from McCrady, expressing the probable numbers of coliform organisms in 100 ml of water which correspond with the various combinations of positive and negative tubes in the series used. When dilutions are used a series of three tenfold dilutions containing some positive and some negative tubes should be chosen and a finite number obtained by multiplying by the appropriate dilution factor.

2. *Membrane Filtration*

The alternative method of counting coliform organisms in water is by filtering a measured volume of the sample through a membrane composed of cellulose esters. All the bacteria present are retained on the surface of the membrane which is then incubated face upwards on a differential selective medium. Colonies form on the surface of the membrane, and all acid-producing colonies are counted as presumptive coliform organisms. Since it is not possible to detect gas production on a membrane, there is an additional presumption that all acid producing colonies also produce gas. *Cl. perfringens*, a possible cause of false positive reactions in tubes of MacConkey broth, does not form colonies on a membrane, and the false positive reactions which may occur in tubes containing mixtures of organisms are also eliminated in the membrane filtration method.

Membranes have the advantage that the conditions of incubation can be easily varied to encourage the growth of attenuated or slow growing organisms. An initial short period of incubation at a low temperature, usually 4 hours at 30°C, can be followed by a further period at a higher temperature, either 35°, 37° or 44 °C. If required the membrane may be easily transferred to a different medium for the second incubation. By these techniques it is possible to obtain, within a total incubation time of 18 hours, direct presumptive coliform and direct *E. coli* counts which do not depend on the use of probability tables.

Counts on membranes are, however, subject to statistical variation and replicate counts of the same water sample will not in general show the same number of organisms. If a count of organisms (C) of greater than twenty is observed approximate 95 per cent confidence limits for the true number of organisms can be calculated as follows:

The upper limit $\approx C + 2 \times (2 + \sqrt{C})$
The lower limit $\approx C - 2 \times (1 + \sqrt{C})$

For example if 100 organisms were observed the true number of organisms in the water being tested would lie between 78 and 124. For counts of less than 20 the following values of the limits may be useful:

Membrane count	95% confident limits	
	Upper	Lower
1	5.6	0.025
5	11.7	1.6
10	18.4	4.8

The media used with membrane filters usually differ in composition from those used in the multiple tube method, because membranes selectively adsorb some substances but not others. There is also a differential adsorbent capacity between the nutrient pad and the membrane. Modifications of MacConkey broth have been used in the United Kingdom but modified Teepol broths are now favoured (Taylor and Burman, 1964; Burman, 1967a). Gultamate media for use with membranes are also under investigation.

Membrane filtration technique

(a) Equipment and materials. The apparatus must be used in conjunction with a filter flask, preferably with a stopcock between the filter and the flask, and a suitable source of vacuum. An electric pump is the most effective and convenient means but a water pump attached to a tap is adequate where there is sufficient water pressure, or a simple hand pump can be used.

All the experimental work on which the recommendations in this report are based was carried out with Oxoid membranes, which can easily be washed and sterilized for re-use, and are at present the cheapest type available in the United Kingdom. Whatman No. 17 filter papers of the same diameter as the membranes should be used as absorbent pads. These can be placed in petri dishes of standard size. Dishes that are only fractionally larger than the membranes will lead to difficulty in manipulation. It is advisable to incubate the dishes and membranes inside another container with a close-fitting lid to prevent excessive drying out. For accurate temperature control at 44°C immersion in a water bath is advisable, using a heavy brass cylindrical container with screw-on lid sealed against a rubber gasket and with an air expansion tube in the lid projecting above the water level.

(b) Preparation and sterilization of materials. The filtration apparatus and spare funnels, the distilled water and funnel stands should all be sterilized in the autoclave each day before use.

As membranes curl up when they are first wetted and heated, for their initial sterilization they must be interleaved with a smooth-surfaced filter paper (Whatman No. 1 is satisfactory) and packed firmly in a petri dish kept tightly closed with rubber bands. The manufacturers recommend that they should then be sterilized by autoclaving at 115°C for 20 minutes. Their filtration characteristics can however easily be damaged by overheating; it is therefore recommended that they should be treated first by immersing the pack of interleaved membranes in boiling distilled water for a few minutes. The membranes and filter papers should then be separated and the membranes again boiled immediately before use freely immersed in distilled water. This method of sterilization is adequate for membranes used for coliform organisms on a selective medium.

The membranes may be re-used several times. For re-use they should be gently washed in running water for several hours. Damaged membranes should be removed and the remainder immersed in boiling 3 per cent v/v hydrochloric acid. The acid must then be brought back to the boil and immediately poured off. The membranes should then be washed in at least three changes of sterile distilled water. A trace of bromocresol purple and sufficient sodium bicarbonate to neutralize any remaining acid should be added to the final rinse. After a final boiling in this last rinse water they are ready for re-use.

The Whatman No. 17 absorbent pads are sterilized in the autoclave for 20 minutes at 121°C and are then placed singly, with aseptic precautions, into sterile petri dishes. The culture medium is dispensed aseptically from sterile stock into each dish. A sterile adjustable automatic pipette provides the simples means of filling a large number of dishes. A slight excess of medium should be used and 2·5-3·0 ml is adequate. The pads may swell on standing and gradually soak up more medium. The excess should be poured off immediately before use. Failure to remove the excess will encourage the spreading of colonies into flat irregular shapes which can become confluent, especially on crowded membranes, and so interfere with colony counting.

(c) Preparation of sample and dilutions. The same technique for the shaking of the sample and preparation of dilutions, if required, should be used as recommended for the multiple tube technique.

(d) Choice of volumes for filtration. The coliform count and the *E. coli* count are made on separate volumes of water. All samples expected to contain less than 100 coliform organisms per 100 ml require the filtration of 100 ml of sample for each test. The volumes of polluted samples should if possible be so chosen that the number of colonies to be counted on the membranes lies between 10 and 100. In an unexpectedly grossly polluted sample, however, it is possible to give an approximate estimate of numbers even if they greatly exceed 100 per membrane. The use of ruled membranes is recommended to facilitate counting.

(e) Filtration procedure. Membranes sterilized as already described are placed in a container of boiling distilled water which is kept just at the boil. The sterilized filtration apparatus is connected to a stopcock, filter flask and vacuum pump.

A sterile membrane is removed from the boiling water bath with flat-ended forceps and placed on the base of the filter apparatus. During all manipulations membranes must always be held by the edge, never in the middle. The funnel and collar are clamped on firmly and the sample of water added to the funnel. Volumes of 50 or 100 ml may be poured from the bottle up to the funnel marks. For 10 ml volumes a sterile pipette should be used. For 1 ml volumes or dilutions, 9 or 10 ml of sterile dilution water should be added first, then 1 ml of sample or dilution with a sterile pipette and the contents mixed by slight rotation of the funnel. The stopcock is opened and when all the water has filtered through the funnel is again removed and placed in its stand. The membrane is removed, the stopcock closed, and the membrane placed on the absorbent pad saturated with culture medium, any surplus medium having been poured off first.

A second membrane is then placed on the apparatus and the process repeated. After each sample the funnel and collar are re-sterilized by immersing in boiling sterile distilled water for about 1 minute. For counts of coliform organisms and *E. coli* on selective media, the funnel need not be sterilized between two filtrations on one sample nor need the base be sterilized during a series of filtrations on several samples, unless unsterile water is accidentally dropped on the base or a damaged membrane is used.

It is however advisable not to alternate known polluted samples and chlorinated samples in one series through the same apparatus. This can be avoided by filtering all the chlorinated samples and those expected to give negative results first, followed by the known polluted samples. Alternatively a separate membrane filter apparatus can be reserved for all chlorinated samples on any one day, using a separate apparatus for the known polluted samples.

(f) Choice of media and incubation procedures. Most of the original work with membrane filters for the coliform group in the United Kingdom was carried out with modifications of MacConkey broth usually preceded by resuscitation on nutrient broth.

Bile salts however have a number of variable properties which influence growth on membranes in addition to the variability of their inhibitory properties. Some samples have a marked effect on colony colour, some encourage colonies to spread and some are apparently more inhibitory on membranes to coliform organisms than to some other organisms. These variabilities have not been entirely overcome even in the standardized bile salts No. 3 sold by some firms. Media containing Teepol as inhibitor are therefore recommended for the enumeration of coliform organisms by membrane filtration. With these media the use of a separate medium for resuscitation is unnecessary.

All the recommendations which follow are based on work carried out at the Metropolitan Water Board (Metropolitan Water Board, 1966a, 1967a). Further simplifications have since been made, so that only one medium is now recommended, namely 0·4 per cent enriched Teepol broth (0·4 ET). An additional refinement in the preparation of this medium is to adsorb the basic solution with cellulose triacetate to remove possible traces of toxic ingredients that might become concentrated on the membrane (Burman, 1967a). Details of preparation of this medium are given in Appendix B8. In devising these membrane techniques the aim was to obtain a count of coliform bacteria and *E. coli* in 18 hours as high as that from a multiple tube technique in 48 hours at 37°C followed by subculture at 44°C for 24 hours, and which would at the same time restrict the count to the arbitrarily defined coliform group of the water bacteriologists.

In the Metropolitan Water Board Laboratories it was found that in unchlorinated waters or in waters that had become polluted after chlorination, provided no chlorine residue was present, membranes incubated on 0·4 ET for 4 hours at 30°C followed by 14 hours at 35°C generally gave higher counts of coliform organisms than the multiple tube method using either MacConkey broth or Gray's glutamate medium incubated for 48 hours at 37°C. Incubation at 37° instead of 35° gave a slightly lower count.

An *E. coli* count on unchlorinated waters equivalent to multiple tube counts in MacConkey broth was obtained by incubating the membranes on medium 0·4 ET for 4 hours at 30°C followed by 14 hours at 44°±0·25°C.

If rapid results are required the membranes may be examined after a total incubation time of 12 hours. If no colonies of any kind are present, a nil count can be assumed. If small colonies of indeterminate colour are present the membranes must be returned to the incubator for the full 18 hours. In a polluted sample a very high proportion of colonies, particularly of *E. coli* can be counted after as little as 10 hours at 44°C.

For the examination of chlorinated samples both for coliform organisms and *E. coli,* a lower initial temperature of incubation and longer total incubation times are required. For coliform organisms membranes should be incubated on 0·4 ET for 6 hours at 25° followed by 18 hours at 35°C. For *E. coli* membranes should be incubated on 0·4 ET for 6 hours at 25° followed by 18 hours at 44°± 0·25°C. These procedures may give a higher proportion of false positive results than the methods recommended for unchlorinated samples, but if their use is restricted to chlorinated samples, which should normally contain no coliform organisms at all, the few colonies that do develop can be confirmed by the procedures recommended for the multiple tube technique.

Until these techniques have been used in numerous other laboratories on a wide variety of water supplies, it should not be assumed that equivalent results will always be obtainable. It is, therefore, essential that, before adopting membrane filtration as a routine procedure in any laboratory or with any particular water supply, an adequate parallel series of tests should be run comparing membranes with multiple tubes, in order to establish their equivalence or the superiority of one or the other.

(g) Incubation and counting techniques. Membranes are placed face upwards on the absorbent pads previously saturated with culture medium, care being taken to exclude air bubbles between membrane and pad. The Petri dishes are then inverted and placed inside another container with a tight fitting lid to prevent evaporation. For membranes incubated at 44° a special heavy brass cylindrical container with screw-on lid and gasket and a breathing tube has been found to be satisfactory. The weight of the cylinder keeps it completely immersed in the water bath with the breathing tube projecting above the water level.

To change the temperature from 25° or 30° to 35° or 44° the whole container of membranes can be transferred from one incubator to another after the appropriate time. As this may often be required at an inconvenient time it is preferable to arrange for an automatic temperature change. This requires an incubator and a water bath with two thermostats connected in parallel and a time switch controlling one thermostat. Incubators should be of the anhydric type as the temperature changes are too slow in a water-jacketed incubator.

After incubation the membranes should be examined with a hand lens under good lighting, arranged at a low angle of incidence so that the colonies throw shadows. All yellow colonies should be counted irrespective of size. Pink or colourless colonies should be ignored. Membranes should be examined and colonies counted within a few minutes of removal from the incubator or bath as colours are liable to change on cooling and standing.

Interpretation of membrane counts. When interpreting the results of membrane counts it must be remembered that neither gas nor indole production can be detected on membranes. On the other hand false results due to anaerobic spore-forming organisms will not occur and with the medium recommended false results due to aerobic spore-forming organisms will not occur either. The differentiation of gas-producing and non-gas-producing organisms is usually of little significance for treatment control purposes; it becomes of significance only in chlorinated samples and in samples collected in the distribution system which should contain no coliform organisms of any kind. From such samples, yellow colonies from membranes at 35°

or 37°C should be subcultured to lactose peptone water to confirm gas production in 48 hours at 37°C, and yellow colonies from membranes at 44°C should be subcultured to lactose peptone water and peptone water to confirm gas and indole production at 44°C in 24 hours.

Limitations of use of membrane filtration. Membranes are unsuitable for use with waters of high turbidity in association with low counts of coliform organisms. In these circumstances the membrane will become blocked before sufficient water can be filtered. Membranes are also unsuitable for water containing few coliform organisms in the presence of many non-coliform organisms which are capable of growth on the media used and are thus liable to cover the whole membrane and interfere with the growth of coliform organisms. If non-gas-producing lactose-fermenting organisms are predominant in the water, membranes will be unsuitable because of the high proportion of false positive results.

Advantages of membranes. The outstanding advantage of the membrane filtration technique is the speed with which results can be obtained, including an *E. coli* count. This enables rapid corrective action to be taken when required, and it also enables plant to be put back into service more quickly when a negative result is obtained. In the laboratory there is a considerable saving in technical labour, and in the quantities of media and glassware required. There is also a considerable saving in cost of materials if the most expensive consumable item, the membrane, is carefully handled and re-used.

Delayed incubation. It has been stressed elsewhere in this report (p. 16) that water samples should be dispatched to the laboratory and examined with the minimum of delay and certainly not more than six hours should elapse between sampling and examination. Where this is difficult to achieve, the problem can be overcome by filtering the sample through a membrane on site or in a local laboratory with limited facilities. The membrane is then placed in the normal way on an absorbent pad saturated with a transport medium

(Panezai, Macklin and Coles, 1965) (see Appendix B9). This is a very dilute medium on which the organisms survive but do not develop visible colonies in three days at room temperature. If polystyrene petri dishes are used they can be despatched by post to the central laboratory where the membranes should be transferred to Teepol medium and incubated for 4 hours at 30°C followed by 14 hours at 35°C or 44°C. Delays of three days have made little difference to counts of coliform organisms and *E. coli*.

Section VIII. Confirmation and Differentiation of Coliform Organisms

The further investigation of presumptive positive reactions can be considered as two problems. First because of the importance of detecting *E. coli* in water supplies a rapid and simple test is required to show whether or not *E. coli* is present; second it is necessary to confirm that other presumptive positive reactions are due to true coliform organisms. Further differentiation is rarely necessary but may be useful in showing that the water is regularly contaminated with the same organisms and so helping to trace the source of the contamination.

Classification of Coliform Organisms

In previous editions of this report the nomenclature adopted by Wilson and his colleagues (1935) was used to classify the various types of organisms in the coliform group. This classification was based on tests for the production of acid and gas from lactose at 44°C, together with indole-production and methyl-red, Voges-Proskauer and citrate-utilization tests, usually known collectively as the IMViC reactions. More recently changes have been introduced in bacterial nomenclature and, since taxonomists do not attach the same importance as do water bacteriologists to certain reactions including the fermentation of lactose, no direct translation into modern terminology is possible for the organisms listed in previous editions of this report. Complete identification in terms of modern nomenclature would require an extensive series of tests such as those described by Edwards and Ewing (1962) or Cowan and Steel

(1965). The relatively simple IMViC tests at present used by water bacteriologists are usually sufficient for identifying any particular coliform organism found without giving it a specific name. As an example, *E. coli*, if typed by the IMViC method, would be recorded as ++--44°+ and the organism from jute packing, formerly known as Irregular VI, would give the reactions --++44°+.

Rapid Detection of E. coli

The detection of *E. coli* depends on the ability of this organism to produce gas from lactose at 44°C. There is ample evidence that in the United Kingdom such gas production is practically specific for *E. coli*. Some other coliform organisms can produce gas at this temperature but few of them are able to produce indole and the indole test, also carried out at 44°C, provides a means of distinguishing most of them from *E. coli*.

With pure cultures, such as can be obtained from membrane counts, 1 per cent lactose peptone water is suitable for showing gas production. However, when subculturing from presumptive positive tubes, which may contain spore-bearing organisms, some inhibitory substance which will prevent the growth of these organisms is required. Brilliant-green bile broth has been extensively used to suppress the growth of spore-bearing organisms. In trials carried out by the PHLS Standing Committee on the Bacteriological Examination of Water Supplies (1968b) 1 per cent lactose ricinoleate broth gave better results than brilliant green bile broth and was less subject to the variability of ox bile or brilliant green.

Each presumptive positive tube should be subcultured to a tube of either 1 per cent lactose ricinoleate broth or brilliant-green bile broth and incubated for 24 hours at 44°C. At the same time a tube of peptone water should be inoculated for production of indole after 24 hours' incubation at 44°C. If confirmation is required with membrane counts of *E. coli* grown at 44°C the colonies may be subcultured to 1 per cent lactose peptone water and peptone water for incubation at 44°C. Although gas and indole production can frequently be detected after 6 hours' incubation the test should not be regarded as negative until after 24 hours' incubation (Taylor, 1955). For the test to work satisfactorily a temperature in the medium of 44°±0·25°C is required. In order to secure this, incubation in a carefully regulated water bath is essential. The variations in temperature in ordinary incubators, even if water jacketed, are too great to permit the maintenance of a constant temperature in the medium. It is useful to include positive and negative control organisms with each batch of tests. Care should be taken in selecting the peptone for the indole tests at 44°C (see Appendix B6).

A most probable number of *E. coli* per 100ml can be calculated from those presumptive positive tubes subcultured which have given a positive reaction for *E. coli* in the same way as the presumptive coliform count is calculated.

Confirmatory Tests for Coliform Organisms

With unchlorinated waters it is usually sufficient to demonstrate the presence of *E. coli* by the method outlined above. With chlorinated waters it is essential to confirm the presence of coliform organisms by subculture to 1 per cent lactose ricinoleate broth or brilliant-green bile broth for incubation at 37°C for 48 hours, in order to exclude false positive results due to aerobic or anaerobic gas-producing, spore-bearing organisms. Production of gas in these media within 48 hours can be taken as sufficient confirmation that coliform organisms are, in fact, present. Further differentiation should only be necessary in special circumstances.

At the same time as these confirmatory tests are carried out it is advisable to plate out each positive tube derived from chlorinated water on a suitable solid medium such as MacConkey agar. When the tubes are examined for gas production the colonial morphology of the organisms present will then be available for inspection, and pure cultures will be easily obtainable should further differential tests be necessary. If differentiation of coliform organisms from unchlorinated water is required the presumptive positive tube may similarly be plated out on MacConkey agar. It is essential to ensure that well separated colonies are obtained.

(a) Appearance of colonies. On MacConkey agar coliform colonies are usually circular in shape, convex or low convex with a smooth surface and entire edge. They are red, but the depth of colour varies considerably. Normally *E. coli* colonies are deep red and non-mucoid but no reliance can be placed on colonial differences as a means of differentiation within the group. For further examination two or three colonies, as far as possible of different appearance, should be selected and subcultured. Should no red or pink colonies develop within 48 hours, at least one colony of the predominant variety should be selected for further investigation, bearing in mind that such atypical colonies may nevertheless produce acid and gas in lactose peptone water within 48 hours. If there is any doubt that the colonies picked are coliform organisms a film should be made and stained by Gram's method. Only colonies consisting of Gram-negative, non-sporing rods should be examined further. Many organisms found in water differ from the coliform group as defined on page 3 of this report solely in their inability to produce gas from lactose at 37°C, although they may readily be able to do so at lower temperatures. Many organisms naturally found in water, including coliform organisms and *Aeromonas* species, have an optimum growth temperature below 37°C but may produce gas from lactose at 37°C. *Aeromonas* species, which resemble coliform organisms but are of no sanitary significance, may be excluded by carrying out an oxidase test on the colonies by Kovacs' (1956) method.* Colonies giving a positive oxidase reaction are not coliform organisms.

(b) Subculturing for differential tests. Each colony selected should be lightly touched with a straight wire and inoculated into one tube of citrate medium for the citrate-utilization test. Two

---

**Oxidase reaction:* On a piece of filter paper in a petri dish place 2 to 3 drops of 1% tetramethyl-*p*-phenylenediamine aqueous solution. Smear the culture under test across the impregnated paper with a platinum (not nichrome) loop. A positive reaction is indicated by the appearance of a dark purple colour on the paper within 10 seconds. Controls: Positive: *Pseudomonas aeruginosa*. Negative: *Escherichia coli*.

tubes of glucose phosphate medium for the methyl-red and Voges-Proskauer tests, one tube of peptone water for indole formation and two tubes of lactose peptone water should also be inoculated from the same colony.

(c) The differential tests. (1) Fermentation of lactose at 37°C. One lactose peptone water culture is incubated for 48 hours at 37°C. Acid and gas production within this time is sufficient to show that a coliform organism has been selected.

(2) Fermentation of lactose at 44°C. The other tube of lactose peptone water is incubated at 44°C. The presence of any amount of gas in the inverted inner tube after 6 to 24 hours' incubation indicates a positive reaction. Absence of gas production after 24 hours' incubation even though growth or acid production is present is regarded as a negative reaction.

The need for accurate temperature control in water baths for this test has already been mentioned when describing the rapid test for *E. coli* from presumptive positive tubes. The bath should be electrically heated and controlled by a thermostat capable of maintaining the temperature at 44°±0.25°C. Thermostatically controlled heating and circulating units are available which are capable of maintaining the temperature at 44°±0.1°C. The indicator thermometer should be checked against a N.P.L. standard. Two control tubes should be included, one inoculated with a known strain of *E. coli* and the other with a strain of coliform organism such as *Klebsiella aerogenes* which does not produce gas from lactose at 44°C. A recording thermometer provides a useful check.

(3) Indole test. After incubation of the peptone water culture at 37°C for at least 24 hours, 0.2-0.3 ml of Kovacs' reagent (Appendix B 7) is added and the tube gently shaken. If the test is positive a deep red colour appears in the upper layer almost immediately.

(4) Methyl-red test. To one of the inoculated tubes of glucose phosphate medium incubated at 30°C for three days, two drops of 0·04 per cent methyl-red solution are added. A magenta red colour

indicates a positive result, a yellow colour a negative result; pink or pale red are best considered as doubtful results.

(5) Voges-Proskauer test. To the other inoculated tube of glucose phosphate medium, also incubated at 30°C for three days, a knife point of creatine and 5 ml of 40 per cent (or a solid pellet of) potassium hydroxide are then added. The tube should be well shaken and observed for a colour change. If positive a pink colour develops; if negative no colour is produced (O'Meara, 1931). The colour may be slow to develop and tubes should not be discarded as negative until 4 hours after the potassium hydroxide has been added.

The α-naphthol modification described by Barritt (1936) is considered too sensitive a test for the differentiation of coliform organisms.

(6) Growth in citrate. The tube of citrate medium (Appendix B 13) receives the first inoculum from the selected colony with the straight wire so that none of the other culture media are carried over into the citrate medium. It is then incubated at 30°C for three days. Opacity is considered evidence of growth. If difficulty is experienced in determining opacity, bromothymol blue may be added in 0·008 per cent concentration as an indicator. Growth is accompanied by a change in colour from a pale green to a bluish green or blue. Simmons' (1926) citrate agar medium can be used instead of Koser's citrate medium.

The Faecal Streptococcus Test

Before describing methods of counting faecal streptococci it is necessary to define the meaning to be given to this term. The species of streptococci normally occurring in human and animal faeces and therefore most likely to be found in polluted water are *Streptococcus faecalis, S. faecium, S. durans, S. bovis* and *S. equinus*. Strains with properties intermediate between these are also common. The term "faecal streptococci" should be used to refer only to these named species and to the intermediate strains which also belong to Lancefield's serological group D. The term "enterococci," often used as a synonym for faecal streptococci, has not been precisely defined, and its use is not recommended. Other streptococci occasionally present in faeces, but not belonging to Lancefield's group D, include *S. mitis* and *S. salivarius*, which originate in the mouth and are swallowed in the saliva. Such strains differ in many of their properties from true faecal streptococci and their presence in water should not necessarily be regarded as evidence of faecal pollution.

The properties of faecal streptococci are fully described in textbooks of bacteriology. Their important characteristics are the ability to grow at 45°C; to grow in the presence of 40 per cent bile and in concentrations of sodium azide which are inhibitory to coliform organisms and most other Gram-negative bacteria. Some species resist heating at 60°C for 30 minutes, will grow at pH 9·6 and in media containing 6·5 per cent sodium chloride.

Technique for Enumerating Faecal Streptococci

The techniques for enumeration depend on the use of sodium azide as a selective inhibitor, and differentiation by growth at 45°C. A multiple tube method may be used which takes five days for a confirmed result, or a membrane filtration method may be used which gives a result in two days. Although faecal streptococci will normally grow at 45°C, a preliminary resuscitation or growth period at 37°C is recommended since some organisms may temporarily lose this ability outside the body (Allen, Pierce and Smith, 1953).

(a) Multiple tube method. Various volumes of water are inoculated into tubes of single or double strength glucose azide broth (Hannay and Norton, 1947)(see Appendix B14) in exactly the same manner as in the multiple tube method for coliform organisms (see p. 19). Inverted inner tubes are not necessary as there is no gas production. The tubes are incubated at 37°C for 72 hours. As soon as acidity is observed a heavy inoculum is subcultured into further tubes of single strength Hannay and Norton's medium, and incubated at 45°C for 48 hours; all tubes showing acidity at this temperature contain faecal streptococci. The most probable number can then be determined from the probability tables (Appendix C). Childs and Allen (1953) avoided the double incubation in azide medium at 37° and 45°C by resuscitating in glucose broth for 2 hours at 37°C and then adding the azide portion of the medium and incubating for 48 hours at 45°C. This method shortens the time and gives significantly higher counts but involves an elaborate procedure after the first 2 hours.

(b) Membrane filtration method. The membrane filtration procedure as described for coliform organisms on pp. 20-26 is used except that a different medium and incubation procedure are required. Membranes cannot be washed and re-used. Membranes previously used for coliform organisms must not be used for streptococci. After filtration, the membrane is placed on a well dried plate of Slanetz and Bartley's (1957) glucose azide agar (see Appendix B15). This is incubated at 37°C for four hours and then at 44° or 45°C for 44 hours. All red or maroon colonies are counted as faecal streptococci (Taylor and Burman, 1964; Mead, 1966). Mead found that initial resuscitation without the inhibitory azide did not increase the recovery of streptococci. Incubation throughout at 37°C will permit the growth of many streptococci which are neither faecal nor of group D.

(c) Confirmatory tests. The presence of streptococci may be confirmed by direct microscopical examination, which will show typical short-chained streptococci, or by subculture on MacConkey agar, which will show small red colonies after 24 or 48 hours' incubation at 37°C.

Mead (1963 and 1964) has described a single rapid test which will differentiate *S. faecalis* from other faecal streptococci. Cultures are plated on a tyrosine sorbitol thallous acetate agar (see Appendix B16) and incubated at 45°C for three days. Differentiation depends on the ability of *S. faecalis* to reduce 2, 3, 5-triphenyltetrazolium chloride (T.T.C.) at pH 6.2, to ferment sorbitol, to produce tyrosine decarboxylase and to grow at 45°C in the presence of 0.1 per cent thallous acetate. Colonies of *S. faecalis* on this medium have a uniform deep maroon colour encircled by clear zones where the tyrosine has been decomposed. As Mead found that *S. faecalis* was particularly associated with man this test when positive suggests that the pollution is of human origin. A negative result does not however exclude human pollution.

REFERENCES

Attenborough, S. J., and Scarr, M. P. (1957). The use of membrane filter techniques for control of thermophilic spores in the sugar industry. *J. Appl. Bacteriol. 20* (3): 460–466.

Ayres, P. A., and Barrow, G. I. (1978). The distribution of *Vibrio parahaemolyticus* in British coastal waters. Report of a collaborative study 1975-6. *J. Hyg. 80,* 281.

Barrow, G. I. (1977). Bacterial indicators and standards of water quality in Britain. In *Bacterial Indicators/Health Hazards Associated with Water,* ASTM SPT 635, A. W. Hoadley and B. J. Dutka (Eds.). American Society for Testing and Materials, Philadelphia.

Barrow, G. I., Miller, D. C., Gray, R. D., and Lowe, G. H. (1978). *Report on a Feasibility Study on the Distribution and Use of Simulated Water Samples for Comparative Bacteriological Analysis*. CEC, Luxembourg. (In press.)

Beling, A., and Maier, K. H. (1954). Die Wasserversorgung und Uberwachung mit Hilfe der membranfilter Methode, Jahresber. *Vom Wasser 21:* 118.

Bonde, G. J. (1977). Bacterial indication of water pollution. In *Advances in Aquatic Microbiology*. M. R. Droop and J. H. Jannasch (Eds.). Academic Press, London.

Brown, M. R. W., and Scott Foster, J. H. (1970). A simple diagnostic milk medium for *Pseudomonas aeruginosa*. *J. Clin. Pathol. 23:* 172–177.

Burman, N. P. (1955). The standardization and selection of bile salts and peptone for culture media used in the bacteriological examination of water. *Proc. Soc. Water Treat. Exam. 4:* 10.

Burman, N. P. (1960). Developments in membrane filtration techniques. 1. Coliform counts on MacConkey broth. *Proc. Soc. Water Treat. Exam. 9:* 60.

Burman, N. P. (1961). Some observations on coli-aerogenes bacteria and streptococci in water. *J. Appl. Bacteriol. 24* (3): 368–376.

Burman, N. P. (1965). Taste and odour due to stagnation and local warming in long lengths of piping. *Proc. Soc. Water Treat. Exam. 14:* 125.

Burman, N. P. (1967a). Developments in membrane filtration techniques. 2. Adaptation to routine and special requirements. *Proc. Soc. Water Treat. Exam. 16:* 40.

Burman, N. P. (1967b). Recent advances in bacteriological examination of water. In *Progress in Microbiological Techniques,* C. H. Collins (Ed.). Butterworth, London, p. 185.

Burman, N. P. (1973). The occurrence and significance of actinomycetes in water supply. In *Actinomycetales,* G. Sykes and F. A. Skinner (Eds.). Academic Press, New York.

Burman, N. P., and Colbourne, J. S. (1977). Techniques for the assessment of growth of micro-organisms on plumbing materials used in contact with potable water supplies. *J. Appl. Bacteriol. 43:* 137—144.

Burman, N. P., Oliver, C. W., and Stevens, J. K. (1969). Membrane filtration techniques for the isolation from water, of coli-aerogenes, *Escherichia coli,* faecal streptococci, *Clostridium perfringens,* actinomycetes and micro-fungi. In *Isolation Methods for Microbiologists,* D. A. Shapton and G. W. Gould (Eds.), No. 3. Academic press, New York, pp. 127—134.

CEC (1977). *Bacteriological Analyses of Drinking Water.* EUR 5694E. Commission of the European Communities--Environment and Quality of Life, Luxembourg.

Chaina, P. N. (1968). A study of bacterial flora, bacteriophages and a bacteriocin-like agent isolated from seawater and sea weeds collected at selected stations in the North Sea. Ph.D. thesis, University of Aberdeen, Aberdeen, Scotland.

Chiori, C. O., Hambleton, R., and Rigby, G. J. (1965). The inhibition of spores of *Bacillus subtilis* by cetrimide retained on washed membrane filters and on the washed spores. *J. Appl. Bacteriol. 28*(2): 322-330.

Cholodny, N. (1928). Contributions to the quantitative analysis of bacterial plankton. *Trav. Sta. Biol. Dniepre 3:* 157.

CMEA (1968). *Standard Methods for the Water Quality Examination for the Member Countries of the Council for Mutual Economic Assistance.* The Ministry of Forestry and Water Management in Co-operation with the Hydraulic Research Institute, Prague, 1968.

Colbourne, J. S. (1978). Personal communication.

Collins, V. G. (1963). The distribution and ecology of bacteria in freshwater. *Proc. Soc. Water Treat. Exam. 12:* 40—73.

Collins, V. G. (1977). Methods in sediment microbiology. In *Advances in Aquatic Microbiology*, vol. 1, M. R. Droop and H. W. Jannasch (Eds.). Academic Press, New York, pp. 219–272.

Collins, V. G., and Willoughby, L. G. (1962). The distribution of bacteria and fungal spores in Blelham Tarn with particular reference to an experimental overturn. *Arch. Mikrobiol. 43:* 294.

Collins, V. G., Jones, J. G., Hendrie, M. S., Shewan, J. M., Wynn-Williams, D. D., and Rhodes, M. E. (1973). Sampling and estimation of bacterial populations in the aquatic environment. In *Sampling--Microbiological Monitoring of Environments*, R. G. Board and D. W. Lovelock (Eds.). Academic Press, New York.

Cranston, P. M., and Calver, J. H. (1974). Quantitative fluorescent microscopy of yeast in beverages. *Food Technol. Aust. 26:* 15.

Daubner, I., and Peter, H. (1974). *Membranfilter in der Mikrobiologie des Wassers*. Walter de Gruyter, Berlin, p. 216.

Emmenegger, T. (1956). Vergleichende Untersuchungen über das Wachstum van *Escherichia coli* auf verschiedenen Endoagar--Nahrboden mit Hilfe der membranfilter Methode, *Arch. Lebensmittelhyg. 47:* 536.

Funder, S., and Johannessen, S. (1957). The membrane filter as an aid in the cultivation and identification of fungi. *J. Gen. Microbiol. 17:* 117–119.

Goetz, A., and Tsuneishi, N. (1951). Application of molecular filter membranes to the bacteriological analysis of water. *J. Am. Water Works Assoc. 43:* 943.

Gray, R. D. (1964). An improved formate lactose glutamate medium for the detection of *Escherichia coli* and other coliform organisms in water. *J. Hyg. Camb. 62:* 495.

Gray, R. D., and Lowe, G. H. (1976). The preparation of simulated water samples for the purpose of bacteriological quality control. *J. Hyg. 76:* 49.

Hannay, C. L., and Norton, I. L. (1947). Enumeration, isolation and study of faecal streptococci from river water. *Proc. Soc. Appl. Bacteriol. 10:* 39.

Jannasch, H. W. (1953). Zur Methode der quantitativen Untersuchungen von Bakterienkulturen in flussigen Mediem. *Arch. Microbiol. Berlin 18:* 425.

Jannasch, H. W. (1954). Beitrag zur Methode der direkten mikroskopischen Untersuchung von Mikroorganismen auf Membranfiltern: über die Aufhellung der Filter. *Zentralbl. Bakteriol. Parasitenkd. Infektionskr. Hyg. I Orig. 161:* 225.

Jannasch, H. W. (1958). Studies on planktonic bacteria by means of a direct membrane filter method. *J. Gen. Microbiol. 18:* 609–620.

Jannasch, H. W., and Jones, G. E. (1959). Bacterial populations in sea water as determined by different methods of enumeration. *Limnol. Oceanogr. 4:* 128.

Jawad, L. A. (1976). On the sanitary significance of *Pseudomonas aeruginosa*. Msc. thesis, University of Dundee, Scotland.

Jessen, O. (1965). *Pseudomonas aeruginosa* and other green fluorescent pseudomonads. A taxonomic study. Copenhagen, Munksgaard.

Jones, J. G. (1972). Studies of freshwater bacteria: association with algae and alkaline phosphatase activity. *J. Ecol. 60:* 59—75.

Jones, J. G. (1974). Some observations on direct counts of freshwater bacteria obtained with a fluorescence microscope. *Limnol. Oceanogr. 19*: 540—543.

Jones, J. G. (1975). Some observations on the occurrence of the iron bacterium *Leptothrix ochracea* in fresh water, including reference to large experimental enclosures. *J. Appl. Bacteriol. 39:* 63—72.

Jones, J. G. (1978). The distribution of some freshwater planktonic bacteria in two stratified eutrophic lakes. *Freshw. Biol. 8:* 127—140.

Jones, J. G., and Simon, B. M. (1975). An investigation of errors in direct counts of aquatic bacteria by epifluorescence microscopy, with reference to a new method for dyeing membrane filters. *J. Appl. Bacteriol. 39:* 317—329.

Levin, M. A., and Cabelli, V. J. (1972). Membrane filter technique for enumeration of *Pseudomonas aeruginosa*. *Appl. Microbiol. 24:* 864.

Lingappa, Y., and Lockwood, J. L. (1962). Chitin media for selective isolation and culture of actinomycetes. *Phytopathologia 52:* 317.

Mead, G. C. (1966). Faecal streptococci in water supplies and the problem of selective isolation. *Proc. Soc. Water Treat. Exam. 15:* 207.

Miller, E. J. (1963). A method for the removal from membrane filters of microorganisms filtered from water and air. *J. Appl. Bacteriol. 26*(2): 211—215.

Moore, B. (1971). The health hazards of pollution. In *Microbial Aspects of Pollution,* G. Sykes and F. A. Skinner (Eds.). Academic Press, London.

Müller, G. (1977). Bacterial indicators and standards for water quality in the Federal Republic of Germany. In *Bacterial Indicators/Health Hazards Associated with Water,* ASTM STP635, A. W. Hoadley and B. J. Dutka (Eds.). American Society for Testing and Materials, Philadelphia.

Mulvany, J. G. (1969). Membrane filter techniques in Microbiology. In *Methods in Microbiology,* vol. 1, J. R. Norris and D. W. Ribbons (Eds.). Academic Press, New York, pp. 205–253.

Niemela, S. (1965). The quantitative estimation of bacterial colonies on membrane filters. *Ann. Acad. Sci. Fenn. Ser. A. IV Biol.,* p. 90.

Ormerod, K. S. (1974). *Review and Comparison of Methods Used in Different Countries for the Analysis of Total Coliform Bacteria, Fecal Coliform Bacteria, Fecal Streptococci and* Clostridium perfringens. [Report NIVA X8-01.] Norwegian Institute for Water Research, Oslo.

Panezai, A. K., Macklin, J. J., and Coles, H. G. (1965). Coli-aerogens and *Escherichia coli* counts on water samples by means of transported membranes. *Proc. Soc. Water Treat. Exam. 14:* 179.

Paton, A. M., and Jones, S. M. (1975). The observation and enumeration of microorganisms in fluids using membrane filtration and incident fluorescence microscopy. *J. Appl. Bacteriol. 38*(2): 199.

Public Health Laboratory Service (1968). Standing Committee on the Bacteriological Examination of Water Supplies. Comparison of MacConkey broth, Teepol broth and glutamic acid media for the enumeration of coliform organisms in water. *J. Hyg. 66:* 167.

Rasumov, A. S. (1932). Die direkte Methode der Zahlung der Bakterien im Wasser und ihre Vergleichung mit der Kochsohen Plattenkulture Methode. *Microbiol. Mosc. 1:* 131.

Report (1969). *The Bacteriological Examination of Water Supplies.* Reports on Public Health and Medical Subjects No. 71. Department of Health and Social Security, Welsh Office, and Department of Environment. Her Majesty's Stationery Office, London.

Rodina, A. G. (1972). *Methods in Aquatic Microbiology.* R. R. Colwell and M. S. Zambruski (Trans./Eds./Rev.). Butterworth, London, and University Park Press, Baltimore, Md.

Slanetz, L. W., and Bartley, C. H. (1957). Numbers of enterococci in water, sewage and feces determined by the membrane filter technique with an improved medium. *J. Bacteriol. 74:* 591–595.

Stanier, R. Y., Palleroni, N. J., and Doudoroff, M. (1966). The aerobic pseudomonads: a taxonomic study. *J. Gen. Microbiol. 43:* 159–271.

Suess, M. J. (1977). Bacterial water quality and standards: the role of the World Health Organization. In *Bacterial Indicators/Health Hazards Associated with Water,* ASTM STP 635, A. W. Hoadley and B. J. Dutka (Eds.). American Society for Testing and Materials, Philadelphia.

Taylor, E. W. (1955). Membrane filtration technique for the bacteriological examination of water. *Metropolitan Water Board, Report on the results of bacteriological, chemical and biological examination of London Waters for the years 1953-54, 36:* 48—53.

Taylor, E. W. (1957). Membrane filtration. *Metropolitan Water Board, Report on the results of bacteriological, chemical and biological examination of London Waters for the years 1955-6, 37:* 13—20.

Taylor, E. W. (1959). Progress with membrane filtration. *Metropolitan Water Board, Report on the results of bacteriological, chemical and biological examination of London Waters for the years 1957-8, 38:* 26—37.

Taylor, E. W. (1961). Further progress with membrane filtration. *Metropolitan Water Board, Report on the results of bacteriological, chemical and biological examination on London Waters for the years 1959-60, 39:* 20—27.

Taylor, E. W. (1963). Further progress with membrane filtration. *Metropolitan Water Board, Report on the results of bacteriological, chemical and biological examination of London Waters for the years 1961-2, 40:* 15—17.

Taylor, E., and Burman, N. P. (1963). La mise en application de la technique de filtration par membrane pour l'examen biologique de l'eau. *Tech. Eau 17*(200): 37.

Taylor, E., and Burman, N. P. (1964). The application of membrane filtration techniques to the bacteriological examination of water. *J. Appl. Bacteriol. 27*(2): 294—303.

Taylor, E., Burman, N. P., and Oliver, C. W. (1953). Use of the membrane filter in the bacteriological examination of water. *J. Appl. Chem. 3:* 233.

Taylor, E., Burman, N. P., and Oliver, C. W. (1955). Membrane filtration technique applied to the routine bacteriological examination of Water. *J. Inst. Water Eng. 9:* 248.

WHO (1971). *International Standards for Drinking Water.* World Health Organization, Geneva, 1971.

Williams, E. D. F. (1969). A submerged membrane filter apparatus for microbiological sampling. *Mar. Biol. 3:* 78.

12

MEMBRANE FILTRATION: LATIN AMERICAN AND CARIBBEAN VIEWS

MARIA THEREZINHA MARTINS Companhia de Tecnologia de Saneamento Ambiental (CETESB), São Paulo, Brazil

I. INTRODUCTION

The use of membrane filters in routine water analysis is still in its infancy in South America and has been emphasized only during the past 4 years. Although the technique has been known for over 15 years in several South American countries, its use has been restricted to a few classroom demonstrations, experimental studies, or occasionally for emergency analysis of drinking water where the bacteriological quality has to be determined as fast as possible. Only recently have membrane filtration procedures been used routinely in drinking water analysis.

To obtain data concerning the use of the membrane filter technique in South and Central America, interviews were held with laboratory personnel and laboratory supply companies, and questionnaires were sent to Latin American and Caribbean laboratories (Appendix), as data and references on membrane filtration are very scarce in these areas. Because of the vastness of the continent and the limited exchange of scientific information that takes place between countries, the task of collecting the necessary data for this chapter was very difficult. The necessary information was not received from some countries.

This chapter describes the present use of the membrane filter technique in South and Central America and the Caribbean area. In this study we have tried to group the countries that have several points in common, such as cultural formation, political and socioeconomic situation, and/or similarity of language.

II. APPLICATION OF THE MEMBRANE FILTER TECHNIQUE IN SOUTH AND CENTRAL AMERICA

A. Central America

In some laboratories in Mexico the membrane filter procedure is routinely used for the bacteriological analysis of water, phytoplankton analysis, chlorophyll concentrations, and in certain chemical analyses, such as toxic metal and dissolved salts techniques (Díaz, 1978; Martinez, 1978). Approximately 15,000 membrane filters are used annually by the main environmental control laboratories in Mexico. Drinking water control laboratories in Mexico use approximately 12,000 membrane filters per annum. The methods used are those found in the 14th edition of the APHA *Standard Methods* (1975).

In Guatemala the membrane filter technique has been used since 1959. The technique is now routinely used in the bacteriological examination of water for total and fecal coliforms, plankton analysis, and the presence of ova and cysts of parasitic organisms (Tabarini, 1975; Calzada, 1974; Tabarini, 1972; García and Tabarini, 1972; Tabarini, 1963). Coliform methodology is that found in the 14th edition of the APHA *Standard Methods* (1975), and approximately 400 membrane filters are used annually in this procedure. Coliform membrane filtration procedures are used in emergency situations, such as during the February 4, 1976, earthquake, when 2000 tests for coliforms were carried out to control the drinking water quality during this disaster (Castagnino and Guzmán, 1976).

Honduras uses the membrane filtration technique for the examination of drinking water following standard North American practices

(American Public Health Association, 1975) and for monitoring total and fecal coliform levels in river water used by water treatment plants (Gomez, 1978; Rosales, 1978). For these procedures, approximately 600 membranes are used annually; however, as there are no commercial sources within the country, there are occasions when the multiple-tube technique must be used because of the depletion of membrane filter stocks.

In El Salvador, membrane filters are used to varying degrees in the quality control of bottled water and soft drinks.

In Nicaragua, the membrane filtration technique is used for total and fecal coliform estimation in water and for sterilization of culture media (Bengoechea, 1978). However, the multiple-tube technique is still very popular, as there appears to be no sound reason to change from traditional methods.

In Barbados, approximately 2000 membrane filters are used each year in water quality studies (Fields, 1978).

In the Virgin Islands (U.S.A.), membrane filtration techniques are used routinely, whereas in Puerto Rico (U.S.A.) multiple-tube techniques are primarily used in water quality studies (Geldreich, (1978). In Costa Rica, membrane filtration procedures are only used in classroom demonstrations (Castro, 1978).

Of the 13 countries consulted on their use of membrane filter technique, only 7 replied. Four of the seven responding countries indicated that they were either routinely using the membrane filtration technique in routine water analysis or were in the process of introducing the technique.

B. South America

Colombian laboratories use the membrane filtration technique to monitor the bacteriological content of drinking water following the APHA *Standard Methods* (1975) procedures (Lojada, 1978). Approximately 2600 membranes are used annually. Only on rare occasions is portable field equipment used to test rural water quality.

Trinidad and Tobago have used the membrane filtration technique for approximately 10 years to test for total and fecal coliforms using the APHA *Standard Methods* (1975) procedures without verification (Millete, 1978). Approximately 14,000 membranes manufactured by Millipore and Oxoid are used annually in water quality testing.

Guyana has used the membrane filter procedure for approximately 2 years, performing approximately 3000 tests per year (Rampersand, (1978).

In Ecuador, approximately 200 membrane filters are used annually on special studies, as the preferred method for water analysis appears to be the multiple-tube method (Guzmán, 1978; Marcos, 1978; Riófrio, 1978; Vera Reyes, 1978). One laboratory in Ecuador has noticed that pads impregnated with media, for the coliform test, tend to deteriorate with storage (Marcos, 1978). Sartorius and Gelman membranes are the preferred brands in Ecuador.

Peru does not use the membrane filtration procedure because of the high cost and difficulty in importing the products. The view is that membrane filtration procedures offer no advantages over conventional methods (Tarazona, 1978).

Uruguayan laboratories use the membrane filtration procedure only in experimental bacteriological studies. However, membrane filters have been used widely for approximately 15 years mainly in chlorophyll, biomass, and plankton analyses, as well as in chemical procedures where filtration is required (e.g., determination of phosphates) (Alciaturi, 1977). The membranes used are cellulose ester, 0.45 μm and 47 mm in diameter.

In Argentina, the membrane filter procedure is used only in special water quality studies, not in routine analysis. It is believed that the procedure has no advantage over conventional methods and the costs are prohibitive (Deambrosi, 1978; Ripoli, 1978).

In 1978, Chile will be introducing the membrane filtration technique for routine bacteriological examination of drinking water (Castillo, 1977; Villalon, 1978). This introduction is based on the studies by Castillo (1977), who compared the membrane filtration technique with the most probable number multiple-tube dilution technique. These studies are based on WHO (1972) standards for the examination of drinking water.

Membrane filtration is also used to control *Pseudomonas aeruginosa* in demineralized and distilled water used by pharmaceutical laboratories and in the cosmetic industry (Castillo and Thiers, 1977). Membranes are also used occasionally to test waters and vegetables for *Salmonella* (Castillo et al., 1977). Aquatic biology research in Chile has used membrane filtration technology in phytoplankton and zooplankton analysis, chlorophyll determination, and in primary productivity studies.

Venezuelan water quality studies are based on the most probable number multiple-tube technique, with the membrane filtration technique being used only for special studies and classroom demonstrations (Geldreich, 1978).

In Brazil, the membrane filtration technique has been used for approximately 10 years, but it is only during the last 4 years that some laboratories have been using this technique routinely. In the majority of cases, prior to switching from the multiple-tube dilution technique to membrane filtration procedures, comparison studies were first carried out (Martins and Sanchez, 1971; Consendy, 1970). Some Brazilian state laboratories still use the membrane filtration technique only for special studies, such as *Salmonella* incidence (Martins and Sanchez, 1975; Barreto, 1977; Teixeira, 1978), fecal streptococci, *Pseudomonas aeruginosa* and *Staphylococcus aureus* levels (Olinda, 1977), and in the concentration of water and effluent samples for enterovirus isolation (Schatzmayr, 1976; Homma and Schatzmayr, 1975; Schatzmayr, 1974; Homma et al., 1973; Christovão et al., 1967). Membranes are also used in chlorophyll determinations, primary productivity studies (Teixeira, 1978) and the isolation of fungi (Kerr, 1977). There is some reported use of membranes in chemical procedures related to water quality. Bacteriological and virological culture media are sometimes sterilized by the membrane filtration procedure. It is believed that approximately

70,000 membrane filters are used annually in Brazil, purchased from Millipore, Gelman, and Sartorius.

Of the 14 South American countries consulted on their use of the membrane filtration technique, only 9 replied to the questionnaire. Two of the countries that did not reply, Bolivia and Paraguay, are known not to use this technique routinely. It is also known that where a country is reported to use the membrane filtration technique, this use is usually not widespread and often occurs only in some state laboratories or capital city laboratories. Thus, it would appear that the technique is not widely applied within most countries. Brazil, for example, has 27 federative units, of which 18 replied to our questionnaire. From these, only three (16.7%)--Rondonia territory, São Paulo, and Cuiaba (capital of Mato Grosso)--reported that the membrane filter was used in routine water quality studies. Four states--Parana, Rio Grande do Sul, Espirito Santo, and Bahia--indicate that they will soon be using this technique routinely, and five states indicated that the procedure was being used in special studies.

C. Summary

From the replies received to the questionnaire, it can be concluded that:

1. Membrane filtration procedures are more commonly used in physicochemical analysis and aquatic biology for such parameters as chlorophyll, phytoplankton, biomass estimation, and primary productivity.
2. In drinking water analysis the membrane filtration technique has not been stressed and, when used, is usually in coliform estimations.
3. The concentration ability of the membrane is utilized in *Salmonella* incidence studies (Brazil and Chile) and in enterovirus studies. Four laboratories are involved in these enterovirus studies: Instituto Oswaldo Cruz in Rio de Janeiro, Companhia de Tecnologia de Saneamento Ambiental and Instituto de Ciencias Biomedicas, Univ. S. Paulo in São Paulo and Centro de Estudos de Saneamento Basico in Rio Grande do Sul. One other laboratory that has been working with virus isolations from water using the membrane filtration technique is the Laboratorio de Obras Sanitarias de la Nación in Buenos Aires, Argentina.

III. WHY THE MEMBRANE FILTER TECHNIQUE IS NOT USED MORE FREQUENTLY

A. Central America

A total of 10 replies to our questionnaire (Appendix) were received from 7 countries within Central America. Four countries (57%, five replies) indicated that the membrane filtration technique was used for water quality studies and three countries (five replies) answered negatively with the following justifications:

1. The membrane filtration technique was much more expensive than the multiple-tube technique (four replies).
2. The membrane filtration technique offered no advantages over conventional methods (two replies).
3. There are no manufacturers' representatives in the country for consultation (one reply).
4. Lack of confidence in the method, and the technique is not as flexible as the multiple-tube technique (one reply).

B. South America

Brazilian replies will be treated separately. Of the nine replies received from South American countries, four supported and used the technique, whereas five had negative views and rarely used membrane filtration, for the following reasons:

1. The membrane filtration procedure was much more expensive than the multiple-tube technique (five replies).
2. The membrane filtration technique was not considered practical or economical (one reply).
3. Problems had been encountered with the procedure, especially in using pads containing dehydrated media (one reply).

Brazil

A total of 21 replies were received to the questionnaire (Appendix). Four laboratories indicated that the technique was used routinely, five used the technique but not routinely, four indicated that they were in the process of establishing the technique, and eight reported that they did not use membrane filtration procedures.

The laboratories that did not use the technique routinely presented the following reasons:

1. No advantages were seen in substituting membrane filter technology for conventional procedures (seven replies).
2. High cost of the membrane filter procedure (four replies).
3. The technique is not practical in waters with high turbidity and high densities of noncoliforms (one reply).
4. The membrane filtration technique has not been officially approved (one reply).
5. Multiple-tube dilution results were considered more accurate than membrane filtration results (one reply).
6. There was no drinking water supply system (one reply).
7. Supplies were difficult to purchase (one reply).
8. Lack of technical assistance (one reply).

C. Summary

In reviewing the reasons given for not using the membrane filtration technique, 68.7% of the negative responses were related to the high cost of the procedure and the failure to see any obvious advantages in it. Some of the replies indicated that there were technical problems with the membrane filter procedure, owing probably to lack of technical assistance from the manufacturers as well as isolation from readily available literature.

The emphasis on the high cost of the membrane filtration technique is a fact of life in South and Central America. Costs of foreign currency and import taxes make imported goods, scientific or otherwise, very expensive and therefore often prohibitive in cost. A factor that further emphasizes this is the low cost of untrained or nonqualified labor in Central and South America. Therefore, such tasks as glassware washing and media preparation, which are major factors in conventional multiple-tube dilution methods, are very inexpensive. Again, many of the laboratories are old and equipped for conventional methods with large stocks of glassware and equipment. To change techniques would be very troublesome and expensive. The majority of laboratories indicating routine use of the membrane filtration technique are either new or have recently been modified and updated. For example, CETESB, a recently formed mixed-economy company (10 years) whose salary structure is different from other institutions, finds that labor is expensive. Therefore, the membrane filtration technique becomes economical when compared to the labor-intensive multiple-tube dilution technique.

D. Brands of Membrane Filters Used in South and Central America

The following membrane filter brands were reported as being used by South and Central American laboratories: Millipore (25 replies), Sartorius (3 replies), Gelman (3 replies), Oxoid (1 reply). According to the questionnaire replies (Appendix), there are no other manufacturers represented in Latin America and in the Caribbean countries.

IV. MEMBRANE FILTRATION PROCEDURES USED

A. General

The majority of countries using this procedure follow the APHA *Standard Methods* (1975) for total coliforms, fecal coliforms, and fecal streptococcus estimations. Many countries verify total coliform counts.

In CETESB laboratories in São Paulo, the membrane filter (MF) technique is used primarily for testing drinking water samples, of which 50,000 are analyzed annually. The following procedure is observed:

1. One hundred milliliters of drinking water is membrane-filtered through 0.45 μm HAWG 047 membranes which are placed on m-Endo agar LES (Difco or BBL) in tight-fitting plastic petri dishes.
2. The plates are incubated for 22-24 hr on trays containing water-soaked absorbent paper at 35 ±0.5°C.
3. Typical colonies are counted with the aid of a dissecting microscope.
4. To verify the coliform colonies, typical colonies are transferred to lactose broth and incubated 24-48 hr at 35 ±0.5°C and all tubes showing gas production to brilliant green lactose bile broth 2% and incubated at 35 ±0.5°C for 48 hr and to EC medium that is incubated at 44.5°C ±0.2°C for 24 hr.

5. By this procedure, total coliforms are verified and fecal coliforms differentiated, thus avoiding duplication of fecal coliform and total coliform tests. Our rationale for this procedure is that positive tests are rare and only in positive tests is it necessary to verify the types of coliforms found in the sample.

The testing for fecal coliforms in drinking water by the MF technique using mFC medium has not been started in the CETESB laboratories. Conventional multiple-tube dilution techniques (American Public Health Association, 1975) have proven to be more consistent than the membrane procedure using mFC broth.

Presently, the CETESB Microbiology Division Laboratory is involved in comparative studies between the multiple-tube dilution technique and the membrane filtration technique for the following organisms: total and fecal coliforms, fecal streptococci, and *Ps. aeruginosa* in fresh water, wastewater, and seawater. The data from these studies are now being analyzed.

B. Salmonella Isolation Procedures from Water and Wastewater

The following procedure is used by CETESB Laboratories:

1. Three to five liters of sample are filtered through a Millipore 0.45 μm, 142 mm diameter membrane filter using a 142 mm stainless steel filter holder and a pressure vessel.
2. After filtration, the membrane filter is cut in half. One portion is placed in selenite broth containing novobiocin (Pessôa and Peixoto, 1971) and incubated at 42.5°C and tested at 24, 48, and 120 hours. The other portion is placed in Rappaport's medium as modified by Hofer (1969) and incubated at 35°C and tested at 24 and 48 hr.
3. Each 24 hours, transfers are made from the enrichment media to Wilson Blair bismuth sulfite agar, brilliant green agar, and XLD agar which are incubated at 35°C for 18-24 hr.
4. Colonies typical of *Salmonella* (10 to 20 per plate) are transferred to Rugai-Lisina medium (Pessôa and Da Silva, 1974), where by motility and biochemical reactions, presumptive identification of *Salmonella* is made.
5. All presumptive *Salmonella* are submitted to serotyping procedures.

C. Concentration of Water Samples for Enterovirus Enumeration

Water samples that have been previously treated with $AlCl_3$ (final concentration 0.0005 M) and 6 N HCl to lower the pH to 3.5, are pressure-filtered through a AP20 fiberglass prefilter and a 0.45 μm membrane filter. To elute the viruses adsorbed by the membrane filter or the prefilter, pH 11.5 glycine is used.

Studies are presently being carried out to assess the ability of membrane filters to concentrate large volumes of water contaminated with Mahoney polio virus. These studies tend to indicate that the percentage recovery of viruses inoculated in raw water, sewage, and polluted seawater is insignificant. This low recovery may be due to several factors, such as loss in viability of the viruses or interference by membrane-coating components. From these studies, we believe that this concentration procedure is advisable only for drinking waters with low turbidity and low levels of organic material.

D. Procedures Used in Chile

Apart from the methods detailed in the APHA *Standard Methods* (1975) for total coliforms, fecal coliforms, and fecal streptococci, the following membrane filtration procedures are also used.

Pseudomonas aeruginosa

1. One hundred to five hundred milliliters aliquots of demineralized or distilled water used in cosmetic or pharmaceutical processes are membrane-filtered through 0.45 μm, 47 mm diameter membranes which are placed into asparagine broth and incubated at 35°C for up to 5 days.
2. Asparagine broths showing growth are seeded onto the surface of:
   a. Cetrimide agar (Difco) and incubated at 42.5°C for 24 to 48 hr.
   b. Nutrient agar and incubated at 35°C for 24-48 hr.
3. Suspected *P. aeruginosa* colonies are identified by the following procedures: colony appearance, fluorescein production, characteristic odour, Gram's test, oxidase test, production of nitrite

or nitrogen from nitrate broth, and oxidation or fermentation of glucose; where doubts still exist, the bacteria are examined using an electron microscope (Castillo and Thiers, 1977).

Salmonella *Isolation from Water and Wastewater*

1. One hundred to five hundred milliliters of water is membrane-filtered using 0.45 μm membranes which are placed in selenite-novobiocin broth with incubation at 42°C and in tetrathionate broth with incubation at 35°C.
2. From these broths, platings are made onto selective media from which typical colonies are identified conventionally (Castillo, (1977).

Salmonella *Isolation from Vegetables*

This procedure was specifically designed for lettuce (Castillo et al., 1977).

1. Lettuce leaves are placed in peptone water for approximately 15 min.
2. The peptone medium is membrane-filtered and the membrane is treated similarly to those used in *Salmonella* isolation from water and wastewater.

V. SUMMARY OF PROBLEMS ENCOUNTERED BY USERS OF MEMBRANE FILTERS

1. One major problem noted was the high cost of this procedure because of import taxes and currency costs.
2. Some membranes, including those recommended by the APHA *Standard Methods* (1975), such as epoxy-coated membranes, are not available in Latin America and the Caribbean.
3. Frequently, material and equipment, delivered at very high prices, are obsolete in other countries (.e.g, kits for field work, incubators, and filter holders of old design which have been replaced in North America with newer improved versions).
4. Delivery of faulty materials presents another problem. CETESB Laboratories received 20,000 membranes (Gelman) that presented problems; they were wrinkled and contained hydrophobic areas.
5. It is common to receive culture media that are toxic or have inhibitory characteristics, as well as out-of-date reagents and culture media.

6. Materials that are not adequately sensitive (Geldreich, 1975), such as Coli-Count Dip Sticks and pads impregnated with dehydrated media for performing coliform counts, are "forced" on the prospective buyer by representatives who seem to be aware of the problems but do not seem to care about the costs and loss to the consumer.

We believe that Latin American and Caribbean suppliers should carefully check their products to avoid offering laboratories outdated and unreliable products. South and Central America and the Caribbean countries are concerned about quality control and emphasize this aspect more and more in order to be confident of their results. However, they also need the help of the suppliers to provide fresh, reliable products to lessen their quality control problems.

APPENDIX: QUESTIONNAIRE SENT TO THE LABORATORIES TO OBTAIN INFORMATION ON MEMBRANE FILTER USAGE

1. Are they used in your laboratory?

   Yes ( ) No ( )

2. Are they used in your country or state?

   Yes ( ) No ( )

3. For how long?______________________________

4. Is field equipment used?

   Rarely ( )

   Frequently ( )

   No ( )

5. For what purpose is MF used in your country?________________

   ____

6. Why is the technique of MF not used?

   High cost ( )

   No need to change conventional techniques ( ) Problems (mention, please)______________________________

7. What type is used?

   Porosity ________________

   Diameter ________________

Brand ______________________

Type ______________________

8. Average annual consumption__________________________________

9. Methodology used

Standard Methods (a) direct count ( )

(b) with verification ( )

Other methods (describe, please)____________________________

____

10. Special analysis, or other uses of MF.

a. Concentration of sample for virus isolation

b. Other determinations

c. Sterilization of culture media ( ) (mention, please)

____

11. Do you have any work published that uses the MF technique?

Yes ( ) (mention, please)_________________________________

No ( )

ACKNOWLEDGMENTS

The author wishes to thank Dr. Gabriela Castillo for her valuable cooperation; Miss Niva Deana, for typing services, and all our colleagues from Latin America and the Caribbean who so willingly replied to the questionnaire, giving us the information necessary to enable us to prepare this work.

REFERENCES

Alciaturi, F. A. (1977). Personal communication. Administración de las Obras Sanitarias del Estado, Montevideo, Uruguay.

American Public Health Association (1975). *Standard Methods for the Examination of Water and Wastewater,* 14th ed. American Public Health Association, Washington, D.C.

Baretto, M. K. (1977). Personal communication. FEEMA, Rio de Janeiro, Brazil.

Bengoechea, J. (1978). Personal communication. Managua, Nicaragua.

Calzada, M. J. F. (1974). Evaluation of the algae constant kinetics of Lake Amatitlán. Guatemala (in Spanish).

Castagnino, W., and Guzmán Ch., G. (1976). The earthquake and the rehabilitation of the Santa Luzia water treatment plant service. XV Meeting of Sanitary Engineering, Guatemala (in Spanish).

Castillo, G. (1977). Comparison between the techniques of most probable number and membrane filter in coliform analysis in drinking water. 2nd Chilean Meeting of Sanitary Engineering, Santiago, Chile (in Spanish).

Castillo, G., and Thiers, R. (1977). *Pseudomonas aeruginosa* in water utilized in the pharmaceutical and cosmetics industries. VII Meeting of Microbiology, Buenos Aires, Argentina (in Spanish).

Castillo, G., Thiers, R., and Cordano, A. M. (1977). Bacteriological quality of irrigation waters and influence on vegetable contamination. International Symposium on Nutrients and Heavy Metals, Santiago, Chile (in Spanish).

Castillo, G. (1977). Personal communication. Universidad de Chile, Santiago, Chile.

Castro, E. O. (1978). Personal communication. Costa Rica.

Christovão, D. de A., Candeias, J. A. N., and Iaria, S. T. (1967). Sanitary conditions of irrigation waters in the gardens of the city of São Paulo. Isolation of enteric viruses. *Rev. Saúde Pública, S. Paulo 1*(1): 12-17 (in Portuguese).

Consendy, E. A. (1970). Comparative studies between the membrane filter procedure and multiple tubes in water bacteriology. *SURSAN, IES, RJ*, No. 54, Brazil (in Portuguese).

Deambrosi, N. (1978). Personal communication, Buenos Aires, Argentina.

Díaz, M. P. D. (1978). Personal communication. Dirección General de Aguas y Saneamiento, Mexico City, Mexico.

Fields, P. A. (1978). Personal communication. Queen Elizabeth Hospital, Barbados.

García, M. L. E., and Tabarini de Abreu, A. (1972). Procedures and previous results of a limnological survey in Lake Amatitlán. Investigation Programs, Escuela Regional de Ingeniería. Facultad de Ingeniería, Guatemala (in Spanish).

Geldreich, E. E. (1975). *Handbook for Evaluating Water in Bacteriological Laboratories*. EPA 670/9-75-006, U.S. Environmental Protection Agency.

Geldreich, E. E. (1978). Personal communication.

Gomez, J. B. (1978). Personal communication. Servicio Nacional de Acueductos y Alcantarillados, Tegucigalpa, Honduras.

Guzmán, T. N. (1978). Personal communication. Instituto Nacional de Higiene, Guayaquil, Ecuador.

Hofer, E. (1969). Über Abänderunger des Rappaport--Nährbodens. *Zentralbl. Bakteriol. Parasitenkd. Infektionskr. Hyg. Abt. 1 Orig. 210:* 419–422.

Homma, A., and Schatzmayr, H. G. (1975). The isolation of viruses from seawater; field applications of a cellulose membrane method. 3rd International Meeting of Virology, Madrid, Spain.

Homma, A., Schatzmayr, H. G., Frias, L. A., and Mesquita, J. A. (1973). Study of viruses found on the beaches of Guanabara Bay: quantification, characterization, methodology and preliminary results. IX Meeting of Brazilian Symposium of Tropical Medicine, Fortaleza, Ceará, Brazil (in Portuguese).

Kerr, E. W. (1977). Personal communication. INPA, Manaus, Amazonas, Brazil.

Lojada, M. O. R. (1978). Personal communication. Bogotá, Colombia.

Marcos, V. J. E. (1978). Personal communication. Empresa Municipal de Agua Potable, Guayaquil, Ecuador.

Martinez, J. A. (1978). Personal communication. Laboratorio del Centro de Investigación y Entrenamiento, Mexico.

Martins, M. T., and Sanchez, P. S. (1971). Comparison between the membrane filter and multiple tube techniques in coliform determination. *Rev. DAE 81:* 97–101 (in Portuguese).

Martins, M. T., and Sanchez, P. S. (1975). Correlation studies between pollution indicators and pathogenic bacteria, previous note. XVII Brazilian Meeting of Sanitary Engineering, ABES. Rio de Janeiro, Brazil (in Portuguese).

Millete, O. (1978). Personal communication. Water and Sewerage Authority, Trinidad y Tobago.

Olinda, F. E. (1977). Personal communication. CAGECE, Ceará, Fortaleza, Brazil.

Pessôa, G. V. A., and Da Silva, E. A. M. (1974). Milieu pour l'identification présomptive rapide des entérobactéries, des aéromonas et des vibrions. *Ann. Microbiol. (Inst. Pasteur) 125-A-A:* 341–347.

Pessôa, G. V. A., and Peixoto, R. S. (1971). Selenite novobiocine broth: a more selective medium for the isolation of *Salmonella* from fecal specimens. *Rev. Inst. A. Lutz, 31*: 1–3 (in Portuguese).

Rampersand, H. (1978). Personal communication. Guyana Water Authority, Georgetown, Guyana.

Ríofrio, H. (1978). Personal communication. Instituto Nacional de Recursos Hidraulicos, Quito, Ecuador.

Ripoli, J. L. (1978). Personal communication. Laboratorio del Ministerio del Bienestar Social, Buenos Aires, La Plata, Argentina.

Rosales, N. M. O. (1978). Personal communication. Honduras.

Schatzmayr, H. G. (1974). Concentration of enteroviruses found in seawater. Methodology and epidemiological significance. Symposium on Virological Topics. V Brazilian Meeting on Microbiology, Univ. Gama Filho, Rio de Janeiro, Brazil (in Portuguese).

Schatzmayr, H. G. (1976). Studies on the isolation of viruses from seawater. Postdoctorate thesis in virology, Univ. Federal Fluminense, Rio de Janeiro, Brazil (in Portuguese).

Tabarini de Abreu, A. (1960, 1969, 1972). Notes on microbiology, pp. 60–73. Escuela Regional de Ingeniería Sanitaria Facultad de Ingeniería, Univ. San Carlos, Guatemala (in Spanish).

Tabarini de Abreu, A. (1963). Use of membrane filters in bacteriological water analysis. *Revista de la Escuela de Farmacia*, Nos. 299, 300, 301, 302, 303, 304, 305, 306. Guatemala (in Spanish).

Tabarini de Abreu, A. (1975). Eutrophication of Lake Amatitlán (Bioessay techniques). Escuela Regional de Ingeniería Sanitaria Facultad de Ingeniería, Univ. San Carlos, Guatemala (in Spanish).

Tarazona, J. H. (1978). Personal communication. Ministerio de Vivienda y Construcción, Lima, Peru.

Teixeira, A. R. (1978). Personal communication. DEMAE--Rio Grande do Sul, Brazil.

Villalon, D. J. (1978). Personal communication. Ministerio de Salud, Chile.

Vera Reyes, M. (1978). Personal communication. Instituto Nacional de Higiene, Guayaquil, Ecuador.

World Health Organization (1972). International Standards for Drinking Water, 3rd ed. World Health Organization, Geneva, Switzerland.

13

MEMBRANE FILTRATION: A JAPANESE VIEW

MINORU MORITA Suntory Limited, Osaka, Japan

I. INTRODUCTION

The first published report in a Japanese journal on the use of the membrane filtration technique occurred in 1957 and described the enumeration of coliform bacteria in seawater (Fujimoto, 1957). Following this, in 1960 Taguchi reported the results of a study comparing the membrane filtration (MF) procedure with the most probable number (MPN) procedure in the enumeration of coliform bacteria in shallow well waters. However, in spite of these early investigations, the use of the MF technique in the field of water microbiology has not yet been popularly accepted in Japan. Two reasons can be offered to explain why the MF technique has not been used extensively in water microbiology in Japan, one based on geology and the other on economics.

Geologically, Japan, an archipelago composed of four main islands and many adjacent smaller islands, is a mountainous country in which abundant rainfall gives birth to more than 100 river systems whose waters run from the mountainous interior to the sea. All of these rivers have short courses and small drainage basins; only three rivers (Shinano, Ishikari, and Tone) exceed 200 miles in length. As a result, these river waters are characterized by low solids content, low hardness, low pH, and low temperatures. These environmental factors are unfavorable for the growth of microorganisms and thereby help keep the waters naturally clean.

However, because of the rapid industrial growth and expansion of economic activities during the last 20 years, environmental pollution has become a serious social problem. In response, the Japanese government enacted a Basic Antipollution Law in 1967, followed by successive enactments of relevant laws such as the Clean Water Law, Clean Air Law, Sea Water Protection Law, and Nature Conservation Law. Based on these regulations, a variety of water quality standards have been created incorporating such indicators as pH, biological oxygen demand (BOD), chemical oxygen demand (COD), suspended solids (SS), dissolved oxygen (DO), and coliform bacteria as indicators of fecal contamination. As mentioned earlier, the upstream river waters are relatively clean and have few microbiological contamination problems, whereas in the downstream and basin areas, hazardous chemical contaminants derived from industrial wastes are causing major pollution problems. Because of the nature of the pollution, Japanese researchers consider BOD and COD measurements to be more important than the other criteria, as they indicate the overall contamination by organic materials. Thus, coliforms are not tested for routinely. For example, according to the official report issued by Osaka Prefectural Office (1977), a total of 7537 analyses related to water quality tests were performed in 1976 by the official agencies in Osaka. Of these, coliforms comprised 29 while 1249 were for pH, 935 for BOD, 1159 for COD, and 1071 for SS.

The second and more decisive factor for limiting the use of the MF technique is that this method is very expensive compared to other methods, as almost all membranes used are imported from other countries and disposed of after a single use. Although the MF procedure has been adopted as a standard method for the estimiation of coliforms in drinking water (Japanese Water Works Association, 1970; Pharmaceutical Society of Japan, 1973, 1977), an agar plate method using deoxycholate medium is preferentially used because the procedure is simple, time saving, and does not require any special equipment. Thus, in Japan, approximately 80% of coliform enumeration is carried out by the deoxycholate plate method, 20% by the MPN method, and less than 1% by MF.

II. MICROBIOLOGICAL TESTS FOR VARIOUS TYPES OF WATER

A. Natural Water

In 1971, the environmental criteria for natural waters, which included rivers, lakes, and seas, were established by the Environmental Agency's enactment of the Basic Antipollution Law. On the basis of these criteria, all water basins throughout the country were classified into several categories, based on their degree of pollution. Table 1 presents environmental criteria for river waters with their recommended uses.

In these criteria, coliform bacteria are used as an indicator of fecal contamination and the coliform density upper limits set for each grade of waters are used mainly for water supply, bathing, and fisheries. Coliform populations are estimated according to the officialy designated MPN method, which uses brilliant green lactose broth (BGLB) (Environmental Agency, 1971).

B. Potable Water

Potable water supplied by waterworks is under the control of the Waterworks Law, and the water quality has to meet quality standards set by the Ministry of Health and Welfare (MHW) (1966). Microbiologically, these waters must not be contaminated with any pathogenic microorganisms and must not contain more than 100/ml bacteria which are able to grow on a standard nutrient agar plate. Thus, the enumeration of coliforms is unnecessary; however, a three-step qualitative procedure may be used to detect coliforms. This test procedures consists of (1) a presumptive test using lactose broth, (2) a confirmation test with BGLB, and (3) a completed test by isolation of colonies followed by Gram's staining and culture in lactose broth.

TABLE 1 Environmental Criteria of River Water Quality

Quality[a] and use	pH	Criteria (daily average) BOD (ppm)	SS (ppm)	DO (ppm)	Coliform (MPN/dl)
AA	6.5-8.5	<1	<25	>7.5	<50
A	6.5-8.5	<2	<25	>7.5	<1000
B	6.5-8.5	<3	<25	>5	<5000
C	6.5-8.5	<5	<50	>5	-
D	6.0-8.5	<8	<100	>2	-
E	6.0-8.5	<10	-[b]	>2	-

[a]AA: First-grade raw water for potable water supply; usable after a simple purification procedure such as filtration.

A: Second-grade raw water for potable water supply; usable after an ordinary purification procedure such as a combination of sedimentation and filtration.
First-grade fishery water; suitable for oligosaprobic organisms.
Bathing water.

B: Third-grade raw water for potable water supply; usable after intensive purification procedure, including suitable pretreatment.
Second-grade fishery water; suitable for oligosaprobic organisms such as salmon and sweetfish.

C: Third-grade fishery water; suitable for β-mesosaprobic organisms.
First-grade industrial water; usable after an ordinary purification procedure such as sedimentation.

D: Second-grade industrial water; usable after an intensive purification procedure such as chemical treatment.
Irrigating water for agriculture.

E: Third-grade industrial water; usable after a specific purification procedure.

[b]Containing no visible detritus.

Source: Environmental Agency (1971).

Two other reference books are used in potable water analysis: *Standard Methods of Analysis for Hygienic Chemists,* issued by Pharmaceutical Society of Japan (PSJ) (1973), and *Standard Methods of Analysis in Waterworks,* by the Japanese Water Works Association (JWWA) (1970). The former is intended to provide standard methods of analysis for foods, food additives, toys, cosmetics, waters, and so on; the latter provides standard analytical procedures for waters used for drinking water supplies.

Microbiological methods for water analysis in PSJ standards were revised in 1977 and a MF procedure for coliforms similar to that described by the American Public Health Association, *Standard Methods for the Analysis of Water and Wastewater*, 13th ed. (1971), was adopted, in addition to the classic Japanese MPN procedure.

JWWA standards (1970) provide the test methods for various types of microorganisms relevant to controlling water quality, such as heterotrophic bacteria, coliform bacteria, enterococci, sulfate-reducing bacteria, actinomycetes, iron bacteria, fungi, and other microorganisms. Media and culture conditions for these microorganisms are summarized in Table 2. Coliform populations are determined by three methods: (1) MPN procedure utilizing lactose broth, (2) agar plate technique using deoxycholate medium, and (3) MF procedure using m-Endo agar.

In 1960, Taguchi, using Millipore membrane filters, compared the MF technique with the MPN technique in estimating coliform populations from shallow well waters. During the study he evaluated media suitable for coliform enumeration by the membrane filtration technique and found that deoxycholate agar gave the best results. In his study of shallow wells polluted to various degrees, he found that coliform counts by the MF procedure were usually 70% of those obtained by the MPN technique. Based on his studies, Taguchi suggested that the MF procedure was suitable for the enumeration of coliforms in relatively unpolluted waters but that for highly contaminated waters such as sewage, the deoxycholate agar plate method was preferable.

The MF procedure has also been used in the collection and quantitation of microorganisms other than bacteria from water samples. In 1971, Koide described a novel MF technique for the microscopic examination of planktonic algae. The algae on the filter were fixed by ethanol and then the filter was coated with Pleurax (a phenolic resin) to make the membrane filter transparent and suitable for microscopic examination. This method was reported to preserve fragile algae without destroying or shrinking the cells.

TABLE 2 Types of Microbiological Tests Described in the JWWA Standard Methods of Analysis in Waterworks

Type of bacteria	Type of test	Medium[a]	Incubation Temp. (°C)	Incubation Time
Total colony count	Quantitative Agar plate	Nutrient agar	35-37	22-26 hr
Heterotrophic bacteria	Quantitative Agar plate	Henrici	24-26	3 days
Coliform	Qualitative			
	Presumption	LB	35-37	45-51 hr
	Confirmation	BGLB	35-37	45-51 hr
	Isolation	EMB or Endo	35-37	22-26 hr
	Quantitative			
	MPN	LB	35-37	45-51 hr
	MF	M-Endo	35-37	22-24 hr
	Agar plate	DOC	35-37	18-20 hr
	Enteric coli test	ECB	44-45	24 hr
	IMViC test			
	Indole	TB	35-37	22-26 hr
	Methyl red	GPPB	35-37	5 days
	Voges-Proskauer	GPPB	35-37	48 hr
	Sodium citrate	CB	35-37	3-4 days
Enterococcus	Qualitative			
	Presumption (microscopic)	LB	35-37	45-51 hr
	Confirmation	EVAB	35-37	45-51 hr
	Quantitative			
	MPN	LB	35-37	45-51 hr
Sulfate-reducing bacteria	Qualitative			
	Confirmation	Sato	30	2-3 days
	Quantitative			
	MPN	Sato	30	2-3 days
Actinomycetes	Quantitative Agar plate	Krainsky or Romano	30	5-7 days
Iron bacteria	Microscopic identification of sample directly, after collection on MF, or after cultivation according to Yagi or KWM procedure.[a]			
Sulfur bacteria	Microscopic identification of sample directly, or after collection on MF.			
Sphaerotilus	Microscopic identification of sample directly, or after cultivation with 50 ppm yeast extract for 2-3 days at 25°C with aeration. Isolation			

TABLE 2 (Continued)

Type of bacteria	Type of test	Medium[a]	Incubation	
			Temp. (°C)	Time
	culture is made on agar plate with Sakurai's medium for 2-3 days at 25°C.			
Fungi	Microscopic identification of sample directly, or after cultivation on Waksman's medium followed by Czapek's agar plates (for *Hyphomycetes*) or with hemp seeds (for *Phycomycetes*).			

[a]See the Appendix.
Source: JWWA (1970).

This procedure has been adopted in the JWWA *Standard Methods* (1970) and also in the PSJ *Standard Methods* (1977). The details of this procedure will be described later.

C. Recreational Water

Water quality standards for bathing waters (Table 3) are described in the 1971 Environmental Standards established by the Environmental Agency, and coliform objectives are defined as less than 1000 per 100 ml based on the MPN technique. Although the MPN procedure using BGLB is mandatory, the deoxycholate agar plate method is also described in the PSJ *Standard Methods* (1973).

Swimming pool water quality is most frequently monitored by the hygienic chemical standards set by PSJ (1973). In these standards "total" bacterial counts and a semiquantitative test for

TABLE 3 Quality Standards of Bathing Waters

Type of water	Criteria (daily average)					
	pH	BOD (ppm)	COD (ppm)	DO (ppm)	SS (ppm)	Coliform (MPN/dl)
River	6.5-8.5	<2	-	>7.5	<25	<1000
Lake	6.5-8.5	-	<3	>7.5	<5	<1000
Sea	7.8-8.3	-	<2	>7.5	-	<1000

Source: Environmental Agency (1971).

coliform bacteria are used as indicators of microbiological contamination. Total bacterial counts are based on a nutrient agar spread plate technique and an upper limit of 200 bacteria per milliliter. The coliform test is performed by inoculating five tubes of lactose broth with 10 ml of test water and incubating for 45-51 hr at 35-37°C. The water passes the test if two or fewer tubes are found to be positive for gas production.

D. Sewage Effluent and Industrial Wastewater

Sewage effluent and industrial wastewater quality are based on the Clean Water Law (1970) and are regulated by the Effluent Quality Standards (Prime Minister's Office, 1971, 1976). In these standards, coliform bacteria are considered to be indicators of the efficiency of sewage treatment plants, rather than fecal contamination. Accordingly, complicated identification tests for coliforms are not required; instead, simple time-saving enumeration procedures are used. Although the MF procedure appears to be the most suitable method, it has not yet been adopted as the standard method for the examination of sewage effluents. Instead, the deoxycholate agar method described in the *Japanese Industrial Standards* (Japanese Standards Association, 1964, 1974) is used as the official standard method (Evironmental Agency, 1974). Official standard methods for the analysis of coliform bacteria in the various waters described above are summarized in Table 4.

III. STANDARD METHODS USING THE MEMBRANE FILTRATION TECHNIQUE

A. Quantitation of Coliform Bacteria

Japanese Water Works Association (JWWA)
Standard Method 1970

If the coliform content is assumed to be greater than five per milliliter, the water sample should be diluted with buffered water,

TABLE 4 Summary of Standard Methods for Analysis of Coliform Bacteria

Type of water	Type of test	Medium[a]	Reference
Natural water	Quantitative		
	MPN	BGLB	EA (1971)
Potable water	Qualitative		
	Assumption	LB	MHW (1966);
	Confirmation	BGLB	JWWA (1970);
	Isolation	EMB or Endo	PSJ (1973)
	Quantitative		
	MPN	LB	JWWA (1970); PSJ (1973)
	MF	M-Endo	JWWA (1970); PSJ (1977)
	MF	LES-Endo	PSJ (1977)
	Agar Plate	DOC	JWWA (1970)
Recreational water	Quantitative		
	MPN	BGLB	MHW (1966)
	MPN	LB	PSJ (1973)
	Agar plate	DOC	PSJ (1973)
Swimming pool water	Qualitative		PSJ (1973)
	Assumption	LB	
	Confirmation	BGLB	
	Isolation	EMB or Endo	
	Semiquantitative		
	MPN	LB	PSJ (1973)
Sewage, wastewater	Quantitative		
	Agar plate	DOC	EA (1974) JSA (1974)

[a]See the Appendix.

made by adding 1.25 ml of 0.1 M KH_2PO_4 (pH adjusted to 7.2 with N NaOH) to 1 liter of normal saline solution. The quantity of water sample to be filtered should be at least 30 ml and should contain 50-200 coliform bacteria.

A sterilized 47-50 mm membrane filter with grids and pore size of 0.45 μm is placed in a standard funnel holder and the sample is filtered under a negative pressure of 20-50 mm Hg. After rinsing the inner surface of the funnel two to three times with sterilized diluting medium, the membrane is transferred onto a m-Endo broth saturated pad in a petri dish. Overall contact of the membrane with the pad should be assured. The petri dish is incubated in an inverted position at 35-37°C for 22-24 hr. After incubation, colonies with dark red metallic sheen are counted.

Pharmaceutical Society of Japan (PSJ) Standard Method 1977

This procedure is identical with the JWWA method except an Endo agar LES is recommended along with the m-Endo broth saturated-pad procedure.

B. Collection of Microorganisms on Membrane Filter for Microscopic Examination

Planktonic Algae (JWWA, 1970; PSJ, 1977)

A membrane filter (cellulose acetate type, 25 mm diameter, 0.45 μm pore size) is placed in an appropriate holder and a water sample of sufficient size and/or dilution to provide 100-300 organisms is filtered, followed by serial rinsing with 2 ml each of 60, 85, and 99% (vol/vol) ethanol. After the rinsing solution is gently filtered off, the membrane filter is removed and placed on a glass slide together with a few drops of Pleurax. The surface of the membrane filter is covered with additional drops of Pleurax and a cover glass. The transparent preparation thus obtained is examined under a microscope.

Iron and Sulfur Bacteria (JWWA, 1970)

These enumerations are based on direct examination and recognition of typical bacterial morphology. After an aliquot of water has been membrane-filtered, the membrane is removed and placed on a glass slide to which a few drops of immersion oil have been added. As the water evaporates, the immersion oil diffuses through the membrane, rendering it transparent. Additional drops of the oil are added to the membrane surface and a cover glass is provided for microscopic examination.

IV. CONCLUSION AND FUTURE PROSPECTS

As reviewed above, the MF technique is not a familiar procedure with Japanese water microbiologists. The MPN procedure is still the sole official standard method for the enumeration of coliform bacteria from environmental water samples. For monitoring coliform populations in sewage effluents and wastewaters, an agar plate method using deoxycholate medium is used as the official standard method. Even though the MF technique has been adopted as one of the standard methods for the estimation of coliforms in potable water, the deoxycholate method is much more frequently used.

Although the MF procedure has been found to be more expensive than the agar plate and MPN procedures, it has an important advantage in that it is possible to collect or concentrate bacteria from any size of water sample. This factor improves the accuracy of the results, especially in examining water samples with low bacterial populations. Accordingly, if the coliform population in a sample is relatively high, such as sewage effluents or polluted water samples, the agar plate method is considered to be more suitable from an economic point.

Although the MF technique is not yet popular with environmental microbiologists, the technique is widely used in the pharmaceutical and biomedical industries. Recognition of the increasing importance of the sterility and safety of drugs, biomedical devices, and cosmetics, has led to the establishment of Good Manufacturing Practices, an official guideline for the manufacturing of medical products, with special emphasis on the sanitary control of the environment. In these industries, the MF technique is routinely used, not only for determining the sterility of the final products, but also for monitoring microbiological contamination during various production stages. The tests are involved primarily with estimating total colony counts in raw materials, containers, and process water lines. If the increase in membrane filter use in these fields could lead to a reduction in membrane filter prices in Japan, it would conceivably accelerate the use of the MF technique by water microbiologists.

APPENDIX: COMPOSITION OF MEDIA (g/l)

BGLB (brilliant green lactose broth): peptone, 10; lactose, 10; powdered bile, 20; brilliant green, 0.013; pH 7.0-7.4

CB (citrate broth): $NaNH_4HPO_4$, 1.5; K_2HPO_4, 1; $MgSO_4 \cdot 7H_2O$, 0.2; sodium citrate, 2.5-3.0

Czapek: sucrose, 30; $NaNO_3$, 2; K_2HPO_4, 1; KCl, 0.5; $MgSO_4 \cdot 7H_2O$, 0.01; agar, 15

DOC (deoxycholate agar): peptone, 10; ferric ammonium citrate, 2; sodium deoxycholate, 1; neutral red, 0.03; pH 7.0-7.4

ECB (enteric coli broth): tryptone, 20; lactose, 5; cholate, 1.5; KH_2PO_4, 1.5; K_2HPO_4, 4; NaCl, 5

EMB (eosin methylene blue agar): peptone, 10; K_2HPO_4, 2; lactose, 10; eosin yellow, 0.4; methylene blue, 0.065; agar, 15; pH 6.6-7.0

Endo: peptone, 10; meat extract, 4; lactose, 10; Na_2SO_3, 1.6; basic fuchsine, 0.1; agar, 15; pH 7.0-7.4

EVAB (ethyl violet-azide broth): tryptone, 20; glucose, 5; K_2HPO_4, 2.7; NaN_3, 0.4; ethyl violet, 0.83; pH 7.0

GPPB (glucose phosphate-peptone broth): proteose peptone, 5; glucose, 5; K_2HPO_4, 5

Henrici: sodium caseinate, 0.5; polypeptone, 0.5; glycerol, 0.5; starch, 0.5; KH_2PO_4, 0.5; agar, 15; pH 7.0-7.5

Krainsky: starch, 10; NH_4Cl, 0.5; K_2HPO_4, 0.5; agar, 15

LB (lactose broth): meat extract, 5; peptone, 10; lactose, 5; bromthymol blue, 0.024; pH 7.0-7.4

LES Endo: casitone or trypticase, 3.5; thiopeptone or thiotone, 3.7; tryptone, 7.5; lactose, 9.4; K_2HPO_4, 3.3; KH_2PO_4, 1; NaCl, 3.7; Na_2SO_3, 1.6; basic fuchsine, 0.8; sodium deoxycholate, 0.1; sodium lauryl sulfate, 0.05; 95% (vol/vol) ethanol, 20; agar, 15

m-Endo: tryptone or polypeptone, 10; thiotone or thiopentone, 5; casitone or trypticase, 5; yeast extract, 1.5; lactose, 12.5; NaCl, 5; K_2HPO_4, 4.375; KH_2PO_4, 1.375; sodium lauryl sulfate, 0.05; sodium deoxycholate, 0.1; Na_2SO_3, 2.1; basic fuchsine, 1.05; 95 (vol/vol) ethanol, 20

Nutrient agar: meat extract, 5; peptone, 10; agar, 15-20; pH 6.4-7.0

Romano: glucose, 20; $NaNO_3$, 2.0; K_2HPO_4, 1.0; $MgSO_4 \cdot 7H_2O$, 0.2; KCl, 0.2; $FeSO_4 \cdot 7H_2O$, 0.01; agar, 20; pH 7.0

Sakurai: peptone, 2; yeast extract, 1; glucose, 0.5; agar, 15; pH 7.0-7.2

Sato: peptone, 5; meat extract, 5; sodium lactate (70%), 5; KH_2PO_4, 0.5; Na_2SO_4, 1.5; $MgSO_4 \cdot 7H_2O$, 1.5; $FeSO_4(NH_4)_2SO_4 \cdot 6H_2O$, 0.1; pH 7.2

TB (tryptophane broth): tryptone or trypticase, 10

Waksman: glucose, 10; peptone, 5; KH_2PO_4, 1; $MgSO_4 \cdot 7H_2O$, 0.5; agar, 20

Yagi's method: Two flasks are filled with 100 ml of water sample, and 50 mg of powdered iron is added to one flask and 50 mg of iron together with 5 mg of peptone are added to the other. Flasks are bubbled with CO_2 until the pH reaches below 6.0, then incubated at 20°C for 2-7 days.

KWM method: Water sample (3-4 ml) is added to a test tube containing 1.5 ml of FeS sludge and 15 ml of medium consisting of 0.1 g of NH_4Cl, 20 mg of $MgSO_4 \cdot 7H_2O$, 50 mg of K_2HPO_4, 120 μg (as Ca) of Ca $(HCO_3)_2$, 60 μg (as Fe^{2+}) of Fe $(HCO_3)_2$ in 100 ml double-strength evaporated tap water, and bubbled with CO_2. The mixture is incubated at 20°C for 2 weeks.

ACKNOWLEDGMENT

The author wishes to thank Mr. Kuranosuke Ishii of the Nippon Millipore Limited, Tokyo, for his valuable advice.

REFERENCES

American Public Health Association, American Water Works Association, Water Pollution Control Federation (1971). *Standard Methods for the Examination of Water and Wastewater*, 13 ed. American Public Health Association, Washington, D.C., pp. 678–688.

Environmental Agency (1971). Environmental criteria relevant to water pollution. *EA Official Announcement, No. 59* (1971).

Environmental Agency (1974, 1975). Methods of analysis for the quality standards of effluent provided by the Secretary of EA. *EA Official Announcement, No. 64* (1974); *No. 4* (1975).

Fujimoto, I. (1957). Studies on night-soil discharge to the ocean (XI); membrane filtration technique as a microbiological test for sea water pollution. *Public Health (Kokumin Eisei), 26:* 231 (cited in Taguchi, 1960).

Japanese Standards Association (1964, rev. 1974). Methods of analysis for industrial waste water. In *Japanese Industrial Standards,* JIS K0102.

Japanese Water Works Association (1970). Microbiological and biological test methods. In *Standard Methods of Analysis in Waterworks.* Japanese Water Works Association, Tokyo, pp. 347–547.

Koide, G. (1971). A novel method for the preparation of microscope specimens and biological estimation of planktonic algae with use of membrane filter. *J. Water Works Assoc. (Suido Kyokai Zasshi) No. 439:* 1–5.

Ministry of Health and Welfare (1966). Ministerial ordinance on quality standards for drinking water. *MHW Ordinance No. 11* (1966).

Osaka Prefectural Office (1977). Statistics of sample analysis relevant to water quality standards. In *Annual Report on Environmental Protection in Osaka Prefecture,* Osaka Prefectural Office, p. 295.

Pharmaceutical Society of Japan (1973). Methods of analysis for water quality. In *Standard Methods of Analysis for Hygienic Chemists--with Commentary.* Kanehara Publishing Co., Tokyo, pp. 681–822.

Pharmaceutical Society of Japan (1977). Methods of analysis for water quality, a revision of *Standard Methods for Analysis for Hygienic Chemists--with Commentary. J. Hyg. Chem. (Eisei Kagaku) 23:* 138–146.

Prime Minister's Office (1971, 1976). Government ordinance on the quality standards of effluent. *PMO Ordinance No. 35* (1971); *No. 37* (1976).

Taguchi, K. (1960). A comparable study on the most probable number method and membrane filtration method in the coliform test of water. *J. Water Works Assoc. (Suido Kyokai Zasshi) No. 304:* 42–48.

14

MEMBRANE FILTRATION: A SOUTH AFRICAN VIEW

W. O. K. GRABOW Council for Scientific and Industrial Research, Pretoria, South Africa

I. INTRODUCTION

Southern Africa has limited freshwater resources (Stander and Funke, 1967). Calculations based on the present rapid growth in population, industry, and agricultural activities indicate that toward the end of the century the demand for potable water will exceed the supplies obtainable by conventional methods (Stander and Clayton, 1977). Another problem is that the limited rainfall is irregular and not evenly distributed over the subcontinent. The eastern parts enjoy a relatively high rainfall, which decreases westward to arid areas in central regions and extreme desert conditions in coastal parts of South West Africa. This situation has enforced intensive research on water conservation, pollution abatement, and reuse (Cillié, 1975; Stander, 1977). A milestone of special significance in the history of water research in South Africa was reached in 1969 when a wastewater reclamation plant was commissioned in Windhoek, South West Africa, to supplement the city's domestic water supplies (Van Vuuren et al., 1971; Sebastian, 1974).

The critical water situation in southern Africa necessitates careful monitoring and control of the quality of natural water resources (Coetzee, 1962; Livingstone, 1969, 1976; Grabow and Prozesky, 1973), wastewater discharges (Coetzee and Fourie, 1965;

Malherbe et al., 1965; Malherbe and Strickland-Cholmley, 1965a, b; Grabow and Nupen, 1972; Grabow et al., 1973; Grabow and Prozesky, 1973; Grabow et al., 1975b, 1976), drinking water supplies (Isaäcson et al, 1974; Hattingh and Nupen, 1976; Grabow, 1977), and special attention is of course given to reclaimed wastewater (Nupen, 1970; Nupen et al., 1974; Barnard and Hattingh, 1975; Nupen and Hattingh, 1975; Grabow and Isaäcson, 1977; Hattingh, 1977). The available information indicates that waterborne outbreaks of diseases seldom occur in southern Africa, which implies that water treatment management and surveillance of the microbiological quality of water supplies are of a high standard. This chapter deals with the role of membrane filtration in the methodology, approach, and views of the microbiologists who have to safeguard the microbiological quality of water in this part of the world, with its own special conditions regarding available water resources, climate, industry, socioeconomical structure, and population composition.

II. WATER QUALITY STANDARDS IN SOUTH AFRICA

South Africa has no statutory standards for potable water. The South African Bureau of Standards (1971) recommends the bacteriological limits in Table 1, which are in agreement with internationally accepted standards (Hattingh, 1977). The following limits have been proposed for drinking water reclaimed from wastewater by means of multiple-barrier processes which conform to specified physicochemical parameters (Grabow et al., 1978a): total bacterial plate count, 100 per milliliter; total coliforms, 0 per 100 ml; enteric viruses, 0 per 10 liters; coliphages, 0 per 10 ml. These proposals are based on internationally accepted standards for drinking water, detailed evaluation of the removal of microorganisms in each individual unit process of the multiple-barrier plants concerned, and elaborate epidemiological studies on the consumers of reclaimed water (Grabow et al., 1969; Nupen, 1976; Grabow, 1977; Grabow and Isaäcson, 1977; Grabow et al., 1978a, b).

TABLE 1 Bacteriological Limits for Potable Water Specified by the South African Bureau of Standards

Organism	Recommended limit	Maximum allowable limit
Coliform organisms (number/100 ml)	[a]	10
E. coli I (number/100 ml)	0	0
Total viable organisms (colonies/ml)	100	Not specified

[a]If any coliform organisms are found in a sample, a second sample must be taken immediately after the tests on the first sample have been completed, must be free from coliform organisms, and not more than 5% of the total number of water samples (from any one reticulation system) tested per year may contain coliform organisms.
Source: South African Bureau of Standards (1971).

The only statutory water quality standard in South Africa specifies that wastewater discharged into certain rivers or their catchment areas must contain no *Escherichia coli* type I bacteria per 100 ml (Department of Water Affairs, 1962). However, the Department of Water Affairs may grant permits to exceed this limit on request motivated by acceptable reasons for not being able to comply with the standard. The Department of Health (Smith, 1969) has recommended the limits in Table 2 as a guideline for the discharge or reuse of purified wastewater. Coetzee (1963) suggested that swimming bath water should have a total bacterial plate count of less than 100 per milliliter, no fecal coliforms per 100 ml and a free residual chlorine concentration of 1 mg/liter.

III. METHODS OFFICIALLY SPECIFIED IN SOUTH AFRICA FOR THE MICROBIOLOGICAL ANALYSIS OF WATER

The South African Bureau of Standards (1971) specifies in its SABS Method 221 that the membrane filtration (MF) technique should be used for the analysis of drinking water. Details are given for apparatus and procedures. Samples should be analyzed within 6 hr after collection by personnel adequately trained and experienced in

TABLE 2 Bacteriological Quality Guidelines for Purified Wastewater Discharge or Reuse Recommended by the Department of Health

Purpose	*E. coli* I/100 ml
Irrigation	
Crops eaten raw	Use of effluent not permitted
Crops unlikely to be eaten raw	1000
Fruit trees, trellised vines	1000
Golf courses, sports fields, etc.	1000
Pasturage grazing	1000
Pasturage nongrazing	No standard
Industrial use	0
Discharge	
River, stream, etc.	1000
Seawave zone	
Bathing beach	1000
Remote from bathing beach	Free of floatable material
Intermediate between 1 and 2	Free of floatable material
Beyond wave zone	Each case considered on its merits

Source: Smith (1969).

microbiological techniques. Membrane filters, preferably marked with a grid, should have a pore size not exceeding 450 nm, and they should have been proved by laboratory tests to provide full bacterial retention and satisfactory speed of filtration, to be stable in use, and to be free of chemicals that retard the growth and development of bacteria. Membranes and filter holders should be sterilized by autoclaving. After filtration, membranes should be incubated for 2 hr at 35 ±1°C on a resuscitation medium which consists of 40 g of peptone, 6 g of yeast extract, 30 g of lactose, and 1000 ml of distilled water sterilized by autoclaving at 115 ±2°C for 15 min. The membranes may either be incubated on appropriate pads saturated with 2.5 ml of medium or on medium solidified by the addition of 1.5% (wt/vol) agar. The membranes should subsequently be transferred to a MacConkey medium prepared as follows. Add 10 g of peptone, 30 g of lactose, 4 g of bile salts, and 5 g of sodium chloride to 1000 ml of distilled water, adjust the pH to ensure a value of 7.2 in the final medium, and steam until all solids are dissolved. After this add 12 ml of 1% (wt/vol) alcoholic solution of bromcresol purple indicator and sterilize the medium by autoclaving. The peptone and bile salts used shall be equivalent to the reference standards held by the South African Bureau of Standards. Appropriate dehydrated culture media may be used in lieu of the media described provided that they conform to the description given and yield equivalent results. Incubation of membranes on the MacConkey medium may be done either on pads saturated with the medium or on medium solidified by the addition of 1.5% (wt/vol) agar. Incubation on MacConkey medium is either at 35 ±1°C for 18-24 hr to obtain a total coliform count or at 44 ±0.25°C for 18-24 hr to obtain a fecal coliform count. Incubation at either temperature should be done in a water bath. Identification of *E. coli* I among fecal coliforms is performed by picking colonies from the membrane and testing their ability to produce indole at 44 ±0.25°C.

The Department of Water Affairs (1969) specifies that a most probable number (MPN) method should be used for the evaluation of coliform bacteria in wastewater effluents. Details are given for procedures, apparatus, and the MacConkey medium to be used. However, no reason is given for the choice of an MPN method. The only reason is probably that the specification was drafted at a time when the MPN method was the only commonly known procedure and the MF technique had not been well established in the country. This situation, where one authority specifies an MPN evaluation and another an MF method for analyses that are generally done in the same laboratory, illustrates that authorities concerned with the standardization of techniques should regularly review their specifications in the light of the continuous development of better methods. An important disadvantage of outdated specifications is that laboratories eventually tend to follow their own discretion in the choice of methods. The ultimate result is that a wide variety of techniques are being used and few laboratories adhere to any standardized techniques.

Apart from coliform methods, the only officially specified technique for analysis of the microbiological quality of water is a pour plate procedure for total viable bacteria (South African Bureau of Standards, 1971).

IV. METHODS USED IN SOUTHERN AFRICA FOR THE BACTERIOLOGICAL ANALYSIS OF WATER

Methods used in southern Africa for the bacteriological analysis of water were reviewed by sending detailed questionnaires to more than 60 laboratories. The 59 laboratories that replied (Table 3) adequately represented all sectors involved in water quality evaluation and included laboratories from national research institutes, various state departments, universities, local authorities, and private enterprises. These laboratories analyzed a total of about 102,300 samples of water per year.

TABLE 3 Laboratories That Participated in a Review of Methods Used for the Bacteriological Analysis of Water in Southern African Countries

Location	Laboratory
Transvaal	
Germiston	Municipal Health
	Municipal Water
Jan Smuts Airport	SA Airways[a]
Johannesburg	Coca-Cola Export Corp.
	Electr. Supply Comm.
	Municipal Health
	SAIMR[b]
Klerksdorp	SAIMR Regional
Nelspruit	SAIMR Regional
Phalaborwa	Water Board
Pietersburg	SAIMR Regional
Potchefstroom	SAIMR Regional
Pretoria	Institute of Pathology
	Municipal Water
	NIWR Technol. Appl.
	NIWR Water Quality
	SA Bureau of Standards
	SA Railway and Harbours
Rustenburg	Municipal

TABLE 3 (Continued)

Location	Laboratory
Vereeniging	Municipal
	Rand Water Board
	SAIMR Regional
Orange Free State	
Bloemfontein	Municipal Health
	SAIMR Regional
Kroonstad	SAIMR Regional
Welkom	SAIMR Regional
South West Africa	
Windhoek	Dept. Water Affairs
	Municipal
	NIWR Regional
	SAIMR Regional
Cape Province	
Bellville	NIWR[c] Regional
Cape Town	Fishing Industry
	Municipal
	SA Bureau of Standards
	State Health
East London	Frere Hospital
	Municipal
George	State Health Regional
Kimberley	Municipal
Port Elizabeth	Municipal
	SAIMR Regional
Upington	State Health Regional
Wellington	Municipal
Worcester	State Health Regional
Natal	
Durban	Addington Hospital
	Drummond Pathologists
	Municipal Health
	Municipal Water
	NIWR Regional
	State Health Regional
Pietermaritzburg	Darvil Sewage
	D. V. Harris Water
	Signal Hill Water
Rhodesia	
Bulawayo	Bactochemical Anal.
	Mpilo Hospital
Salisbury	Municipal
	Public Health

TABLE 3 (Continued)

Location	Laboratory
Transkei	
Umtata	Transkei Water
Lesotho	
Maseru	Queen Elizabeth II Hospital

[a]SA, South African.
[b]SAIMR, South African Institute for Medical Research.
[c]NIWR, National Institute for Water Research.

A. Methods for Coliform Bacteria

Membrane Filtration vs. Most Probable Number Evaluation

Coliform analyses were done by means of MF in 22 (37%) of the 59 laboratories, 28 (48%) applied the MPN technique, and 9 (15%) used either of the two methods, depending on conditions, such as the quality of the water under investigation (Table 4). Although the laboratories that exclusively used the MF technique were in the minority, they included almost all the major and more sophisticated laboratories involved in research or large-scale routine analyses of potable water supplies to cities. A large number of laboratories of small local authorities that have limited association with research or development applied the MPN procedure exclusively. This is

TABLE 4 Application of Membrane Filtration, Most Probable Number, or Both Methods by a Representative Number of Laboratories in Southern Africa[a]

	MF	MPN	MF or MPN	Total
Number of laboratories	22	28	9	59
% of total	37	48	15	
Number of samples analyzed annually	43,780	25,590	32,930	102,300
% of total	43	25	32	

[a]Laboratories listed in Table 3.

illustrated by the data in Table 4, which show that only 25% of the total number of water samples analyzed annually were done by means of MPN evaluations exclusively; 43% of the samples were analyzed by MF methods only and 32% by either of the two methods.

Only two (8%) of the 24 laboratories that used MPN evaluations exclusively had ever compared their methods to MF. The reasons given by one of them for adhering to the MPN method were that it was cheaper, took less time to perform (although results were not available as soon as with MF), had been the traditional laboratory method, and was the technique officially specified by the Department of Water Affairs (1969). The second laboratory decided to do so because it feared that its field staff, which had limited training, might not be able to cope with MF, whereas they turned out reliable and satisfactory results with a traditional MPN procedure. These were also among the most important reasons given by the other laboratories that used MPN evaluations exclusively. Among the 28 laboratories that perform MPN evaluations only, 7 (25%) maintained that they preferred the method because it was cheaper, 17 (61%) stated that it was the traditional laboratory procedure and that staff with limited training can produce more reliable results with MPN evaluations, 3 (11%) maintained that it was the method specified by the Department of Water Affairs (1969), and 6 (21%) explained that although results were not available as soon as with MF tests, MPN evaluations took less time to perform. Rhodesian laboratories pointed out that membranes had to be imported and that they were not always freely available. Two laboratories indicated that they avoided problems with long-distance transportation of samples by submitting to the field staff containers half-filled with double-strength MacConkey medium. These containers had only to be filled with water and returned to the laboratory for a qualitative positive or negative evaluation of coliforms in 50 ml or 10 ml of the water concerned. None of the laboratories that used MPN evaluations exclusively were aware of any occasion on which their method failed to give the desired indication of water quality; neither did any of them express concern about the fact that their MPN evaluations yielded results after only 48-96 hr, whereas corresponding results were available within 18 hr by means of MF procedures.

Among the 9 laboratories that used both MPN and MF methods, 5 (56%) indicated that they applied MF for all samples of water except those with high turbidities, which clog membranes. This implies that 30 (51%) of the 59 laboratories that participated in this investigation used MF for the analysis of drinking water supplies. According to 5 (56%) of the laboratories that used both methods, the earlier availability of results was an important advantage of MF. Two laboratories reported that the time required for performing MF tests was less than that for MPN evaluations, whereas one laboratory stated the opposite. One laboratory thought that MPN evaluations were more suitable for staff with limited training, another pointed out that MPN evaluations were useful as purely qualitative tests, and two maintained that MF tests yielded more accurate results than did MPN evaluations. One laboratory pointed out that volumes of water much larger than the normal 100 ml could be analyzed by means of MF without additional media, whereas this is not possible with MPN evaluations. The same laboratory also mentioned that MF had the advantage of giving a direct indication of numbers of bacteria, whereas MPN tests are based on statistical evaluations subject to inherent errors.

Among the 22 laboratories that used MF procedures exclusively, 9 (41%) had experimentally compared the method to MPN evaluations and all found MF techniques to be superior. The published results of Pretorius and Coetzee (1965) show that MF yielded higher counts of *E. coli* I than did MPN evaluations in all qualities of water tested, except heavily polluted water such as raw sewage. They concluded that MF saves time and materials. Melmed (1966) reported the following advantages for MF: speed and simplicity of operation, results obtained within 18 hr, a direct count is made and not a statistical evaluation, the dispersion from the mean of results obtained by MF is less than for results obtained by MPN evaluations, tests can be done without inconvenience to staff on 4 days out of a 5-day work week, the cost of the test per sample is cheaper than that of MPN evaluations and is still further reduced if membranes are reused, and there is saving in the time needed for media preparation. According to Melmed (1966), the limitations of MF were clogging of membranes by highly turbid samples, occasional difficulties in selecting a suitable volume for filtration, and growth of coliforms may be obscured by that of non-lactose-fermenting bacteria or spreading colonies. However, the problem of spreading colonies could be limited by careful control of moisture on the surface of agar plates and the use of standardized bile salt for the preparation of the MacConkey medium used. The laboratories that used MF exclusively without having experimentally compared the method to MPN evaluations, had decided in favor of MF on the basis of results of comparative evaluations published in the literature.

Eighteen (82%) of the 22 laboratories that used MF only considered the fact that results were available within 18 hr as compared to the 48-96 hr of MPN evaluations to be an important advantage of MF techniques, 18 (82%) of the laboratories maintained that MF was more accurate, 10 (45%) regarded MPN evaluations as more cumbersome, and 4 (18%) estimated MF tests to be cheaper. Another advantage of MF, which was regarded as important by 4 (18%) of the laboratories, was that the growth of non-lactose-fermenting bacteria on membranes provides useful additional information on the quality of the water under investigation, and colonies can easily be picked from membranes for further individual characterization. Three (14%) of the laboratories pointed out that the petri dishes used in MF take up less incubator space than do the racks with tubes used in MPN evaluations.

Membrane Filtration Methods

Media. Six (19%) of the 31 laboratories that performed MF tests used a MacConkey medium which they prepared themselves according to the formula specified by the South African Bureau of Standards (1971). Eight (26%) of the laboratories used Oxoid MacConkey membrane broth, which is identical to the medium specified by the Bureau of Standards (Oxoid Manual, 1976). The other 16 laboratories (52%) used MacConkey media that differed from the Bureau of Standards' medium in the concentration of certain constituents. Ten of these laboratories prepared a medium that contained 20 g instead of 10 g of peptone, 10 g instead of 30 g of lactose, and 5 g instead of 4 g of Oxoid bile salts No. 3 per liter; two used MacConkey purple broth (Oxoid Manual, 1976) which contains 20 g of peptone, 10 g of lactose, 5 g of bile salts, and 0.01 g instead of 0.12 g of bromcresol purple per liter; one used a Metropolitan Water Board agar, which contains 9 g of bile salts per liter; one used 20 g of peptone, 10 g of lactose, 1.5 g of bile salts, no sodium chloride instead of 5 g, and 12 g instead of 15 g of agar per liter; one used 20 g of peptone, 10 g of lactose, 5 g of bile salts, and 13.6 g of agar per liter; and one used Difco MacConkey agar, which differs extensively from the other media and contains 17 g of Bacto-peptone, 3 g of proteose peptone, 10 g of lactose, 1.5 g of bile salts, 5 g of sodium chloride, 13.5 g of agar, 0.03 g of neutral red, and 0.001 g of crystal violet (Difco Manual, 1963).

Fifteen (48%) of the 31 laboratories that applied MF had compared the media they used to other media. In the case of 11 of these laboratories, the comparisons were limited to different varieties of MacConkey media. Four laboratories had compared their MacConkey media to Teepol, m-Endo, and mFC media. One found that Teepol media yielded inconsistent results with many false positives, and another reported "promising results" for a Teepol medium. Two laboratories maintained that the green metallic sheen of coliform colonies on m-Endo medium was not as clearly distinguishable as the bright yellow colonies on their MacConkey medium. In our own laboratory we have performed a detailed comparative study on MacConkey, Teepol, LES Endo agar, and mFC media. The MacConkey medium consistently yielded the lowest counts for both total and fecal coliforms in water of various qualities. Counts of these organisms were generally somewhat higher on the Teepol medium and considerably higher on LES Endo agar and mFC media, respectively. Although we agree that the definition of coliform colonies on MacConkey medium was not matched by any of the other media, the recognition of these colonies on LES-Endo agar did not prove an acceptable objection to the medium. Even if doubtful colonies were disregarded, total coliform counts were generally still higher on LES Endo agar than on MacConkey medium. The reason for the differences in counts is currently under investigation. A report containing information on whether this is due to excessive false positives on the one hand, or selection excluding certain coliforms on the other, should be released in due course.

Resuscitation Procedures. Ten (32%) of the 31 laboratories included an enrichment step in their MF tests. Nine of them used Oxoid Resuscitation Membrane Broth, which meets the requirements of the medium specified by the South African Bureau of Standards (1971). Four laboratories incubated membranes for both total and fecal coliforms for 2 hr at 37°C on this medium. One reported incubation for 1 hr at 36°C, another incubation for 1 hr at 35°C, and a third one kept the plates for 15 min at room temperature. Two laboratories incubated total coliform membranes for 1 hr at 37°C and fecal coliform membranes for 2 hr at the same temperature. One laboratory resuscitated fecal coliforms by preincubation of membranes on the final MacConkey medium for 2 hr at 37°C prior to transfer to 44.5°C, without using a separate enrichment medium. Melmed (1966) found no significant differences in comparative evaluations of coliform counts with or without preliminary incubation on separate enrichment media. Another laboratory maintained that enrichment on Oxoid Resuscitation Membrane Broth increased counts of both total and fecal coliforms in river water by 15-24%. We have investigated the advantages of a double-layer enrichment medium for fecal coliforms similar to that recommended by Rose et al. (1975). The bottom layer consisted of the MacConkey agar generally used in our laboratory (Grabow and Nupen, 1972), and the top layer was prepared as recommended by Rose et al. (1975). Membranes were put directly onto this medium and after incubation for 2 hr at 37°C, the plates were transferred to 44.5°C for another 16 hr. Although this procedure generally yielded higher counts than the ordinary method in which membranes are directly incubated on MacConkey agar without enrichment or preincubation, fecal coliforms were rarely if ever isolated by the enrichment procedure from water that proved to be free of these organisms using the ordinary method. The question of whether, in fact, there is a need for the higher counts and whether any enrichment procedures are necessary remains to be answered.

Agar Media vs. Saturated Pads. Saturated pads proved popular for enrichment procedures. Eight of the nine laboratories who incubated membranes on a separate resuscitation medium used pads saturated with Oxoid Resuscitation Membrane Broth, and one laboratory saturated pads with the same medium containing 0.75% (wt/vol) agar to prevent dehydration. Twenty-four (77%) of the total number of 31 laboratories who performed MF tests preferred agar media for the final incubation of membranes. The reasons given for the preference of agar media were that saturated pads tended to dry out during incubation (5 laboratories), and that agar media were less cumbersome (20 laboratories), time consuming (15 laboratories), and liable to contamination (2 laboratories). Two laboratories pointed out that personnel with limited training had difficulties in handling saturated pads. Among the seven laboratories (23%) that used saturated pads, five gave no particular reason for their choice; one maintained that solid media tended to dry out more easily than did saturated pads, and another stated that saturated pads yielded better growth. In comparative evaluations Coetzee and Pretorius (1965) found that counts on agar media were slightly higher than on saturated pads, and they concluded that the increase in cost due to the incorporation of agar is offset by the saving in labor and the higher counts. Higher counts on agar media were confirmed by two other laboratories, whereas a third found no significant difference.

Membrane Filters and Filtration Equipment. The choice of membranes was determined primarily by such factors as cost, availability, convenience in handling, and suitability for reuse. The general opinion indicated, for instance, that Sartorius membranes handled very well and could be reused but that they were expensive, Millipore membranes tended to curl up during decontamination in boiling water, and Gelman membranes were cheap but brittle and could not be reused. Detailed comparative studies of counts on different membranes have not been reported by any of the laboratories. One reason was that, according to the literature, there was no significant difference in the efficiency of the locally available membranes (Green et al., 1975; Lin, 1977; Tobin and Dutka, 1977). In addition, 26 (84%) of the 31 laboratories had experience with different membranes and apart from considerations such as those mentioned earlier, none of these laboratories noticed any difference in efficiency. The membranes used in the 31 laboratories were Gelman (14 laboratories), Oxoid (9 laboratories), Millipore (8 laboratories), Sartorius (7 laboratories), and Selectron S & S (1 laboratory). At the time of this investigation, Gelman GN-6 membranes were popular mainly because they were available at a reasonable price. Sixteen (52%) of the laboratories did not reuse membranes. However, reuse depends to a large extent on the membranes used because some membranes, particularly Gelman, are not suitable for reuse. Obviously, new presterilized membranes save time and limit the possibility of contamination and using damaged or clogged membranes. Twelve (67%) of the 18 laboratories that reused membranes or did not use presterilized membranes decontaminated membranes by autoclaving at 121°C for 15 min; the others boiled the membranes in water for periods that varied from 10 to 30 min. The number of times that membranes were reused varied from once only to 10 times for Oxoid and Sartorius membranes in one laboratory and up to 15 times for Oxoid membranes in another laboratory.

A variety of filtration units were used. The most popular ones were Sartorius (9 laboratories), Gelman (7 laboratories), Millipore (6 laboratories), and Gallenkamp (4 laboratories). Although there are differences in the composition and operation of these units, satisfactory service was obtained from all of them. These units were decontaminated by autoclaving (14 laboratories), immersion in boiling water (11 laboratories), flaming (6 laboratories), sterilization in a dry-air oven (3 laboratories), ultraviolet irradiation (1 laboratory), or flushing with sterile distilled water (1 laboratory).

Incubation Temperatures and Incubators. The incubation temperature for total coliform counts was not regarded as critical; 84% of the laboratories reported incubation at 37°C and the others at 35°C. Only 10% of the laboratories used water baths for the incubation of total coliform counts. The others used conventional incubators, many even without circulating air facilities. We have compared coliform counts obtained by incubation of parallel tests in circulating air incubators set at 35°C and 37°C and found no significant differences. The incubation temperature for fecal coliform counts was considered more important. Eleven laboratories reported incubation temperatures of 44.0°C, 10 reported 44.5°C, and one 45.0°C. Twenty (65%) of the laboratories used water baths for the incubation of fecal coliform plates, whereas the others used circulating air incubators. None of the laboratories had investigated the effect of temperature variability on fecal coliform counts.

Petri Dishes. Twenty (65%) of the laboratories used disposable plastic dishes, whereas the others used glass dishes. Some of the laboratories that generally used glass dishes kept a supply of disposable dishes for emergencies. The diameter of the glass dishes used varied from 50 to 100 mm. One laboratory used 50 mm diameter metal dishes with tight-fitting lids. Among the laboratories that used disposable dishes, 12 (60%) used 85 mm diameter dishes with loose-fitting lids, 5 (25%) used 65 mm dishes with loose-fitting lids, and 3 (15%) used 50 mm dishes with tight-fitting lids (the latter dishes were used mainly in field work). The dishes with tight-fitting lids were not made locally and had to be imported, with the result that they were very expensive. Another disadvantage of these dishes is that their handling is cumbersome and time consuming. None of the laboratories reported any problems in using dishes with loose-fitting lids for either total or fecal coliform counts. We find it surprising that so many laboratories used 85 mm dishes. In our experience the 65 X 15 mm dishes have ample space for the conventional 47 mm membranes, and they have the advantage of being cheaper; requiring less medium, which reduces costs; and taking up less incubator space.

Transportation of Water Samples. Sixteen (52%) of the 31 laboratories rarely, if ever, analyzed samples more than 3 hr after they had been taken. Thirteen (42%) of the laboratories frequently analyzed samples between 4 and 24 hr after collection. All but one of these laboratories transported samples cooled to temperatures between about 4 and 10°C. One laboratory analyzed cooled samples up to 48 hr after they were taken. Three laboratories used a Millipore portable water laboratory kit for field work. Only one laboratory filtered samples on site and transported filtered membranes on a transportation medium for eventual incubation on MacConkey agar in the laboratory. Melmed (1966) presented evidence that storage of samples for more than 1 hr at room temperature resulted in increased or decreased fecal coliform counts, depending on factors such as organic pollution of the water, total bacterial numbers, and temperature.

B. Membrane Filtration Methods for Bacteria Other Than Coliforms

Fifteen laboratories employed MF for the evaluation of bacteria other than coliforms. Nine laboratories performed enterococcus counts, eight by means of Difco m-Enterococcus agar and one by means of Oxoid Slanetz and Bartley medium containing 0.04% (wt/vol) potassium tellurite. Six laboratories used a qualitative presence or absence MF procedure for *Pseudomonas aeruginosa*. The test consisted of incubating membranes through which 100 ml of water had been filtered in Drake's medium followed by confirmation on acetamide medium (Grabow, 1977). This method proved much more sensitive than the direct MF method of Levin and Cabelli (Grabow, 1977). Eight laboratories evaluated *Clostridium perfringens* counts by incubating membranes face down on Wilson and Blair glucose sulfite iron medium. The membranes were covered with a layer of nutrient agar to establish anaerobic conditions (Grabow and Nupen, 1972). After incubation, colonies with a black halo were picked for confirmation of *C. perfringens* by means of the Nagler test (Cruickshank et al., 1975). In our experience, this method is unsatisfactory and better techniques for the evaluation of *C. perfringens* should be investigated. *Staphylococcus aureus* counts were evaluated by incubating membranes on mannitol salt agar (7 laboratories), Staphylococcus Medium No. 110 (2 laboratories) or Merck Vogel-Johnson Agar (1 laboratory). *Staphylococcus aureus* was confirmed by either a plasma coagulase test (Livingstone, 1969) or a DNase test (Grabow, 1977). Salmonellas were isolated from water by incubation of membranes on deoxycholate citrate agar in one laboratory and by enrichment of membranes on selenite brilliant green medium followed by incubation on SS agar in another laboratory. One laboratory performed shigella tests by incubating membranes on SS agar.

C. Cost of Membrane Filtration Compared with That of Most Probable Number Evaluation

Both methods require certain basic laboratory facilities and apparatus, such as incubators, sterilization equipment, pipettes, and glassware, which need not be taken into account in a cost comparison. The same applies to procedures required for the eventual identification and characterization of bacteria, such as IMViC testing and gram staining. The cost differences between MF and MPN evaluations are concerned primarily with media, filtration equipment, membranes, petri dishes, and some glassware. The following evaluations are based on prices in South Africa in May 1978.

The MPN method specified by the Department of Water Affairs (1962) requires a presumptive test with five tubes for each of three dilutions. Each tube contains 10 ml of single-strength MacConkey medium. The total volume of 150 ml single-strength MacConkey medium would cost about R0.28 if commercial dehydrated powder medium were used. If waters with low bacterial counts, such as drinking supplies, were to be tested, analysis of a single 100 ml sample of water requiring 100 ml of double-strength medium would cost about R0.37. In addition, a fecal coliform count requires subinoculation of all positive presumptive tubes into similar tubes for incubation at 44.5°C. If 80% of the presumptive tubes were to be positive, this would involve another R0.22. Although it is not mentioned in the specifications of the Department of Water Affairs, a total coliform count should also be confirmed by subinoculation (American Public Health Association, 1976), which implies that the commercial MacConkey powder medium required for an MPN evaluation of either total or fecal coliforms would cost about R0.50. If the more convenient tablet form of dehydrated medium were to be used, the cost would be about R1.46 (U.S. $1.68) per test. An MPN evaluation on fivefold quantities of 100 ml, 10 ml, and 1 ml of water would cost substantially more. In addition, the MPN test requires 350 ml test jars, test tubes, and durham tubes, which are not needed for MF. Since this glassware is reused, the cost per test in capital expenditure and labor required for washing and sterilization is difficult to estimate.

MF tests require filter holders and a suction device, which are capital items not required for MPN evaluations. The price of a good-quality filter funnel was about R50.00 and that of a manifold with three filter funnels about R135.00 (U.S. $156). A suitable negative pressure for filtration may be obtained by means of a water jet pump (R5.45), a hand vacuum pump (R48.00), or an electric pressure/vacuum pump (R230.00). These items have an almost unlimited lifetime and in the long run the expenditure will be much less than that of the additional glassware and associated labor required for MPN evaluations. In addition, this equipment is being used for so many other purposes that it should be regarded as standard in any microbiological laboratory. The resuscitation step specified by the South African Bureau of Standards (1971) would cost about R0.01 for 2.5 ml of broth (commercial tablets), R0.04 for an absorbent pad, and R0.03 for a disposable petri dish (65 mm diameter) per single analysis. Final incubation on MacConkey broth (commercial tablets) would cost R0.01 for 2.5 ml of medium, R0.04 for an absorbent pad, and R0.03 for a disposable petri dish. The cost of an Oxoid membrane filter was about R0.31. A single analysis would thus cost R0.47 and a complete test on three dilutions R1.41 (U.S. $1.63). If agar medium would be used for the final incubation instead of saturated pads, 15 ml of the same medium containing 1.5% (wt/vol) agar would be required per plate (65 mm diameter) at a cost of R0.09, which is considerably more than the R0.05 for a pad saturated with 2.5 ml of broth.

The cost of R1.41 for running expenses of a MF test, which is already less than R1.46 of an MPN evaluation, represents about the most expensive way of performing the test. Most laboratories omit the resuscitation step, which is acceptable for general purposes (American Public Health Association, 1976). This would reduce the cost per test to R1.17. The cost can be further reduced extensively by reusing membranes and petri dishes, which are the most expensive items in MF. An important advantage of MF is that the cost remains the same for any volume of water that will pass through a membrane filter, whereas the cost for MPN evaluations increases extensively for volumes of more than 1 ml. In addition, we feel that any analysis on numbers of bacteria in water should be done at least in triplicate, which increases the cost difference between MF and MPN evaluations even more. The foregoing costs of MF and MPN evaluations can, of course, be reduced by preparing media from basic ingredients in the laboratory. However, commercial media save time and they are generally of good and homogeneous quality.

D. Application of Membrane Filtration for Purposes Other than Conventional Bacterial Counts

MF equipment is being used for many purposes other than conventional counting of bacteria. Twenty-six (44%) of the 59 laboratories that participated in this review used MF and filters with appropriate pore size for evaluating organisms such as algae, yeasts, and parasite ova in water; for decontamination of heat-labile media, serum, antibiotic solutions, and chemical reagents; for the detection of fungal contaminants in jet aircraft fuels; for microbiological studies on dairy products and foodstuffs; for sterility tests on antibiotics; for certain studies on the chemical quality of water and air pollution; and for determination of suspended solids in water and crystal sizes of salts in seawater. MF proved most useful in qualitative and quantitative studies on bacteria resistant to antimicrobial drugs in water. In these studies samples of water are membrane-filtered and the membranes incubated on appropriate media containing any desired single drug or combination of drugs for the selection of resistant bacteria (Grabow and Prozesky, 1973; Grabow et al., 1974). MF is also being used in microbial genetics laboratories for the concentration and selection of desired recombinants in transduction (Coetzee, 1974) and conjugation (Coetzee, 1975) experiments. In view of the wide variety of applications, MF facilities should be regarded as standard in any microbiological laboratory.

E. Need for Better Methods

Among the 28 laboratories that did not use MF techniques for counting bacteria, 8 (29%) expressed the desire to improve on the methods they were using; 4 thought that they should change to MF and the other 4 had no particular suggestions. The remaining 20 laboratories (79%) were of the opinion that their MPN methods had never failed to prove or disprove the microbiological safety of water, so there was no need for changing the traditional laboratory procedures.

Seventeen (55%) of the 31 laboratories that used MF only or both MF and MPN tests felt that they would like to improve on the methods they were using. Four laboratories expressed concern about the recovery of stressed bacteria, four thought that available procedures for field work could be improved, one was interested in practical methods for obtaining coliform results sooner than the 18 hr required by MF tests, and one hoped for a reduction in the cost of MF tests. Many laboratories had no particular suggestions but indicated that various aspects of bacteriological evaluations involving MF, such as media, sample preparation, and filtration procedures, could probably be improved to increase sensitivity, reliability, and reproducibility. One laboratory proposed interlaboratory comparisons. Fourteen laboratories were happy with the MF techniques they were using, as they had never failed for their purpose.

V. DISCUSSION

A. Membrane Filtration vs. Most Probable Number Evaluation

The general opinion and experience of laboratories in southern Africa indicate that MF is the method of choice for the bacteriological analysis of drinking water as well as most other waters. Although views of overseas laboratories on the preference for methods differ (McCarthy et al., 1961; Geldreich et al., 1965; Dutka and Tobin, 1976; Lanz and Hartman, 1976; Lin, 1976; Mowat, 1976), water quality authorities generally agree that MF counts are at least as reliable as MPN evaluations and that MF has various important advantages (Department of Health and Social Security, 1970; World Health Organization, 1970, 1971; American Public Health Association, 1976; Aurand et al., 1976; Canada Department of Environment, 1976). In a recent publication, the World Health Organization indicates a preference for MF (World Health Organization, 1976). Current objections to MF are concerned principally with observations that MPN evaluations yield higher coliform counts on waters with a high incidence of stressed organisms, such as chlorinated wastewater effluents (Lanz and Hartman, 1976; Lin, 1976; Mowat, 1976; Bissonnette et al., 1977; Green et al., 1977). However, this concern may not be altogether justified. Injury and recovery of bacteria is a complex matter that is not well understood (Hurst, 1977). None of the authors concerned offers any supporting information on the health significance of stressed bacteria. It seems likely that pathogens injured to the same extent as stressed coliforms not capable of recovery during direct exposure to selective media may also not be able to survive the adverse conditions of the human stomach and small intestines or other defense mechanisms. This would imply that stressed organisms have limited health significance. Even if concern about stressed bacteria should be justified, there is ample evidence that inclusion of appropriate enrichment or adaptation steps will increase the sensitivity of MF to similar or higher levels than that of MPN evaluations (Rose et al., 1975; Lanz and Hartman, 1976; Lin, 1976; Green et al., 1977; Stuart et al., 1977). Purified wastewater effluents are rarely chlorinated in South Africa, and consequently the problem of chlorine-injured bacteria in discharges does not concern most laboratories in this country.

Another important reason for differences in counts obtained by MF and MPN evaluations is the inherent statistical error of MPN evaluations. The 95% confidence limit for the five-tube three-dilution MPN method yields values between 31 and 289% of the actual number of bacteria (Geldreich et al., 1965; Department of Health and Social Security, 1970). It is also known from both theoretical and experimental data that MPN estimates are positively biased (Lin, 1976; Green et al., 1977), and Thomas (1955) gave a factor of 0.851 to correct for the positive bias in five-tube tests. This error implies that the more bacteria in a sample of water, the more MPN evaluations estimate the count too high. MPN evaluations should therefore be blamed for differences in evaluations on highly polluted waters and not MF techniques (Lin, 1976).

The large number of laboratories in southern Africa that still use MPN tests seems to be the result of shortcomings in communication, education, and training. Many of these laboratories are just not aware of alternative methods and their advantages. Authorities concerned with water quality, such as the Department of Health and Department of Water Affairs as well as the Bureau of Standards, may have obligations in this regard.

B. Advantages of Membrane Filtration

The most important advantages of MF may be summarized as follows:

1. MF gives a direct count, whereas MPN evaluations are based on statistical estimates with inherent errors.

2. MF yields coliform results within 18 hr and MPN evaluations only after 48 to 96 hr.
3. MF tests for total coliforms yield valuable additional information on the qualitative and quantitative presence of organisms, such as *Aeromonas* and *Pseudomonas* species, which may grow in distribution systems (Burman and Colbourne, 1977; Van der Kooij, 1977) and certain units of advanced treatment processes (Grabow et al., 1978a).
4. Colonies can easily be picked from membrane filters for further characterization, whereas the isolation of organisms from MPN tubes is cumbersome and unreliable.
5. Organisms such as *Clostridium perfringens* (Department of Health and Social Security, 1970) and coliphages (Schiemann et al., 1976) may interfere with MPN evaluations of coliform bacteria, whereas they have no effect on MF tests.
6. MF tests are cheaper than MPN evaluations.
7. The performance of MF tests is less cumbersome and time consuming than that of MPN evaluations.
8. The petri dishes used in MF take up less incubator space than the racks with tubes required for MPN evaluations.
9. MF is more convenient under field conditions and useful MF field test kits have been developed (World Health Organization, 1976).

MPN evaluations remain useful for analyses on low numbers of bacteria in highly turbid waters, sludges, mud, or solid wastes. One disadvantage of MF tests for total coliforms is that the growth of exceptionally large numbers of certain non-lactose-fermenting bacteria present in some waters may obscure the growth of coliform colonies on membranes. However, the reliability of MPN evaluations for such waters is also questionable (Van der Kooij, 1977). Presence-absence tests with liquid media may prove useful for routine qualitative evaluation of drinking water supplies under certain conditions (Clark and Vlassoff, 1973).

C. Methods and Media for Membrane Filtration

The experience of laboratories in southern Africa allows recommendations on tests for coliform bacteria only. Although the available information indicates that MacConkey media have never failed their purpose, m-Endo and mFC media yield higher counts. The time may have arisen for these laboratories to seriously consider changing to the latter media, which at present probably are the most commonly used in the world (American Public Health Association, 1976; Aurand et al., 1976; Canada Department of Environment, 1976; World Health Organization, 1976). MacConkey media are also outdated in MPN evaluations, and better results are being obtained with glutamate media (Public Health Laboratory Service Standing Committee on the Bacteriological Examination of Water Supplies, 1968; Department of Health and Social Security, 1970; Alivisatos and Papadakis, 1975; Kampelmacher et al., 1976) and probably also with brilliant green lactose bile broth (American Public Health Association, 1976; Canada Department of Environment, 1976).

Resuscitation procedures generally increase MF counts. However, the need for inclusion of these steps, which are time consuming and increase costs, is not well motivated. Experience of local laboratories seems to support the view (American Public Health Association, 1976; Canada Department of Environment, 1976) that enrichment or adaptation procedures are not necessary for general purposes.

The majority of laboratories agree that the many advantages of agar media compared to saturated pads justify their higher cost (McCarthy et al., 1961; Canada Department of Environment, 1976).

A number of different brands of membrane filters yield satisfactory results. However, poor results have been reported for some brands and membranes should be selected with care (Lin, 1977; Tobin and Dutka, 1977). The same is true for filter devices.

The incubation temperature for total coliform counts does not seem to be critical and circulating air incubators generally yield satisfactory results. No significant differences in results have been reported for temperatures varying between 35°C and 37°C. The incubation temperatures of 44 ±0.25°C (Department of Health and Social Security, 1970; South African Bureau of Standards, 1971) or 44.5 ±0.2°C (American Public Health Association, 1976; Canada Department of Environment, 1976) are not clearly motivated. Indications are that the growth of coliforms other than fecal coliforms is suppressed at 42.5°C (Deutscher Verein von Gas- und Wasserfachmännern, 1976) and the Canada Department of Environment (1976) regards results obtained at 43°C as equivalent to those obtained at 44.5 ±0.2°C. However, the general consensus is that the incubation temperature for fecal coliform counts is important and a water bath with a temperature variation not exceeding 0.2°C set at 44.5°C is preferred for incubation.

Disposable plastic petri dishes (65 X 15 mm) with loose-fitting lids yield satisfactory results. Even for fecal coliform counts, tight-fitting lids are not necessary if sufficient humidity to prevent dehydration of media is maintained, as is the case when plates are incubated in watertight containers in a water bath. Glass dishes may prove more economical.

The survival of bacteria in water samples is subject to a wide variety of variables, and there is no doubt that laboratory analyses should be completed as soon as possible after collection. However, a delay of 6 hr is generally acceptable for uncooled samples (Lonsane et al., 1967; South African Bureau of Standards, 1971; American Public Health Association, 1976) and samples cooled to about 4°C generally yield acceptable results even after 24 hr (American Public Health Association, 1976; Kampelmacher et al., 1976; Standridge and Lesar, 1977).

D. The Value of Bacteria as Indicators of Virus Survival in Water Treatment Processes

Serious concern has been expressed about the possible dangers of viruses in water (Berg, 1973; Phillips, 1974; Gerba et al., 1975; Melnick, 1976; Shuval, 1977; Grabow et al., 1978a, b). However, there are indications that the problem of viruses in water tends to be exaggerated. Authoritative estimates indicate that each year some 500 million people are affected by incapacitating waterborne or water-associated diseases, and that as many as 10 million of these die, and an estimated 25% of the world's hospital beds are occupied because of unwholesome water (World Health Organization, 1976). The great majority of people in developing countries do not have access to reasonably uncontaminated water (Feachem, 1977), and even in the United States a survey revealed that 20 to 30% of drinking water supplies did not comply with conventional standards (McCabe et al., 1970). Notwithstanding this situation, there is no conclusive evidence for waterborne transmission of any virus other than the type A hepatitis virus (Grabow, 1968, 1976, 1977; Mosley, 1967; Craun and McCabe, 1973; Grabow et al., 1975a; Cooper, 1977; Kraus, 1977; Tate and Trussell, 1977). Apart from occasional hepatitis A outbreaks (Mosley, 1967), indications of possible association with water of infections by viruses such as polio (Grabow, 1968; Berg, 1977), adeno (Foy et al., 1968) and gastroenteritis viruses (Appleton and Pereira, 1977; Berg, 1977), and unconfirmed theoretical virus hazards (Grabow, 1968; Gerba et al., 1975; Cooper, 1977), waterborne diseases are caused by bacteria and parasites (Geldreich, 1972; Craun and McCabe, 1973; Craun, 1977).

Concern about viruses in water is often motivated by statements that viruses are more resistant to disinfection processes than bacteria because some viruses proved more resistant to chlorine than *E. coli* (Bates et al., 1977; Grabow et al., 1978b). However, this assumption is not justified because both laboratory experiments (Friberg and Hammarström, 1956; Engelbrecht et al., 1974) and field studies (Grabow et al., 1978a) showed that many bacteria, especially gram-positive organisms, bacterial spores, and acid-fast bacilli, are more resistant to chlorine than any viruses tested so far. In fact, *E. coli* proved one of the most sensitive bacteria, and even other members of the coliform group, such as *Klebsiella pneumoniae*, were found more resistant to chlorination (Bagley and Seidler, 1977). Many bacteria, notably enterococci, also proved more resistant than enteric viruses to high-lime treatment (Grabow et al., 1978b). Available data on the relative resistance and behavior of bacteria and viruses confirm views that bacteria may serve as reliable indicators for the removal of viruses in water treatment processes provided that the appropriate bacterial tests are applied (Engelbrecht et al., 1974; Grabow, 1977; Grabow et al., 1978a, b).

Experience over many years has proved the value of coliform tests for general purposes (Mack. 1977), but under certain conditions other bacteria, such as acid-fast bacilli (Engelbrecht et al., 1974), enterococci (Cohen and Shuval, 1973; Grabow et al., 1978b), or *Pseudomonas aeruginosa* (Grabow, 1977; Hoadley, 1977) may be more suitable. The simple and cheap total bacterial plate count is still the most sensitive tool in microbiological water quality evaluation and it is unfortunate that this test is so often neglected (Geldreich, 1973; Environmental Protection Agency, 1975; Grabow, 1977; Müller, 1977; Ptak and Ginsburg, 1977; Grabow et al., 1978a).

The conventional coliform test is often discredited because enteric viruses have been isolated from 379-3785 liter (100-1000 gal) concentrates of water that yielded negative coliform tests per 100 ml (Berg, 1973; Phillips, 1974; Gerba et al., 1975; Melnick, 1976; Shuval, 1977). Even if the presence of a single virus particle in that volume of treated water should impose an acceptable health hazard, there is no reason to believe that bacteriological methods may not be capable of supplying evidence that the desired level of purification has been attained (Mack, 1977). We have found that 3 liters of Pretoria tap water readily passes through conventional membrane filters. If this quantity is passed through each of 10 membranes, the sensitivity of the conventional coliform test would be increased 300 times, while the test remains cheap and simple with results available within 18 hr. MF evaluations of total bacterial counts would even be much more sensitive.

Bacteria are also indirectly valuable in water quality evaluation because the viruses (bacteriophages) that infect organisms, such as *E. coli*, proved useful indicators of virus survival and behavior in water purification processes (Grabow et al., 1978a, b).

E. Need for Better Methods

The basic requirement in water quality evaluation is practical, simple, and cheap methods, giving a reliable indication of the microbiological quality of water within a reasonable period of time.

Research workers in sophisticated laboratories with highly skilled labor and vast financial resources (Berg, 1973; Gerba et al., 1975; Melnick, 1976; Shuval, 1976), should consider the position of the public health worker in the elementary laboratory of a small community or developing country who stands in the front line of combat against waterborne diseases, before proposing standards such as the absence of enteric viruses in hundreds or thousands of liters of drinking water. Although the value for research purposes of the excellent achievements in concentrating viruses from those quantities of water is fully appreciated, much more information is necessary to justify the consideration (American Public Health Association, 1976) of their application in conventional water quality evaluation. Waterborne diseases have never been due to shortcomings in currently available methods for microbiological quality evaluation or water treatment technology, but always to the choice of treatment processes or their control (Craun and McCabe, 1973; Grabow, 1977).

The conclusion that currently used fecal coliform tests only represent "the tip of the iceberg" (Dutka and Kuchma, 1977) reflects the experience and literature reviewed in this chapter, which show that, should the need arise, the sensitivity of available bacteriological methods can be increased extensively without notably depriving them of their practical features. Although there is no indication that the wide variety of methods used by the 59 laboratories in southern Africa to analyze 102,300 samples of water each year has ever failed the basic purposes of water quality evaluation, the situation illustrates that the standardization of methods is the most important immediate need. Standardization is necessary to improve reliability, reproducibility, and comparability, and only results obtained by well-defined methods are acceptable for research on further improvements. The need for increasing the sensitivity of available methods, as well as the choice of tests and standards for particular purposes, will depend on the results of continued critical evaluation of relationships between indicator organisms and pathogens in water environments and purification processes, as well as epidemiological data on relationships between water quality and the health of consumers.

ACKNOWLEDGMENTS

The kind cooperation of the laboratories listed in Table 3 is appreciated. Thanks are due to Professors L. S. Smith and O. W. Prozesky for their valuable comments and suggestions.

REFERENCES

Alivisatos, G. P., and Papadakis, J. A. (1975). MacConkey and glutamate media in the bacteriological examination of seawater. *J. Appl. Bacteriol. 39*: 287–293.

American Public Health Association (1976). *Standard Methods for the Examination of Water and Wastewater,* 14th ed. American Public Health Association, Washington, D.C., pp. 875–981.

Appleton, H., and Pereira, M. S. (1977). A possible virus aetiology in outbreaks of food-poisoning from cockles. *Lancet 1*: 780–781.

Aurand, K., Hässelbarth, U., Müller, G., Schumacher, W., and Steuer, W. (1976). *Die Trinkwasser-Verordnung*. Erich Schmidt Verlag, Berlin, pp. 197–204.

Bagley, S. T., and Seidler, R. J. (1977). Significance of fecal coliform-positive *Klebsiella*. *Appl. Environ. Microbiol. 33*: 1141–1148.

Barnard, J. J., and Hattingh, W. H. J. (1975). Health aspects of the re-use of wastewater for human consumption. Presented at a Water Research Centre Colloquium entitled *Drinking Water Quality and Public Health,* High Wycombe, England, November, 4–6.

Bates, R. C., Shaffer, P. T. B., and Sutherland, S. M. (1977). Development of poliovirus having increased resistance to chlorine inactivation. *Appl. Environ. Microbiol. 34:* 849–853.

Berg, G. (1973). Reassessment of the virus problem in sewage and in surface and renovated waters. *Prog. Water Technol. 3*: 87–94.

Berg, G. (1977). Microbiology--detection, occurrence, and removal of viruses. *J. Water Pollut. Control Fed. 49*: 1290–1299.

Bissonnette, G. K., Jezeski, J. J., McFeters, G. A., and Stuart, D. G. (1977). Evaluation of recovery methods to detect coliforms in water. *Appl. Environ. Microbiol. 33*: 590–595.

Burman, N. P., and Colbourne, J. S. (1977). Techniques for the assessment of growth of micro-organisms on plumbing materials used in contact with potable water supplies. *J. Appl. Bacteriol. 43*: 137—144.

Canada Department of Environment (1976). *Methods for Microbiological Analysis of Waters, Wastewaters and Sediments.* Canada Centre for Inland Waters, Burlington, pp. I.5—II.15.

Cillié, G. G. (1975). Review paper. Current technology and advances in true re-use of effluents. In *Radiation for a Clean Environment,* Proceedings of a Symposium, Munich, Germany, March, 17—21. International Atomic Energy Agency, Vienna, pp. 29—43.

Clark, J. A., and Vlassoff, L. T. (1973). Relationships among pollution indicator bacteria isolated from raw water and distribution systems by the presence-absence (P-A) test. *Health Lab. Sci. 10*: 163—172.

Coetzee, J. N. (1974). High frequency transduction of kanamycin resistance in *Proteus mirabilis. J. Gen. Microbiol. 84*: 285—296.

Coetzee, J. N. (1975). Chromosome transfer in *Proteus mirabilis* mediated by a hybrid plasmid. *J. Gen. Microbiol. 86:* 133—146.

Coetzee, O. J. (1962). The possible use of oysters as integrated bacteriological samples of a specific area to detect the presence of pathogens in sea water. *Public Health, Johannesburg 62* (August): 21—25.

Coetzee, O. J. (1963). Bacteriological criteria for swimming bath water. *Public Health, Johannesburg 63* (June): 29—31.

Coetzee, O. J., and Fourie, N. A. (1965). The efficiency of conventional sewage purification works, stabilization ponds and maturation ponds with respect to the survival of pathogenic bacteria and indicator organisms. *J. Inst. Sewage Purif. 3*: 210—215.

Coetzee, O. J., and Pretorius, T. (1965). A modification to the membrane filtration technique applied to the routine determination of *Escherichia coli* bacteria in water. *Public Health, Johannesburg 65* (August): 383—385.

Cohen, J., and Shuval, H. I. (1973). Coliforms, fecal coliforms, and fecal streptococci as indicators of water pollution. *Water Air Soil Pollut. 2*: 85—95.

Cooper, R. C. (1977). Health considerations in use of tertiary effluents. *J. Environ. Eng. Div. Am. Soc. Civ. Eng. 103*: 37—47.

Craun, G. F. (1977). Waterborne outbreaks. *J. Water Pollut. Control Fed. 49*: 1268—1279.

Craun, G. F., and McCabe, L. J. (1973). Review of the causes of waterborne-disease outbreaks. *J. Am. Water Works Assoc. 65*: 74–84.

Cruickshank, R., Duguid, J. P., Marmion, B. P., and Swain, R. H. A. (1975). *Medical Microbiology*, 12th ed., vol. 2. Churchill Livingstone, Edinburgh, pp. 473–476.

Department of Health and Social Security (1970). *The Bacteriological Examination of Water Supplies.* Reports on Public Health and Medical Subjects No. 71. Her Majesty's Stationery Office, London, pp. 1–52.

Department of Water Affairs (1962). Regional standards for industrial effluents (No. R. 553). In *Government Gazette Extraordinary*, vol. 4. Government Printer, Pretoria, South Africa, pp. 9–14.

Department of Water Affairs (1969). Regional standards for industrial effluents (No. R. 3208). In *Government Gazette*, vol. 50. Government Printer, Pretoria, South Africa, pp. 1–8.

Deutscher Verein von Gas- und Wasserfachmännern (1976). Zum Nachweis von Escherichia coli und coliformen Bakterien nach der Trinkwasserverordnung vom 31. Januar 1975. *Gas Wasserfach, Wasser Abwasser 117*: 170–176.

Difco Manual (1963). *Difco Manual of Dehydrated Culture Media and Reagents for Microbiological and Clinical Laboratory Procedures.* Difco Laboratories, Inc., Detroit.

Dutka, B. J., and Kuchma, S. R. (1977). Fecal coliform density estimates, tip of the iceberg. *Abstr. Annu. Meet. Am. Soc. Microbiol.*, p. 272.

Dutka, B. J., and Tobin, S. E. (1976). Study on the efficiency of four procedures for enumerating coliforms in water. *Can. J. Microbiol. 22*: 630–635.

Engelbrecht, R. S., Foster, D. H., Greening, E. O., and Lee, S. H. (1974). *New Microbial Indicators of Wastewater Chlorination Efficiency*, Environmental Protection Technology Series EPA-670/2-73-082. U.S. Environmental Protection Agency, Washington, D.C., pp. 1–65.

Environmental Protection Agency (1975). National interim primary drinking water regulations. *Fed. Regist. 40*: 59566-59587.

Feachem, R. G. (1977). Water supplies for low-income communities: resource allocation, planning and design for a crisis situation. In *Water, Wastes and Health in Hot Climates*, R. Feachem, M. McGarry, and D. Mara (Eds.). Wiley, London, pp. 75–95.

Foy, H. M., Cooney, M. K., and Hatlen, J. B. (1968). Adenovirus type 3 epidemic associated with intermittent chlorination of a swimming pool. *Arch. Environ. Health 17*: 795–802.

Friberg, L., and Hammarström, E. (1956). The action of free available chlorine on bacteria and bacterial viruses. *Acta Pathol. Microbiol. Scand. 38*: 127–134.

Geldreich, E. E. (1972). Water-borne pathogens. In *Water Pollution Microbiology*, R. Mitchell (Ed.). Wiley, New York, pp. 207–241.

Geldreich, E. E. (1973). Is the total count necessary? Presented at the AWWA First Water Quality Technology Conference, Cincinnati, Ohio, December 2–4.

Geldreich, E. E., Clark, H. F., Huff, C. B., and Best, L. C. (1965). Fecal-coliform-organism medium for the membrane filter technique. *J. Am. Water Works Assoc. 56*: 208–214.

Gerba, C. P., Wallis, C., and Melnick, J. L. (1975). Viruses in water: the problem, some solutions. *Environ. Sci. Technol. 9*: 1122–1126.

Grabow, W. O. K. (1968). Review paper. The virology of waste water treatment. *Water Res. 2:* 675–701.

Grabow, W. O. K. (1976). Progress in studies on the type A (infectious) hepatitis virus in water. *Water S. A. 2*: 20–24.

Grabow, W. O. K. (1977). South African experience on indicator bacteria, *Pseudomonas aeruginosa* and R^+ coliforms in water quality control. In *Bacterial Indicators/Health Hazards Associated with Water*, ASTM STP 635, A. W. Hoadley and B. J. Dutka (Eds.). American Society for Testing and Materials, Philadelphia, pp. 168–181.

Grabow, W. O. K., and Isaäcson, M. (1977). Microbiological quality and epidemiological aspects of reclaimed water. Presented at an IAWPR International Conference on Advanced Treatment and Reclamation of Wastewater, Johannesburg, South Africa, June 13–17. *Prog. Water Technol. 10*: 329–335.

Grabow, W. O. K., and Nupen, E. M. (1972). The load of infectious micro-organisms in the waste water of two South African hospitals. *Water Res. 6*: 1557–1563.

Grabow, W. O. K., and Prozesky, O. W. (1973). Drug resistance of coliform bacteria in hospital and city sewage. *Antimicrob. Agents Chemother. 3*: 175–180.

Grabow, W. O. K., Grabow, N. A., and Burger, J. S. (1969). The bactericidal effect of lime flocculation/flotation as a primary unit process in a multiple system for the advanced purification of sewage works effluent. *Water Res. 3*: 943–953.

Grabow, W. O. K., Middendorff, I. G., and Prozesky, O. W. (1973). Survival in maturation ponds of coliform bacteria with transferable drug resistance. *Water Res. 7*: 1589–1597.

Grabow, W. O. K., Prozesky, O. W., and Smith, L. S. (1974). Review paper. Drug resistant coliforms call for review of water quality standards. *Water Res. 8*: 1–9.

Grabow, W. O. K., Prozesky, O. W., Appelbaum, P. C., and Lecatsas, G. (1975a). Absence of hepatitis B antigens from feces and sewage as a result of enzymatic destruction. *J. Infect. Dis. 131*: 658–664.

Grabow, W. O. K., Prozesky, O. W., and Burger, J. S. (1975b). Behavior in a river and dam of coliform bacteria with transferable or nontransferable drug resistance. *Water Res. 9*: 777–782.

Grabow, W. O. K., Van Zyl, M., and Prozesky, O. W. (1976). Behaviour in conventional sewage purification processes of coliform bacteria with transferable or non-transferable drug-resistance. *Water Res. 10*: 717–723.

Grabow, W. O. K., Bateman, B. W., and Burger, J. S. (1978a). Microbiological quality indicators for routine monitoring of wastewater reclamation systems. Presented at the IAWPR 9th International Conference, Stockholm, Sweden, June 12–16. *Prog. Water Technol. 10:* 317–327.

Grabow, W. O. K., Middendorff, I. G., and Basson, N. C. (1978b). The role of lime treatment in the removal of bacteria, enteric viruses and coliphages in a wastewater reclamation plant. *Appl. Environ. Microbiol. 35:* 663–669.

Green, B. L., Clausen, E., and Litsky, W. (1975). Comparison of the new Millipore HC with conventional membrane filters for the enumeration of fecal coliform bacteria. *Appl. Microbiol. 30*: 697–699.

Green, B. L., Clausen, E. M., and Litsky, W. (1977). Two-temperature membrane filter method for enumerating fecal coliform bacteria from chlorinated effluents. *Appl. Environ. Microbiol. 33*: 1259–1264.

Hattingh, W. H. J. (1977). Reclaimed water: a health hazard? *Water S.A. 3:* 104–112.

Hattingh, W. H. J., and Nupen, E. M. (1976). Health aspects of potable water supplies. *Water S. A. 2*: 33–46.

Hoadley, A. W. (1977). Potential health hazards associated with *Pseudomonas aeruginosa* in water. In *Bacterial Indicators/ Health Hazards Associated with Water,* ASTM STP 635, A. W. Hoadley and B. J. Dutka (Eds.). American Society for Testing and Materials, Philadelphia, pp. 80–114.

Hurst, A. (1977). Bacterial injury: a review. *Can. J. Microbiol. 23*: 935–944.

Isaäcson, M., Clarke, K. R., Ellacombe, G. H., Smit, W. A., Smit, P., Koornhof, H. J., Smith, L. S., and Kriel, L. J. (1974). The recent cholera outbreak in the South African gold mining industry. *S. Afr. Med. J. 48*: 2557–2560.

Kampelmacher, E. H., Leussink, A. B., and Van Noorle Jansen, L. M. (1976). Comparative studies of methods for the enumeration of coli-aerogenes bacteria and *E. coli* in surface water. *Water Res. 10*: 285–288.

Kraus, M. P. (1977). Bacterial indicators and potential health hazards of aquatic viruses. In *Bacterial Indicators/Health Hazards Associated with Water,* ASTM STP 635, A. W. Hoadley and B. J. Dutka (Eds.). American Society for Testing and Materials, Philadelphia, pp. 196–217.

Lanz, W. W., and Hartman, P. A. (1976). Timed-release capsule method for coliform enumeration. *Appl. Environ. Microbiol. 32*: 716–722.

Lin, S. D. (1976). Membrane filter method for recovery of fecal coliforms in chlorinated sewage effluents. *Appl. Environ. Microbiol. 32*: 547–552.

Lin, S. D. (1977). Comparison of membranes for fecal coliform recovery in chlorinated effluents. *J. Water Pollut. Control Fed. 49*: 2255–2264.

Livingstone, D. J. (1969). An appraisal of sewage pollution along a section of the Natal coast. *J. Hyg. 67*: 209–223.

Livingstone, D. J. (1976). An appraisal of sewage pollution along a section of the Natal coast after the introduction of submarine outfalls. *J. Hyg. 77*: 263–266.

Lonsane, B. K., Parhad, N. M., and Rao, N. U. (1967). Effect of storage temperature and time on the coliforms in water samples. *Water Res. 1*: 309–316.

McCabe, L. J., Symons, J. M., Lee, R. D., and Robeck, G. G. (1970). Survey of community water supply systems. *J. Am. Water Works Assoc. 62*: 670–687.

McCarthy, J. A., Delaney, J. E., and Grasso, R. O. (1961). Measuring coliforms in water. *Water Sewage Works 108*: 238–243.

Mack, W. N. (1977). Total coliform bacteria. In *Bacterial Indicators/Health Hazards Associated with Water,* ASTM STP 635, A. W. Hoadley and B. J. Dutka (Eds.). American Society for Testing and Materials, Philadelphia, pp. 59–64.

Malherbe, H. H., and Strickland-Cholmley, M. (1965a). Quantitative studies on viral survival in sewage purification processes. In *Transmission of Viruses by the Water Route,* G. Berg (Ed.). Interscience, New York, pp. 379–387.

Malherbe, H. H., and Strickland-Cholmley, M. (1965b). Survival of viruses in the presence of algae. In *Transmission of Viruses by the Water Route,* G. Berg (Ed.). Interscience, New York, pp. 449–458.

Malherbe, H. H., Strickland-Cholmley, M., and Geyer, S. M. (1965). Viruses in abattoir effluents. In *Transmission of Viruses by the Water Route*, G. Berg (Ed.). Interscience, New York, pp. 347–354.

Melmed, L. N. (1966). Membrane filtration used for *Escherichia coli* counts in sewage works effluents and the effect of sample storage on these counts. *J. Proc. Inst. Sewage Purif. 3*: 272–279.

Melnick, J. L. (1976). Viruses in water. In *Viruses in Water*, G. Berg, H. L. Bodily, E. H. Lennette, J. L. Melnick and T. G. Metcalf (Eds.). American Public Health Association, Washington, D.C., pp. 3–11.

Mosley, J. W. (1967). Transmission of viral diseases by drinking water. In *Transmission of Viruses by the Water Route*, G. Berg (Ed.). Interscience, New York, pp. 5–23.

Mowat, A. (1976). Most probable number versus membrane filter on chlorinated effluents. *J. Water Pollut. Control Fed. 48*: 724–728.

Müller, G. (1977). Bacterial indicators and standards for water quality in the Federal Republic of Germany. In *Bacterial Indicators/Health Hazards Associated with Water*, ASTM STP 635, A. W. Hoadley and B. J. Dutka (Eds.). American Society for Testing and Materials, Philadelphia, pp. 159–167.

Nupen, E. M. (1970). Virus studies on the Windhoek waste-water reclamation plant (South-West Africa). *Water Res. 4*: 661–672.

Nupen, E. M. (1976). Viruses in renovated waters. In *Viruses in Water*, G. Berg, H. L. Bodily, E. H. Lennette, J. L. Melnick, and T. G. Metcalf (Eds.). American Public Health Association, Washington, D.C., pp. 189–195.

Nupen, E. M., and Hattingh, W. H. J. (1975). Health aspects of reusing wastewater for potable purposes--South African experience. Presented at a Conference on Wastewater Reclamation Research Needs, University of Colorado, Boulder, Colorado, March 20–22.

Nupen, E. M., Bateman, B. W., and McKenny, N. C. (1974). The reduction of viruses by the various unit processes used in the reclamation of sewage to potable waters. In *Virus Survival in Water and Wastewater Systems*, J. F. Malina, Jr., and B. P. Sagik (Eds.). Water Resources Symposium No. 7, Center for Research in Water Resources, University of Texas, Austin, Tex., pp. 107–114.

Oxoid Manual (1976). *The Oxoid Manual of Culture Media, Ingredients and Other Laboratory Services*. Oxoid Limited, Basingstoke, England.

Phillips, II, W. J. (1974). The direct reuse of reclaimed wastewater: pros, cons, and alternatives. *J. Am. Water Works Assoc. 66*: 231–237.

Pretorius, T., and Coetzee, O. J. (1965). A comparison of the membrane filter and the tube dilution method (most probable numbers) for the enumeration of *E. coli*. *Public Health, Johannesburg 65*: 387–406.

Ptak, D. J., and Ginsburg, W. (1977). Bacterial indicators of drinking water quality. In *Bacterial Indicators/Health Hazards Associated with Water,* ASTM STP 635, A. W. Hoadley and B. J. Dutka (Eds.). American Society for Testing and Materials, Philadelphia, pp. 218–221.

Public Health Laboratory Service Standing Committee on the Bacteriological Examination of Water Supplies (1968). Comparison of MacConkey broth, Teepol broth and glutamic acid media for the enumeration of coliform organisms in water. *J. Hyg. 66*: 67–81.

Rose, R. E., Geldreich, E. E., and Litsky, W. (1975). Improved membrane filter method for fecal coliform analysis. *Appl. Microbiol. 29*: 532–536.

Schiemann, D. A., Manley, J. L., and Arnold, M. O. (1976). Coliphage interference in the recovery of coliform bacteria. *J. Water Pollut. Control Fed. 48*: 533-539.

Sebastian, F. P. (1974). Purified wastewater--the untapped water resource. *J. Water Pollut. Control Fed. 46*: 239–246.

Shuval, H. I. (1976). Water needs and usage. The increasing burden of enteroviruses on water quality. In *Viruses in Water,* G. Berg, H. L. Bodily, E. H. Lennette, J. L. Melnick, and T. G. Metcalf (Eds.). American Public Health Association, Washington, D.C., pp. 12–26.

Shuval, H. I. (1977). Health considerations in water renovation and reuse. In *Water Renovation and Reuse*, H. I. Shuval (Ed.). Academic Press, New York, pp. 33–72.

Smith, L. S. (1969). Public health aspects of water pollution control. *Water Pollut. Control 68*: 544–549.

South African Bureau of Standards (1971). *Specification for Water for Domestic Supplies,* SABS 241-1971, S. A. Bureau of Standards, Pretoria, South Africa, pp. 1–13.

Stander, G. J. (1977). Modern thinking in wastewater management. Presented at an IAWPR International Conference on Advanced Treatment and Reclamation of Wastewater, Johannesburg, South Africa, June 13–17. *Prog. Water Technol. 10:* 9–15.

Stander, G. J., and Clayton, A. J. (1977). Planning and construction of wastewater reclamation schemes as an integral part of water supply. In *Water, Wastes and Health in Hot Climates*, R. Feachem, M. McGarry, and D. Mara (Eds.). Wiley, London, pp. 383–391.

Stander, G. J., and Funke, J. W. (1967). Conservation of water by reuse in South Africa. *Chem. Eng. Prog. Symp. Ser. 63*: 1–12.

Standridge, J. H., and Lesar, D. J. (1977). Comparison of four-hour and twenty-four-hour refrigerated storage of nonpotable water for fecal coliform analysis. *Appl. Environ. Microbiol. 34*: 398–402.

Stuart, D. G., McFeters, G. A., and Schillinger, J. E. (1977). Membrane filter technique for the quantification of stressed fecal coliforms in the aquatic environment. *Appl. Environ. Microbiol. 34*: 42–46.

Tate, C. H., and Trussell, R. R. (1977). Developing drinking water standards. *J. Am. Water Works Assoc. 69*: 486–498.

Thomas, H. A., Jr. (1955). Statistical analysis of coliform data. *Sewage Ind. Wastes 27:* 212–222.

Tobin, R. S., and Dutka, B. J. (1977). Comparison of the surface structure, metal binding, and fecal coliform recoveries of nine membrane filters. *Appl. Environ. Microbiol. 34*: 69–79.

Van der Kooij, D. (1977). The occurrence of *Pseudomonas* spp. in surface water and in tap water as determined on citrate media. *Antonie van Leeuwenhoek; J. Microbiol. Serol. 43*: 187–197.

Van Vuuren, L. R. J., Henzen, M. R., Stander, G. J., and Clayton, A. J. (1971). The full-scale reclamation of purified sewage effluent for the augmentation of the domestic supplies of the city of Windhoek. In *Advances in Water Pollution Research*, vol. 1, S. H. Jenkins (Ed.). Pergamon Press, Oxford, pp. I–32/1-9.

World Health Organization (1970). *European Standards for Drinking Water*, 2nd ed. World Health Organization, Geneva, pp. 17–20.

World Health Organization (1971). *International Standards for Drinking Water*, 3rd ed. World Health Organization, Geneva, pp. 18–21.

World Health Organization (1976). *Surveillance of Drinking Water Quality*. WHO Monograph Series No. 63. World Health Organization, Geneva, pp. 11–125.

15

MEMBRANE FILTRATION: A NEW ZEALAND VIEW

B. H. PYLE[*] *Ministry of Works and Development, Hamilton, New Zealand*

I. INTRODUCTION

In New Zealand's temperate climate, the relatively high annual rainfall, which has an even seasonal distribution in most places (Tomlinson, 1976), assures an abundance of water. With growing urbanization, agricultural, and industrial development, the country has become increasingly dependent on this resource (McKenzie, 1976). Water supply and wastewater disposal facilities, hydroelectric, geothermal, and thermal power production have kept pace with this growth over the last few decades. The close proximity of most of the population to inland and coastal waters has encouraged extensive use of natural waters for recreation and as a source of food.

These activities necessitate the maintenance and improvement of water quality. Regular testing of water supplies for potable and industrial use is undertaken and the quality of surface, underground, and coastal waters is subject to an increasing range and intensity of monitoring programs. There are about 100 laboratories throughout the country with the capability of bacteriological analysis of water

---

**Present affiliation:* Lincoln University College of Agriculture, Canterbury, New Zealand.

samples. A survey of the activities of such laboratories (Mulcock, 1977) indicated that over 30,000 samples were tested for bacteria content in 1976. Most samples were tested for coliform bacteria, and a small proportion were tested for streptococci or other bacteria.

In this chapter, the use of membrane filter (MF) tests for coliform bacteria in New Zealand water laboratories is discussed.

II. WATER QUALITY ADMINISTRATION AND STANDARDS IN NEW ZEALAND

A. Potable Water Supplies

The New Zealand Department of Health is responsible for surveillance and grading of potable public water supplies as part of its commitments embodied in the Health Act (1956). Public health laboratories operated by or on behalf of the Department perform the analyses.

There are no statutory standards for the bacteriological quality of potable water, but the Department of Health applies the latest international recommendations (WHO, 1971) when considering the quality of a supply (Till, 1978). Methods specified by the World Health Organization (WHO, 1963, 1971) are generally used, the most probable number (MPN) technique being the most common. Many municipalities and industrial concerns have water testing laboratories for day-to-day monitoring of their supplies. The absence of coliform bacteria from the majority of samples is the common bacteriological criterion applied to potable supplies used for drinking and food processing purposes.

B. Natural Waters

Administration and coordination of activities relating to the management of natural waters is the responsibility of the National Water and Soil Conservation Authority (Water and Soil Conservation Act, 1967). Local administration and surveillance is performed by 20 Regional Water Boards whose areas correspond to the main catchment districts. The Water and Soil Conservation Act (1967) and amendments incorporate 10 schedules for different classes of natural water, the class being determined to promote in the public interest the conservation and the best use of the water concerned. Standards based on the number of coliform bacteria are included in four of the schedules, but no methods are specified. Of the schedules with coliform standards, the second and third relate to inland (fresh) water and the fifth and sixth to coastal (saline) waters (Table 1).

So-called "total" and "fecal" coliform bacteria standards are specified, although there is no known method for detecting the total number of coliform bacteria in a sample (New Zealand Microbiological Society (NZMS), 1976; Loutit, 1979). The term "total coliform" is considered to be equivalent to the term "presumptive coliform," and the term "fecal coliform" is taken to indicate that the organisms have a higher probability of being of fecal origin and hence indicating fecal contamination (NZMS, 1976).

TABLE 1 Summary of Bacteriological Standards from the Schedules of the New Zealand Water and Soil Conservation Act (1967) and Amendments

Schedule	Class	Use[a]	"Total" coliforms[b]	Fecal coliforms[b]
2nd	B (fresh)	Protected water supplies, treated before use	10,000	2,000
3rd	C (fresh)	Contact recreation	-	200
5th	SA (saline)	Edible-shellfish growing	70	-
6th	SB (saline)	Contact recreation	-	200

[a]Ministry of Works and Development, 1973.
[b]Each schedule states that "based on not fewer than five samples taken over not more than a 30 day period, the median value" of the bacteria content per 100 cm^3 must not exceed that given above.

III. METHODS USED IN NEW ZEALAND FOR COLIFORM DETECTION IN WATER

For more than 10 years, New Zealand microbiologists involved in water testing have been concerned about the number of methods in use for coliform detection, the reliability of the methods used, and the background of those employed to carry out the tests. In 1973, the New Zealand Microbiological Society held a symposium on coliform bacteria during its annual conference. The results of a survey in use for coliform detection reported at that symposium (Noonan, 1974) confirmed that the lack of statutory methods for water analysis had resulted in the adoption of many different methods. Some of these methods were taken from standard handbooks, but almost all had been modified in some way and some were not related closely to any standard procedure.

Matters raised at the symposium resulted in a committee being formed by the Microbiological Society to "investigate, and make recommendations on, the possibility of standardising methods for assessing numbers and types of coliform bacteria in New Zealand" (NZMS, 1976).

A. Survey of Methods Used

The survey of methods (Noonan, 1974) included 23 water testing laboratories. Of these, 14 used the MPN evaluation only, 5 used both the MPN and MF techniques, and 4 used only the MF method. These laboratories were using methods from a variety of sources, the most common at that time being the American Standards Methods (APHA, 1971). International Standards (WHO, 1963, 1971) and the British methods (Ministry of Health, 1957, 1969). Most laboratories using MF employed the American Standard Methods (APHA, 1971). Six laboratories gave no source for the method used; five of these used a MPN method and one used a MF method.

B. Recommended Reference Method

The report of the New Zealand Microbiological Society's Coliform Committee (NZMS, 1976) defined the terms relating to coliform testing and recommended the adoption of a specified MPN test as the reference method. The committee's recommendations have been adopted by the Water Resources Council and circulated to regional water boards.

The MPN technique was recommended because it is applicable to samples from many environments, including water, wastewater, soil, feces, and food. Most important, it can be applied to almost every sample of water or wastewater likely to be encountered. The widespread use of this method in New Zealand offered better prospects for standardization than alternative techniques such as MF, which has been found (Noonan, 1974) to be less popular. The report (NZMS, 1976) stated that the membrane filtration technique "should only be used where experimental comparison with the multiple tube method shows that conditions would be satisfactory for the use of such a method."

A summary of the committee's (NZMS, 1976) definitions and recommended procedures follows. A five-tube, three-dilution procedure was specified.

Coliform Organisms

These were defined as the group that "includes all the aerobic and facultatively anaerobic non-spore-forming, rod-shaped bacteria which ferment lactose with the production of acid and gas within 48 hours ±2 hours at 37°C ±0.5°C."

Presumptive Coliform Test

In this preliminary screening test, measured inocula are added to a suitable medium that contains lactose and indicator, such as minerals-modified glutamate medium (Ministry of Health, 1969), and incubated at 37 ±0.5°C for 48 ±2 hr. The production of acid and gas constitutes a positive result.

Confirmed Coliform Test

The positive presumptive tubes are subcultured in a more selective medium, specified as lauryl sulfate tryptose broth (LSTB) (APHA,

1975), to be incubated at 37 ±0.5°C for 48 ±2 hr. Gas production from lactose confirms the presence of coliform bacteria.

Completed Coliform Test

It was noted that this series of tests is intended to establish that the bacteria isolated in the presumptive and confirmed tests fulfill the morphological and biochemical definition of a coliform when in pure culture. The stated requirements were:

1. The growth of typical colonies on suitably buffered Levine's eosin methylene blue agar.
2. Demonstration that gram-negative, non-spore-forming rod-shaped bacteria are present when a typical colony from test (1) is cultured on nutrient agar.
3. Organisms from the same colony used in test (2) should produce acid and gas in a lactose-containing broth.

Fecal Coliform Test

Fecal coliform organisms were defined as those that fulfill the definition of coliform organisms and in addition are capable of fermenting lactose with gas production in 24 ±2 hr at 44.5 ±0.2°C. Positive tubes from the presumptive coliform test should be inoculated into LSTB and incubated for 24 ±2 hr at 44.5 ±0.2°C.

Escherichia coli

The committee adopted the definition according to the latest edition of *Bergey's Manual* (Buchanan and Gibbons, 1974). The typical characteristics of this species were specified as those of the fecal coliform group, and in addition the ability to produce indole from tryptophane at 44.5°C with IMViC reactions + + - -. The procedures specified to estimate the numbers of *Escherichia coli* were to subculture positive presumptive coliform tubes into two tubes of brilliant green bile broth (BGLB), one to be incubated at 37 ±0.5°C for 48 ±2 hr, the other at 44.5 ±0.2°C for 24 ±2 hr, and into peptone water to test for indole production after incubation at 44.5 ±0.2°C for 24 ±2 hr.

IV. COMPARATIVE EVALUATION OF METHODS AND QUALITY CONTROL

The New Zealand Microbiological Society recommended that the MF procedure should be used routinely only after satisfactory comparison with the multiple-tube MPN method. For this recommendation to be adopted, procedures for such comparisons, for verification, and for control checks are needed. No complete procedures are given in the method handbooks commonly used in New Zealand (Ministry of Health, 1969; WHO, 1971; APHA, 1975).

A. MF:MPN Comparisons

According to Geldreich et al. (1967), the results of MF coliform tests should fall consistently within the 95% confidence limits of the confirmed coliform MPN. The tests should be run over a period of at least 3 months, and 100 samples known to contain coliforms should be tested. The APHA (1975) recommended parallel testing "to demonstrate applicability and to familiarize the worker with the procedures involved" but gave no indication of how this was to be done or on how many samples. More detailed requirements were stated by Geldreich (1975), who recommended that the MF "total" coliform test should be compared with the multiple-tube MPN test over a 3-month period using samples from a variety of sources.

Based on these recommendations, parallel testing and verification procedures have been proposed in relation to the New Zealand reference MPN method (Pyle and Davis, 1979). Coliform colonies isolated by MF are to be tested for gas production in LSTB at 37.0 ±0.5°C within 48 ±2 hr. Positive tubes would then be subcultured by streaking onto solid buffered Levine's eosin methylene blue agar (APHA, 1975). A typical colony growing within 24 hr at 37°C should be streaked onto a nutrient agar slope. Demonstration by staining and microscopy of gram-negative non-spore-forming rod-shaped bacteria provides a positive result. An inoculum from the same colony should be capable of acid and gas production in a lactose broth.

Similarly, fecal coliform colonies may be subinoculated into LSTB and incubated at 44.5 ±0.2°C for 24 ±2 hr. Growth and production of gas constitute a positive result. When confirmation that fecal coliform isolates obtained by either MF or MPN have the defined characteristics of coliform bacteria (NZMS, 1976) is required, positive LSTB tubes from the MPN test or MF verification may be subjected to the completed test for coliforms.

It was proposed that when laboratories initially attempt MF determinations for coliforms, tests for "total" and fecal coliforms should be run in parallel with multiple-tube MPN tests on at least 100 samples over a period of at least 3 months. The MF total counts should be adjusted according to the proportion of isolates that fulfill the requirements of the completed coliform test, and 80% of the adjusted counts should lie within the 95% confidence limits of the corresponding completed MPN values. The MF fecal counts should be adjusted according to the proportions of isolates for which gas production at the elevated temperature was confirmed, and 80% of the adjusted counts should be within the MPN confidence limits. When samples from sources not examined by the laboratory previously which may not be amenable to MF are presented for testing, one-fifth of the samples examined over a 3-month period should be subjected to such parallel testing. Subsequently, one-tenth of the total and fecal coliform MF colonies from at least two water types representative of those being examined routinely should be subjected to tests verifying gas production in appropriate media at the appropriate temperature, and at least two samples should be examined in parallel by the reference MPN method with each batch of tests.

B. Control Tests

As well as parallel tests to compare MF with the reference MPN method, control tests are required, especially when there is a change of operator, materials, or equipment. The APHA (1975) gives methods for culture preparation and specifies the use of *Enterobacter aerogenes* as a control organism for the MF total coliform test and *E. coli* for use with the MF fecal coliform test. To ensure that both the MF and MPN procedures in use adequately differentiate coliform and noncoliform organisms, the use of both positive and negative isolates would be preferable. A typical negative organism has not been stated in the routine method handbooks, but *Serratia marcescens* has been used (Fifield and Schaufus, 1958; Schiff et al., 1970). Some strains of this organism have since been found to be enteropathogenic (Sattar et al., 1972).

C. Interlaboratory Comparability

Apart from comparing results within a laboratory, in a small country such as New Zealand and indeed elsewhere, it is desirable to compare results obtained by different laboratories. Often, two or more laboratories may be involved in testing samples from the same area, and consequently regular comparison of results is essential. Where no particular methods are specified by legislation and there is a range of methods in use, cooperating laboratories must either adopt identical methods or otherwise attempt to correlate results. Even when two laboratories use the same methods, the results achieved may vary significantly.

Exchange of paired samples between laboratories located near each other is possible for these purposes, but on a national scale the use of proficiency test specimens would be more suitable. Methods for preparation of such samples by lyophilization (Cada, 1975) and chemical preservation (Brodsky et al., 1978) have been proposed.

Efforts to standardize laboratory operation have been made by the Testing Laboratory Registration Council of New Zealand (TELARC, 1975). The Council was established to develop a system for registration of competent testing laboratories and to promote good laboratory practice by all laboratories in New Zealand.

V. DISCUSSION

A. Membrane Filtration vs. Most Probable Number Determination

The relative merits of these methods have been ascertained from the literature, the comments of New Zealand microbiologists and water quality testing staff, and from personal experience (Pyle and Davis, 1979).

Advantages of Membrane Filtration

The MF technique is based on a count of the viable coliform bacteria detected in a given amount of water (Geldreich et al., 1967), providing an estimate of the number of coliforms present with smaller confidence limits than the MPN (Ministry of Health, 1969; Geldreich, 1975). The sensitivity is good, because large amounts of sample can be tested (Thomas et al., 1956) provided that the sample is relatively free of particulate material or interfering organisms. The results of MF tests are obtained more rapidly with more significant and accurate data than are results using the MPN procedure (Geldreich et al., 1967; Peterson, 1974).

There is less work overall for the MF test than for the MPN when preparation, analysis, and cleaning are taken into account. The filter plates require less incubator space than the MPN tubes and there may be cost savings in terms of materials (Peterson, 1974). The use of several filter units at once, although requiring a larger capital outlay, may permit a significant increase in sample throughput.

Soluble inhibitors present in the sample may be separated from the bacteria during the process of filtration and rinsing (Peterson, 1974). Resuscitation by preincubation of filters on an enrichment medium for a few hours enhances the recovery of stressed cells (McCarthy et al., 1961). Further, to facilitate collection of data from sites remote from the laboratory, filtration may be performed in the field and the incubation delayed by placing the filter on a transport holding medium (APHA, 1975). The use of portable incubators is not recommended unless tests have been undertaken to ensure that their accuracy and stability are satisfactory under the conditions of use.

Disadvantages of Membrane Filtration

The main disadvantage of the MF method is that whereas the MPN procedure is based on the premise that it is a fundamental characteristic of coliform bacteria that they ferment lactose with the production of acid and gas (APHA, 1975; NZMS, 1976), the MF procedure detects only the production of an acid or aldehyde. These products do not prove conclusively that lactose fermentation with gas production has occurred. Further, the metallic sheen produced on Endo-type media is the result of an aldehyde complex formed with basic fuchsin and sodium sulfite (Rose, 1966); the sheen is not produced clearly by some coliform colonies, and some operators may have difficulty in determining its presence or absence (Thomas et al., 1956).

The relatively low workload of the MF procedure may be increased by the requirement that several volumes or dilutions of each sample must be examined. Since the acceptable ranges of counts are 20-80 colonies per filter with Endo-type media and 20-60 per filter with mFC medium (APHA, 1975), serial dilutions within decimal ranges are required. For example, instead of a decimal series of three dilutions (1, 1/10, 1/100), a series of five dilutions (1, 1/4, 1/16, 1/64, 1/256) is required to cover the same range of expected cell concentration if a membrane count within the 20-80 colony limit is to be obtained. The time taken to count colonies on several filters, which is often necessary when noncoliform organisms are present, may exceed the time taken to check for positive tubes for the MPN test. Although the overall labor input may not be important, the amount of work at a critical time, for example inoculation or counting, may limit the number of tests that can be performed. A skilled operator may be able to process almost as many samples for MPN evaluation as for MF at such critical stages.

Some Causes of Failure in Membrane Filtration

Failure of the MF procedure or lack of comparability with the MPN method have often been found to result from nonobservance of simple basic procedural details or alteration of methods without adequately checking the effects of this.

One of the difficulties in New Zealand is that importation of some equipment and materials may be delayed due to the preference for surface transport, which reduces costs. It has sometimes been found that the materials or equipment in use in water testing laboratories were not the most suitable but were the only ones available. It may be necessary to change from one type of filter or medium to another during the course of the investigation because of this.

Membrane Filters. It is now well established that membrane filters must meet certain specifications (APHA, 1975; Green et al., 1975; Sladek et al., 1975; Standridge, 1976), and the products of several manufacturers have been shown to be suitable (Lin, 1977; Tobin and Dutka, 1977).

It has been found that filters autoclaved in the laboratory sometimes become distorted and brittle despite attempts to control the process. The use of filters that have been prepacked and sterilized by the manufacturer is therefore preferred, because the sterilization process is likely to have undergone precise monitoring and control. Filters should carry a certification that the sterilization technique has not altered the properties of the membranes (APHA, 1975).

Procedures. Manuals such as that of the APHA (1975) explain the methods for MF clearly and in detail. Nevertheless, modifications of such procedures have been observed in New Zealand laboratory practice. Some of these modifications would be likely to adversely affect the results of the tests being done. Attention has been drawn to the effects of such procedural modifications elsewhere (Pyle, 1981).

Selection of Media and Incubation Procedures. It is essential that the method used be applicable to the type of water sample being tested. For example, a medium developed for testing potable water containing few bacteria may be quite unsatisfactory for river water samples with a rich bacterial biota.

Some operators fail to recognize the limitation that the MF test cannot be applied to unsuitable samples: those that contain excessive particulate material, excessive noncoliform populations or heavy metal ions, and variable chlorinated effluents (Geldreich, 1975). This appears to be more prevalent in laboratories using only the MF technique. Where the MPN is in use as a reference, it can also be applied to samples not suitable for MF.

B. Need for New and Better Methods

Current techniques for detecting indicator bacteria in water are commonly criticized because they are too difficult and time consuming and the number of samples that should be tested to provide an adequate appraisal of water quality is very large (Loutit, 1979). In many cases, decisions regarding the number and type of samples to be tested are made by nontechnical personnel. Sometimes the technician carrying out the work is tempted to use shortcuts to complete the required number of analyses, thus sacrificing accuracy. When the real workload is realized, the option to curtail all bacteriological analysis may be taken rather than the more realistic option to reduce the number of tests done and to ensure that they are done well.

To be acceptable, new methods should be simple, requiring as little time as possible, and automated if possible. With automated methods, the equipment should be as inexpensive as possible or should be readily available in existing laboratories.

If a bacterial indicator is to be used, it is desirable that the methods used for its recovery, identification, and enumeration determine the taxonomic characteristics of that indicator. If an isolation procedure is used that does not directly make use of taxonomic characteristics of the organism concerned, it is essential that the organisms which it isolates have the correct taxonomic characteristics. Although the MF method does not determine gas production by coliforms, correlation between positive reactions in MF tests and coliform confirmations has been demonstrated (McCarthy et al., 1961; Geldreich et al., 1965; Geldreich et al., 1967; Rose et al., 1975; Pyle and Davis, 1977). The possibility that a large proportion of nonfecal bacteria may occur in some samples and be included in results as false positives must be taken into account. For example, organisms such as *E. aerogenes* biotype 2 are capable of lactose fermentation at 44.5°C (Hendricks, 1970) and the test for indole production at 44°C (WHO, 1971) may not eliminate all nonfecal organisms (Loutit, 1979). New methods must be developed that overcome these inadequacies of the existing methods.

Development of the membrane transfer technique proposed by Dufour and Cabelli (1975) for seawater samples could be worthwhile in relation to this. That procedure incorporates in situ substrate tests, the membrane filters being transferred sequentially from the isolation medium to media that enable reactions in the urease, oxidase, and indole tests to be observed. The method obviates the need for manual transfer of individual colonies, and enables differentiation of *Escherichia, Klebsiella,* and *Enterobacter-Citrobacter* organisms.

Both the MF and MPN procedures are based on the determination of the number of viable units in a bacterial suspension that will grow under the conditions of the test to produce particular reactions. Recent research (Morgan, 1978) has indicated that in Waikato River (New Zealand) water, a proportion of bacteria are attached to particulate material larger than 8 μm diameter and are not dispersed by the normal test procedures of shaking the sample and preparing dilutions (APHA, 1975). Simple methods to assist dispersion of samples being tested by the current methods are required where such phenomena occur. Factors affecting the adsorption of bacteria to allophane clay particles, which are common in some New Zealand waters, have been described (Cooper, 1977).

The assumption that colonies on membrane filters have originated from a single bacterium or even bacteria of a single species is invalidated where there are clumps of bacteria. Synergistic metabolism which may occur in MPN tubes may also occur in MF colonies, yielding false-positive results (Schiff et al., 1970).

An alternative to the introduction of better methods for sample dispersion is the development of methods to determine the presence of individual cells, even in clumps, by specific staining and direct examination or by determining the specific activity of cells or the concentration of certain cell components. Some progress has been made toward this alternative: for example, the fluorescent antibody technique (Lieber and Martin, 1965; Abshire and Guthrey, 1973), radiometry (Bachrach and Bachrach, 1974), phage enumeration (Kenard and Valentine, 1974), enzyme activity (Khanna, 1974; Moran and Witter, 1976), and electrochemical detection (Wilkins and Boykin, 1976).

VI. CONCLUSIONS

The MF technique is suitable for coliform detection and enumeration in some water samples. It must be used in conjunction with the MPN procedure so that analysis of all water and wastewater samples encountered will yield satisfactory results. Both procedures need to be used so that the results of parallel MF and MPN tests may be compared.

For these reasons, some water testing laboratories in New Zealand in which the MPN method is in routine use may never implement the MF test because it is regarded as complementary rather than alternative to the MPN method. Others using only the MF procedure are being encouraged to introduce the MPN method as well, as a reference method and to enable results for all samples to be compared.

Recommendation of a MF method as the sole standard procedure in New Zealand is unlikely because of its limited applicability, but where MF is applicable, the American Standard Methods (APHA, 1975) have been commended most strongly because of their general applicability and current acceptance (Noonan, 1974). Although new and better methods are needed, the use of even the most basic components of the present MPN method on a large number of samples would be preferable to exhaustive testing of a few samples (NZMS, 1976). Continued use of the current indicator bacteria tests is warranted until more suitable alternatives are available.

ACKNOWLEDGMENTS

The author wishes to thank the Commissioner of Works for approval to publish this information and the following for their editorial assistance and comments: Professor A. P. Mulcock, Associate Professor M. Loutit, Dr. M. J. Noonan, Dr. H. W. Morgan, Dr. M. D. Cooke, Dr. M. E. U. Taylor, Mr. D. Till, and Miss J. C. A. Davis.

REFERENCES

Abshire, R. L., and Guthrie, R. K. (1973). Fluorescent antibody as a method for the detection of faecal pollution: *Escherichia coli* as indicator organisms. *Can. J. Microbiol. 19*: 201—206.

APHA (American Public Health Association). (1971). *Standard Methods for the Examination of Water and Wastewater*. 13th ed. APHA/AWWA/WPCF, Washington, D.C.

APHA (American Public Health Association). (1975). *Standard Methods for the Examination of Water and Wastewater*. 14th ed. APHA/AWWA/WPCF, Washington, D.C.

Bachrach, V., and Bachrach, Z. (1974). Radiometric method for the detection of coliform organisms in water. *Appl. Microbiol. 28*: 169—171.

Brodsky, M. J., Ciebin, B. W., and Schiemann, D. A. (1978). Simple bacterial preservation medium and its application to proficiency testing in water bacteriology. *Appl. Environ. Microbiol. 35*: 487—491.

Buchanan, R. E., and Gibbons, N. E. (1974). *Bergey's Manual of Determinative Bacteriology*. 8th ed. Williams & Wilkins, Baltimore, Md., 1268 p.

Cada, R. L. (1975). Proficency test specimens for water bacteriology. *Appl. Microbiol. 29*: 255—259.

Cooper, A. B. (1977). Interactions between *Escherichia coli* and allophane. M.Sc. thesis, University of Waikato, Hamilton, New Zealand.

Dufour, A. P., and Cabelli, V. J. (1975). Membrane filter procedure for enumerating the component genera of the coliform group in seawater. *Appl. Microbiol. 29*: 826–833.

Fifield, C. W., and Schaufus, C. P. (1958). Improved membrane filter medium for the detection of coliform organisms. *J. Am. Water Works Assoc. 50*: 193–196.

Geldreich, E. E. (1975). *Handbook for Evaluating Water Bacteriological Laboratories*, 2nd ed. EPA 670/9-75-006. Environmental Protection Agency, Cincinnati, Ohio.

Geldreich, E. E., Clark, H. F., Cuff, C. B., and Best, L. C. (1965). Fecal-coliform-organism medium for the membrane filter technique. *J. Am. Water Works Assoc. 57*: 208–214.

Geldreich, E. E., Jeter, H. L., and Winter, M. S. (1967). Technical considerations in applying the membrane filter procedure. *Health Lab. Sci. 4*: 113–125.

Green, B. L., Clausen, E., and Litsky, W. (1975). Comparison of the new Millipore HC with conventional membrane filters for the enumeration of fecal coliform bacteria. *Appl. Microbiol. 30*: 697–699.

Health Act (1956). Reprinted Act (with amendments incorporated), as on January 1, 1973. *New Zealand Statutes 1972, 2*: 1449–1556.

Hendricks, C. W. (1970). Formic hydrogenlyase induction as a basis for the Eijkman fecal coliform concept. *Appl. Microbiol. 19*: 441–445.

Kennard, R. P., and Valentine, R. S. (1974). Rapid determination of the presence of enteric bacteria in water. *Appl. Microbiol. 27*: 484–487.

Khanna, P. (1974). Rapid enumeration of water bacteria. *Water Res. 8*: 311–315.

Lieber, M., and Martin, A. J. (1965). Detection of coliform organisms by the fluorescent-antibody method. *J. Am. Water Works Assoc. 57*: 99–106.

Lin, S. D. (1977). Comparing ten membranes for a two step membrane filtration method for fecal coliform recovery in chlorinated effluents. *J. Water Pollut. Control Fed. 49*: 2255–2264.

Loutit, M. (1979). *Bacteria and Water and Waste Water Quality.* In *A Review of Some Methods for the Biological Assessment of Water Quality with Special Reference to New Zealand.* Standing Biological Working Party of the Water Resources Council. *Water and Soil Technical Publication 18*: 41-46.

McCarthy, J. A., Delaney, J. E., and Grasso, R. J. (1961). Measuring coliforms in water. *Water Sewage Works* 108: 238–243.

McKenzie, D. W. (1976). Landforms and resources. In *New Zealand Atlas*. I. McL. Wards (Ed.). NZ Government Printer, Wellington, New Zealand, pp. 71–75.

Ministry of Health (1957). *The Bacteriological Examination of Water Supplies*. Reports on Public Health and Medical Subjects, No. 71, 3rd ed. Ministry of Health, Housing and Local Government (Great Britain). Her Majesty's Stationery Office, London.

Ministry of Health (1969). *The Bacteriological Examination of Water Supplies*. Reports on Public Health and Medical Subjects. No. 71, 4th ed. Ministry of Health, Housing and Local Government, Dept. of Health and Social Security, Welsh Office, Dept. of Environment (Great Britain). Her Majesty's Stationery Office, London.

Ministry of Works and Development (1973). *What You Should Know About Water Use*. Published by Ministry of Works and Development for National Water and Soil Conservation Organisation, Wellington, New Zealand. WASCO 17, rev. December 1973.

Moran, J. E., and Witter, L. D. (1976). An automated rapid method for measuring fecal pollution. *Water Sewage Works 123*: 66–67.

Morgan, H. W. (1978). Personal communication. Senior Lecturer, Biological Sciences, University of Waikato, Hamilton, New Zealand.

Mulcock, A. P. (1977). *Report on Survey of Biological Methods of Water Testing*. Report to Biological Standing Working Party of Water Quality Research Committee, Water Resources Council, National Water and Soil Conservation Organisation, New Zealand.

Noonan, M. J. (1974). *Survey of Methods Used for the Enumeration of Coliform Organisms in New Zealand*. Lincoln College, University College of Agriculture, Canterbury, New Zealand.

NZMS (1976). Definition of terms concerning coliform bacteria and recommended methods for their detection. NZ Microbiological Society's Committee on Coliform Bacteria. *N. Z. J. of Sci. 19*: 215–219.

Peterson, J. (1974). Comparison of the MF technique and MPN technique for the estimation of coliforms in water. *Public Health Lab. 32*: 182–193.

Pyle, B. H. (1981). A Review of Methods for Coliform Detection and Enumeration in Water and Wastewater. In *A Review of Some Biological Methods for the Assessment of Water Quality with Special Reference to New Zealand, Vol. 2*. Standing Biological Working Party of the Water Resources Council. *Water and Soil Technical Publication*. In press.

Pyle, B. H., and Davis, J. C. A. (1978). *Methods for Coliform Enumeration in Water*. Draft Report to Biological Standing Working Party of Water Quality Research Committee, Water

Resources Council, National Water and Soil Conservation Organisation, New Zealand. (In press.)

Pyle, B. H., and Davis, J. C. A. (1978). Unpublished data.

Rose, R. E. (1966). Effective use of Millipore membrane filters for water analysis. *Water Sewage Works 113*: R-281—R-284.

Rose, R. E., Geldreich, E. E., and Litsky, W. (1975). Improved membrane filter method for fecal coliform analysis. *Appl. Microbiol. 29*: 532—536.

Sattar, S. A., Synek, E. J., Westwood, J. C. N., and Neals, E. (1972). Hazard inherent in microbial tracers: reduction of risk by the use of *Bacillus stearothermophilus* spores in aerobiology. *Appl. Microbiol. 23*: 1053—1069.

Schiff, L. J., Morrison, S. M., and Mayeaux, J. V. (1970). Synergistic false-positive coliform reaction on M-Endo MF medium. *Appl. Microbiol. 20*: 778—781.

Sladek, K. J., Suslavich, R. V., Sohn, B. I., and Dawson, F. W. (1975). Optimum structures for growth of coliform and fecal coliform organisms. *Appl. Microbiol. 30*: 685—691.

Standridge, J. H. (1976). Comparison of surface pore morphology of two brands of membrane filters. *Appl. Microbiol. 31*: 316—319.

TELARC (1975). *Biological Testing--Requirements for Registration.* Testing Laboratory Registration Council of New Zealand.

Thomas, H. A., Woodward, R. L., and Kabler, P. W. (1956). Use of molecular filter membranes for water potability control. *J. Am. Water Works Assoc. 48*: 1391—1402.

Till, D. G. (1978). Personal communication. Senior Bacteriologist, National Health Institute, Department of Health, Wellington, New Zealand.

Tobin, R. S., and Dutka, B. J. (1977). Comparison of the surface structure, metal binding and fecal coliform recoveries of nine membrane filters. *Appl. Environ. Microbiol. 34*: 69—79.

Tomlinson, A. I. (1976). Climate. In *New Zealand Atlas,* I. McL. Wards (Ed.). NZ Government Printer, Wellington, New Zealand, pp. 82—89.

Water and Soil Conservation Act (1967). Reprinted Act (with amendments incorporated), as on January 1, 1974. *New Zealand Statutes, 1973, 2:* 1703—1804.

WHO (1963). *International Standards for Drinking Water,* 2nd ed. World Health Organization, Geneva.

WHO (1971). *International Standards for Drinking Water,* 3rd ed. World Health Organization, Geneva.

Wilkins, J. R., and Boykin, E. H. (1976). Analytical notes--electrochemical method for early detection and monitoring of coliforms. *J. Am. Water Works Assoc. 68*: 257—263.

16

EFFECT OF INJURY ON THE RECOVERY OF BACTERIA ON MEMBRANE FILTERS

*ALFRED W. HOADLEY** *Georgia Institute of Technology, Atlanta, Georgia*

I. INTRODUCTION

When a cell's adaptive capacity is overtaxed during exposure to stress, damage or injury results. This damage may be sublethal, permitting recovery, or if more intense or prolonged, it may cause death of the cell. The boundaries separating lethal from sublethal injury are difficult to define. The pathologist may establish injury on the basis of ultrastructural disorganization or impaired biochemical function. In bacteria, injury may be studied at either the ultrastructural or functional level, although in practice it is most often examined in terms of function. It may be manifested as a failure to form colonies on defined minimal or selective media while retaining the ability to form colonies on rich media. It is the differential between counts on rich and selective media that is a matter of concern to the applied bacteriologist.

Interest in injury to bacteria subjected to environmental stress is relatively recent, although early investigations of factors affecting the recovery of bacteria in microbiological media were directed at the broad problem. Several investigators, such as

**Present affiliation:* World Health Organization, Dacca, Bangladesh

Knox et al. (1948), Heinmets (1953), and Heinmets et al. (1954), addressed the subject directly, and Heinmets in 1953 stated that "if suitable means are applied to reverse the reversible fraction of cellular injury, provided that the residual injury is in the range of tolerance, the cell may survive. Such a recovery phenomenon would not be restricted necessarily to irradiation injury, but possibly may apply in principle to all chemical and physical injuries produced within the cell." However, it was not until the 1960s that strong general interest in the practical aspects of injury to bacteria emerged.

Interest in cellular injury to bacteria arose at this time primarily among food microbiologists, but also emerged among sanitary microbiologists working with water. Important questions regarding the viability of bacterial cells exposed to stress were surfacing. The phenomenon was most readily apparent to food microbiologists because it was clear that portions of bacterial populations subjected to irradiation, heating, or freezing failed to produce colonies on certain defined minimal media or on selective media. This left questions regarding the safety of preserved food products that had been found apparently free of significant contamination by conventional bacteriological techniques. It is unacceptable that counts of toxigenic, pathogenic, or indicator bacteria upon which judgments of quality are made should not include injured bacteria which, given a favorable environment, may recover their normal capacities. This point was stressed by Postgate and Hunter (1963a), who pointed out that "viability on the richest medium available may be lower than that obtainable were it possible to devise a test environment on which all survivors . . . grew without a lag." Speck and Cowman (1969) have suggested that freeze-injured salmonellae may be as pathogenic as uninjured cells, and Collins-Thompson et al. (1973) have shown that following repair, heat-injured *Staphylococcus aureus* were able to synthesize enterotoxin B just as did uninjured control cells.

It is the purpose of the present chapter to consider some agents of environmental stress. sites of cellular injury, and agents that may influence the response of bacterial cells to stress, with particular emphasis on the application of membrane filter techniques. It is only with an understanding of these that methods can be applied properly, data interpreted properly, and improvements in methodologies achieved.

II. RESPONSE TO ENVIRONMENTAL STRESS

Characteristically, injury in response to stress results in a failure of some portion of a bacterial population to multiply in or on defined minimal or selective media. It is manifested when counts on these media are compared with counts on rich media, the difference between the two counts representing the portion of the population injured but able to effect repair when provided with appropriate environmental conditions. The response of a bacterial population to environmental stresses may reflect (1) the physiological state of the cells, (2) the severity of the stress, (3) the duration of the stress, and (4) the environment during exposure to the stress (thermal, nutrient, osmotic, toxic). The manifestation of that response reflects in addition (1) conditions surrounding cells prior to inoculation into culture media and (2) nutrients, inhibitors, and temperatures present in the recovery medium.

If counts are made simultaneously on both nonselective or rich media and selective media during the course of exposure to an environmental stress, the discrepancy between counts on the enriched and selective media may be seen to increase with increasing time of exposure (Fig. 1). The same phenomenon may at times be observed when counts on membrane filters incubated on rich media are compared with counts on spread plates made from the same medium (Table 1). This is important because it suggests that the longer the exposure to stress, the less reliable are counts made on selective media.

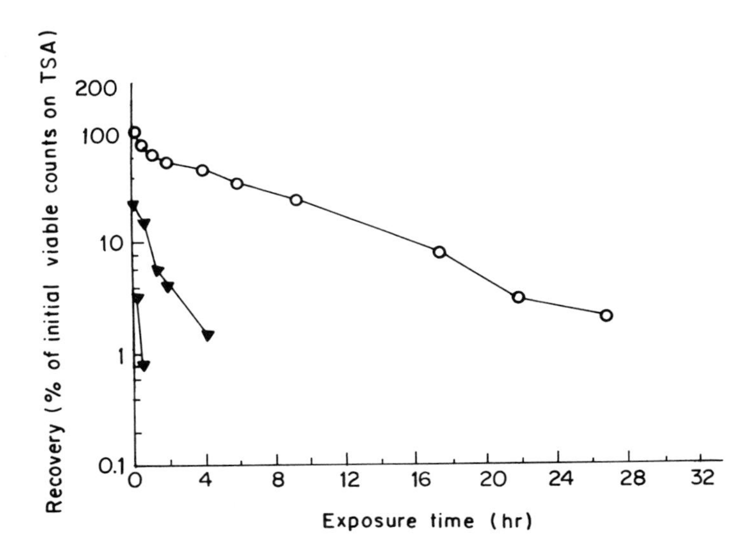

FIGURE 1 Recovery of *P. aeruginosa* suspended in phosphate buffer at 20°C. Plated on TSA (O), Drake's medium (▽), and cetrimide agar (▼). (After Hoadley and Cheng, 1974).

TABLE 1 Recovery of *E. coli* ATCC 27622 from Chlorinated Secondary Sewage on Trypticase Soy Agar[a]

Chlorine contact time (min)	Counts/ml		
		Membrane filter	
	Spread plate	TSA[b]	TSB[c]
1	33,670	33,570	30,600
5	31,630	28,070	22,370
10	20,700	14,630	4,800
20	500	45	1
30	30	2	

[a]Temperature, 21°C; pH 6.9; chlorine dosage, 2 mg/liter; chlorine residual after 30 min, 0.4 mg/liter.
[b]TSA, trypticase soy agar.
[c]TSB, trypticase soy broth.
Source: Hoadley (1977).

That the physiological state of a cell population may exert a great influence upon its response to environmental stress was demonstrated elegantly by Favero et al. (1971) and Carson et al. (1972) who showed that *Pseudomonas aeruginosa* can grow in distilled water and that naturally occurring cells grown in distilled water were more resistant to disinfecting agents than were cells grown on laboratory media (Fig. 2). Furthermore, whereas naturally occurring cells grew rapidly when inoculated into distilled water, populations of subcultured cells declined. In some distilled waters, populations of subcultured cells recovered eventually and multiplied to levels equal to those of naturally occurring cells (Fig. 3).

The repair of injured cells under favorable conditions explains the discrepancy between counts of stressed bacteria made on rich and selective media. It can also be called upon to explain, at least in part, numerous other phenomena, such as the appearance of bacterial contaminants in apparently sterile items, aftergrowth in highly polluted streams or soil, apparent declines in viable counts during the phase of adjustment (lag phase), or apparent absence of a phase of adjustment.

In their earlier works on thermal injury to *Staphylococcus aureus* and *Streptococcus faecalis*, Iandolo and Ordal (1966) and Clark et al. (1968) demonstrated repair following heating of cells suspended in phosphate buffer. Unheated cells suspended in trypticase soy broth (TSB) multiplied without lag and could be counted on either trypticase soy agar (TSA) or trypticase soy agair containing 6% sodium chloride (TSAS). A period of adjustment was evident in counts of heated cells suspended in TSB when made on TSA. That the lag was attributable to repair was evident when counts were made simultaneously on TSAS.

Bissonnette et al. (1975) applied the approach used by Ordal and his associates to investigate injury to *S. faecalis* and *E. coli* suspended in membrane filter chambers in stream environments. Cultures were suspended in chambers containing sterilized stream waters

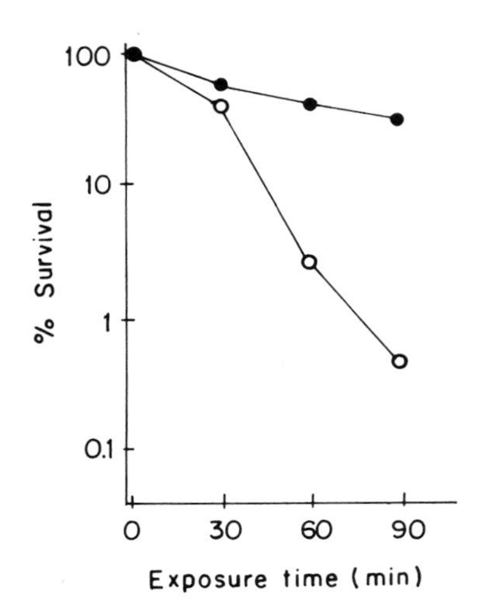

FIGURE 2 Recovery of naturally occurring (●) and subcultured (o) *P. aeruginosa* exposed to 0.25% acetic acid in mist therapy unit water at 25°C. (After Carson et al., 1972.)

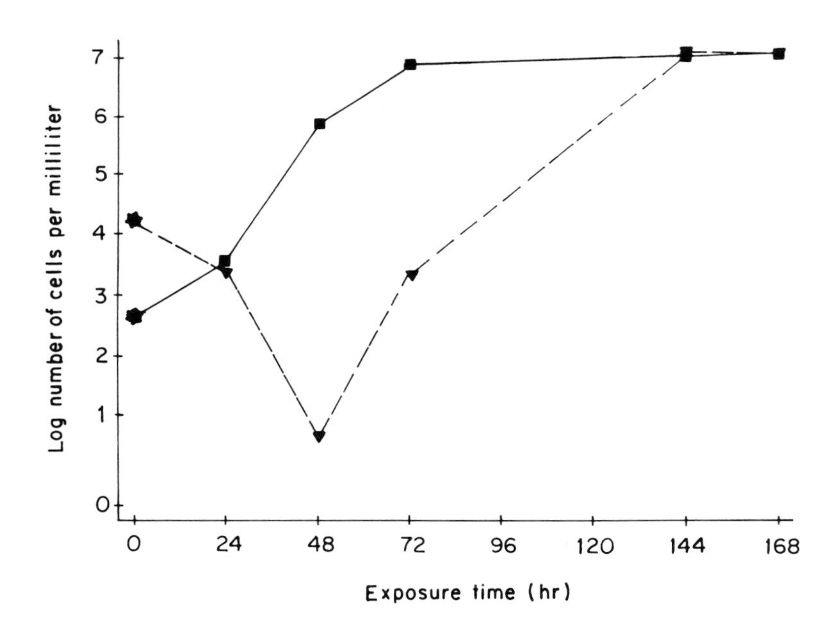

FIGURE 3 Recovery of *P. aeruginosa* inoculated into mist therapy unit water. Naturally occurring cells (■); subcultured cells (▼). [After M. S. Favero, L. A. Carson, W. W. Bond, and N. J. Peterson (1971). *Science 173*: 836–838. Copyright 1971 by the American Association for the Advancement of Science.]

and placed in the streams where they remained for varying periods of time. Suspensions of *E. coli* cells not exposed to the stream environment grew after only a short lag period when inoculated into trypticase soy broth supplemented with yeast extract (TSY broth). Cells that had remained in the stream for a period of 2 days exhibited a lag when inoculated into TSY broth and counted on trypticase soy agar supplemented with yeast extract (TSY agar). This lag again could be explained as a period of repair of injury evident when counts on deoxycholate lactose agar (DLA) were compared with counts on TSY agar.

Injury and repair were demonstrated by Collins-Thompson et al. (1973), who subjected a strain of *S. aureus* suspended in phosphate buffer to heating for periods up to 15 min. Injury was demonstrated by counts on TSA and TSA containing 7.5% sodium chloride (TSAS). After 15 min of heating in potassium phosphate buffer, more than 99.9% of the population forming colonies on TSA failed to do so on TSAS. To permit repair, cells were transferred to an enriched broth medium after 15 min of exposure to the heat stress. Cells were removed periodically and plated on TSA and TSAS. Over a period of 6 hr the counts on TSAS increased until they equaled counts on TSA. That counts on TSA remained unchanged during the period of repair indicates again that increasing counts on TSAS can be attributed to repair and not multiplication.

In water a similar decline in counts on selective or inhibitory media may be followed by recovery without transfer to an enriched medium. This has been demonstrated by Hoadley and Cheng (1974), who suspended a strain of *E. coli* in sterile stream water and compared counts on TSA and membrane filters incubated on M-FC medium (Fig. 4). It has been demonstrated also by Koellner (1975), who suspended a strain of *P. aeruginosa* originally isolated from Lake Erie in autoclaved stream water and compared counts made on TSA and membrane filters incubated on M-PA medium (Fig. 5).

It should be noted that while rich media generally have been employed for the repair of injured cells or for the enumeration of

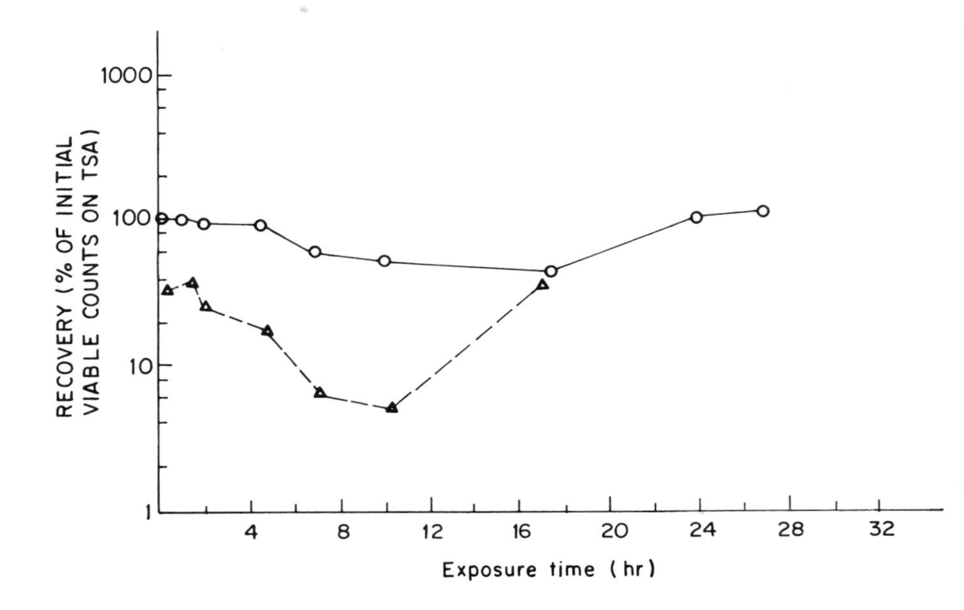

FIGURE 4 Injury and repair of *E. coli* during exposure to stream water. Counts on TSA (○) and mFC medium (△). (After Hoadley and Cheng, 1974.)

both injured and uninjured cells, a rich environment may be neither necessary nor desirable. Heinmets (1953) achieved substantial increases in counts when irradiated cells were incubated for 24 hr in phosphate buffer containing 0.05 M pyruvate, but achieved moderate improvements even when cells were incubated in buffer alone. Hurst et al. (1976) achieved far better repair of heat-stressed *S. aureus* in skim milk containing 1% milk solids than in milk containing 10% milk solids (or 0.1% solids). Gomez et al. (1973) demonstrated that while recovery on a minimal medium of heat-stressed *Salmonella typhimurium* grown on a rich medium resembled its recovery on a rich medium, recoveries of injured cells originally grown on the minimal medium were greater on plates of minimal medium. Although aquatic microbiologists are well aware of the advantages of dilute media for the recovery of aquatic bacteria, "minimal medium recovery" of enteric bacteria in water has not been extensively investigated.

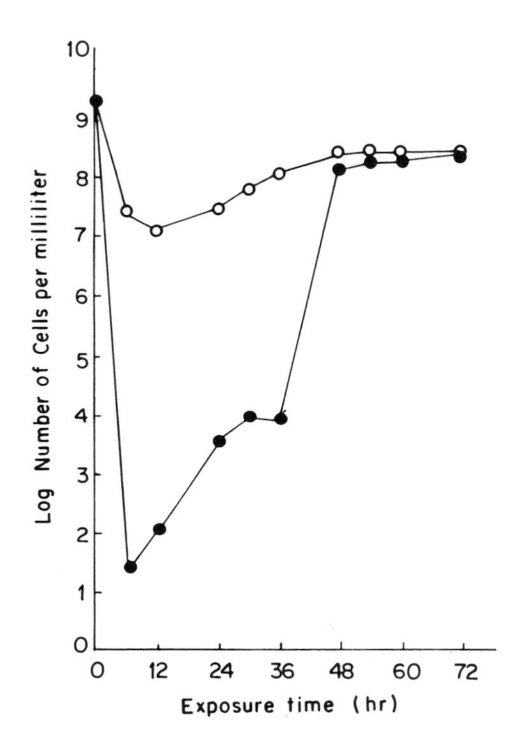

FIGURE 5 Injury and repair of *P. aeruginosa* during exposure to stream water. Counts on TSA (O) and mPA medium (●). (Unpublished data kindly provided by J. Koellner, Ministry of the Environment, London, Ontario, Canada.)

It may, under certain circumstances, affect recoveries, especially in view of the observations of Favero et al. (1971).

III. AGENTS OF ENVIRONMENTAL STRESS

A variety of agents and environmental conditions may cause injury or death in bacteria. These may include physical agents, such as irradiation; chemical agents, including disinfectants; biological agents, such as viruses or predaceous bacteria or amoebae; and environmental conditions, such as temperature or the availability of oxygen or nutrients. They may cause damage to cell walls, cell membranes, ribosomes, chromosomes, or enzymes, resulting in loss of function. Many of these have been reviewed in several papers (Busta, 1976; Hurst, 1977) and in the important volume edited by Gray and Postgate (1976).

In the present section, agents of environmental stress are reviewed in brief.

A. Temperature

Injury to bacteria caused by heat is of obvious interest in food microbiology since elevated temperatures are employed to kill vegetative bacteria and spores during the processing of foods. It is perhaps of particular significance in the context of the present paper that exposure to cold may cause injury which can be demonstrated in the same way. Cold shock, which occurs when cells are exposed to sudden chilling, has been demonstrated by a number of workers (e.g., see review by MacLeod and Calcott, 1976).

Exponential death of heated bacteria is often assumed to be a consequence of the denaturation of a single critical cellular protein. Deviations from linear survival curves are thought to be caused by variations in the sensitivity of the cells composing the population, to clumping of cells, or to a threshold that must be reached before damage is sufficient to cause death.

That the activity of enzymes may be reduced during heating was demonstrated by Bluhm and Ordal (1969) and by Tomlins et al. (1971). In their experiments, Bluhm and Ordal showed that sublethal heating caused reductions in the O_2 uptake rates of *S. aureus* provided with a variety of substrates. Respiratory activity was recovered following repair when cells were provided some substrates, but not when they were provided with ribose or succinate. It was shown in cell-free extracts that the specific activity of fructose diphosphate aldolase, lactate dehydrogenase, and butanediol dehydrogenase were reduced by heating and that activity returned during recovery of cells in TSB. In the later paper of Tomlins et al. (1971), it was shown that a loss of malate dehydrogenase and lactate dehydrogenase activity was reversible and that recovery of their activities did not depend upon protein synthesis. Oxoglutarate dehydrogenase activity failed to return during recovery, however.

Tomlins and Ordal (1971) showed that RNA synthesis accounted for the major part of the biosynthetic activities of *S. typhimurium* during recovery, and that recovery was dependent on this synthesis, as well as ATP and synthesis of protein, perhaps related to RNA synthesis. Strange and Shon (1964) confirmed a relationship between the rates of death and of RNA degradation in *Aerobacter (Enterobacter) aerogenes*. These investigators observed also that addition of magnesium to the medium in which cells were heated decreased both rates. The loss of magnesium from *S. aureus* during heating was demonstrated by Hurst et al. (1974). Hurst and Hughes (cited in Hurst, 1977) showed that preventing this loss during heating prevented loss of RNA and repair of injury occurred in the presence of actinomycin D. That injury without RNA degradation can occur also in *S. typhimurium* was demonstrated by Lee and Goepfert (1975). Thus, RNA degradation may well be secondary to loss of magnesium required to maintain the integrity of the ribosomes.

It may be noted that the loss of intracellular materials from cells and possibly their utilization also as a source of energy for repair during and after heating can result in the loss of endogenous sources of energy (Pierson and Ordal, 1971). This can cause decreases in respiration rates of more than 50% in heat-stressed *S. typhimurium*. The loss of endogenous substrate may cause a decrease in the rate of active uptake. Pierson and Ordal increased the uptake of methyl-α-D-glucopyranoside by the addition of succinate to *S. typhimurium*.

Single-stranded breaks in DNA caused by heating were demonstrated by Bridges et al. (1969), who concluded that sensitivity to heat was in general correlated with sensitivity to γ-radiation. Gomez and Sinskey (1973) also found that increases in single-strand breakage accompanied declines in viable counts. However, Sedgwick and Bridges (1972) concluded, on the basis of experiments with strains of *E. coli* possessing and lacking DNA polymerase I, that damage to DNA depended upon the strain and the physiological state of the bacteria and was not related to the direct effect of heat on the DNA. Rather, their results were consistent with the release or activation of nucleases. Because when heated *S. typhimurium* grown in a minimal medium were incubated in a minimal medium prior to incubation on TSY agar counts were higher than when incubated in TSY broth prior to plating, Gomez and Sinskey (1973) concluded that damage occurred when heated cells were exposed to TSY broth. This raises questions regarding the value of preenrichment of stressed bacteria under certain circumstances.

Exposure to low temperatures may cause similar effects. Exposure to cold may cause damage to enzymes. Thus, Speck and Cowman (1969) showed that storage of *S. lactis* at 3°C caused aggregation and inactivation of a proteinase associated with the cell membrane. The enzyme could be reactivated in the presence of ferrous ions, cysteine, and glutathione.

Leakage of nucleotides, amino acids, and low-molecular-weight proteins have been reported to accompany cold shock (Strange and Dark, 1962; Strange and Ness, 1963; Smeaton and Elliott, 1967), and the increase in permeability of *A. aerogenes* was demonstrated by Strange and Postgate (1964). Leder (1972) has reported extreme sensitivity to osmotic composition during cold shock, which he attributes to crystallization of lipids within the membrane and resultant creation of hydrophilic channels which would facilitate loss of permease accumulated substrates.

B. Freezing

Gunderson and Rose (1948) and Hartsell (1951) observed that recoveries of coliforms and salmonellas from frozen foods were greater on rich media than on selective media. Straka and Stokes (1957) were the first to distinguish among unharmed, injured, and dead cells in frozen samples and their relation to time and temperature of storage. Their investigations of the survival of *E. coli* and three *Pseudomonas* species were the first of a group of papers including studies of *Shigella sonnei* (Nakamura and Dawson, 1962) and

A. aerogenes (Postgate and Hunter, 1963b) which established the damaging effect of freezing on bacteria and the importance of this effect in the assessment of the bacteriological quality of frozen foods. Strange and Postgate (1964) and Moss and Speck (1966) began to explore mechanisms of injury associated with freezing. Later, Ray et al. (1972), Ray and Speck (1972, 1973a,b) and Warseck et al. (1973) investigated the repair process.

Freezing was shown by Strange and Postgate (1964) to cause an increase in the permeability of *A. aerogenes* to RNase. The loss of peptides from cells into the medium was observed by Moss and Speck (1966) to parallel loss in viability of frozen *E. coli*. Increased permeability was indicated by loss of peptides and some carbohydrate, amino acids, and RNA residues. This was further supported by the observation of Ray et al. (1972) that cells of *Salmonella anatum* injured by freezing were highly sensitive to lysozyme, whereas unfrozen cells were not. Similar sensitivity of *E. coli* to lysozyme and surface-active agents was observed by Ray and Speck (1972).

The repair of freeze-injured *S. anatum* and *E. coli*, investigated by Ray et al. (1972) and by Ray and Speck (1972), indicated that repair depended on ATP production, as it was prevented by uncouplers of oxidative phosphorylation in phosphate buffer but not in minimal broth except when cyanide and azide were employed. Inhibitors of the synthesis of proteins, nucleic acids, and mucopeptides were without effects on repair.

That enzymes may be affected by freezing was demonstrated by Speck and Cowman (1969), who reported the irreversible aggregation or polymerization of a proteinase associated with the cell membrane of *S. lactis*.

C. Radiation

Radiation may cause injury or death to bacteria. This has been a matter of interest to microbiologists concerned with the preservation of foods employing γ-irradiation and with the assessment of the safety of these foods. The microbiologist concerned with the bacterial flora of waters, particularly surface waters, should also have an interest in radiation damage. That damage to coliforms caused by visible light occurs in surface and marine waters has been demonstrated by light and dark bottle studies on English estuaries. There is a rich literature on aquatic bacteria and bacteria found on leaf surfaces, suggesting that pigmentation and high guanine and cytosine content protect these bacteria from harmful effects of light and, in turn, that light is indeed inimical to species not possessing these characteristics. Light should be especially damaging to the allochthonous flora of surface waters, which includes enteric pathogens and indicator bacteria. It is interesting that Grabow et al. (1973; 1975) have suggested on the basis of their investigations of pond systems and surface waters that plasmids bearing determinants for resistance to ultraviolet radiation may enhance the survival of enteric bacteria carrying them.

It should be noted, furthermore, that ultraviolet light may be used to sterilize water in special situations, and γ-radiation has been used to reduce bacterial populations in sewage sludges, especially where these are applied to the land (Hess and Lott, 1971). The use of γ-radiation for the treatment of sludges is currently the subject of intense investigation.

The literature on the effects of ionizing and nonionizing radiation is extensive, and the subject cannot be examined thoroughly in this review. The present discussion will be limited, therefore, to a brief review of the effects of ultraviolet radiation and visible light on bacteria.

Ultraviolet Radiation

Injury to bacteria exposed to ultraviolet radiation and its repair was demonstrated in 1949 by Kelner, in 1953 by Heinmets, and in 1958 by Alper and Gillies. Ultraviolet radiation is absorbed strongly by nucleic acids, and the DNA is the most important site of damage. When exposed to ultraviolet light, cyclobutane-type dimers are formed between adjacent pyrimidine residues (Setlow, 1966).

Ultimately, damage to the DNA causes death of the cell. This is clear in view of the enhancement of repair by light, which was so well demonstrated by Kelner.

Light causes the splitting of pyrimidine dimers. Rupert (1962a, b) described evidence of an enzyme that was bound in the dark to ultraviolet-irradiated DNA. Exposure to light was required for the dissociation of the complex releasing free enzyme and repaired DNA.

Repair may be effected also by excision described by Kelly et al. (1969) or by mechanisms involving the RecA+ gene. In excision repair, DNA polymerase exhibiting endonuclease activity makes incisions, releasing a short piece of single-stranded DNA containing the dimer. The polymerase then repairs the gap.

Visible Light

It has been demonstrated nicely both in the laboratory and in the field that visible light can cause death of bacteria. The mechanisms, however, differ from the mechanisms of injury caused by ultraviolet or other ionizing radiations as well as from the mechanisms of injury caused by other agents considered in this chapter. One consequence of this difference is that repair of damage caused in the presence of oxygen by visible light occurs to a very much smaller extent, and that caused by near ultraviolet to a smaller extent than that occurring as a result of exposure to ultraviolet radiation (Webb and Lorenz, 1970). Uretz (1964) was unable to demonstrate any change in the photosensitivity of *E. coli* by assay in minimal media, nor was he able to demonstrate recovery during storage in the dark or photoreactivation. Thus, the extent to which visible light is a matter of concern during bacteriological investigations of surface waters is not clear.

D. Starvation

The survival of bacteria that have been deprived of exogeneous sources of energy or other essential nutrients depends upon numerous factors relating to their previous history, nutritional status, and population density, and to the characteristics of the environment (see Postgate and Hunter, 1962; Postgate and Hunter, 1963a; Dawes, 1976). During starvation, cells depend upon endogeneous substrates which must provide energy to maintain their integrity. Most investigations of starvation have not been concerned with injury per se. Rather, they have dealt with survival measured as viable counts on rich media and with cellular damage. Injury manifested in discrepancies between counts on rich, selective, or minimal media have not been of prime interest, although Postgate and Hunter (1962) called attention to the phenomenon.

Dawes (1976) has pointed out that it has not been possible to attribute the death of bacteria to the loss of any specific cell component during starvation, although changes in certain of these do occur. The depletion of protein and RNA was observed to accompany the starvation of *P. aeruginosa* by MacKelvie et al. (1968), as were levels of constitutive enzymes involved in glucose oxidation. Thomas and Batt (1969a) observed the hydrolysis of RNA and protein in *S. lactis* and release of the products during starvation. Although addition of exogenous energy sources enhanced resynthesis, they found no evidence that this favored survival. The presence of magnesium ion was found to prolong survival and to protect the integrity of ribosomes (Thomas et al., 1969). Only RNA was found to decrease in starved *Zymomonas anaerobia* and *Z. mobilis* investigated by Dawes and Large (1970). Magnesium ion prevented RNA breakdown and also increased viability, although Burleigh and Dawes (1967) failed to observe an increase in survival associated with reduced RNA degradation in *S. lutea*.

In bacteria able to store glycogen, the substrates of endogenous metabolism vary according to initial growth conditions, and consequently to the composition of cells. Strange et al. (1961) showed that the composition of *Aerobacter (Enterobacter) aerogenes* cells varied in response to growth conditions and that degradation of cellular components during suspension in buffered saline similarly depended upon these conditions. Cells grown in tryptic meat broth lost RNA at a rapid rate; those grown in tryptone glucose medium lost carbohydrate at a high rate. Bacteria in the stationary phase containing glycogen survived well. The depletion of RNA and protein were not immediately related to loss of viability and they appeared to be present in excess. These observations and the initial lag preceding the decline in viable cells suggested that utilization of all these reserves permitted cells to survive during starvation.

Similarly, Ribbons and Dawes (1963) and Dawes and Ribbons (1965) demonstrated that *E. coli* grown in a glucose ammonium salts medium and a glucose-tryptone medium stored glycogen which it used initially during endogenous metabolism, after which ammonia was released. Cells grown on tryptone medium stored no glycogen and released ammonia when starved. When *S. lutea* was grown on peptone medium (Ribbons and Dawes, 1963), the amino acid and peptide pools were depleted, but lipid and carbohydrate contents remained unchanged. When grown in glucose-peptone medium, however, polyglucose was utilized concurrently. However, Burleigh and Dawes (1967) found that *S. lutea* with a high carbohydrate content survived less well than cells containing low concentrations. That poly-β-hydroxybutyrate (PHB) may also serve as a reserve enhancing survival was demonstrated by Sobek et al. (1966), who showed that this stored material was utilized during endogenous metabolism and that cells containing large amounts of PHB survived longer than cells containing small amounts.

That changes may occur in the enzyme composition of cells during starvation was suggested by Postgate and Hunter (1962), who reported a decline in the dehydrogenase activity of starved *A. aerogenes* which paralleled the decline in viable counts. Strange (1966) observed a rapid loss of protein and β-galactosidase activity in starving *E. coli* cells which exceeded the rate at which viable cells declined. MacKelvie et al. (1968) also suggested the possibility of the loss of constitutive enzymes involved in the oxidation of glucose by *P. aeruginosa* during starvation. They noted an ability of this species to reestablish inducible protein synthesizing systems in spite of greatly reduced ribosomal complement, however.

E. Survival in Aerosols

Whereas aerosols are of general interest to those concerned with the assessment of bacterial populations in water and waterborne wastes, they are of only marginal concern with respect to membrane filter techniques and the subject will be considered only in a cursory manner here. A more thorough review is to be found in the paper by Strange and Cox (1976).

Damage to aerosolized bacteria may be related to a number of factors. Relative humidity generally exerts a major influence over survival of aerosolized bacteria such as *E. coli* at relative humidities greater than about 50% (Cox, 1966, 1969, 1970). At high relative humidities, survival declines and becomes highly unstable. Although instability at high relative humidity may be related to water content, Cox (1970) has suggested that this is not strictly so. At low relative humidities, oxygen concentration becomes important in the survival of aerosolized bacteria (Hess, 1965; Cox, 1970). At high relative humidities, however, survival is greater in air than in anaerobic environments (Cox, 1970).

It is of interest in reference to the subject of this chapter that mechanisms for the repair of radiation damage do not appear to be involved in the repair of damage caused by aerosolization (Cox et al., 1971).

That pollutants of the atmosphere apparently may affect survival of bacteria in aerosols was demonstrated by Druett and May (1968), who showed that *E. coli* survived very much better under enclosed conditions than in free air. These investigators showed that a substance which they designated "open air factor" rapidly disappeared from air introduced into enclosed areas, was highly lethal to *E. coli*, and suggested that it might consist of an ozone-olefin complex such as those associated with phytotoxic smogs.

Damage to bacteria in aerosols was investigated first by Anderson and Dark (1967), who noted a loss of potassium following collection. Hambleton (1971) demonstrated damage to the surface of *E. coli*, which became sensitive to lysozyme. Magnesium, zinc, and ferric ions enhanced repair, presumably by stabilizing stressed outer wall layers. Repair required a source of carbon and energy, but not synthesis of protein, RNA, or mucopeptide. The sensitivity of *Klebsiella pneumoniae* to lysozyme, DNase, and RNase was shown by Maltman and Webb (1971) to be related to the medium from which the cells were aerosolized, relative humidity, time of exposure as an aerosol, and the method by which the cells were rehydrated. Effects on the cell membrane reducing the accumulation of methyl-α-D-glucopyranoside and isoproply-thio-β-D-galactopyranoside, but increasing permeability to β-D-galactopyranoside were observed by Benbough et al. (1972). These effects apparently were related to loss of components rather than damage to permeases because leaked substances restored, at least in part, the activity of the cells.

F. Survival in Water

The survival of bacteria in water and its measurement depend upon many factors. These may include characteristics of strains, phase of growth and rate of growth prior to suspension in water, nutritional status, density of population, prior stresses on the population, nature of the aqueous environment, and counting technique. Injury in relation to bacterial survival in water has been investigated primarily in connection with studies of diluents, but to a small extent also in connection with investigations of stream waters.

Diluents

The behavior of bacteria in diluents has concerned bacteriologists since the 1920s. Winslow and Brooke in 1927 observed that bacteria were protected by 0.005% nutrient broth or 0.003% meat extract incorporated in dilution water, and concluded that the effect was not attributable to their utilization as nutrients. Sodium chloride and sugars failed to provide protection. Ballentyne (1930) found that *E. coli* and *S. typhi* survived longer in distilled water than in 0.85% saline solution. They also observed a protective effect of nutrient broth and low temperatures. In 1932, Butterfield compared distilled water; phosphate buffer; phosphate buffer containing calcium, magnesium, and iron salts; and various tap waters for their value as diluents. Phosphate buffer and buffer containing salts yielded the best results. Gunter (1954) investigated the effects of buffer concentration on the survival of various bacteria and yeasts. Although buffer concentrations from 0.001 to 0.5 M did not influence the survival of *E. coli*, *P. fluorescens* was highly sensitive to concentration, 0.1 M being optimum. Supplementation with magnesium salts improved recoveries of *P. fluorescens* over a period of 3 hr. In 1957, Straka and Stokes reported investigations of a variety of diluents, and recommended the use of 0.1% peptone, which afforded consistent and full protection to the mixed flora of foods. Although various other substances afforded protection also, bacteria died rapidly in distilled water and phosphate buffer. The recommendation that 0.1% peptone be used as a diluent has been given support by numerous later workers, although some (King and Hurst, 1963; Vanderzant and Krueger, 1968) warned that growth may occur in this diluent.

Interest in diluents for use in procedures for the enumeration of stressed bacteria on selective media has led to the development of modifications or alternatives to commonly used diluents. Ray and Speck (1973b) improved recoveries of *E. coli* on selective media when 1% peptone, 1% nonfat dried milk solids, or 1% $MgSO_4 \cdot 7H_2O$ were added to dilution water. Weiler and Hartsell (1969) recommended 0.1% trypticase.

Injury to bacteria in diluents was investigated by Hoadley and Cheng (1974), who found that *P. aeruginosa* and to a lesser extent *E. coli* were sensitive to the stress of suspension in commonly employed diluents. Injury to *P. aeruginosa* preventing recovery on selective media occurred not only in distilled water and phosphate buffer but in 0.1% peptone as well, although repair eventually occurred in peptone. Injury followed by repair occurred also in *E. coli* suspended in 0.1% peptone. The rapid decline of populations of *P. aeruginosa* in distilled water and phosphate buffer was consistent with the initial decline in distilled water observed by Favero et al. (1971). Growth in distilled water was not observed, although growth during 5-10 day periods has been reported by Botzenhart and Röpke (1971) and Favero et al. (1971).

That exposure to heat leaves cells of *P. fluorescens* more susceptible to injury during dilution has been demonstrated by Gray et al. (1977). These investigators observed injury and repair in 0.1% peptone resembling that observed by Hoadley and Cheng (1974). They showed also that 0.1% peptone lacked constituents present in a glutamate salts defined medium which protected the bacteria against injury in the diluent. It was concluded that heat-induced lesions were produced at the cell surface, and in the absence of cations the cells were unable to maintain their integrity. The addition of 0.01 M magnesium sulfate to diluent was found to minimize the injury to cells during dilution.

Surface Waters

Although numerous authors have attempted to identify factors affecting the survival of fecal bacteria in surface waters, there appear in the literature few reports of investigations of injury to bacteria in surface waters and its implications. Evidence that injury preventing coliforms from forming colonies on membrane filters incubated on m-Endo broth was reported by McCarthy et al. (1961), who observed that most probable numbers of total coliforms in certain surface waters exceeded counts on membrane filters. Since then, various authors have been able to improve recoveries of indicator bacteria on membrane filters by manipulations of the procedure (see Sec. IV). These observations indicate that injured cells are indeed present in surface waters.

Hoadley and Cheng (1974) examined the recovery of *E. coli* suspended in stream water on TSA and on membrane filters incubated on mFC medium. Whereas counts on TSA decreased to about 60% of their initial value and subsequently began to increase, counts on mFC medium decreased to nearly 5% of their initial value, suggesting injury. After about 10 hr, counts on mFC medium increased until they equaled counts on TSA, suggesting repair. Injury to *P.*

aeruginosa with subsequent recovery was also demonstrated. In view of the demonstration of growth of *P. aeruginosa* in distilled waters, distilled waters supplemented with inorganic salts, and in water from mist therapy units (Favero et al., 1971; Botzenhart and Röpke, 1971), repair and growth in stream water is not surprising.

Injury to both *E. coli* and *S. faecalis* was demonstrated in situ by Bissonnette et al. (1975), who investigated recoveries of populations in membrane filter chambers immersed in streams over 4 day periods. Although repair was not demonstrated in the field, cells injured during exposure to the stream environment for 2 days and then inoculated into TSY broth recovered their ability to form colonies on desoxycholate lactose agar.

Injury to *E. coli, S. faecalis,* and *P. aeruginosa* in some tap waters was demonstrated by Hoadley and Cheng (1974), who found that recoveries of *E. coli* on deoxycholate lactose agar and *S. faecalis* on KF medium and m-Enterococcus medium were very poor compared to recoveries on TSA.

There appear to be in the literature no reports of the nature of injury to cells exposed to stream environments or factors affecting injury to and repair of such cells.

G. Chemical Agents

In 1958, McKee et al. showed that counts of coliforms in chlorinated settled sewage on membrane filters were lower than most probable numbers. This observation has been made many times since and has led a number of investigators to employ modifications of membrane filter techniques that allow improved recoveries (see Sec. IV). Braswell and Hoadley (1974) showed that this discrepancy between counts on membrane filters and rich media could be attributed to injury (see Table 2).

Injury-preventing colony formation by *E. coli* subjected to the stress of chlorination was demonstrated first by Heinmets et al. (1954). These investigators observed that whereas stressed cells failed to form colonies on m-9 medium when plated immediately after exposure, incubation in 0.1% pyruvate for 24 hr prior to plating

TABLE 2 Recovery of *E. coli* ATCC 27622 from Chlorinated Secondary Sewage on Trypticase Soy Agar and on M-FC Medium

Chlorine contact time (min)	Counts/ml			
	Expt. 1[a]		Expt. 2[b]	
	TSA[c]	M-FC	TSA	M-FC
1	16,000	15,400	40,000	14,000
5	1,600	17	26,200	1,400
10	1,000	0	9,600	460
20	10	0	1,000	0
30	6	0	100	0

[a]Temperature, 21°C; pH 7.0; chlorine dosage, 3 mg/liter; chlorine residual after 30 min, 0.75 mg/liter.
[b]Temperature, 21°C; pH 7.0; chlorine dosage, 2 mg/liter; chlorine residual after 30 min, 0.35 mg/liter.
[c]TSA, trypticase soy agar.
Source: After Braswell and Hoadley (1974).

permitted repair of injury. Milbauer and Grossowicz (1959) showed that recoveries of *E. coli* exposed to chlorine on nutrient agar were greater than on a minimal agar, and Maxcy (1970) showed that when *E. coli* were exposed to chlorine, counts on violet red bile agar were lower than counts on standard plate count agar and the proportion of viable cells injured increased with time of exposure. That quaternary ammonium compounds also cause injury in *E. coli, S. aureus,* and *S. faecalis* was demonstrated by Scheusner et al. (1971a, b).

Phenolics, quaternary ammonium compounds, and other surface-active compounds are frequently observed to cause leakage of cellular components. Salton in 1951 exposed *S. aureus, S. faecalis, E. coli,* and *S. pullorum* to cetyltrimethylammonium bromide (CTAB) and observed the release of materials absorbing at 260 nm, suggesting that free purines and pyrimidines were being lost from cells. Inorganic phosphorus was also lost from all species, and glutamic acid from the gram-positive species. It is well known that *P. aeruginosa* is highly resistant to agents such as benzalkonium chloride, chlorohexidine, and cetrimide. The envelopes of this species are rich in magnesium, calcium, and lipids, and treatment with EDTA enhances susceptibility to these disinfectants (Haque and Russell, 1974, 1976). EDTA has a lesser potentiating effect on the killing of *E. coli*, *Enterobacter cloacae*, and *K. aerogenes*, the envelopes of which contain less magnesium and calcium.

Exposure to the surface-active tertiary amine N-dodecylidiethanolamine affects the cytoplasmic membrane, causing the release of potassium ion and 260 nm absorbing material from *E. coli*. Phenolic disinfectants have been shown to cause release of glutamic acid from *E. coli*.

Venkobachar et al. (1975) investigated the effects of chlorine on enzyme activity in *E. coli*. They showed that dehydrogenase activity and survival were related to chlorine dosages. They showed also that succinic dehydrogenase activity was reduced in crude extracts exposed to chlorine, whereas ATPase and catalase were not sensitive to chlorine. Succinic dehydrogenase is said to have sulfhydryl groups at active sites, whereas the other enzymes are not rich in thiol groups. The inactivation of succinic dehydrogenase exposed to low chlorine dosages was reversible by reduced glutathione. This and investigation of the oxidation of those groups in crude extracts suggested that disulfide bonds were produced at low dosages but that these were further oxidized to sulfonyl groups at higher concentrations. Subsequently, these authors (1977) have suggested that oxidative phosphorylation is inhibited as well.

That ATPase may be inhibited by other disinfectants has been demonstrated by Harold et al. (1969), who found that ATPase in *S. faecalis* was inhibited by chlorhexidine. Chlorhexidine is known also to cause aggregation of proteins in the cytoplasm (Hugo and Longworth, 1966).

That DNA may be involved during disinfection with oxidizing agents is indicated in results of Hamelin and Chung (1975a, b), who reported mutations in *E. coli* exposed to ozone; of Wlodkowski and Rosenkranz (1975), who reported base substitution-type mutations in *S. typhimurium* exposed to hypochlorite; and of Shih and Lederberg (1976a), who reported small increases in numbers of *trp+* revertants in *Bacillus subtilis* exposed to monochloramine. Shih and Lederberg (1976b) demonstrated single-strand breaks in DNA in vitro and in vivo and reduced endonuclease activity. At high chloramine dosages, double-strand breaks occurred as well. It is of interest that while sodium thiosulfate stops most DNA degradation, the process continues slowly even after addition.

IV. RECOVERY OF STRESSED BACTERIA ON MEMBRANE FILTERS

Once it was recognized that counts of indicator bacteria in certain aquatic environments made on membrane filters could be substantially lower than most probable numbers determined simultaneously, efforts were made to improve recoveries on membrane filters by providing for the resuscitation of injured cells, usually prior to application of membranes to selective differential media. The first to recognize the need to provide a period of enrichment or resuscitation to permit "adjustment" or "weaker" organisms were McCarthy et al. (1961), who had observed that most probable numbers of total coliforms in certain surface waters, but not sewage or highly polluted river waters, were higher on the average by a factor of 1.53 than counts on membrane filters incubated on M-Endo broth.

They were able to improve the recoveries on membrane filters by incubating filters first on lauryl tryptose broth for 2 ±1/2 hr at 35°C and then transferring the membrane to an m-Endo medium that had been reduced in strength and to which agar had been added. By making these changes in the procedure, recoveries were improved so that on the average, recoveries of total coliforms on membrane filters equaled those achieved by the most probable number procedure, and the full advantages of the membrane filter technique were realized. The approach was to be used later by workers enumerating fecal streptococci in surface waters and sewage and fecal streptococci, total coliforms, and fecal coliforms in chlorinated effluents.

Rose and Litsky (1965) were next to apply an enrichment step to the recovery of indicator bacteria on membrane filters. These authors were able to improve the recovery of fecal streptococci from river waters and sewage on m-Enterococcus agar by incubating membranes first on a peptone-yeast extract Casitone broth and incubating for 3 hr (or up to 18 hr if required for the convenience of laboratory personnel) prior to transfer to M-Enterococcus agar which was incubated in the usual manner. The incorporation of an enrichment step increased recoveries of fecal streptococci by a factor of 2.44, on the average.

More recently, Lin (1973, 1974, 1976) employed preenrichment of membrane filters to improve recoveries of indicator bacteria from chlorinated effluents. Lin (1973) first applied the technique of McCarthy et al. (1961) to the recovery of total coliforms in secondary effluents chlorinated in the laboratory. Recoveries achieved after preenrichment were approximately 1.5 times those achieved by incubation on M-Endo broth without preenrichment, and, on the average, equaled most probable numbers. Similarly, Lin (1974) applied the method of Rose and Litsky (1965) to the recovery of fecal streptococci from effluent samples chlorinated in the laboratory. He found that preenrichment of membranes on peptone-yeast extract Casitone broth increased recoveries by a factor of 1.6,on the average. However, he demonstrated further that preenrichment of membranes on brain-heart infusion broth containing bile increased recoveries by a factor of 2.14 over incubation on m-Enterococcus agar without the benefit of preenrichment.

Lin (1976) applied preenrichment also to improve recoveries of fecal coliforms from chlorinated sewage effluents on membrane filters. The best recoveries were achieved when filters were incubated first on phenol red lactose broth at 35°C for 4 hr prior to transfer to mFC agar and subsequent incubation at 44.5°C for 18 hr. When conducted in this way, counts on membrane filters were, on the average, equal to most probable numbers.

Rose et al. (1975) were able to achieve improved recoveries of injured fecal coliforms from river waters, marine waters, raw sewage, and chlorinated sewage without having to transfer membranes from the preenrichment by overlaying the mFC agar with a layer of lactose broth containing agar. Membranes were placed on the solid surfaces of the recently poured overlying layer and incubated at 35°C for 2 hr. This permitted resuscitation at the lower temperature and in the absence of inhibiting components of the mFC medium. After incubation at 35°C, the temperature was increased to 44.5°C and components of the mFC medium diffused through the surface layer. Recoveries increased by an average of 93% over those on mFC agar, and were higher by a factor as high as 38 in chlorinated sewage.

Recently, Stuart et al. (1977) modified the double-layer procedure of Rose et al. (1975), by the addition of sodium acetate, glycerol, thioglycolate, glutathione, and sodium thiosulfate to both layers, adding yeast extract to the surface layer, and increasing the strength of the mFC agar layer. The reducing agents were added to inactivate any residual chlorine and to reduce cellular components oxidized by exposure to chlorine. Glycerol and acetate, which do not cause catabolite repression of lactose utilization, were added to enhance repair of injury. Recoveries of fecal coliforms from effluents chlorinated in the laboratory were greater by factors varying from 1.3 to 11.7 when compared to those on mFC agar. Recoveries were in general comparable to most probable numbers.

More recently, Green et al. (1977) employed preincubation of membrane filters at 35°C for 5 hr on mFC agar followed by incubation at 44.5"C for 18 hr to allow recovery of injured fecal coliforms. Preincubation of the membrane filters increased recoveries from 16% to 79% of most probable numbers in chlorinated secondary effluents.

V. SPECIAL TECHNIQUES EMPLOYING MEMBRANE FILTERS

Membrane filters have been employed in techniques designed to test for injury or death of stressed bacteria. However, the ability of injured bacteria to form colonies on membranes incubated on rich media generally have not been evaluated, and the results of these tests may underestimate injured, and thus surviving populations (refer to Table 1).

Goff et al. (1972) and Claydon (1975) employed membrane filters to differentiate injured and noninjured populations and to segregate revived injured from uninjured bacteria. Samples (milk) were filtered and the filters incubated for 48 hr at 32°C on TSB or TBS containing 3% sodium chloride (TSBS). Injury was indicated by lower counts on TSBS. When membranes incubated on TSBS were transferred to TSB and incubated for an additional 48 hr, however, counts increased to approximately those in filters initially incubated on TSB. When colonies appearing initially on TSBS were marked by perforating with a sterile needle and filters transferred to TSB and incubated further, colonies of resuscitated injured cells could be identified and picked.

Prickett and Rawal (1972) employed a membrane filtration technique for the evaluation of quaternary ammonium compounds as disinfectants. The test was performed by adding 0.05 ml of a suspension of test organisms (*S. typhi* or *S. aureus*) to the surfaces of test filters followed by 0.5 ml of the disinfectant solution to be tested. The cell suspension and disinfectant solution were mixed and allowed to stand for 10 min, after which 20 ml of letheen broth was added to neutralize the quats and filtered. The membrane was then washed with 50 ml of sterile distilled water and incubated on nutrient agar at 37°C for 24 hr. Later, Prince et al. (1975) applied the technique employing 1 ml of cell suspension (containing 10^6-10^8 bacteria) which was spread over the surface of the filter. The membrane was then covered with 20 ml of disinfectant solution which was removed by filtration at 2.5 or 8 min. The membrane was washed twice with 10 ml of nutrient broth and incubated on nutrient agar containing 10% horse blood for 24 or 48 hr at 37°C. These investigators employed a variety of bacterial species, including *P. aeruginosa, S. aureus, E. coli, Proteus vulgaris,* and *E. aerogenes*. Rawal (1976) applied the test to cells (1 ml suspension containing approximately 200 *P. aeruginosa* organisms) mixed with 3 ml of disinfectant solution in a reaction bottle. After 8 min, 1 ml volumes of the mixture were transferred to membrane filters drop by drop and washed with 30 ml of a recovery broth. The filters were incubated on nutrient broth as before.

The use of membrane filters in such tests permits the effective removal of disinfectant and is said to be especially useful for quaternary ammonium compounds, which may not be effectively neutralized. It should be adaptable to larger volumes of mixtures of cells and disinfectants as well.

VI. CONCLUSIONS

Allochthonous bacteria, including enteric pathogens and indicator bacteria of concern to readers of this volume, must be subject to numerous stresses when they enter aquatic environments. These include thermal and nutritional stresses, radiation, and biological interactions that cause damage to cells. Response to stress must be influenced by physiological state as cells enter the water and the organic and inorganic components of the water itself. The physiological state of cells entering water in turn depends upon prior nutrition, growth phase, and chemical stresses such as exposure to disinfectants. Thus, although it may be possible to understand the effects of individual stresses and their interactions on microorganisms, it would be impossible to know how these are acting in a water treatment plant or distribution system, a well water, a surface water, a marine water, a stabilization pond, or raw or treated effluent. What seems clear, however, is that injury does indeed occur.

Injured cells may not multiply on media customarily employed for the enumeration of bacteria of sanitary significance in water, and in particular, those employed in membrane filter techniques. Populations of injured but viable cells unable to form colonies under the conditions used to enumerate bacteria in water may exceed the uninjured population manyfold, particularly in some drinking waters and chlorinated effluents.

It is unacceptable that counts of toxigenic, pathogenic, or indicator bacteria upon which judgments of water quality are made not include injured bacteria. It is, furthermore, only with an understanding of agents causing environmental stress, sites of cellular injury, and factors that may influence the response of bacterial cells to stress and their repair that methods can be applied properly and improved and that data can be interpreted properly.

Although very few investigations have been undertaken to understand the response of allochthonous bacteria exposed to aquatic environments, techniques have been developed to permit repair and subsequent colony formation on membrane filters, while retaining the selective and diagnostic characteristics of the procedures. Preenrichment and preincubation at moderate temperatures have been employed most often, but in recent years double-layer techniques have been developed as well. With these techniques, recoveries have been greatly increased, specificity and selectivity retained, and in some cases the differential properties of the tests have been improved.

REFERENCES

Alper, T., and Gillies, N. E. (1958). "Restoration" of *Escherichia coli* strain B after irradiation: its dependence on suboptimal growth conditions. *J. Gen. Microbiol. 18*: 461–472.

Anderson, J. D., and Dark, F. A. (1967). Studies on the effects of aerosolization on the rates of efflux of ions from populations of *Escherichia coli* strain B. *J. Gen. Microbiol. 46*: 95–105.

Ballentyne, E. N. (1930). On certain factors influencing the survival of bacteria in water and in saline solutions. *J. Bacteriol. 19*: 303–320.

Benbough, J. E., Hambleton, P., Martin, K. L., and Strange, R. E. (1972). Effect of aerosolization on the transport of α-methyl glucoside and galactosides into *Escherichia coli*. *J. Gen. Microbiol. 72*: 511–520.

Bissonnette, G. K., Jezeski, J. J., McFeeters, G. A., and Stuart, D. G. (1975). Influence of environmental stress on enumeration of indicator bacteria from natural waters. *Appl. Microbiol. 29*: 186–194.

Bluhm, L., and Ordal, Z. J. (1969). Effect of sublethal heat on the metabolic activity of *Staphylococcus aureus*. *J. Bacteriol. 97*: 140–150.

Botzenhart, K., and Ropke, S. (1971). Lebensfahigkeit und Vermehrung von *Pseudomonas aeruginosa* in anorganischen Salzlosungen. *Arch. Hyg. Bakteriol. 154:* 509—516.

Braswell, J. R., and Hoadley, A. W. (1974). Recovery of *Escherichia coli* from chlorinated secondary sewage. *Appl. Microbiol. 28*: 328—329.

Bridges, B. A., Ashwood-Smith, M. J., and Munson, R. J. (1969). Correlation of bacterial sensitivities to ionizing radiation and mild heating. *J. Gen. Microbiol. 58*: 115-124.

Burleigh, I. G., and Dawes, E. A. (1967). Studies on the endogenous metabolism and senescence of starved *Sarcina lutea*. *Biochem. J. 102*: 236—250.

Busta, F. F. (1976). Practical implications of injured microorganisms in food. *J. Milk Food Technol. 39*: 138—145.

Butterfield, C. T. (1932). The selection of a dilution water for bacteriological examinations. *J. Bacteriol. 23:* 355—368.

Carson, L. A. Favero, M. S., Bond, W. W., and Petersen, N. J. (1972). Factors affecting comparative resistance of naturally occurring and subcultured *Pseudomonas aeruginosa* to disinfectants. *Appl. Microbiol. 23*: 863—869.

Clark, C. W., Witter, L. D., and Ordal, Z. J. (1968). Thermal injury and recovery of *Streptococcus faecalis*. *Appl. Microbiol. 16*: 1764—1769.

Claydon, T. J. (1975). A membrane-filter technique to test for the significance of sublethally injured bacteria in retail pasteurized milk. *J. Milk Food Technol. 38*: 87—88.

Collins-Thompson, D. L., Hurst, A., and Kruse, H. (1973). Synthesis of enterotoxin B by *Staphylococcus aureus* strain S6 after recovery from heat injury. *Can. J. Microbiol. 19*: 1463—1468.

Cox, C. S. (1966). The survival of *Escherichia coli* sprayed into air and into nitrogen from distilled water and from solutions of protecting agents, as a function of relative humidity. *J. Gen. Microbiol. 50*: 139—147.

Cox, C. S. (1969). The cause of loss of viability of airborne *Escherichia coli* K12. *J. Gen. Microbiol. 57*: 77—80.

Cox, C. S. (1970). Aerosol survival of *Escherichia coli* B disseminated from the dry state. *Appl. Microbiol. 19*: 604—607.

Cox, C. S., Bondurant, M. C., and Hatch, M. T. (1971). Effects of oxygen on aerosol survival of radiation sensitive and resistant strains of *Escherichia coli* B. *J. Hyg. Camb. 69*: 661—672.

Dawes, E. A. (1976). Endogenous metabolism and the survival of starved prokaryotes. In *The Survival of Vegetative Microbes*, T. R. G. Gray and J. R. Postgate (Eds.). Cambridge University Press, Cambridge, England, pp. 19—53.

Dawes, E. A., and Large, P. J. (1970). Effect of starvation on the viability and cellular constituents of *Zymomonas anaerobia* and *Zymomonas mobilis*. *J. Gen. Microbiol. 60*: 31–42.

Dawes, E. A., and Ribbons, D. W. (1965). Studies on the endogenous metabolism of *Escherichia coli*. *Biochem. J. 95*: 332–343.

Druett, H. A. and May, K. R. (1968). Unstable germicidal pollutant in rural air. *Nature 220*: 395–396.

Favero, M. S., Carson, L. A., Bond, W. W., and Petersen, N. J. (1971). *Pseudomonas aeruginosa*: growth in distilled water from hospitals. *Science 173*: 836–838.

Goff, J. H., Claydon, T. J., and Iandolo, J. J. (1972). Revival and subsequent isolation of heat-injured bacteria by a membrane filter technique. *Appl. Microbiol. 23*: 857–862.

Gomez, R. F., and Sinskey, A. J. (1973). Deoxyribonucleic acid breaks in heated *Salmonella typhimurium* LT-2 after exposure to nutritionally complex media. *J. Bacteriol. 115*: 522–528.

Gomez, R. F., Sinskey, A. J., Davies, R., and Labuza, T. P. (1973). Minimal medium recovery of heated *Salmonella typhimurium* LT2. *J. Gen. Microbiol. 74*: 267–274.

Grabow, W. O. K., Middendorff, I. C., and Prozesky, O. W. (1973). Survival in maturation ponds of coliform bacteria with transferable drug resistance. *Water Res. 7:* 1589-1597.

Grabow, W. O. K., Prozesky, O. W., and Burger, J. S. (1975). Behavior in a river and dam of coliform bacteria with transferable and nontransferable drug resistance. *Water Res. 9:* 777-782.

Gray, R. J. H., Ordal, Z. J., and Witter, L. D. (1977). Diluent sensitivity in thermally stressed cells of *Pseudomonas fluorescens*. *Appl. Environ. Microbiol. 33*: 1074–1078.

Gray, T. R. G., and Postgate, J. R. (Eds.) (1976). The Survival of Vegetative Microbes. Cambridge University Press, Cambridge, England.

Green, B. L., Clausen, E. M., and Litsky, W. (1977). Two-temperature membrane filter method for enumerating fecal coliform bacteria from chlorinated effluents. *Appl. Environ. Microbiol. 33*: 1259–1264.

Gunderson, M. F., and Rose, K. D. (1948). Survival of bacteria in precooked fresh-frozen foods. *Food Res. 13:* 254–263.

Gunter, S. E. (1954). Factors determining the viability of selected microorganisms in inorganic media. *J. Bacteriol. 67*: 628–634.

Hambleton, P. (1971). Repair of wall damage in *Escherichia coli* recovered from an aerosol. *J. Gen. Microbiol. 69*: 81–88.

Hamelin, C., and Chung, Y. S. (1975a). Characterization of mucoid mutants of *Escherichia coli* K-12 isolated after exposure to ozone. *J. Bacteriol. 122:* 19-24.

Hamelin, C., and Chung, Y. S. (1975b). The effect of low concentrations of ozone on *Escherichia coli* chromosome. *Mut. Res. 28:* 131-132.

Haque, H., and Russell, A. D. (1974). Effect of chelating agents on the susceptibility of some strains of gram-negative bacteria to some antibacterial agents. *Antimicrob. Agents Chemother. 6*: 200—206.

Haque, H., and Russell, A. D. (1976). Cell envelopes of gram negative bacteria: composition, response to chelating agents and susceptibility of whole cells to antibacterial agents. *J. Appl. Bacteriol. 40*: 89—99.

Harold, F. M., Baarda, J. R., Baron, C., and Abrams, A. (1969). $D10_9$ and chlorhexidine: inhibitors of membrane-bound ATPase and of cation transport in *Streptococcus faecalis*. *Biochem. Biophys. Acta 183*: 129—136.

Hartsell, S. E. (1951). The longevity and behavior of pathogenic bacteria in frozen foods: the influence of plating media. *Am. J. Public Health 41*: 1072—1077.

Heinmets, F. (1953). Reactivation of ultraviolet inactivated *Escherichia coli* by pyruvate. *J. Bacteriol. 66*: 455—457.

Heinmets, F., Taylor, W. W., and Lehman, J. J. (1954). The use of metabolites in the restoration of the viability of heat and chemically inactivated *Escherichia coli*. *J. Bacteriol. 67*: 5—12.

Hess, G. E. (1965). Effects of oxygen on aerosolized *Serratia marcescens*. *Appl. Microbiol. 13*: 781—787.

Hess, E., and Lott, G. (1971). Klarschlamm aus der Sicht des Veterinarhygienikers. *Gas Wasser Abwasser 51*: 62.

Hoadley, A. W. (1977). Effects of injury on the recovery of indicator bacteria on membrane filters. In *Recovery of Indicator Organisms Employing Membrane Filters*, R. H. Bordner, C. F. Frith, and J. A. Winter (Eds.). U.S. Environmental Protection Agency Report EPA-600/9-77-024.

Hoadley, A. W., and Cheng, C. M. (1974). The recovery of indicator bacteria on selective media. *J. Appl. Bacteriol. 37*: 45—57.

Hugo, W. B., and Longworth, A. R. (1966). The effect of chlorohexidine on the electrophoretic mobility, cytopasmic constituents, dehydrogenase activity and cell walls of *Escherichia coli* and *Staphylococcus aureus*. *J. Pharm. Pharmacol. 18:* 569-578.

Hurst, A. (1977). Bacterial injury: a review. *Can. J. Microbiol. 23*: 935—944.

Hurst, A., Hughes, A., Collins-Thompson, D. L., and Shah, B. G. (1974). Relationship between loss of magnesium and loss of salt tolerance after sublethal heating of *Staphylococcus aereus*. *Can. J. Microbiol. 20*: 1153—1158.

Hurst, A., Hendrey, G. S., Hughes, A., and Paley, B. (1976). Enumeration of sublethally heated staphylococci in some dried foods. *Can. J. Microbiol. 22*: 677—683.

Kelly, R. B., Atkinson, M. R., Huberman, J. A., and Kornberg, A. (1969). Excision of thymine dimers and other mismatched sequences by DNA polymerase of *Escherichia coli*. *Nature 224*: 495—501.

Kelner, A. (1949). Photoreactivation of ultraviolet-irradiated *Escherichia coli*, with special reference to the dose-reduction principle and to ultraviolet-induced mutation. *J. Bacteriol. 58*: 511—522.

King, W. L., and Hurst, A. (1963). A note on the survival of some bacteria in different diluents. *J. Appl. Bacteriol. 26*: 504—506.

Knox, W. E., Stumpf, P. K., Green, D. E., and Auerbach, V. H. (1948). The inhibition of sulfhydryl enzymes as the basis of the bactericidal action of chlorine. *J. Bacteriol. 55*: 451—458.

Koellner, J. (1975). Unpublished data.

Leder, I. G. (1972). Interrelated effects of cold shock and osmotic pressure on the permeability of the *Escherichia coli* membrane to permease accumulated substrates. *J. Bacteriol. 111*: 211—219.

Lee, A. C., and Goepfert, J. M. (1975). Influence of selected solutes on thermally induced death and injury in *Salmonella typhimurium*. *J. Milk Food Technol. 38*: 195—200.

Lin, S. (1973). Evaluation of coliform tests for chlorinated secondary effluents. *J. Water Pollut. Control Fed. 45*: 498—506.

Lin, S. (1974). Evaluation of fecal streptococci tests for chlorinated secondary sewage effluents. *J. Environ. Eng. Div. Am. Soc. Civ. Eng. 100*: 253—267.

Lin, S. (1976). Membrane filter method for recovery of fecal coliforms in chlorinated sewage effluents. *Appl. Environ. Microbiol. 32*: 547—552.

McCarthy, J. A., Delaney, J. E., and Grasso, R. J. (1961). Measuring coliforms in water. *Water Sewage Works 108*: 238—243.

McKee, J. E., McLaughlin, R. T., and Lesgourgues, P. (1958). Application of molecular filter techniques to the bacterial assay of sewage. III. Effects of physical and chemical disinfection. *Sewage Ind. Wastes 30*: 245—252.

MacKelrie, R. M., Campbell, J. J. R., and Gronlund, A. F. (1968). Survival and intracellular changes of *Pseudomonas aeruginosa* during prolonged starvation. *Can J. Microbiol. 14*: 639–645.

MacLeod, R. A., and Calcott, P. H. (1976). Cold shock and freezing damage to microbes. In *The Survival of Vegetative Microbes*, T. R. G. Gray and J. R. Postgate (Eds.). Cambridge University Press, Cambridge, England, pp. 81–109.

Maltman, J. R., and Webb, S. J. (1971). The action of hydrolytic enzymes and vapor rehydration on semidried cells of *Klebsiella pneumoniae*. *Can. J. Microbiol. 17*: 1443–1450.

Maxcy, R. B. (1970). Non-lethal injury and limitations of recovery of coliform organisms on selective media. *J. Milk Food Technol. 33*: 445–448.

Milbauer, R., and Grossowicz, N. (1959). Reactivation of chlorine-inactivated *Escherichia coli*. *Appl. Microbiol. 7*: 67–70.

Moss, C. W., and Speck, M. L. (1966). Release of biologically active peptides from *Escherichia coli* at subzero temperatures. *J. Bacteriol. 91*: 1105–1111.

Nakamura, M., and Dawson, D. A. (1962). Role of suspending and recovery media in the survival of frozen *Shigella sonnei*. *Appl. Microbiol. 10:* 40-43.

Pierson, M. D., and Ordal, J. Z. (1971). The transport of methyl-α-D-glucopyranoside by thermally stressed *Salmonella typhimurium*. *Biochem. Biophys. Res. Commun. 43*: 378–383.

Postgate, J. R., and Hunter, J. R. (1962). The survival of starved bacteria. *J. Gen. Microbiol. 29*: 233–263.

Postgate, J. R., and Hunter, J. R. (1963a). The survival of starved bacteria. *J. Appl. Bacteriol. 26*: 295–306.

Postgate, J. R., and Hunter, J. R. (1963b). Metabolic injury in frozen bacteria. *J. Appl. Bacteriol. 26*: 405.

Prickett, J. M., and Rawal, D. D. (1972). Membrane filtration method for the evaluation of quaternary ammonium disinfectants. *Lab. Pract. 21*: 425–428.

Prince, J., Deverill, C. E. A., and Ayliffe, G. A. J. (1975). A membrane filter technique for testing disinfectants. *J. Clin. Pathol. 28*: 71–76.

Rawal, D. D. (1976). Microbiological evaluation of disinfectants by membrane filtration capacity test. *Microbios Lett. 3*: 213–215.

Ray, B., and Speck, M. L. (1972). Metabolic process during the repair of freeze-injury in *Escherichia coli*. *Appl. Microbiol. 24*: 585–590.

Ray, B., and Speck, M. L. (1973a). Discrepancies in the enumeration of *Escherichia coli*. *Appl. Microbiol. 25*: 494–498.

Ray, B., and Speck, M. L. (1973b). Enumeration of *Escherichia coli* in frozen samples after recovery from injury. *Appl. Microbiol. 25*: 499–503.

Ray, B., and Janssen, D. W., and Busta, F. F. (1972). Characterization of the repair of injury induced by freezing *Salmonella anatum*. *Appl. Microbiol. 23*: 803–809.

Ribbons, D. W., and Dawes, E. A. (1963). Environmental and growth conditions affecting the endogenous metabolism of bacteria. *Ann. N.Y. Acad. Sci. 102:* 564–586.

Rose, R. E., and Litsky, W. (1965). Enrichment procedure for use with the membrane filter for the isolation and enumeration of fecal streptococci in water. *Appl. Microbiol. 13*: 106–108.

Rose, R. E., Geldreich, E. E., and Litsky, W. (1975). Improved membrane filter method for fecal coliform analysis. *Appl. Microbiol. 29:* 532-536.

Rupert, C. S. (1962a). Photoenzymatic repair of ultraviolet damage in DNA. I. Kinetics of the reaction. *J. Gen. Physiol. 45*: 703–724.

Rupert, C. S. (1962b). Photoenzymatic repair of ultraviolet damage in DNA. II. Formation of an enzyme-substrate complex. *J. Gen. Physiol. 45*: 725–741.

Salton, M. R. J. (1951). The adsorption of cetyltrimethylammonium bromide by bacteria, its action in releasing cellular constituents, and its bactericidal effects. *J. Gen. Microbiol. 5*: 391–404.

Scheusner, D. L., Busta, F. F., and Speck, M. L. (1971a). Injury of bacteria by sanitizers. *Appl. Microbiol. 21*: 41–45.

Scheusner, D. L., Busta, F. F., and Speck, M. L. (1971b). Inhibition of injured *Escherichia coli* by several selective agents. *Appl. Microbiol. 21*: 46–49.

Sedgwick, S. G., and Bridges, B. A. (1972). Evidence for indirect production of DNA strand scissions during mild heating of *Escherichia coli*. *J. Gen. Microbiol. 71*: 191–193.

Setlow, R. B. (1966). Cyclobutane-type pyrimidine dimers in polynucleotides. *Science 153*: 379–386.

Shih, K. L., and Lederberg, J. (1976a). Chloramine mutagenesis in *Bacillus subtilis*. *Science 192*: 1141–1143.

Shih, K. L., and Lederberg, J. (1976b). Effects of chloramine on *Bacillus subtilis* deoxyribonucleic acid. *J. Bacteriol. 125*: 934–935.

Smeaton, J. R., and Elliott, W. H. (1967). Selective release of ribonuclease inhibitor from *Bacillus subtilis* cells by cold shock treatment. *Biochem. Biophys. Res. Commun. 26*: 75–81.

Sobek, J. M., Charba, J. F., and Foust, W. N. (1966). Endogenous metabolism of *Azotobacter agilis*. *J. Bacteriol. 92*: 687–695.

Speck, M. L., and Cowman, R. A. (1969). Metabolic injury to bacteria resulting from freezing. In *Freezing and Drying of Microorganisms*, T. Nei (Ed.). University Park Press, Baltimore, Md., pp. 39–51.

Straka, R. P., and Stokes, J. L. (1957). Rapid destruction of bacteria in commonly used diluents and its elimination. *Appl. Microbiol. 5*: 21–25.

Strange, R. E. (1966). Stability of β-galactosidase in starved *Escherichia coli*. *Nature 209*: 428–429.

Strange, R. E., and Cox, C. S. (1976). Survival of dried and airborne bacteria. In *The Survival of Vegetative Microbes*, T. R. G. Gray and J. R. Postgate (Eds.). Cambridge University Press, Cambridge, England, pp. 111–154.

Strange, R. E., and Dark, F. A. (1962). Effect of chilling on *Aerobacter aerogenes* in acqueous suspension. *J. Gen. Microbiol. 29*: 719–730.

Strange, R. E., and Ness, A. G. (1963). Effect of chilling on bacteria in aqueous suspension. *Nature 197*: 819.

Strange, R. E., and Postgate, J. R. (1964). Penetration of substances into cold-shocked bacteria. *J. Gen. Microbiol. 36*: 393–403.

Strange, R. E., and Shon, M. (1964). Effects of thermal stress on viability and ribonucleic acid of *Aerobacter aerogenes* in aqueous suspension. *J. Gen. Microbiol. 34*: 99–114.

Strange, R. E., Dark, F. A., and Ness, A. G. (1961). The survival of stationary phase *Aerobacter aerogenes* stored in aqueous suspension. *J. Gen. Microbiol. 25*: 61–75.

Stuart, D. G., McFeters, G. A., and Schillinger, J. E. (1977). Membrane filter technique for the quantification of stressed fecal coliforms in the aquatic environment. *Appl. Environ. Microbiol. 34*: 42–46.

Thomas, T. D., and Batt, R. D. (1969). Degradation of cell constituents by starved *Streptococcus lactis* in relation to survival. *J. Gen. Microbiol. 58*: 347–362.

Thomas, T. D., Lyttleton, P., Williamson, K. I., and Batt, R. D. (1969). Changes in permeability and ultrastructure of starved *Streptococcus lactis* in relation to survival. *J. Gen. Microbiol. 58*: 381–390.

Tomlins, R. I., and Ordal, Z. J. (1971). Requirements of *Salmonella typhimurium* for recovery from thermal injury. *J. Bacteriol. 105*: 512–518.

Tomlins, R. I., Pierson, M. D., and Ordal, Z. J. (1971). Effect of thermal injury on the TCA cycle enzymes of *Staphylococcus aureus* MF 31 and *Salmonella typhimurium* 7136. *Can. J. Microbiol. 17*: 759–765.

Uretz, R. B. (1964). Sensitivity to acridine sensitized photoinactivation in *Escherichia coli* B, B/r, and BS-1. *Radiat. Res. 22*: 245 (Abstr.).

Vanderzant, C., and Krueger, W. F. (1968). Effect of certain variations in diluent and dilution procedure on survival of *Pseudomonas* species grown in various media. *J. Milk Food Technol. 31*: 65–71.

Venkobachar, C., Iyengar, L., and Prabhakara Rao, A. V. S. (1975). Mechanism of disinfection. *Water Res. 9*: 119–124.

Venkobachar, C., Iyengar, L., and Prabhakara Rao, A. V. S. (1977). Mechanism of disinfection: effect of chlorine on cell membrane functions. *Water Res. 11*: 727–729.

Warseck, M., Ray, B., and Speck, M. L. (1973). Repair and enumeration of injured coliforms in frozen foods. *Appl. Microbiol. 26*: 919–924.

Webb, R. B., and Lorenz, J. R. (1970). Oxygen dependence and repair of lethal effects of near ultraviolet and visible light. *Photochem. Photobiol. 12*: 283–289.

Weiler, W. A., and Hartsell, S. E. (1969). Diluent composition and the recovery of *Escherichia coli*. *Appl. Microbiol. 18*: 956–957.

Winslow, C. E. A., and Brooke, O. R. (1927). The viability of various species of bacteria in aqueous suspensions. *J. Bacteriol. 13*: 235–242.

Wlodkowski, T. J., and Rosenkranz, H. S. (1975). Mutagenicity of sodium hypochlorite for *Salmonella typhimurium*. *Mut. Res. 31*: 39-42.

17

ULTRASTRUCTURAL AND OTHER FACTORS INFLUENCING THE SUITABILITY OF MEMBRANE FILTERS FOR ENUMERATING FECAL COLIFORMS

BERNARD J. DUTKA National Water Research Institute, Canada Centre for Inland Waters, Burlington, Ontario, Canada

RICHARD S. TOBIN Health and Welfare Canada, Ottawa, Ontario, Canada

I. INTRODUCTION

Many investigators have reported the results of comparative studies on the recovery of total and fecal coliform bacteria from pure culture studies and natural water samples by a variety of commercially available membrane filters. The results of these studies (Table 1) indicate that there are significant differences between various brands of membranes in their abilities to recover bacteria from both pure cultures and natural waters.

Some studies (Sladek et al., 1975; Standridge, 1976; Tobin and Dutka, 1977; Sladek et al., 1977) have indicated that differences in membrane filter surface pore morphology may be the key factor in explaining differences in recovery of bacteria from liquid samples. Surface pore morphology is the three-dimensional structure of the openings on the surface of the membrane which has been best demonstrated by scanning electron microscope studies. From the results of these studies it appears that the larger the pore size, the better the recovery rate because of:

1. Increased flow rate of sample through the filter, which may lessen the potential of bactericidal substances absorbing to the filter surface.

TABLE 1 Summary of Previous Comparative Studies on the Recovery of Bacteria from Water by Different Brands of Membrane Filters

Number	Investigator(s)	Source (sample/ culture) of coliforms	Type of membrane filters compared	Observations
1	Presswood and Brown (1973)	*Escherichia coli* type I, isolated by mFC method from river water and domestic sewage	Gelman (GN-6), autoclaved Millipore (HAWG 047S0) ethylene oxide-sterilized	Gelman filters recovered significantly more fecal coliform bacteria than Millipore filters.
2.	Dùtka et al. (1974)	River water/ Burlington Canal water	Gelman (GN-6), autoclaved, GNA Millipore (HAWG 047A0), autoclaved-MA Millipore (HAWG 047S0), ethylene oxide-sterilized-ME Sartorius (11406), autoclaved-SA Sartorius (13706), ethylene oxide-sterilized-SE	*March study:* Gelman superior to both types of Millipore and Sartorius for the recovery of total coliforms; for fecal coliform recovery, Gelman were superior to Sartorius but equivalent to Millipore. *June study:* Gelman superior to Millipore (MA) and Sartorius (SE) and equivalent to Millipore (ME) and Sartorius (SA) for total coliform recovery; all filters were equivalent for fecal coliform recovery.
3.	Harris (1974)	Unchlorinated water	Gelman Millipore	Gelman superior to Millipore for fecal coliform recovery.

4.	Hufham (1974)	*E. coli* (ATCC 11775) *Enterobacter aerogenes* (ATCC 13048) An mFC isolate IMViC type I from lake water	Gelman (GA-6), autoclaved Millipore (HAWG 047S0), ethylene oxide-sterilized	Gelman equivalent to Millipore for total coliform recovery but superior for fecal coliform enumeration.
5.	Schaeffer et al. (1974)	Natural water	Gelman Millipore	Gelman superior to Millipore for total coliforms but equivalent for fecal coliform recovery.
6.	Brodsky and Schiemann (1975)	River water samples	Millipore (HAWG 047S0), ethylene oxide-sterilized Sartorius (11456), ethylene oxide-sterilized Johns-Manville (045M047SG), ethylene oxide-sterilized Johns-Manville (045M047AG), autoclave sterilized Gelman (GN-6), autoclave sterilized	With EC broth cultures of water samples, Johns-Manville superior to Sartorius but not different from Millipore for fecal coliform recovery; with EC broth cultures, no differences among Johns-Manville, Millipore, and Sartorius for the recovery of total coliforms; using river water samples, Johns-Manville superior to Sartorius but equivalent to Millipore for total coliform recovery; no differences among Johns-Manville, Millipore, and Sartorius for recovery of fecal coliform from river water samples, Gelman equivalent to Millipore for fecal coliform recovery from river water samples.

TABLE 1 (Continued)

Number	Investigator(s)	Source (sample/ culture) of coliforms	Type of membrane filters compared	Observations
7.	Green et al. (1975)	Nonchlorinated sewage effluent, surface waters receiving sewage, farm run-off, and industrial effluents	Millipore (HA), ethylene oxide-sterilized Millipore (HC), ethylene oxide-sterilized Gelman (GN-6), autoclave sterilized Schleicher and Schuell (B-9, 67S4, 47S3), autoclave sterilized Schleicher and Schuell (5053), ethylene oxide-sterilized Sartorius nitrocellulose, autoclaved and ethylene oxide-sterilized Johns-Manville (416C92), ethylene oxide-sterilized	Ranked membranes for the enumeration of fecal coliform in order of decreasing recovery as follows: Millipore HC $>$ Gelman $>$ Johns-Manville $\simeq$ Sartorius $>$ Millipore HA $>$ Schleicher and Schuell.

2. The cradling of the bacterium within the pore and thus enveloping the bacterium with readily available nutrient.
3. The media enveloped bacterium is protected from the stress of surface evaporation (Sladek et al., 1975).

To further investigate the relationship between pore size and increased bacterial recovery, we studied the bacterial recovery, pore structure, and flow rates in a variety of membrane filters. We particularly wanted to compare membranes sterilized by different procedures and, where possible, improved versions of regular membranes sold by the same manufacturer for use in bacteriological water quality studies. Another aspect that was explored was the capacity of the various membranes to bind metals. If significant differences were noted in binding or sorption ability, this might help to explain some of the resuscitation differences shown by various membranes. To this end, nine membrane filters of five major manufacturers were tested.

II. METHODS

A. Bacterial Suspension

An EC-positive *Escherichia coli*, freshly isolated from Hamilton Bay, Ontario, waters was used to test the nine membrane filters. A 35°C overnight broth culture (tryptic soy broth) was used to inoculate 20 liters of the following suspending fluids (to give a final concentration of 200-250 bacteria per 100 ml) at 4°C:

1. Lake Ontario water, membrane filter (0.2 μm)-sterilized
2. Hamilton Bay water, membrane filter (0.2 μm)-sterilized
3. Burlington tap water (total residual chlorine measured by amperometric titration, 0.25 mg/liter)

B. Membranes Tested

The following 47 mm, 0.45 μm (except Millipore HC) gridded membranes were used in this study:

Millipore HAW047S0, Lot No. 086193 (ethylene oxide-sterilized by manufacturer, mixed esters of cellulose acetate and nitrate

Millipore HAWG047A0, Lot No. 109858 (autoclave-sterilized by laboratory), mixed cellulose acetate and nitrate

Millipore HCWG047S0, Lot No. 273922 (ethylene oxide-sterilized by manufacturer), mixed cellulose acetate and nitrate

Johns-Manville 045M047AG, Lot No. 421F239 (autoclave-sterilized by laboratory), cellulose acetate

Johns-Manville 045M047SG, Lot No. 429J396 (ethylene oxide-sterilized by manufacturer), cellulose acetate

Sartorius SM11406, Lot No. 304382709 (ethylene oxide-sterilized by manufacturer), cellulose nitrate

Sartorius SM13806, no lot number given--new membrane (ethylene oxide-sterilized by manufacturer), cellulose nitrate

Oxoid Nuflow N47/45, Lot No. 3356 (ethylene oxide-sterilized by manufacturer), cellulose acetate

Gelman GN-6, Lot No. 80951 (autoclave-sterilized by manufacturer), mixed cellulose acetate and nitrate

Gelman GN-6, Lot No. 81173 (autoclave-sterilized by manufacturer), mixed cellulose acetate and nitrate

Gelman membrane 80951 was used only in the chlorinated tap water study; Gelman membrane 81173 was used in the other two studies.

The Millipore membrane filter HCWG047S0 was the only membrane not listed at 0.45 μm; it is stated to have 2.4 μm surface openings and 0.7 μm absolute retention pore size and is recommended by the manufacturer for fecal coliform estimations.

C. Testing of Membranes

Inoculated 20 liter samples in glass carboys were thoroughly mixed and a 1900 ml aliquot removed for each test. The order of filtration was randomized for each of the five replicates; each replicate comprised 18 filtrations of 100 ml (nine grid side up and nine grid side down). The length of time to filter each sample was recorded, as was the nature of the edge of the filtering area after filtration. For consistency, the same membrane filtration unit was used throughout the experiment. The membranes were incubated on mFC (Difco) agar at 44.5°C for 22-24 hr.

D. Electron Micrographs

Filters were coated with approximately 20 nm each of carbon and gold in a Bendix evaporator. Micrographs were taken on a AMR 1000 Scanning Electron Microscope operated at 20,000 V at magnifications up to $10{,}000_x$. Measurements of the diameters of the largest surface pores were made on enlarged photographs of each membrane filter.

E. Trace Metals Analysis

One hundred milliliters of deionized, distilled water, prefiltered Hamilton Bay water (0.2 μm filter), and Lake Ontario water (0.2 μm filter) were passed through each of the membranes. Filters were dissolved in 6 ml of concentrated nitric acid with gentle heating to dryness. The residue was taken up in 2 ml of 6 N HCl by evaporating to near dryness, diluted to 25 ml with distilled water. The metals Cu, Zn, Ni, Pb, and Cd were determined by atomic absorption spectrophotometry.

F. Nonionic Surfactants

Fifty milliliters of distilled water was passed through each membrane to remove loosely bound nonionic surfactants. The surfactant (and polyethylene glycol) in the filtrate was quantified by the bismuth-active substance method as described previously (Anthony and Tobin, 1977).

G. Statistical Analysis

Analysis of variance for one-way design, revised June 24, 1969, by Health Sciences Computing Facility, UCLA, was used in this study.

III. RESULTS

A. Bacterial Counts and Flow Rates

From preliminary experiments, it appeared that there were differences in bacterial counts with the fecal coliform membrane-filtration method when filters were used conventionally (grid side up), as compared with grid side down. To determine the relationships of recoveries with surface pore size, we performed all experiments in both manners. Table 2a presents the results of the studies performed with

TABLE 2 Membrane Filter Flow Rates and *E. coli* Recovery from Three Types of Waters

(a) Filtration and incubation grid side up

Membrane	Count range, five replicates	Mean number of colonies/ 100 ml		Significance of F	Flow rate (sec/100 ml)	Flow rate rank		Significance
Normal chlorinated tap water (0.25 mg total residual chlorine/liter)								
Johns-Manville SG	28-158	85			12.4	3		A-B[a]
Johns-Manville AG	37-164	83			13.7	4	B[b]	A-C[a]
Sartorius 13806	38-168	83			11.6	2		A-D[a]
Millipore HC	28-151	82	A[b]	A-B[a]	4.2	1	A[b]	B-C[a]
Gelman	34-153	82			17.5	6	C	B-D[a]
Oxoid	34-140	81			23.4	9	D	C-D[a]
Sartorius 11406	24-143	66			15.6	5	C	
Millipore SO	17-117	49	B		20.8	7	D	
Millipore AO	11-113	44			22.1	8		
Membrane-filtered Lake Ontario water								
Millipore HC	150-266	229	A	A-B[a]	4.7	1	A	A-B[a]
Gelman	200-245	218		A-C[a]	14.0	6	C	A-C[a]
Johns-Manville SG	184-216	192		A-D[a]	10.0	2		A-D[a]
Johns-Manville AG	148-218	183	B	B-C[a]	10.1	3	B	
Sartorius 13806	111-257	177		B-D[a]	10.2	4		
Sartorius 11406	69-154	107	C	C-D[a]	11.3	5		
Millipore AO	17- 53	68			14.6	7		
Oxoid	14- 50	42	D		17.2	9	C	
Millipore SO	10- 63	28			15.7	8		

Membrane-filtered Hamilton Bay water

Membrane	Count range, five replicates	Mean number of colonies/100 ml	Group	Comparison	Flow rate (sec/100 ml)	Flow rate rank	Group	Comparison
Millipore HC	215-245	240	A	A-B[a]	4.8	1	A	A-B[a]
Gelman	191-277	233	A	A-C[a]	11.2	5		A-C[a]
Sartorius 13806	133-270	216	A	A-D[a]	9.4	2	B	B-C[a]
Johns-Manville AG	135-222	178	B	B-C[c]	10.6	4	B	
Johns-Manville SG	131-209	175	B	B-D[a]	10.3	3	B	
Sartorius 11406	105-186	147	B	C-D[a]	11.8	6	B	
Millipore AO	43-164	107	C		15.4	8	C	
Millipore SO	24- 85	58	D		13.2	7	B	
Oxoid	34- 82	51	D		18.1	9	C	

[a]Significant difference at 1% level.
[b]Groups marked by or enclosed by same letter not significantly different at 5% level.
[c]Significant different at 5% level.

(b) Filtration and incubation grid-side down

Membrane	Count range, five replicates	Mean number of colonies/100 ml	Flow rate (sec/100 ml)	Flow rate rank
Normal chlorinated tap water (0.25 mg total residual chlorine/liter)				
Millipore HC	65-172	113	6.0	1
Johns-Manville AG	48-178	109	15.2	4
Johns-Manville SG	38-170	97	11.0	2
Sartorius 11406	31-140	82	16.6	5
Oxoid	32-132	77	27.5	8
Millipore AO	19-149	68	20.6	7
Millipore SO	22-141	66	18.3	6
Sartorius 13806	11-122	65	12.2	3
Gelman	11-128	49	33.0	9

TABLE 2(b) (Continued)

Membrane	Count range, five replicates	Mean number of colonies/100 ml	Flow rate (sec/100 ml)	Flow rate rank
Membrane-filtered Lake Ontario water				
Millipore HC	186-274	227	4.9	1
Johns-Manville AG	128-217	179	11.9	5
Sartorius 11406	110-187	146	11.6	4
Oxoid[a]	71-143	113	17.1	9
Sartorius 13806	34-104	75	10.9	3
Johns-Manville SG[a]	8- 87	44	9.8	2
Millipore AO[a]	19- 41	32	13.9	7
Millipore SO[a]	5- 15	11	14.7	8
Gelman[a]	6- 9	8	13.4	6
Membrane-filtered Hamilton Bay water				
Millipore HC	198-265	236	4.7	1
Sartorius 11406	142-196	164	11.5	6
Johns-Manville AG	135-180	155	11.4	5
Millipore AO	37-174	124	14.8	8
Oxoid[a]	108-143	122	17.2	9
Sartorius 13806[a]	60-160	107	8.7	2
Johns-Manville SG[a]	39- 97	66	10.6	3
Millipore SO	15- 59	38	13.0	7
Gelman[a]	6- 59	26	10.6	4

[a]Membranes showing significant difference (5% level) on recovery rates from grid-up membranes.

the membranes used conventionally (i.e., grid side up). In most cases the bacteria counts decreased with time in suspending waters. The greatest effect was seen with chlorinated tap water, where recoveries were highest on the first replicate of all nine membranes and lowest on the last (fifth replicate). The least such effect was found when bacteria were suspended in Lake Ontario water, where there was no consistent relationship between recovery and replication order. It is apparent that there are large differences among the membranes in their ability to recover *E. coli* from water. The Millipore HC and Gelman GN-6 recovered the highest number of *E. coli* and were not statistically different from one another. In the Hamilton Bay water test, Sartorius 13806 was not significantly different from these two, although it recovered fewer *E. coli* suspended in the Lake Ontario water. In the chlorinated tap water, the recoveries of bacteria were generally low, and differences between membranes were minimal.

The flow rate patterns were fairly consistent in that the membranes with the fastest flow rates also had the highest recovery rates, with Millipore HC having the fastest flow rate and Oxoid the slowest. With one exception, in chlorinated tap water, Oxoid, Millipore AO, and Millipore SO had the lowest recoveries and slowest flow rates. Recovery and flow rate experiments using the reverse side of the membrane (grid down) are presented in Table 2b. Millipore HC and Johns-Manville AG membranes recovered bacteria equally well on either side, with no differences in flow rate.

The Gelman membranes (on their reverse side) always produced the lowest count of all membranes tested. Bacterial recoveries were 27 and 9 times greater on the grid-side-up filtrations with Lake Ontario and Hamilton Bay waters, respectively.

The two Sartorius membranes produced opposite bacterial recovery effects, with membrane 11406 producing more colonies when used grid side down, whereas the newer membrane, 13806, produced more colonies grid side up. The Oxoid membranes were also found to recover more bacteria on their reverse side.

TABLE 3 Growth Pattern on Reverse Side of Millipore HC Membranes

Membrane number	Colonies/100 ml[a]		
	Counting side	Reverse side	Corrected count
Sample filtered grid up			
1	245	3	248
2	252	2	254
3	244	7	251
4	242	9	251
5	215	10	225
Mean	240	6	246
Sample filtered grid down			
1	212	32	244
2	249	9	258
3	255	12	267
4	265	7	272
5	198	19	217
Mean	236	16	252

[a]Colonies growing top of the membrane were counted (filtering side) as well as those growing beneath the membrane (reverse side).

Only Millipore HC membranes produced bacterial colonies between the membrane and the medium. Table 3, using Hamilton Bay water, shows that the average number of bacteria passing through the membranes and growing below them was 6 when filtrations were performed routinely, and 16 when filtrations were performed grid side down.

B. Wetting Characteristics

During the study it was noted that some of the membranes were wetted under the edge of the filter funnel in an irregular pattern or completely to the edge of the membrane. Initially, concern was felt that bacteria could be transported under the edge of the filter funnel. A record of the type of edge found on the filtering area of each membrane is presented in Table 4. Some membranes consistently produced the same type of edge throughout the study (i.e.,

TABLE 4 Nonionic Surfactant Content and Wetting Characteristics of Membrane Filters

Membrane	Nonionic surfactant[a] (μg)	Edge of filtering area[b]	
		Grid up	Grid down
Millipore HC	29	Round sharp	Round sharp, irregular
Millipore A0	67	Irregular	Irregular
Millipore S0	22	Round sharp	Round sharp
Sartorius 13806	12	Round sharp irregular	Irregular, no edge
Sartorius 11406	21	No edge	Irregular, no edge
Johns-Manville AG	214	Irregular	Irregular
Johns-Manville SG	160	No edge	Irregular, no edge
Oxoid	10	Irregular, no edge	Irregular, no edge
Gelman	35	Irregular	Irregular

[a]Measured as bismuth-active substance eluted by 50 ml of distilled water (Anthony and Tobin, 1977), expressed as micrograms of $NPEO_{10}$ (nonylphenol ethoxylate).
[b]After filtration, the area of the membrane under the filter funnel rim was examined for wetting characteristics and classified as having no edge wetting (round sharp), partial edge wetting (irregular), or complete edge wetting (no edge).

Millipore S0, Millipore A0, Johns-Manville AG and Gelman), whereas the other membranes showed some variations. Membranes of the same manufacturer did not produce similar results, which was surprising, as the only stated difference between some of these membranes was the type of sterilization. The amount of nonionic surfactant (or bismuth iodide-active substance) released in 50 ml of water passing through the membrane varied from a low of 10 μg to a high of 214 μg per filter. There was no obvious relationship between the amount of water-extractable surfactant and edge-wetting characteristics (Table 4). Furthermore, in these studies, no bacteria were found growing outside the main filtration area.

C. Scanning Electron Microscopy

All membranes had a large percentage of open area on the surface, with large variations in the size and spacing of pores. Figures 1 to 8 illustrate typical areas on the grid side at low (a) and high (b) magnification and the ungridded side at low (c) and high (d) magnification. Most membranes showed some basic similarity in the surface texture with the notable exceptions of (a) the grid side of the Oxoid Nuflow (Fig. 7a and b), which has a sheetlike appearance; (b) the gridded side of Johns-Manville AG (Fig. 5a and b), which has a "filamentous" appearance; (c) the ungridded side of the

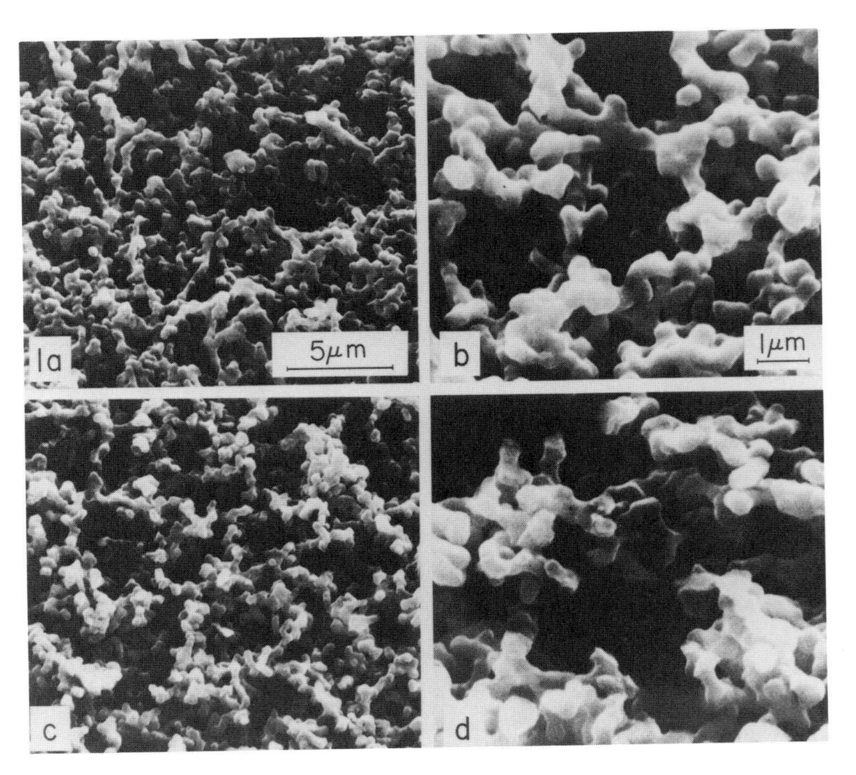

FIGURE 1 Scanning electron micrographs of the upper gridded (a, b) and the reverse, nongridded (c, d) surface of Millipore HC membranes at low (a, c) and high (b, d) magnifications.

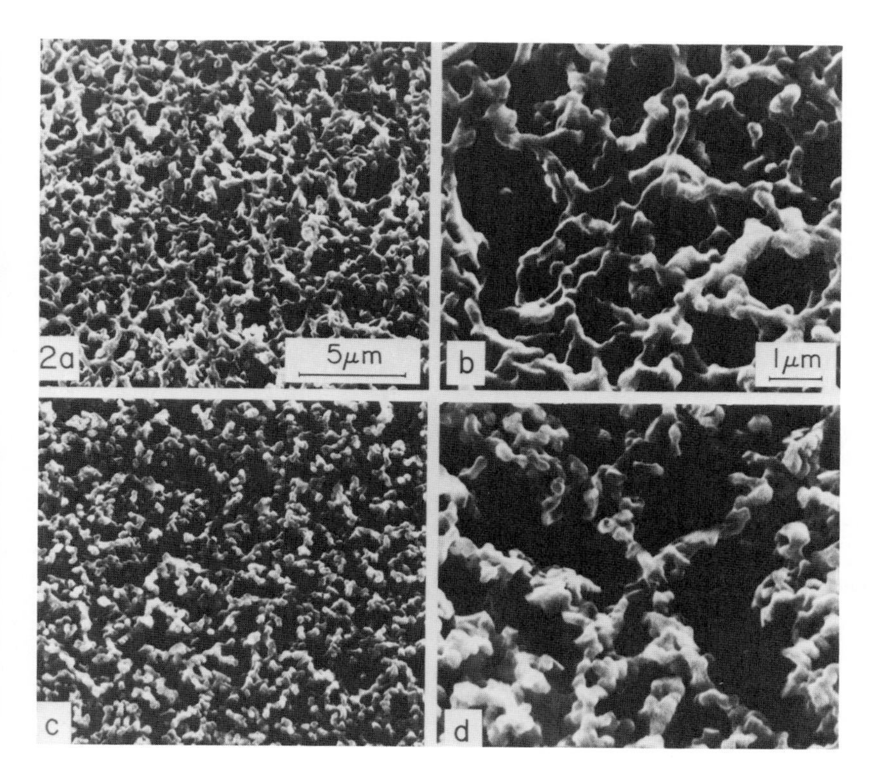

FIGURE 2 Scanning electron micrographs of the upper gridded (a, b) and the reverse, nongridded (c, d) surface of Millipore S0 (nearly identical to A0) membranes at low (a, c) and high (b, d) magnifications.

Johns-Manville AG (Fig. 5c and d), which has large, flat, apparently nonporous areas interspersed with "valleys" of open porous areas; and (d) the ungridded side of Sartorius 13806 (Fig. 3c and d), which is covered with variable-size globules, probably the coloring agent used in this green filter to facilitate colony recognition.

The other main differences are the size and distribution of pores. Although the majority of surface pores on most of the membranes are relatively small, it is the larger pores that may be most important in bacterial recoveries. The average diameter of the

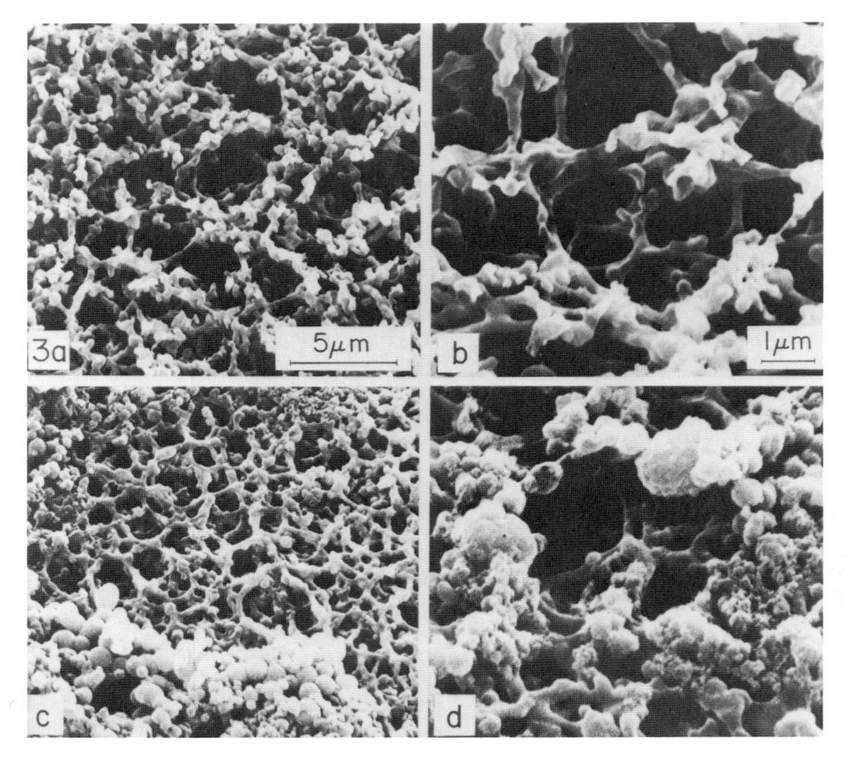

FIGURE 3 Scanning electron micrographs of the upper gridded (a, b) and the reverse, nongridded (c, d) surface of Sartorius 13806 membrane at low (a, c) and high (b, d) magnifications.

largest surface pores was measured from enlarged photographs of the membrane surfaces. Surface pore sizes (in micrometers) and ranking of the grid side are:

Johns-Manville SG (4.5) > Johns-Manville AG (3.7)> Millipore HC (3.2) ≥ Sartorius 13806 (3.2) > Gelman (3.1) > Sartorius 11406 (2.9) > Millipore AO (2.1) > Millipore SO (1.8)

Oxoid (3.1 μm) is not ranked because of the unusual surface structure. Surface pore sizes (in micrometers) and ranking of the ungridded side are:

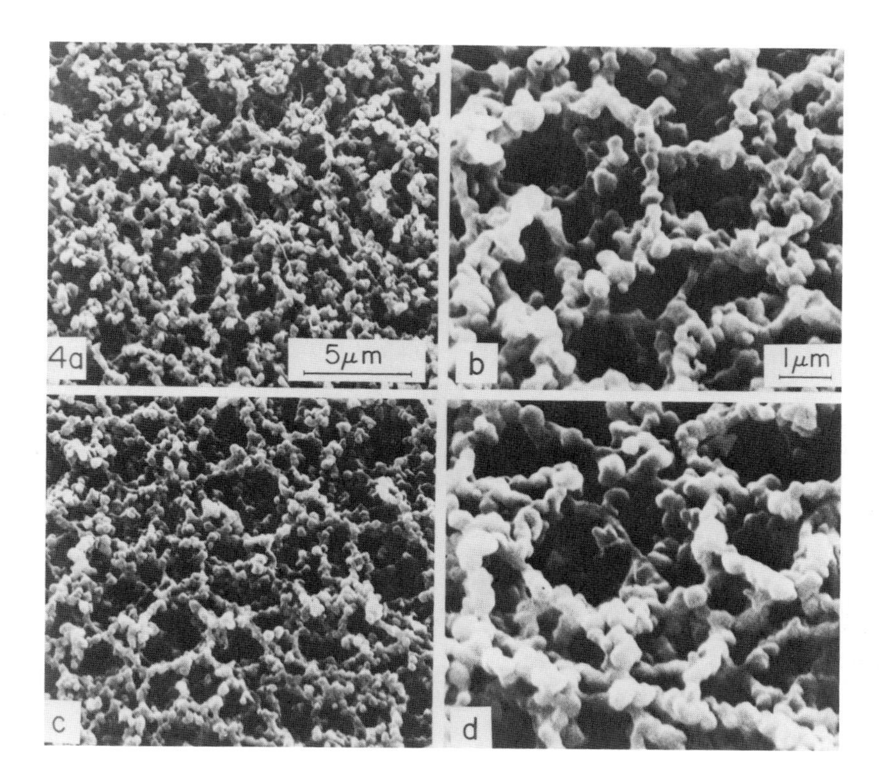

FIGURE 4 Scanning electron micrographs of the upper gridded (a, b) and the reverse, nongridded (c, d) surface of Sartorius 11406 membranes at low (a, c) and high (b, d) magnifications.

Millipore HC (3.5) > Johns-Manville SG (3.0) > Millipore SO (2.9) ≥ Millipore AO (2.9) > Sartorius 11406 (2.8) > Oxoid (2.3) > Sartorius 13806 (1.9) > Gelman (1.1).

Johns-Manville AG (3.8 μm) is not ranked because much of the ungridded surface is smooth and apparently nonporous. Only the interspersed open areas could be measured.

Experience has shown that some membranes have a certain amount of variability in the pore structure from one lot to another. Several different sizes and lot numbers of Gelman membranes were examined for their surface pore structure. The differences were almost exclusively limited to the undersides (ungridded side when

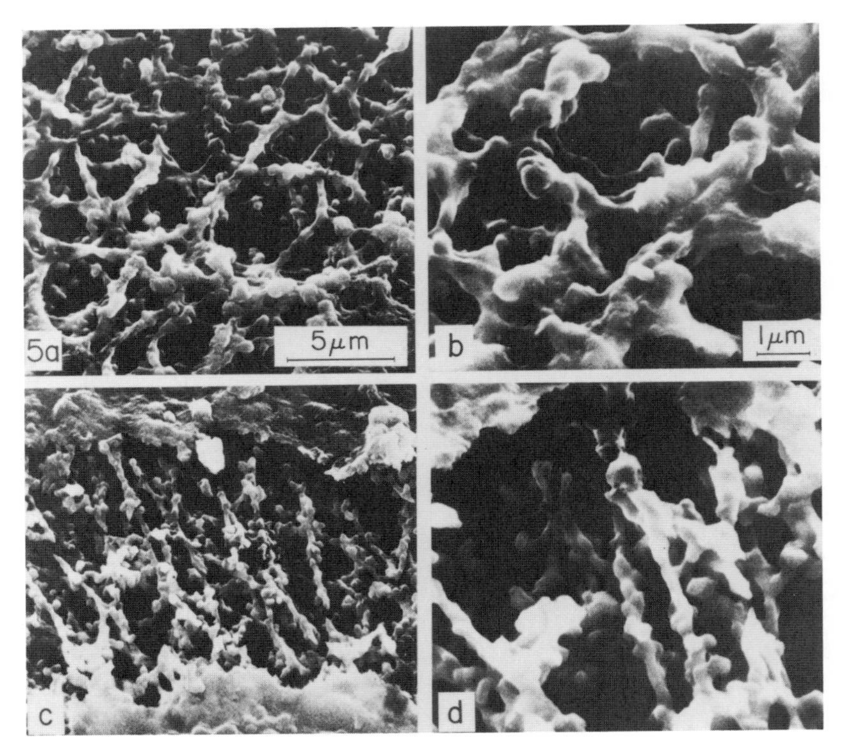

FIGURE 5 Scanning electron micrographs of the upper gridded (a, b) and the reverse, nongridded (c, d) surface of Johns-Manville AG membranes at low (a, c) and high (b, d) magnifications.

gridded membranes were tested), and these are shown in Fig. 9. These differences may affect flow rates but are not believed to influence bacterial colony formation when the filtration is performed conventionally (i.e., grid side up).

D. Trace Metals Analysis

The ability of membranes to bind metals from solution was determined. Environmental water samples with particulates >0.2 μm removed were filtered through each of the membranes which were subsequently analyzed. To determine the contribution of the membranes themselves to these measurements, another set of analyses was performed on untreated membranes after washing with 100 ml deionized

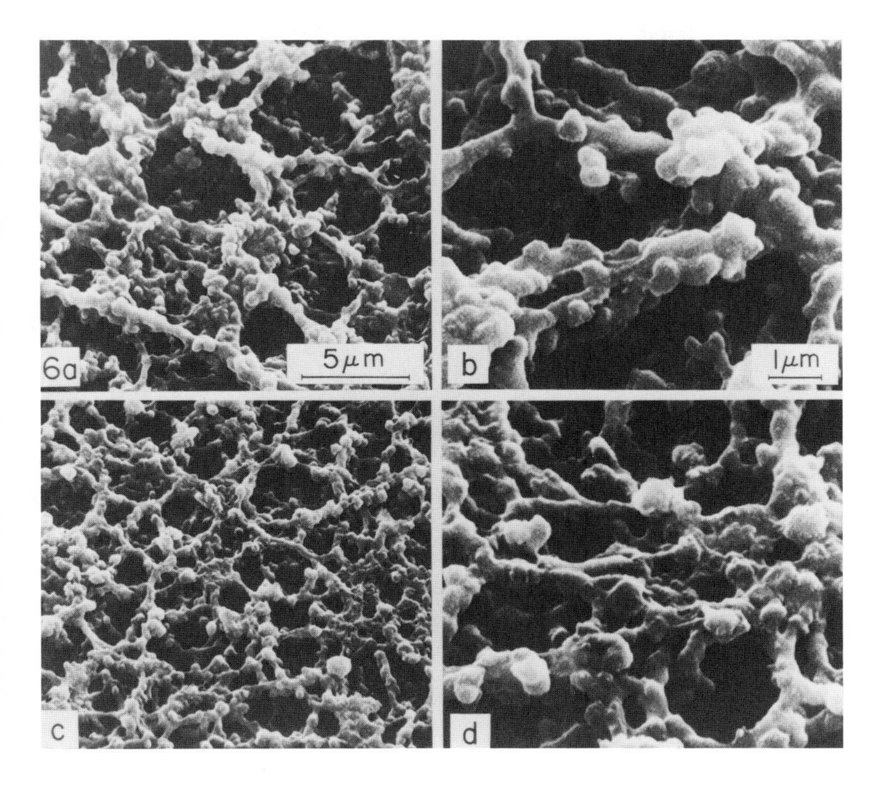

FIGURE 6 Scanning electron micrographs of the upper gridded (a, b) and the reverse, nongridded (c, d) surface of Johns-Manville SG membranes at low (a, c) and high (b, d) magnifications.

distilled water. Most of the washed membranes were low in metals (Table 5) except the green Sartorius 13806, which contained 8.5 μg of copper and 1.0 μg of zinc. The Gelman membrane contained 3.3 μg of zinc and 2.5 μg of lead. Filters analyzed after filtering Lake Ontario or Hamilton Bay water (Table 5) showed that copper binding occurred in all three Millipore membranes, Johns-Manville AG, Oxoid, and Gelman. More copper binding occurred in the Johns-Manville SG from the Lake Ontario water than from the Hamilton Bay water. Neither nickel nor lead was significantly bound, nor was cadmium (not shown), which was less than 0.3 μg per membrane in every case. The nickel content of the washed filters did not originate from the

TABLE 5 Analysis of the Trace Metal Content of Membrane Filters by Atomic Absorption Spectrometry

Membrane	Metal content (μg/filter)[a]											
	Copper			Zinc			Nickel			Lead		
	DW[b]	LO	HB	DW	LO	HB	DW	LO	HB	DW	LO	HB
Millipore HC	0.3	0.5	1.0	0.3	0.3	0.5	1.0	0.5	<0.3	2.0	<1.3	<1.3
Millipore AO	1.3	1.5	2.0	0.3	0.3	1.3	2.8	1.0	<0.3	<1.3	<1.3	<1.3
Millipore SO	0.5	0.8	1.0	0.3	0.3	4.3	7.0	2.0	0.5	<1.3	<1.3	<1.3
Sartorius 13806	8.5	7.8	8.5	1.0	0.5	0.3	1.0	0.5	<0.3	1.3	<1.3	1.8
Sartorius 11406	<0.3	0.3	0.3	1.5	1.5	1.8	0.5	4.0	<0.3	<1.3	<1.3	1.3
Johns-Manville AG	0.8	1.0	1.8	0.5	0.8	0.8	0.3	0.5	<0.3	<1.3	<1.3	<1.3
Johns-Manville SG	1.0	8.5	1.8	1.0	1.3	1.3	1.5	0.5	<0.3	<1.3	<1.3	<1.3
Oxoid	<0.3	0.3	1.3	0.5	0.5	0.8	0.5	0.3	0.5	<1.3	<1.3	<1.3
Gelman	0.3	0.5	1.0	3.3	3.3	3.5	1.0	0.5	<0.3	2.5	<1.3	1.8

[a]Filters were analyzed after a 100 ml of deionized, distilled water wash (DW) or after filtration of 100 ml of 0.2 μm filter-sterilized Lake Ontario water (LO) or Hamilton Bay water (HB).
[b]Metal content (in μg/liter) of water used for wash: Cu, 2.0; Zn, 5.0; Ni, 2.0; Pb, <1.0.

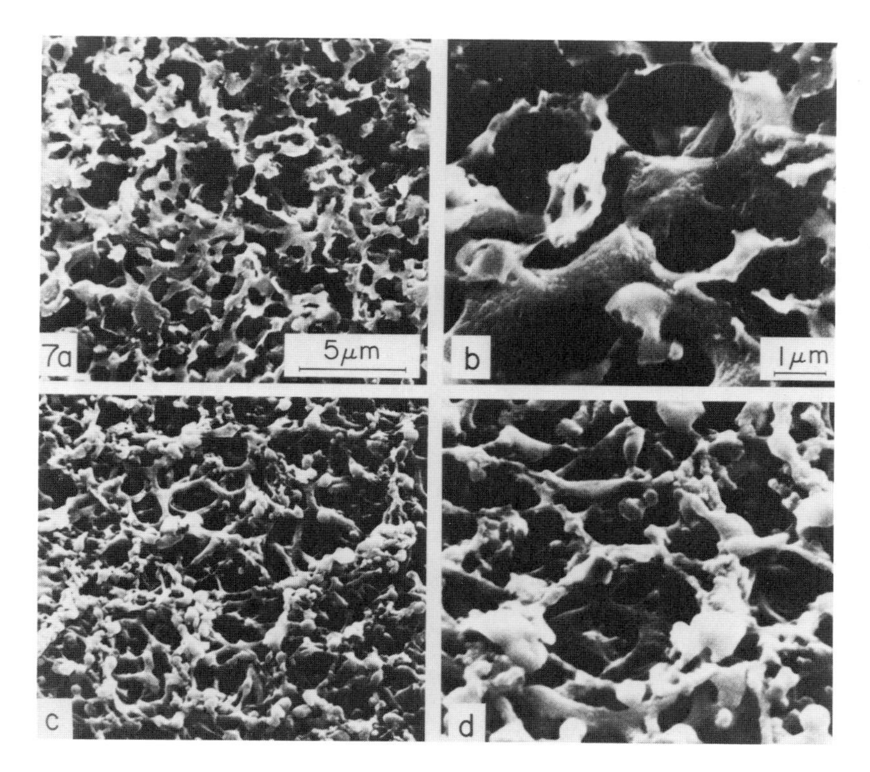

FIGURE 7 Scanning electron micrographs of the upper gridded (a, b) and the reverse, nongridded (c, d) surface of Oxoid Nuflow membranes at low (a, c) and high (b, d) magnifications.

deionized distilled water, which contained 0.2 μg/100 ml. The lower nickel values after filtration of environmental samples may reflect the leaching out of endogenous nickel by the higher-ionic-strength water.

IV. DISCUSSION

It is readily apparent from this study that there are significant differences among brands of membrane filters and also within brands. These differences were found in all aspects examined: resuscitation ability, flow rates, type of edge of filtration area, pore structure, surfactant content, and heavy metal binding.

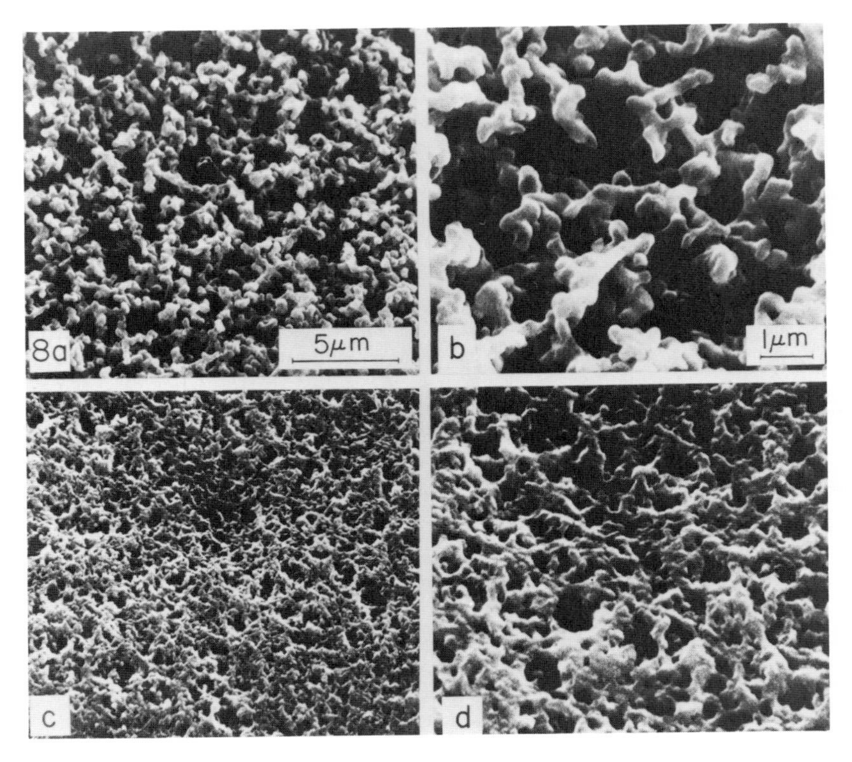

FIGURE 8 Scanning electron micrographs of the upper gridded (a, b) and the reverse, nongridded (c, d) surface of Gelman GN-6 membranes at low (a, c) and high (b, d) magnifications.

It was found that where the suspending medium (i.e., chlorinated tap water) produced a strong stress on the organism, differences in recovery rates of the membranes tended to diminish, with only the Millipore A0 and S0 membranes producing significantly less colonies than the other membranes (Table 2). This "equalizing" effect was observed by Lin (1976) when comparing fecal coliform recoveries from chlorinated sewage effluents with Millipore HC and HA membranes. To improve recovery of chlorine-stressed microorganisms, a two-layer medium has been advocated (Rose et al., 1977), as has the use of a 35°C preincubation period. Under less stressful conditions, greater

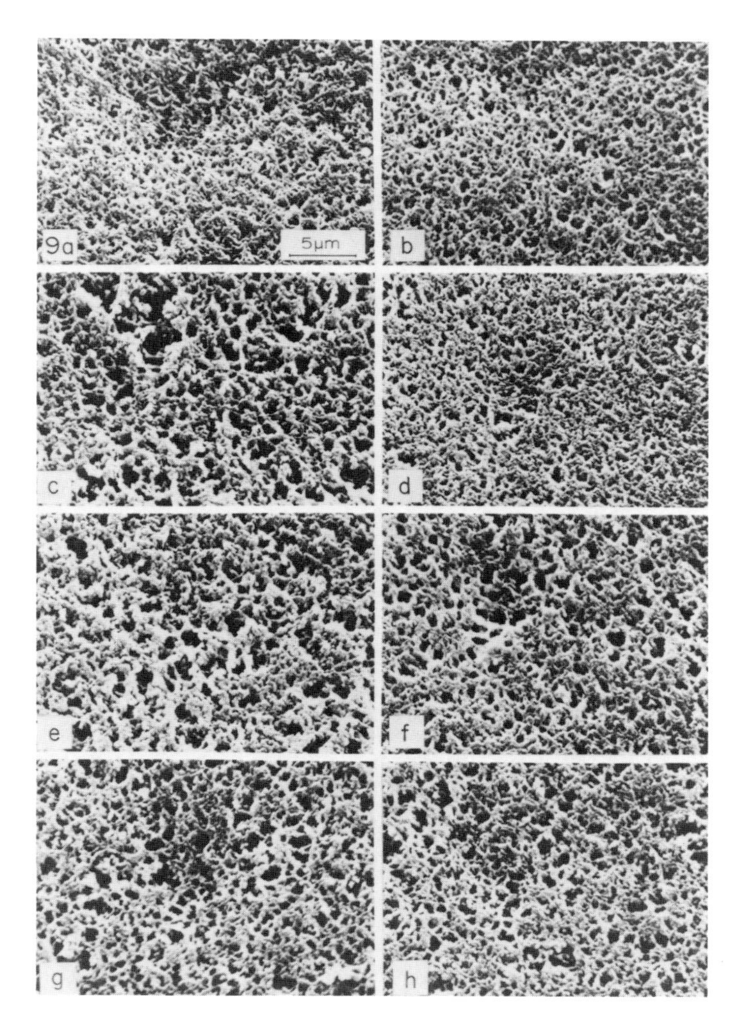

FIGURE 9 The underside (nongridded side) of several 0.45 μm Gelman membranes was examined to determine the consistency in pore structure. The membranes and their lot numbers are: (a) 47 mm, gridded, Lot No. 80951; (b) same box as (a), another membrane; (c) 47 mm, gridded, Lot No. 81173; (d) 13 mm, nongridded, Lot No. 81031; (e) same box as (d), another membrane; (f) 47 mm, nongridded, Lot 81281; (g) 25 mm, nongridded, Lot 80689; (h) same box as (g), another membrane.

variations in recovery were noted, with the most productive membranes, Millipore HC and Gelman, exceeding the least productive membranes by four to eight times.

Although most membranes recovered similar numbers of organisms when used grid side up or grid side down, some were found to produce significantly more colonies when used conventionally (e.g., Sartorius 13806, Johns-Manville SG, and Gelman), and the Oxoid tended to produce more colonies when used grid side down.

It has been postulated that differences in bacterial recoveries on membrane filters depend upon the surface pore morphology (Sladek et al., 1975). An optimum surface structure, it is hypothesized, would allow most bacterial cells to be cradled below the surface and submerged in nutrient solution maintained by capillary action. Suboptimal depth could lead to desiccation or nutrient starvation, or in the event of evaporation and concentration of nutrients, plasmolysis and death. In the experiments on bacterial recovery from three different water sources, the five membrane filters with the best recoveries (Millipore HC, Gelman, Sartorius 13806, and Johns-Manville SG and AG) are those with the largest pore size openings. When the membranes were used on their ungridded side, the three membranes with the best recoveries (Millipore HC, Johns-Manville AG and SG) were also those with the largest pores (although Johns-Manville AG had an unusual structure and was not listed in the ungridded-side surface pore rankings). It can also be seen from Table 2 that membranes with the greatest recoveries and largest pore size (Figs. 1 to 8) were usually those with the fastest flow rates. Thus, the hypothesis that flow rates and bacterial recovery are related has been strengthened by the results of this study. Furthermore, it can be seen that flow rates are directly related to pore structure; the larger the pores, the faster the flow rates.

When surface pore size was related to bacterial recoveries (regardless of filter side used) in either Lake Ontario or Hamilton Bay waters, optimal recovery was invariably obtained from those membranes that had a maximum pore opening of about 3.2 μm (Fig. 10). Several

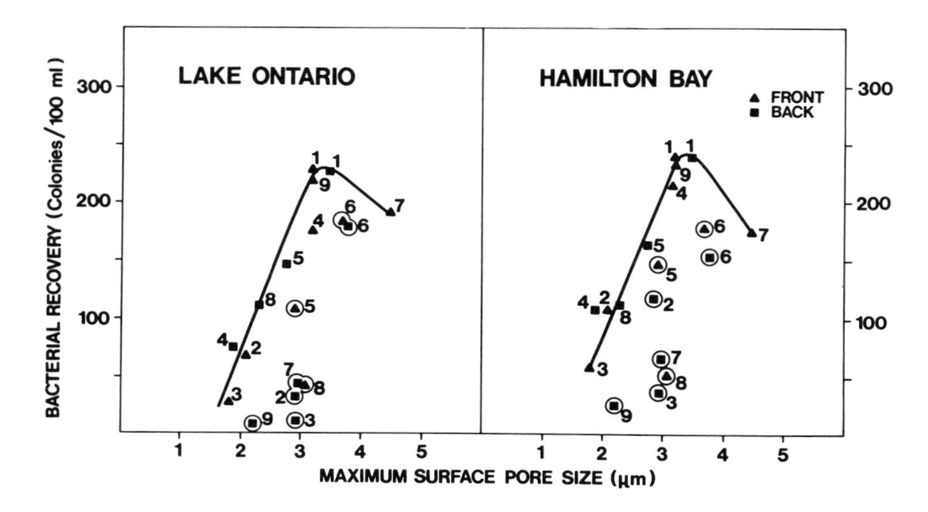

FIGURE 10 Fecal coliform recoveries as related to membrane maximum surface pore size. Circled symbols indicate that the membrane does not have the "ideal" configuration of relatively wide surface pores with finer mesh of pores below for a funneling effect. The numbers indicate the corresponding membranes: 1, Millipore HC; 2, Millipore AO; 3, Millipore SO; 4, Sartorius 13806; 5, Sartorius 11406; 6, Johns-Manville AG; 7, Johns-Manville SG; 8, Oxoid; 9, Gelman GN-6.

of the membrane surfaces (grid up: Johns-Manville AG, Sartorius 11406, Oxoid; grid down: Johns-Manville SG, Oxoid, Millipore AO and SO) recovered fewer bacteria than was predicted by this relationship. In each case, it was evident that the surface pores were more shallow or had less fine subjacent support than those fitting the relationship. Thus, factors such as depth, which is difficult to determine quantitatively, must be considered when predicting recoveries based on membrane morphology.

These studies indicate that optimal bacterial recoveries are obtained on those membranes that have a maximum surface pore size of about 3.2 μm and have relatively deep pores with a finer mesh below. The membranes that were found to have this structure were Millipore HC (both sides), Gelman (grid side), Sartorius 11406 (ungridded side), and to a lesser extent, Sartorius 13806 (grid side). Considering only the use of grid sides, the choice of an appropriate membrane from those tested may be limited to Millipore HC or Gelman, with Sartorius 13806 or Johns-Manville SG or AG as second choice. Of these, the flow rate of the Millipore HC is over twice as fast as any of the others and would presumably be most resistant to clogging and perhaps metal binding.

The binding of heavy metals to membrane filters may be an important consideration where they are present in large quantities. Because of potential cytotoxic or other effects, membranes that do not appreciably bind metals may be desired. Of the five membranes with maximal bacterial recoveries (Table 2a), it may be important to consider the relatively high copper content of the Sartorius 13806 or the zinc content of the Gelman membrane. These are constant for all types of water filtered, however, and do not appear to hinder population estimation. The Johns-Manville SG membranes seemed to bind variable amounts of copper from different types of water, but the remainder of these membranes displayed relatively minor metal sorption. This metal binding may be insignificant except in waters containing extremely high metal concentrations.

Wetting agents are used in the manufacturing process and may be removed to some extent before packaging. The wetting of the edge of the membranes under the filter lip was not correlated with surfactant content, indicating that this edge wetting is due to different lattice structures, facilitating lateral transfer. No bacteria were found growing in this area of the membrane, so loss of bacteria past the filter funnel lip is not considered likely. Wetting agents are required to assist in proper wetting of the membrane during use, but an excess may be undesirable. In no case was there any obvious problem with nonionic surfactants in this study, but in nonmicrobiological studies in which samples are being processed for nonionic surfactant determinations, the presence of significant surfactant concentrations must be avoided. For these purposes, membranes with low amounts of nonionic surfactants are preferable, such as Oxoid (Table 4). This particular brand of membrane has also the lowest capacity to bind nonionic surfactants from aqueous solution (Tobin, 1976) and thus is well suited to studies where nonionic surfactants are to be measured.

When sterilization procedures and bacterial recovery rates were compared using membranes grid up, in only one instance was a difference noted--Millipore AO producing a significantly higher count than Millipore SO. Comparing membranes used grid side down, Johns-Manville AG produced significantly greater numbers of colonies than did Johns-Manville SG on two occasions, as did Millipore AO over Millipore SO. These findings again tend to verify earlier reports that sterilization procedures do affect recovery rates. However, if one examines the electron micrographs (Figs. 1 to 8) of these membranes, it can be seen that most of the ethylene oxide-sterilized membranes have a different pore structure than autoclaved membranes. It would seem that some manufacturers make different membranes for different sterilization procedures or that there is a great deal of variation in the pore structure within a given brand.

Two relatively new membranes were tested in this study: Millipore HC and Sartorius 13806. Both were found to have improved bacterial recovery compared to the regular membranes issued by their respective manufacturers, an indication that some manufacturers recognize the need to further refine membrane filters for particular purposes, such as the fecal coliform procedure.

The data presented in this study underscore the complexity of the task of selecting a membrane filter for bacterial enumeration studies. They emphasize the need for membrane standardization for bacterial enumeration studies and the variable results that may be obtained depending on the membrane filter used.

ACKNOWLEDGMENTS

We thank Scott Kuchma for technical assistance, Tom Bistricki for advice and guidance with the electron microscopy, and Doug Sturtevant and John Gamble of Water Quality Laboratories for metal analyses.

REFERENCES

Anthony, D. H. J., and Tobin, R. S. (1977). Immiscible solvent extraction scheme for biodegradation testing of polyethoxylate nonionic surfactants. *Anal. Chem. 49*: 398–401.

Brodsky, M. H., and Schiemann, D. A. (1975). Influence of coliform source on evaluation of membrane filters. *Appl. Microbiol. 30*: 727–730.

Dutka, B. J., Jackson, M. J., and Bell, J. B. (1974). Comparison of autoclave and ethylene oxide-sterilized membrane filters used in water quality studies. *Appl. Microbiol. 28*: 474–480.

Green, B. L., Clausen, E., and Litsky, W. (1975). Comparison of the new Millipore HC with conventional membrane filters for the enumeration of fecal coliform bacteria. *Appl. Microbiol. 30*: 697–699.

Harris, F. L. (1974). Coliform recoveries on membrane filters, p. 4. In *Analytical Quality Control Laboratory Newsletter*, J. B. Anderson (Ed.), July. U.S. Environmental Protection Agency, Washington, D.C.

Hufham, J. B. (1974). Evaluating the membrane fecal coliform test by using *Escherichia coli* as the indicator organism. *Appl. Microbiol. 27*: 771–776.

Lin, S. D. (1976). Evaluation of Millipore HA and HC membrane filters for the enumeration of indicator bacteria. *Appl. Environ. Microbiol. 32*: 300–302.

Presswood, W. C., and Brown, L. R. (1973). Comparison of Gelman and Millipore membrane filters for enumerating fecal coliform bacteria. *Appl. Microbiol. 26*: 332–336.

Rose, R. E., Geldreich, E. F., and Litsky, W. L. (1977). A layered membrane filter medium for improved recovery of stressed fecal coliforms. In *Proceedings of the Symposium on the Recovery of Indicator Organisms Employing Membrane Filters*, R. H. Bordner, C. F. Frith, and J. A. Winter (Eds.). U.S. Environmental Protection Agency, Cincinnati, Ohio, pp. 101–104.

Schaeffer, D. J., Long, M. C., and Janardan, K. G. (1974). Statistical analysis of the recovery of coliform organisms on Gelman and Millipore membrane filters. *Appl. Microbiol. 28*: 605–607.

Sladek, K. J., Suslavich, R. V., Sohn, B. I., and Dawson, F. W. (1975). Optimum membrane structures for growth of coliform and fecal coliform organisms. *Appl. Microbiol. 30*: 685–691.

Sladek, K. J., Suslavich, R. V., Sohn, B. I., and Dawson, F. W. (1977). Optimum membrane structures for growth of fecal coliform organisms. In *Proceedings of the Symposium on the Recovery of Indicator Organisms Employing Membrane Filters*, R. H. Bordner, C. F. Frith, and J. A. Winter (Eds.). U.S. Environmental Protection Agency, Cincinnati, Ohio, pp. 46–53.

Standridge, J. E. (1976). Comparison of surface pore morphology of two brands of membrane filters. *Appl. Environ. Microbiol. 31*: 316–319.

Tobin, R. S. (1976). Unpublished data

Tobin, R. S., and Dutka, B. J. (1977). Comparison of the surface structure, metal binding, and fecal coliform recoveries of nine membrane filters. *Appl. Environ. Microbiol. 34*: 69–79.

18

IN SITU STUDIES USING MEMBRANE FILTER CHAMBERS

*GORDON A. McFETERS and DAVID G. STUART** *Montana State University, Bozeman, Montana*

I. DEVELOPMENT OF MEMBRANE FILTER CHAMBERS

It has long been realized that the relationship between data obtained in the laboratory and what is actually occurring in the field is tenuous, highly questionable, and often misleading when various parameters associated with the actual survival of microorganisms in natural aquatic environments are being studied.

Developments in equipment, methods, and techniques to deal with this problem by studying the behavior of microorganisms under natural conditions in the field were perfected around the turn of the century with the work of Jordan et al. (1904) using sacs of permeable membrane material to expose bacterial populations to natural waters.

By 1935, cells fabricated of brass were being fitted with permeable-membrane sidewalls to be used for this purpose (Beard and Meadowcraft, 1935). The brass was heavily coated with paraffin to prevent oligodynamic phenomena, and rubber gaskets were used to seal the membranes to the brass cells to prevent the loss of organisms.

**Present affiliation:* The Baker Company, Inc., Sanford, Maine.

Techniques and methods for preparing populations of organisms to be used, exposing them to the natural environment, sampling exposed populations, and enumerating the survivors developed by these early workers have been used and modified by others over the years as discussed by Schultz and Gerhardt (1969) and Vasconcelos and Swartz (1976).

The need for a diffusion chamber that could be used in situ was generated in our laboratory during a study of the bacteriological water quality of two adjacent mountain watersheds, one open and the other closed to public entry (Stuart et al., 1971). Total coliform counts in water from the closed watershed were found to be up to five times those measured in the open area. Although these data could be explained largely by a larger concentration of big game animals, resulting in more contamination within the closed watershed, there was some question as to whether coliform bacteria might be growing, or surviving longer, in the waters of the closed area. This hypothesis was supported by water chemistry data which indicated that such parameters as conductivity, alkalinity, nitrate, and phosphate were also higher in the waters of the closed watershed.

To test this hypothesis in situ, there was need for a chamber or cage that could be used to hold a captive population of bacteria in the streams while exposing those bacteria to the ambient conditions of that aquatic environment. Although similar studies have been conducted utilizing filter paper (Jordan et al., 1904; Beard and Meadowcraft, 1935) and dialysis sac materials (Slanetz and Barley, 1965; Zobel, 1936; Metcalf and Stiles, 1967), it was deemed essential to have conditions inside the chamber reflect instream conditions as accurately and reliably as possible and to minimize the cryptic growth phenomena discussed below under applications. Theory, rationale, and techniques for this approach using membrane filter materials have been discussed by Schultz and Gerhardt (1969). Although this review dealt with dialysis culture of bacteria in terms of medical applications (i.e., inside the bodies of experimental animals), the concepts are applicable to aquatic studies.

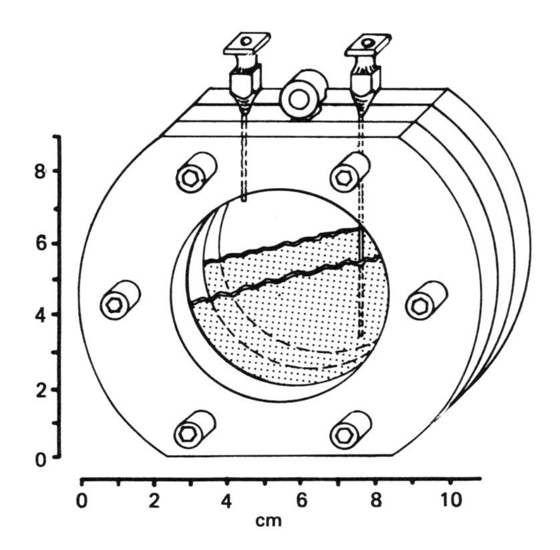

FIGURE 1 Schematic drawing of the MSU-VME membrane diffusion chamber. (From McFeters and Stuart, 1972.)

Therefore, a plexiglass cell with membrane filter sidewalls (Fig. 1) was designed (McFeters and Stuart, 1972). Dialysis tubing material which is permeable to large molecules was ruled out as the sidewall material to be used in this study.

The objective was to utilize a membrane with as large a pore size as possible, to maximize exchange between the environment and the inside of the chamber and still retain the bacteria being studied. Just as important to achieving this objective of maximum exchange was the design criterion of a high surface area-to-volume ratio. This consideration lead to the rather flat and narrow chamber with about 25 ml capacity (Fig. 2). Another major design criterion, ease of sampling, was satisfied by the installation of two syringe needles as sampling ports, with one of the needles having a plastic tube attached that reached the lower portion of the cavity (Fig. 1).

Upon fabrication and testing of the chamber (McFeters and Stuart, 1972), it was found that with a membrane surface area-to-volume ratio of 2.84 (HAWP304F0 membranes, Millipore Corp.),

FIGURE 2 Picture of the MSU-VME diffusion chamber. (From McFeters and Stuart, 1972.)

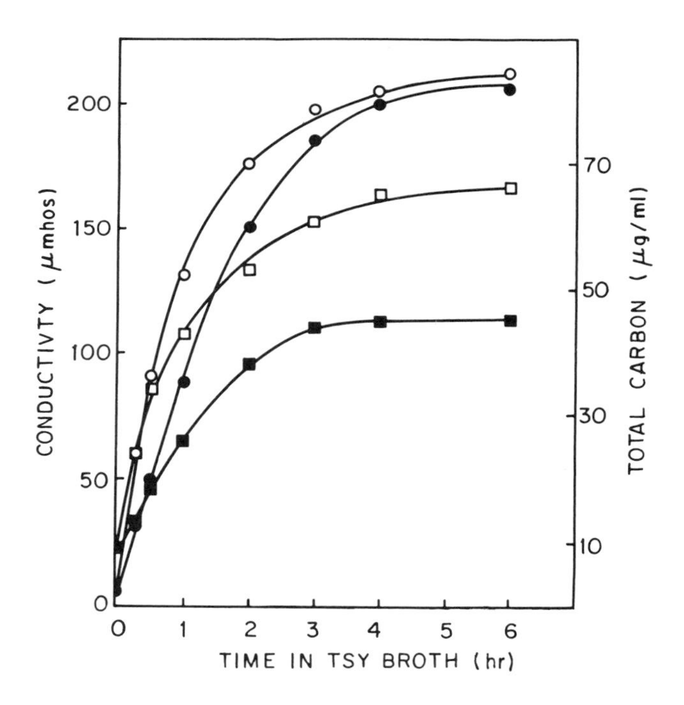

FIGURE 3 Comparative uptake of carbon-containing compounds (□,■) and ions responsible for conductivity (o,●) into a membrane dialysis chamber (o,□) and a dialysis tubing sac (●,■). The exclosures were filled with distilled water and submerged in natural water enriched with Dextran 500 and mixed continuously. At the indicated times, samples were withdrawn from inside each exclosure and analyzed for total carbon and conductivity. The water temperature was 23°C. (From McFeters and Stuart, 1972.)

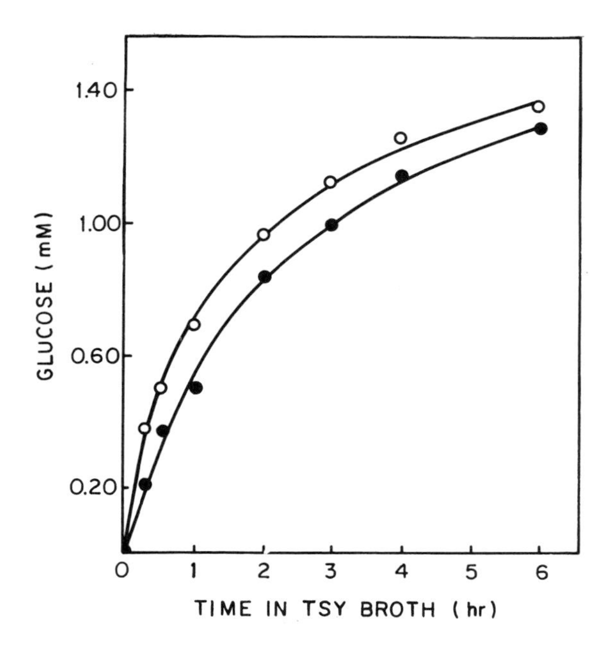

FIGURE 4 Comparative uptake of glucose into a membrane dialysis chamber (O) and a dialysis tubing sac (●). The exclosures were filled with distilled water and submerged in distilled water containing glucose with continuous mixing. Samples were removed from the exclosure and analyzed for glucose concentration at timed intervals. The water temperature was 22.5°C. (From McFeters and Stuart, 1972.)

diffusion of salts and dextran (Fig. 3) and glucose (Fig. 4) was more rapid in the chamber than in dialysis tubing sacs. It was also shown that whereas survival data obtained in the lab revealed no differences between the two streams under study, marked differences were observed in data obtained in the field (Fig. 5). Thus, it was demonstrated that the membrane filter diffusion chamber technique was sensitive enough to measure the difference in behavior of the same organism exposed to the aquatic environments of adjacent mountain streams. Further work showed that such differences can be measured when the same organism is exposed to different sites on the same stream (Fig. 6). Additionally, variations in behavior of different organisms exposed at the same stream site were measurable (Fig. 7).

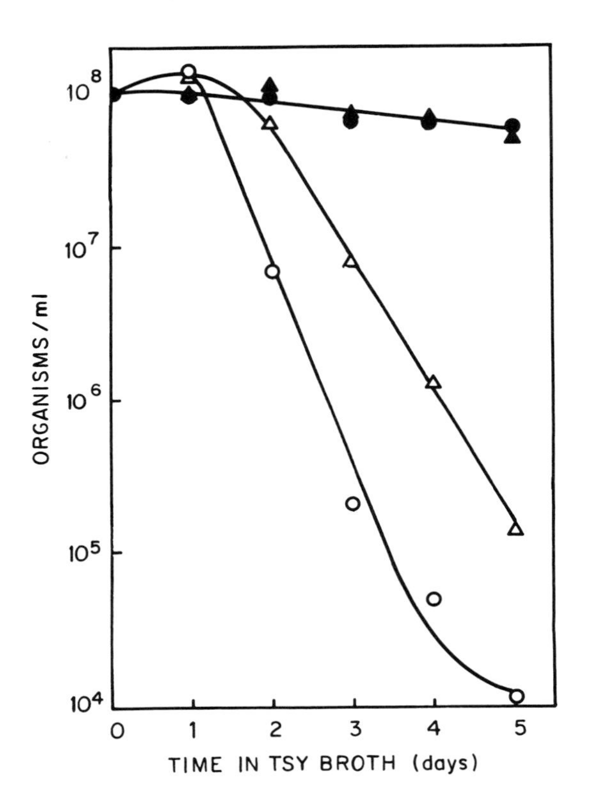

FIGURE 5 Survival of *E. coli* MH 3427 in situ and in the laboratory. Washed cells that were placed in membrane dialysis chambers were immersed in Bozeman Creek (△) and Middle Creek (O) and in the laboratory simulator in the water from Bozeman Creek (▲) and Middle Creek (●). Water samples were brought to the laboratory daily to be used in the simulator. Samples were removed from the chambers daily for bacterial enumeration. (From McFeters and Stuart, 1972.)

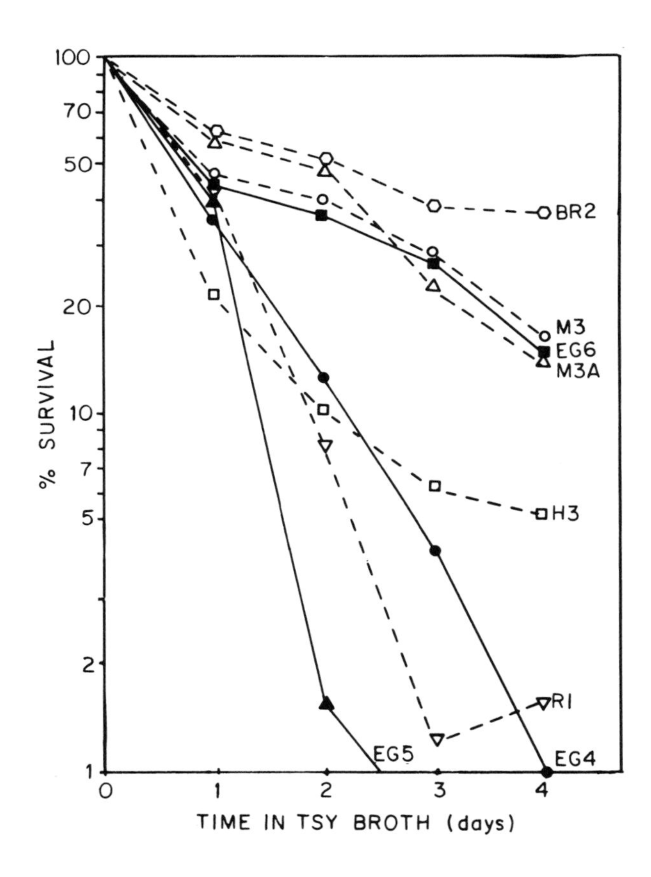

FIGURE 6 Comparative survival of *E. coli* C320MP25 in membrane filter chambers that were immersed in various streams within the Gallatin valley, Montana.

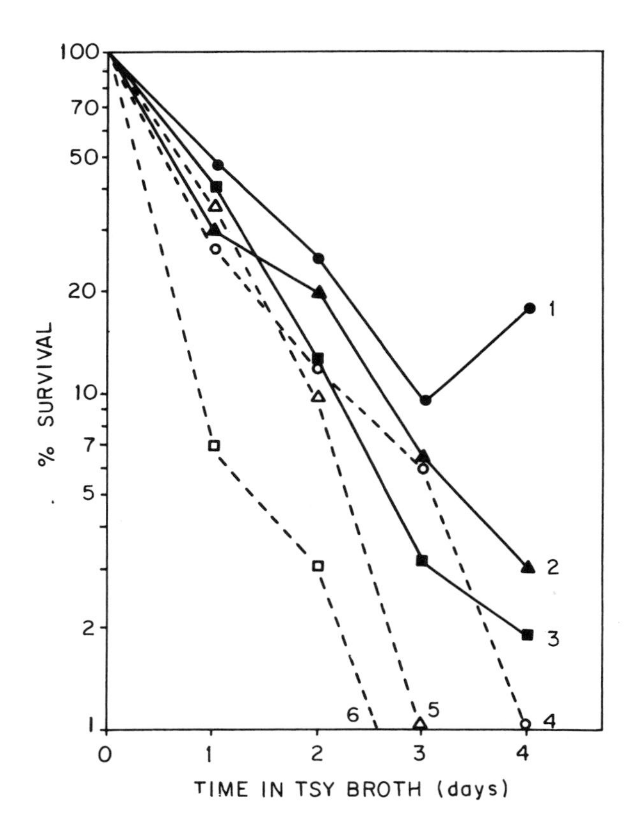

FIGURE 7 Comparative survival of *K. pneumoniae* 1881 (1, ●), *E. aerogenes* EC2072 (2, ▲), *E. coli* C320MP25 (3, ■), *S. faecium* BS895 (4, o), *S. faecalis* RS1009 (5, △), and *S. bovis* C35E17 (6, □) in membrane filter chambers during a 4 day exposure period located at site EG6. Calculations of survival were determined by counts obtained on TSY pour plates.

II. USE OF MEMBRANE FILTER DIFFUSION CHAMBERS

A. Chambers

Membrane filter chambers can be fabricated to the researcher's specifications at a university or local machine shop. The MSU-VME chamber may be purchased from Valley Machine and Engineering, Rt. 2, Box 486, Belgarde, Montana 59714 (unpolished: 25 ml, $40; 100 ml, $50, and unpolished: 20 ml, $45; 100 ml, $55). The size of the chamber will be determined by the intended application.

Although the original MSU-VME chambers were fabricated out of preshrunk plexiglass, some warpage problems were encountered upon repeated autoclaving. Since encountering these difficulties, the chambers were made of polycarbonate for awhile, but the material currently being used is Plex-II.

The larger chambers require more bolts to provide adequate sealing pressure.

B. Membrane Material

Membrane material is available from several manufacturers in sheets ranging in size from 8 to 12 in. on a side. A variety of materials and pore sizes are available:

Millipore Corporation, Ashby Road, Bedford, Massachusetts 01730.
Membranes of mixed cellulose acetate and nitrate, nylon, Teflon, polytetrafluoroethylene, and polyvinyl chloride are available with varying pore sizes in each. Pore sizes in the mixed cellulose acetate and nitrate filters run from 0.025 to 8.0 μm.

Nuclepore Corporation, 7035 Commerce Circle, Pleasanton, California 94566
Nuclepore filters are made of a polycarbonate film and are available in pore sizes from 0.015 to 12 μm.

Gelman Instrument Company, 600 South Wagner Road, Ann Arbor, Michigan 48106
Gelman provides filters made from cellulose triacetate, mixed cellulose esters, cellulose acetate, regenerated cellulose, polyvinyl chloride, aromatic polymer (HT), and so on. Pore sizes range from 0.2 to 5.0 μm. They also provide filter material with a nylon web support.

The specific material to be ordered depends on many variables concerning the proposed application. Some examples are:

1. Pore size. This will be determined by such criteria as size of organism being studied and quality of exposure water.
2. Strength of the membrane material. Applications involving larger chambers and rough water require stronger membranes. Nuclepore membranes are tough; Gelman provides supported membranes to solve this problem.
3. Transparency. Use of chambers with photosynthetic organisms can be facilitated by using polished chambers and Nuclepore membranes, which are transparent in the large pore sizes.
4. Biodegradability and refractiveness. Some environments are hard on membranes; for example, hot-springs environments can degrade certain kinds of membrane material quite rapidly. A word of caution is appropriate also for saltwater applications of cellulose-based membranes. Studies have shown that problems can arise in as little as 5 days of exposure.

C. Sterilization

The cut membrane disks may be sterilized with ultraviolet light for 10 min on each side. The plexiglass chambers withstand autoclave conditions for 10 min but may warp if longer times are used. Plastic caps that fit the needles may be autoclaved. After sterilizing, the chamber parts and membranes are assembled aseptically using flamed forceps to handle the membranes that were hydrated and washed with sterile dilution buffer. The retainer bolts are secured with a small wrench until firm.

The cell suspensions are placed in the chambers with large sterile syringes. The needle with the tubing attached should be used for filling and sample withdrawal, and should be marked so that when the membrane sidewalls are in place it can be identified. When samples are withdrawn with a sterile 1 ml syringe, the syringe is pumped 20 times to resuspend cells that may have settled to the bottom of the chamber. Care should be taken not to break the tip of glass syringes in the needle hubs.

III. APPLICATIONS AND MODIFICATIONS OF MEMBRANE DIFFUSION CHAMBERS

A. Applications

General Microbiological Considerations

The applications of membrane diffusion chambers to the study of aquatic microbiological questions of population dynamics should be carried out within the confines of certain microbiological constraints. This is particularly the case if the experimental results are to be related to the unique environmental conditions of some particular aquatic system under investigation. Starting with this approach, the worker can design the experimental protocol to satisfy the specific criteria that are deemed important and avoid some of the problems associated with the use of diffusion chambers that can contribute to artifactual results.

The use of excessive population densities of microorganisms in membrane diffusion chambers is the most frequently observed factor that can lead to unrealistic results. Postgate and Hunter (1962) studied the effect of population density on the survival of heterotrophic bacteria under batch starvation conditions within flasks and concluded that the death rate became progressively slower when the bacterial population densities ranged from 10^5 to 10^8 cells per milliliter. In some instances, limited growth was even observed following the initial decline in viable bacteria. This phenomenon is termed cryptic growth and it has been established that the death of 50 *Enterobacter aerogenes* organisms will support one call division of that bacterium (Postgate and Hunter, 1962). On the other hand, chamber experiments are less susceptible to this process if the porosity of the membrane that is used in the chamber is properly selected because much of the debris and potential nutrients liberated from the death and lysis of cells will diffuse into the water surrounding the chamber.

This property represents one of the most compelling reasons for using chambers with porous membrane walls rather than enclosed flasks when conducting studies of bacterial population dynamics in aquatic environments. The importance of this consideration in the design of meaningful survival experiments seems apparent regardless of the methodology employed, but many previously published reports ignore this point and none has given a maximum population density range that can be used to avoid this problem. An upper limit of 10^6 bacteria per milliliter is currently employed in our laboratories for most studies. However, lower densities are preferable if they can be accommodated within the limits dictated by the experimental design. If, on the other hand, greater numbers of microorganisms are needed or they are present in the environmental situation to which the experiment is addressed, higher populations of cells may be used with the knowledge that biological colloids such as bacterial suspensions provide a protective influence that is proportional to the density of the mixture (Postgate and Hunter, 1962). In any event, the investigator should be aware that the population density of organisms used does influence behavioral characteristics such as the survival of bacteria when planning and interpreting data obtained from experiments where chambers are employed.

Another microbiological consideration of importance relates to the cultural history of heterotrophic bacteria destined for chamber experiments. Bacteria grown in more complete media are capable of greater survival than are cells from the same stock culture that are grown on a minimal medium (Strange et al., 1961). Additionally, the longevity of *E. aerogenes* under conditions of starvation has been shown to be greater when harvested from the late logarithmic growth phase rather than the midstationary phase of growth in a defined medium. More complete media also yield bacteria that are more able to survive starvation. These findings suggest that, at least in *E. aerogenes*, the potential to withstand environments in which the concentration of organic nutrients is very dilute is directly related to the endogenous macromolecular reserves that accumulate during protracted growth under optimal conditions. Therefore, in the design and interpretation of chamber experiments, particularly those in which the survival of heterotrophic bacteria is followed under minimal nutrient conditions, the investigator should be cognizant that the bacterial response is influenced by previous cultural conditions.

Predation is another factor that may be of importance in performing chamber experiments to examine the dynamics of microbial populations. Bacterial predators such as protozoans and bdellovibrios may or may not pass through the membrane, depending upon the type of membrane that is used, whereas bacteriophage almost certainly can diffuse into the chambers. As a result, chamber experiments can be designed to examine the role of the larger predators in altering bacterial numbers by adding unsterile natural water or sediments to control chambers. Comparable experiments to examine the role of bacteriophage in bacterial persistance are more difficult to devise using chambers.

From the preceding it can be seen that factors such as population density, cultural manipulations, and proper controls are critical in planning studies where suspensions of bacteria are examined following exposure in aqueous environments in chambers. These considerations are important if one wishes to relate the experimental results to some aquatic system of interest. As described earlier in this chapter, the rationale that is basic to the use of membrane diffusion chambers in aquatic microbiological research is an improved degree of interaction between natural aquatic ecosystems and the organisms under investigation. It would truly be an ironic misfortune if microbiologists in their application of this methodology fail to take into consideration a few relatively simple, but important microbiological principles.

Studies of Bacterial Population Dynamics by the Use of the Membrane Diffusion Chamber

In this section we give an overview of a limited number of bacterial survival studies using the chambers. These applications are representative of experiments designed to investigate bacterial population dynamics in natural waters that have been carried out in our laboratories in the past few years. In so doing, we hope to demonstrate the utility of chamber methodology in such experiments.

Early Survival Studies Carried Out in Two Adjacent Mountain Streams. The first application of membrane diffusion chambers in our laboratories entailed a comparative survival study of a water-isolated *Escherichia coli* strain in two adjacent streams (McFeters and Stuart, 1972). Figure 5 illustrates how the use of chambers in the stream yielded results that were quite different from those obtained in the laboratory, both from the standpoint of the slope of the survival curves and in distinguishing the influence of the streams on the survival of that bacterium. The excessive bacterial population that was used in that study might be pointed out as an experimental defect that was not fully appreciated at that time. However, the comparative data agreed qualitatively with the other supportive information obtained in that study.

Comparative Survival Studies of Indicator Bacteria and Enteric Pathogens in Water. Our purpose in this series of experiments was to compare the persistence of bacterial pathogens to various indicator bacteria in natural water (McFeters et al., 1974). This necessitated the use of a reliable source of flowing unchlorinated water that could be used for our experiments and which would later be disinfected prior to discharge in case leakage of pathogens should occur. These requirements were satisfied by a well located at a sewage treatment plant where the water flowed into a tub wherein the chambers were immersed and then directly through the treatment plant. Table 1 gives representative survival data that were obtained using pure cultures of indicators and bacterial pathogens that may be transmitted by the waterborne route. Such experiments represent an important part of the information base that is required in epidemiological studies that are needed for the waterborne diseases. Further experiments examined the persistence of mixed natural populations of indicator organisms with exposure time in water using the chambers (Fig. 8). These results demonstrated the influence of aquatic exposure on the fecal coliform/fecal

TABLE 1 Comparative Die-Off Rates (Half-Time)[a] of Fecal Indicator Bacteria and Enteric Pathogens

Bacteria	Half-time (hr)	Number of strains analyzed
Indicator bacteria		
Coliform bacteria (average)	17.0	29
Enterococci	22.0	20
Coliform from raw sewage	17.5	
Streptococci from raw sewage	19.5	
Streptococcus equinus	10.0	1
S. bovis	4.3	1
Pathogenic bacteria		
Shigella dysenteriae	22.4[b]	1
S. Sonnei	24.5[b]	1
S. flexneri	26.8[b]	1
Salmonella enteritidis ser. paratyphi A	16.0[b]	1
S. enteritidis ser. paratyphi D	19.2[b]	1
S. enteritidis ser. typhimurium	16.0[b]	1
S. typhi	6.0	2
Vibrio cholerae	7.2	3
S. enteritidis ser. paratyphi B	2.4	1

[a]The half-time was determined graphically from Figs. 2 and 4 as the time required for a 50% reduction in the initial population.
[b]The half-time was determined graphically from Fig. 4 as the time required for a 50% reduction in the population at 24 hr.

streptococcus ratio (FC/FS) when bacteria from humans, cattle, and elk were used. As predicted, based on knowledge from the different indicator bacterial composition from these sources and the relative persistence of different indicator species obtained from pure culture survival experiments done earlier, it was shown that the FC/FS in the case of humans decreased and that of cattle increased, whereas the ratio from elk remained low and stable. Although these findings were not surprising, they did corroborate the predictions made earlier and as such could be related to the use of the FC/FS

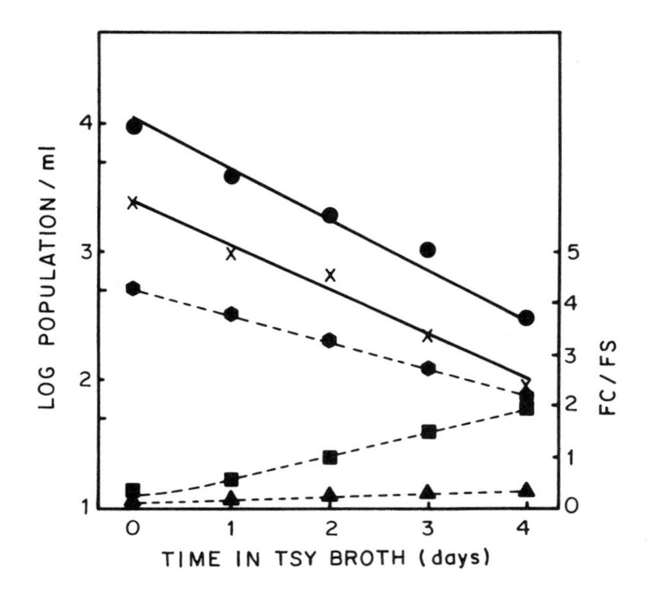

FIGURE 8 Survival of mixed natural populations of indicator bacteria. Bacterial suspensions were placed in membrane chambers that were immersed in well water. Fecal coliform bacteria (●) and fecal streptococci (X) from raw sewage (solid lines) were enumerated daily. The resulting FC/FS (dashed lines) was determined for raw sewage (●) as a human source and from bovine (■) and elk (▲) fecal material. (From McFeters et al., 1974.)

ratio in helping identify the origin of point source contamination in natural waters in the field.

Survival of Acidophilic Bacteria in an Alkaline Environment. In an effort to explain the biogeochemical role of acidophilic iron oxidizing bacteria in nonacid aquatic environments associated with western coal strip mines, survival experiments were carried out using chambers in alkaline mine ponds (Olson and McFeters, 1976). Cultures were loaded into chambers that were immersed in the pond and their persistence was followed with time. The results (Fig. 9) demonstrated that these bacteria did not survive well in that environment when the water temperature was warm and were, therefore, probably transients originating in an acidic microenvironment within the mine-water system.

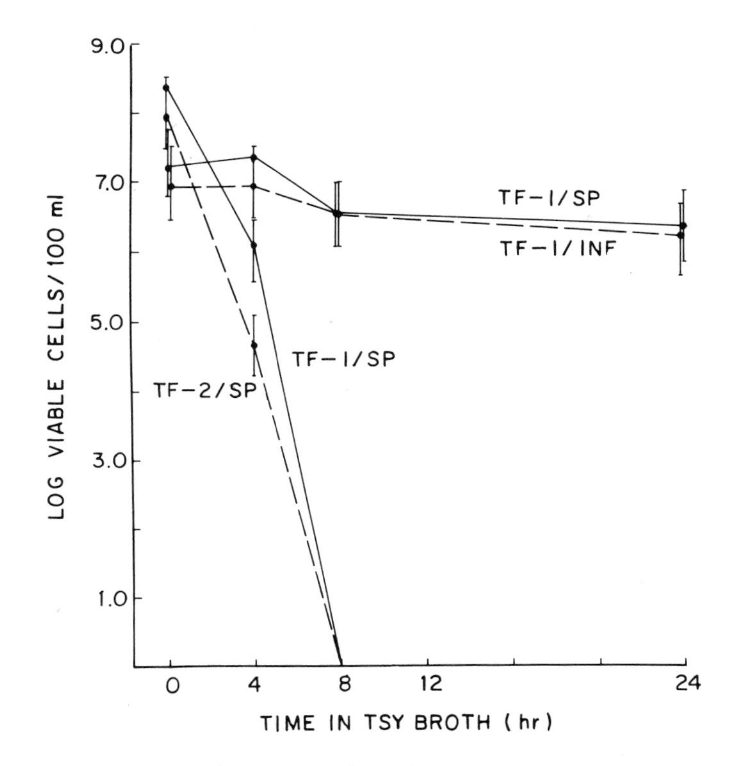

FIGURE 9 Survival of *Thiobacillus ferrooxidans* alkaline settling pond and influent waters using membrane chambers. Bars indicate the 95% confidence limits. The upper lines represent bacterial survival data obtained in October (water temperature 10°C) and the lower lines, an experiment performed in June (water temperature 25°C).

Investigations into the Persistence of Less Commonly Studied Waterborne Pathogens in Natural Waters. These experiments were initiated because information regarding the persistence of *Leptospira* spp. and *Yersinia enterocolitica* in natural aquatic ecosystems was nonexistent. This and other unanswered questions have contributed in large measure to the paucity of epidemiological information concerning diseases that are caused by these organisms.

Specially designed small chambers were constructed for the studies of leptospiral survival using membranes with 0.1 µm mean pore diameter. The results (Fig. 10) indicated that *Leptospira* spp. are capable of long survival in natural waters when compared with other pathogens or the currently employed indicator bacteria (Roberts, 1977).

Pure cultures of *Y. enterocolitica* were loaded into chambers and then placed in natural surface waters and flowing chlorinated tap water, along with a chamber containing a typical fecal coliform. In natural stream water these bacteria persisted much longer than

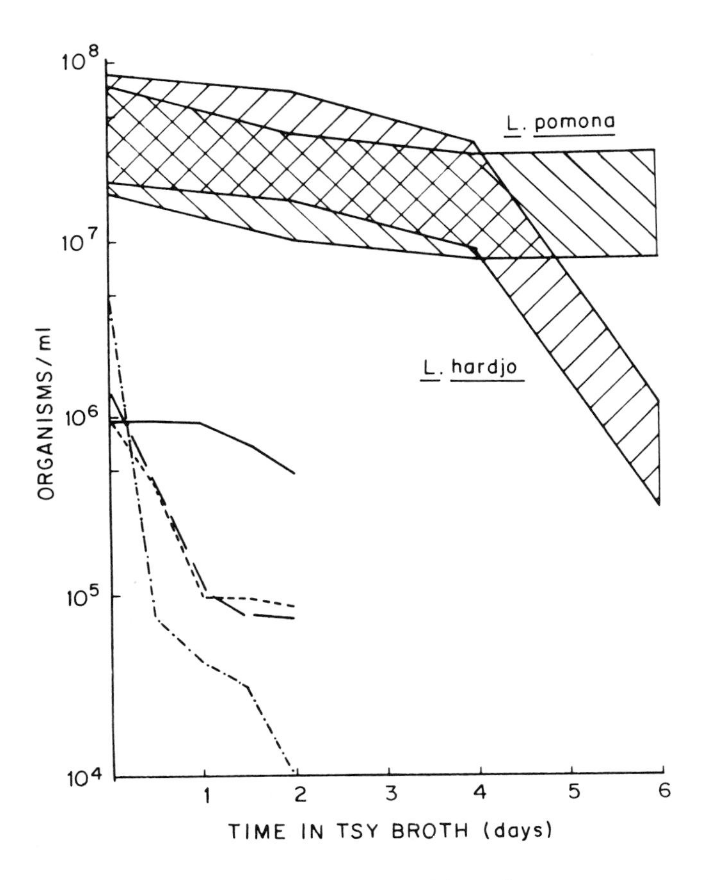

FIGURE 10 Comparison of *L. hardjo* and *L. pomona* survival to *Salmonella enteritidis* ser. paratyphi B (·-·-), *Vibrio cholerae* (— — —), *Salmonella enteritidis* ser. paratyphi D (———), and *Salmonella typhi* (-----) survival in continuous flowing well water. The *L. pomona* (\\\) and *L. hardjo* (///) areas represent the maximum and minimum viability range based on direct cell count and motile cell percentage of the direct count.

the indicator bacteria; but in the chlorinated tap water (residual chlorine, 1-2 mg/l) both organisms died at the same rate within 24 hr.

In Situ Studies of Bacterial Physiological Behavior Using Membrane Diffusion Chambers

Membrane diffusion chambers have been used to study the phsyiological injury of bacteria that has been described in natural waters. This work (Bissonnette et al., 1975, 1977) demonstrated that indicator bacteria such as *E. coli* and *Streptococcus faecalis* progres-

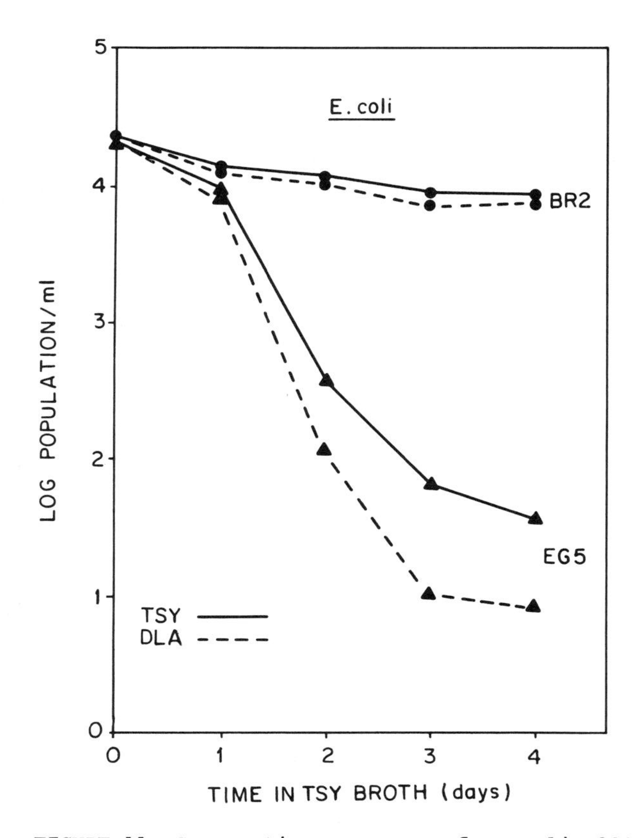

FIGURE 11 Comparative recovery of *E. coli* C320MP25 in membrane filter chambers located at sites BR2 (●) and EG5 (▲) over a 4 day exposure period. Samples were surface-overlay-plated using TSY (——) and DLA (- - - -) agar.

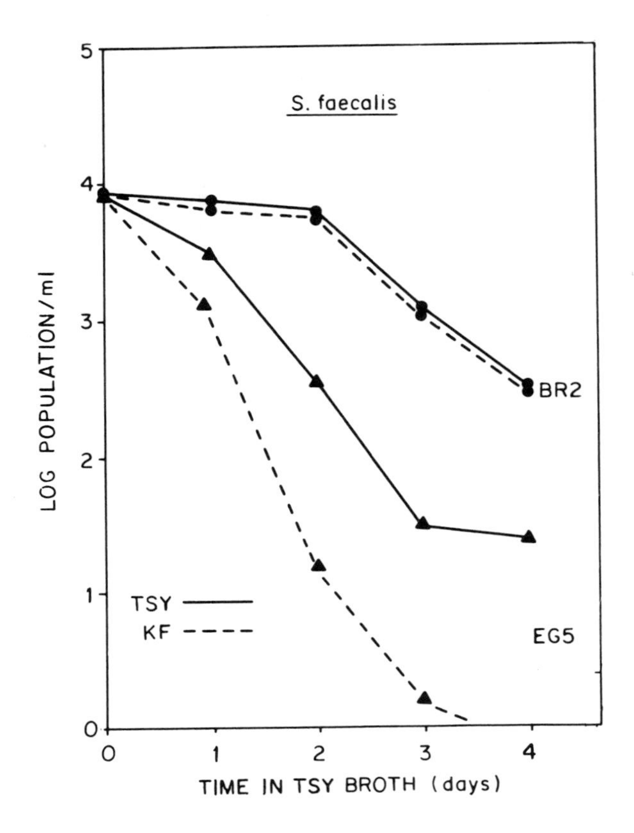

FIGURE 12 Comparative recovery of *S. faecalis* RS1009 in membrane filter chambers located at sites BR2 (●) and EG5 (▲) during a 4 day exposure period. Samples were pour-plated using TSY (——) and KF (- - - -) agar.

sively lose the ability to grow and form colonies on selective media as they age in some natural waters (Figs. 11 and 12) and that a rich nutrient environment and moderate temperatures allow the repair of such injury within 4 hr (Figs. 13 and 14). These studies indicated that this occurrence is widespread and that injured cells may comprise greater than 90% of the total population of indicator bacteria. The results of these in situ studies, which were made possible through the use of the membrane diffusion chambers, have been related to natural waterborne populations of indicator bacteria and problems associated with the enumeration of these injured cells.

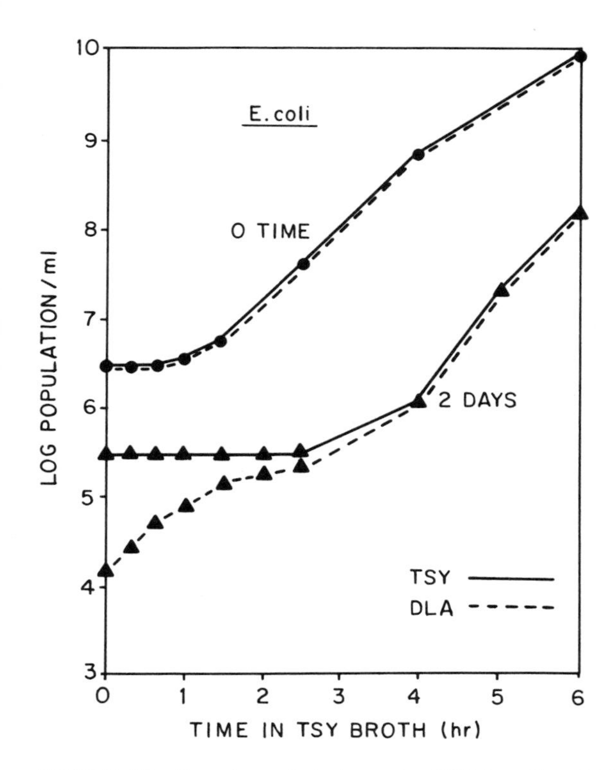

FIGURE 13 Repair of injury in TSY broth of *E. coli* C320MP25 cells having been exposed to the stream environment for 2 days. Control, or 0 hr cells (●) and 2 day exposed cells (▲) were enumerated over a 6 hr growth period in TSY broth with TSY (——) and DLA (- - - -) surface overlay plates. (From Bissonnette et al., 1975.)

As a result, new methods have been proposed to more accurately and completely count debilitated cells from natural and chlorinated waters (Stuart et al., 1977; Green et al., 1977). However, chambers fitted with standard membranes (Millipore Corp.) did not work well in chlorinated waters with less than 0.5 mg/liter of residual chlorine, probably because the low concentration of chlorine reacted with the membrane and, as a result, did not freely diffuse into the chamber. These findings indicate that experiments where chambers loaded with microorganisms are suspended in solutions of highly reactive solutions as well as particulate or colloidal suspensions should be approached with great caution.

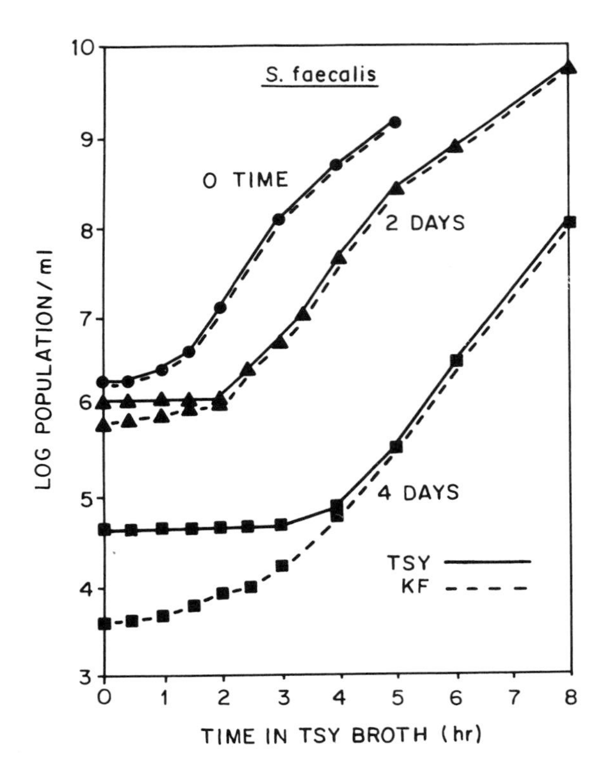

FIGURE 14 Repair of injury in TSY broth for *S. faecalis* RS1009 cells having been exposed to the stream environment of site EG6 for 2 and 4 days. Control or 0 hr cells (●), 2 day exposed cells (▲), and 4 day exposed cells (■) were enumerated over an 8 hr growth period in TSY broth using TSY (——) and KF (- - - -) media with membrane filtration procedures.

Viral Studies Using Membrane Diffusion Chambers

Membrane diffusion chambers have been employed in studies of viral inactivation in surface waters. That in situ experiments of this type have an advantage over similar work done in the laboratory was demonstrated before the development of chambers by Metcalf and Stiles (1967) using dialysis sacs. Because of this, O'Brien and Newman (1977) used membrane diffusion chambers fitted with sidewalls of dialysis tubing material to follow the in situ persistence of polio and coxsackie viruses in the Rio Grande. The relative merits of using the chambers as compared with sacks of dialysis tubing in this case might be debated since dialysis tubing sidewalls were used in the chambers. However, the greater surface area-to-volume ratio and ease of sampling with the chambers are clear advantages. On the other hand, the small relative size of the microorganisms under investigation in this case dictates the use of a filtration medium of low porosity, such as dialysis tubing, to retain the viruses within the chamber. Nonetheless, this necessary trade-off does result in an experimental environment for the viruses that is like the surrounding ambient water in many respects. As a result, a measure of comparability is gained by the use of the chamber over similar experiments performed within enclosed vessels or water in the laboratory.

In Situ Studies with Algae Using Membrane Diffusion Chambers

Algae are commonly used in laboratory bioassays to determine the nutrient status of natural waters or the presence of toxic chemicals (Environmental Protection Agency, 1971). Although this test is designed to be carried out in the laboratory and yields results that have been effectively related to actual field conditions, there could conceivably be circumstances where the test should be performed in situ. Membrane diffusion chambers have been tested for this use (Olson and McFeters, 1976). In this application, chambers of 100 ml capacity that were polished to allow greater light penetration were fitted with a translucent filtration medium (Nuclepore N500CPR), then filled with a suspension of the test alga and immersed in the body of water under investigation. The chambers were suspended within the flotation collar in an attitude that was nearly horizontal. The side of the chambers that contained the ports were maintained slightly higher than the rest of the structure and a tube was connected to the vent port. This arrangement was essential to vent the oxygen that resulted from algal metabolism.

Samples were withdrawn at timed intervals and the algal growth was followed fluorometrically. The preliminary data available at this time indicate the potential utility of this technique as a field algal assay procedure (Olson and McFeters, 1976).

Use of Membrane Diffusion Chambers in Extreme Environments

The chambers have been used in extreme thermal environments to only a limited extent (Ramaley, 1978). From that experience in Yellowstone National Park, Wyoming, it was learned that some of the cellulose-based filtration media that are available commercially will disintegrate within 24 hr in an alkaline hot spring at 85°C. Therefore, when planning chamber experiments in aquatic environments with extremes of pH or temperature, the stability of the membranes should be tested.

B. Modifications

Chamber Volume

The original chamber that was developed and used in our laboratories had an approximate volume of 25 ml. A larger 100 ml capacity chamber was later developed in response to a specific need. Both of these chamber sizes are available from the manufacturer with or without polished edges. The polished models are better suited to applications where photosynthetic organisms are to be used. As mentioned earlier, a modified 5 ml capacity chamber was designed and constructed by Tom Roberts in our laboratories for use with leptospiral survival studies.

Physical Alterations and Alternative Designs

Screens and Exclosures. The use of external screens or exclosures are important modifications when working with pathogens or in locations where the membrane breakage is likely to occur as a result of minor tampering. Chambers used in ponds located on university campuses or in parks have proven to be highly susceptible to perforation and some protection can be important. This type of modification, in the form of stainless steel wire screens covering the outer membrane surfaces (Fig. 15), was suggested by Fliermans and Gordon (1977) for deep-water studies, and they also included an

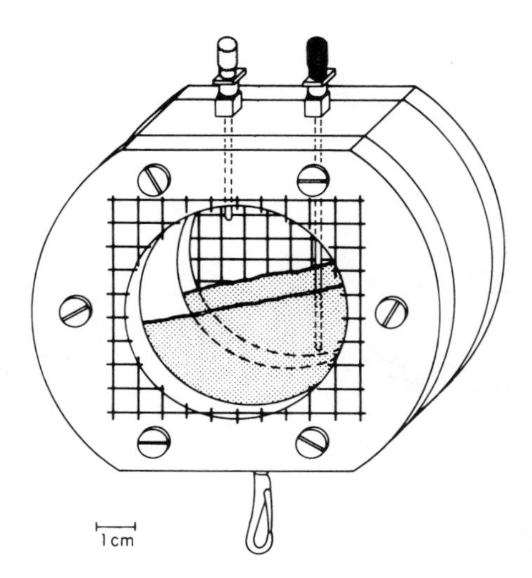

FIGURE 15 Modified membrane diffusion chamber. (From Fliermans and Gordon, 1977.)

enlarged central spacer to accommodate a larger volume and a metal clip on the bottom for an anchor or to attach other chambers. The screens may be secured by the structural bolts that hold the chamber together. In some applications it might be easier to protect the chambers by placing them in wire baskets.

Mixers. Some workers have suggested that diffusion chambers used in microbiological studies should be stirred continually to prevent bacterial settling as the experiment progresses (Sieburth, 1976). This design feature was incorporated into a highly modified chamber with the addition of a battery-operated magnetic mixer unit by Vasconelos and Swartz (1976). The details of the apparatus are shown in Fig. 16. Other chambers that incorporate both internal and external mixers will be described later.

The use of membrane diffusion chambers in the open sea presents problems associated with the growth of microorganisms on the membranes and chamber surfaces with a subsequent reduction in diffusion. This type of fouling prompted environmental microbiologists interested in marine applications to incorporate mixing devices into their chambers to help solve the problem. Additionally, mixers can maintain homogeneous cell suspensions and increase diffusion across

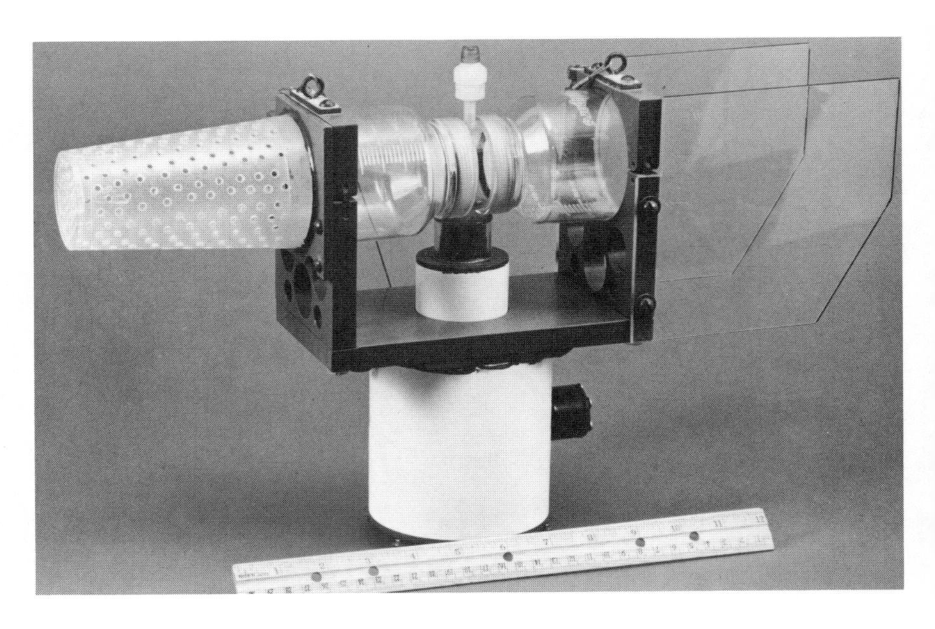

FIGURE 16 Diffusion chamber of Vasconcelos and Swartz (1976) with stirring unit.

the membranes. These considerations become critical in marine studies using chambers, but in most freshwater applications the need for mixers is much less obvious if the worker is careful to ensure that there is a homogeneous and representative suspension before a sample is withdrawn for analysis. However, if agitators are used in chambers, some attention should be given to the speed of the mixer and the influence of the resultant increased shear forces on the test organisms.

Other Designs. A rather large (800-1000 ml) tubular chamber was proposed in 1974 to follow the growth of various organisms when submerged in natural aquatic environments (Schlichting et al., 1974). Because the surface-to-volume ratio of this device is prohibitively low and the use of dialysis material as membranes is recommended, these chambers lack diffusion characteristics that would make them useful for microbiological applications. On the other hand, this chamber might be well suited for use with small fish or other macroaquatic organisms if a porous material such as nylon mesh were used in place of the dialysis membranes.

A number of unique chambers have been designed and constructed by Sieburth (1976) to meet the special requirements of studies in the deep sea. Mixers that are external to the membranes are to prevent the growth of marine organisms that inhibit porosity while air is supplied to the interior of the chamber to maintain a uniform suspension. Another model has magnetic-driven mixers within the chamber as well as on the outside of both membranes (Figs. 17 and 18). Larger versions that are to accommodate as much as 20 liters are being constructed for specific deep-sea studies.

FIGURE 17 Photograph of two chambers devised by J. M. Sieburth. Chamber A may be aerated through a port and chamber B is mixed both internally and externally.

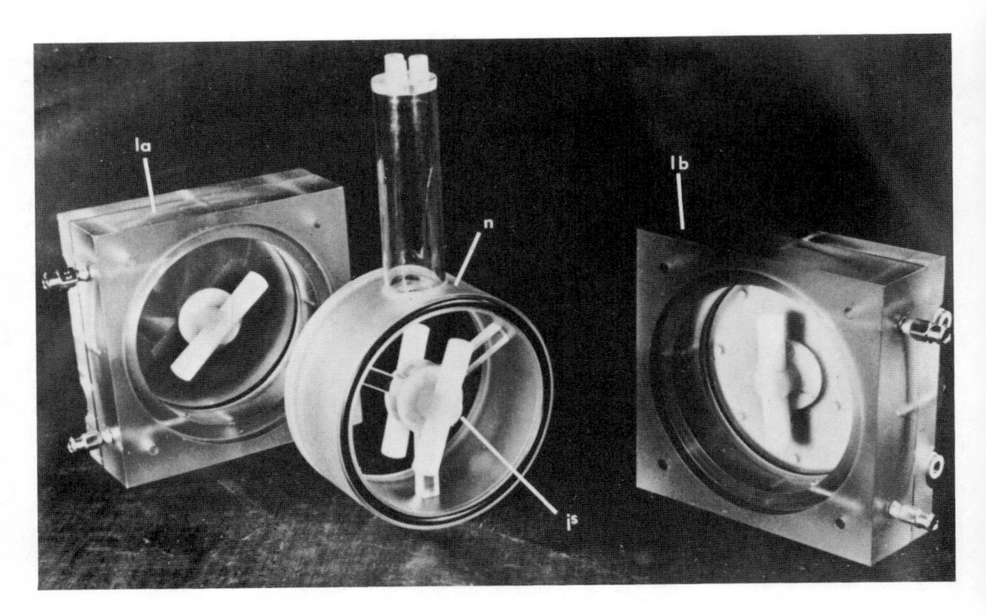

FIGURE 18 Photograph of a disassembled chamber devised by J. M. Sieburth. The internal and external magnetic mixer bars may be seen.

Chambers Adapted to Sediment Studies. The basic MSU-VME membrane diffusion chamber design was slightly modified for an investigation of bacterial predation in the presence of sediments by Marshall and Roper (personal communication). Their modification replaced hypodermic needles, used for sample withdrawal, with 10 mm diameter sampling ports that were plugged with rubber stoppers. In this way, suspensions of sediment could be added or removed with a special pipette. Their results, soon to be published, clearly demonstrate that sediments protect *E. coli* from predation.

Unique membrane chambers have been designed to follow microbiological processes in sediments (Winfrey and Zeikus, 1977). One multichambered device may be loaded with various experimental mixtures of microorganisms and sediments and then driven into a natural sediment (Fig. 19). Samples may be removed at intervals for analysis.

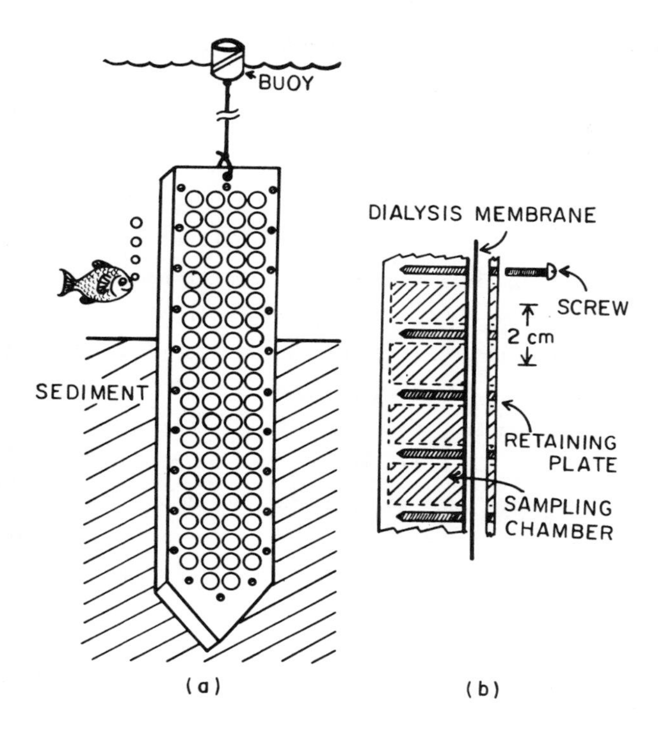

FIGURE 19 Interstitial water sampler: (a) schematic diagram showing sampler in place; (b) cross section showing assembly. (From Winfrey and Zeikus, 1977.)

IV. SUMMARY AND CONCLUSIONS

Plastic diffusion chambers with interchangeable membrane sidewalls have been developed for use in aquatic microbiological applications. The high surface area-to-volume ratio, ease of sampling, and general convenience of this type of device have led to widespread acceptance and diverse applications in a variety of environmental studies. In this way a "captive" population may be retained and studied while in a natural aquatic environment. The microorganisms are exposed in situ to many of the actual ambient physical and chemical factors that influence the natural microbial community.

In the design of experiments that will be ultimately related to some natural ecosystem or community, certain basic microbiological considerations that relate to population density, the cultural background of the cell suspension used, and the presence or absence of bacterial predators are noteworthy. The cell suspension should imitate, if possible, that observed in the natural environment and should not exceed 10^6 organisms per milliliter unless there are compelling reasons to do so. Also, experimental bacterial suspensions cultured for extended periods of time in more favorable nutrient conditions are capable of greater survival.

The chambers have been applied to various studies on population dynamics where pure and mixed bacterial suspensions were observed. In addition, these devices were used to characterize injury that occurs in heterotrophic bacteria when exposed to water and to follow the survival of viruses in water. Preliminary experiments have also demonstrated the utility of these chambers in monitoring algal growth in situ.

The basic design of the chamber has been modified in a number of ways to satisfy certain specific experimental needs. Protective screens and mixers have been included to maintain homogeneous cell suspensions and prevent fouling in certain environments. Also, modifications have made the device useful in the study of microbial processes in sediments and the deep-sea environment.

ACKNOWLEDGMENTS

We thank Tom Wandishin for help in the design and construction of the chambers and John Schillinger, Gary Bissonnette, and Jim Jezeski for their collaborative efforts. The following are also acknowledged for their technical assistance and many helpful discussions: Elva Steinbruegge, Sandra Dempster, Doniella Slanina, Anne Camper, Susan Olson, Greg Olson, Tom Roberts, and Susan Turbak.

This work was supported in part by funds from the U.S. Department of the Interior authorized under the Water Resources Research Act of 1964, Public Law 88-379, and administered through the Montana

University Joint Water Resources Research Center (Grants OWRR B-035 MONT, B-040 MONT, and 14-34-001-6207).

REFERENCES

Beard, P. J., and Meadowcroft, N. F. (1935). Survival and rate of death of intestinal bacteria in seawater. *Am. J. Public Health 25*: 1023–1026.

Bissonnette, G. K., Jezeski, J. J., McFeters, G. A., and Stuart, D. G. (1975). Influence of environmental stress on enumeration of indicator bacteria from natural waters. *Appl. Microbiol. 29*: 186–194.

Bissonnette, G. K., Jezeski, J. J., McFeters, G. A., and Stuart, D. G. (1977). Evaluation of recovery methods to detect coliforms in water. *Appl. Environ. Microbiol. 33*: 590–595.

Environmental Protection Agency (1971). *Algal assay procedure: bottle test*. National Eutrophication Research Program, U.S. Environmental Protection Agency, Corvallis, Ore. 82 pp.

Fliermans, C. B., and Gordon, R. W. (1977). Modification of membrane diffusion chambers for deep-water studies. *Appl. Environ. Microbiol. 33*: 207–210.

Greene, B. L., Clausen, E. M., and Litsky, W. (1977). Two-temperature membrane filter method for enumerating fecal coliform bacteria from chlorinated effluents. *Appl. Environ. Microbiol. 33*: 1259–1264.

Jordan, E. O., Russell, H. L., and Zeil, H. R. (1904). The longevity of the typhoid bacillus in water. *J. Infect. Dis. 1*: 641–689.

McFeters, G. A., and Stuart, D. G. (1972). Survival of coliform bacteria in natural waters: field and laboratory studies with membrane-filter chambers. *Appl. Microbiol. 24*: 805–811.

McFeters, G. A., Bissonnette, G. K., Jezeski, J. J., Thompson, C. A., and Stuart, D. G. (1974). Comparative survival of indicator bacteria and enteric pathogens in well water. *Appl. Microbiol. 27*: 823–829.

Metcalf, J. C., and Stiles, W. C. (1967). Survival of enteric viruses in estuary waters and shellfish, pp. 439–448. In *Transmission of viruses by the water route*, G. Berg (Ed.). Wiley-Interscience, New York.

O'Brien, R. T., and Newman, J. S. (1977). Inactivation of polioviruses and coxackie viruses in surface water. *Appl. Environ. Microbiol. 33*: 334–340.

Olson, G., and McFeters, G. (1976). Unpublished data.

Olson, G., Turbak, S., and McFeters, G. (1976). Bioassays related to the effects of coal strip mining and energy conversion on the aquatic microflora. In *Proceedings of the Symposium on Terrestrial and Aquatic Ecological Studies of the Northwest*, R. A. Soltero (Ed.). EWSC Press, Cheney, Wash. 397 pp.

Postgate, J. R., and Hunter, J. R. (1962). The survival of starved bacteria. *J. Gen. Microbiol. 29*: 233-263.

Ramaley, R. (1978). Personal communication.

Roberts, T. (1977). M.S. thesis, Montana State University, Bozeman, Mont.

Schlichting, H. E., Pendeville, G. N., and Guiry, M. D. (1974). New techniques for biological monitoring of water quality. *Biocontrol Tech. 1*: 1–13.

Schultz, J. S., and Gerhardt, P. (1969). Dialysis culture of microorganisms: design, theory, and results. *Bacteriol. Rev. 33*: 1–47.

Sieburth, J. (1976). Personal communication.

Slanetz, L. W., and Barley, C. H. (1965). Survival of fecal streptococci in seawater. *Health Lab. Sci. 3*: 142–148.

Strange, R. E., Dark, F. A., and Ness, A. G. (1961). The survival of stationary phase *Aerobacter aerogenes* stored in aqueous suspensions. *J. Gen. Microbiol. 25*: 61–76.

Stuart, D. G., Bissonnette, G. K., Goodrich, T. D., and Walter, W. G. (1971). Effects of multiple use on water quality of high mountain watersheds: bacteriological investigations of mountain streams. *Appl. Microbiol. 22*: 1048–1054.

Stuart, D. G., McFeters, G. A., and Schillinger, J. E. (1977). Membrane filter technique for the quantification of stressed fecal coliforms in the aquatic environment. *Appl. Environ. Microbiol. 34*: 42–46.

Vasconcelos, G. J., and Swartz, R. G. (1976). Survival of bacteria in seawater using a diffusion chamber in situ. *Appl. Environ. Microbiol. 31*: 913–920.

Winfrey, M. R., and Zeikus, J. G. (1977). Effect of sulfate on carbon and electron flow during microbial methanogenesis in freshwater sediments. *Appl. Environ. Microbiol. 33*: 275–281.

Zobel, C. E. (1936). Bacterial action of seawater. *Proc. Soc. Exp. Biol. Med. 34*: 113–116.

19

HYDROPHOBIC GRID-MEMBRANE FILTERS: THE (ALMOST) PERFECT SYSTEM

ANTHONY N. SHARPE Health and Welfare Canada, Ottawa, Ontario, Canada

I. INTRODUCTION

The hydrophobic grid-membrane filter (HGMF) was first described by Sharpe and Michaud (1974) as an interesting means of improving the numerical operating range of conventional membrane filter (MF) analyses. In a later paper (Sharpe and Michaud, 1975), a fuller description of the HGMF and the mathematic of its behavior was given. Graphs showing the ability of these early HGMFs to yield a 1:1 relationship between count and inoculum, with both pure cultures and river water samples, at levels up to 3 X 10^4 organisms per filter were provided. Hendry and Sharpe (1976) later reported that apparent coliform counts from lake waters were four times higher on HGMFs than on the corresponding MFs. In 1978, Sharpe and Michaud described HGMFs capable of yielding linear recoveries at inoculum levels up to 9 X 10^4 per filter. Sharpe et al. (1978) suggested how the HGMF could be used in quantitative food microbiology, particularly in the automation of analyses. Later work has been directed toward defining the filterability of food suspensions and factors affecting it (Sharpe, Peterkin, and Dudas, 1979; Peterkin and Sharpe, 1980) and the use of HGMFs in specific food analyses (Sharpe, Peterkin, and Malik, 1979).

Two features characterize the HGMF. First, it is capable of great accuracy and numerical operating range. Second, the nature of its unique performance highlights some important aspects of microbial data acquisition and is worth consideration by anyone interested in the development of enumerative methods. HGMFs are now available from QA Laboratories, Ltd., 135 The West Mall, Etobicoke, Ontario, Canada.

The development of methods for counting microorganisms has preoccupied microbiologists for many years and will presumably continue to do so. The HGMF is one such development. In common with all other techniques, it has both advantages and disadvantages, but there are situations where its advantages appear to be overwhelming. It is an excellent device for the propagation, enumeration, and isolation of microbial colonies. If one believes that the enumeration of colonies is a desirable means of assessing incipient microbiological spoilage or hazard, the HGMF method should be of interest, being potentially the most accurate way of making such measurements.

In the following pages I describe the nature of the HGMF and the reasons for its potential superiority. The claims of this chapter must be on the grounds of interest or provocativeness because, to my knowledge, only a handful of laboratories are currently using the device.

II. WHAT IS THE HGMF?

Exactly as its name says, the HGMF consists of a membrane filter (MF) on which has been prepared a grid of hydrophobic lines. After the filter has been inoculated (in exactly the same way as an ordinary MF), these lines serve as barriers to the lateral growth of colonies (Fig. 1), confining the growth of each colony to the grid cell in which the original colony-forming unit (CFU) fell. This seemingly simple step has far-reaching consequences.

A. Ideal and Real Enumeration Systems

An MF or petri dish containing only one colony is an almost perfect counting system. Provided that we are sure there are no other colonies (invisibly small, say), we know that:

FIGURE 1 Colonies of *Escherichia coli*, approximately 1 mm high, growing on a 1024 grid cell hydrophobic grid-membrane filter on VRB agar.

1. The probability is high that the colony formed from one CFU.
2. The probability is low that the colony formed from two or more CFU.

For with the great area of the MF to choose from, it is most improbable that a plurality of CFU will land so close together that their colonies are indistinguishable.

When the MF contains two colonies, we are not quite so sure. And as the colonies become even more numerous, our confidence in the accuracy of our count decreases even further, for we can be sure that, even if all the CFU managed to multiply, there is an increasing probability of two or more giving rise to only one countable colony. If we had sufficient data about the distribution of colony diameters on each MF, their relative colors, and so on, and the ability of our eyes at resolving close growths, we could calculate the probable error and correct for it. Of course, we never do, so that we must make a trade between the statistical error of the numbers themselves (decreasing with increasing numbers), and this error caused by overlapping of colonies (increasing). For MFs this usually means an upper counting limit of 80 colonies per circle. When colonies tend to be large, as in the case of fecal coliforms, the limit is often only 60 (American Public Health Association, 1971).

The desirability of having no less than 20 colonies per MF to prevent statistical fluctuations defeating our careful pipetting leaves each MF capable of handling a concentration range of 4:1 at best. Faced, in practice, with much larger concentration ranges, we have two choices. We can use very much larger MFs (impractical) or make sequential dilutions and inoculate MFs from each dilution (laborious).

It should be apparent that our lack of information about the size distribution of colonies, and the resolving ability of our eyes, is the root of the conventional MF's or petri dish's inadequacy. If we could make proper corrections for overlap losses, we could extend their working ranges enormously. But we cannot, for the error varies from circle to circle in an unquantifiable way. We always know that the true count is higher than the value obtained, but we are never sure by how much.

The HGMF removes this inadequacy. By bringing uniformity to the sizes of colonies and by regimenting them into a large, but predetermined number of compartments, it allows overlap errors to be (1) reduced and (2) quantified in very simple mathematical terms, with the result that the HGMF can actually mimic the recovery efficiency of an ideal system. It will be shown later how any HGMF count can immediately be converted into a figure representing the number of colonies that would have formed on an ideal MF (i.e., one of infinite area).

B. Why the HGMF Works

The grid lines break up the HGMF into a multitude of tiny identical MFs--the grid cells--which act as independent growth habitats. The precise positions of CFU in grid cells are unimportant, affecting neither the positions nor the sizes of the resultant colonies. Similarly, two CFU landing close together must either fall into separate grid cells (whence they will yield two colonies) or in the same grid cell (whence they will yield a single colony). With fixed dimensions, the probabilities of each occurrence are fixed and easily calculable. If the MF material and the pattern of barriers are uniform, every grid cell filters the same volume of suspension and therefore has the same probability of being inoculated with one or more CFU. The resulting pattern of colonies after incubation is determined, therefore, by the laws of statistics rather than by which organisms form the largest colonies, begin to develop first, are away from the edge, and so on.

At this point we should introduce a nicety by changing the term "colony-forming unit" (CFU) to "growth unit" (GU) because, under some circumstances, the growths in HGMF grid cells may not accurately fit the microbiological definition of colonies. Near the top end of its operating range, each growth in the HGMF grid cell is likely to have arisen from a plurality of growth units. The same is true of conventional MFs or petri dishes, of course, but the implications are usually conveniently ignored.

The mathematic of inoculation of an HGMF is simple, being the same as if we were trying to break the windows of a house by throwing stones, blindfold, or counting whether disintegrations occurred in successive seconds from a weak radioactive source. If the HGMF consists of N grid cells, every growth unit (GU) has a probability 1/N of falling into any particular grid cell. When a certain number of GU have been filtered, there are well-defined probabilities of finding grid cells containing zero, one, two, ...GU. Each grid cell containing one or more GU will produce only a single growth during incubation, and will therefore contribute only one count to the total number of positive grid cells. Grid cells containing zero GU will not, of course, contribute to the count.

If x grid cells are found to be positive, it can be shown that the most probable number of growth units (MPNGU) inoculated onto the HGMF is given by

$$\mathrm{MPNGU} = -N \log_e \frac{N - x}{N} \quad (1)$$

Herein lies the ability of the HGMF to mimic a perfect recovery system. Like any MF or plating method, the count of growths ("colonies") is less than the number of GU inoculated, because of the probability of overlapping. Unlike these others, however, overlapping is regulated and turned to advantage by allowing the true inoculum to be estimated over a very wide operational range indeed. An HGMF consisting of a pattern of 100 X 100 grid cells has a maximum quotable MPNGU of 92,000, for example (see Table 1). Its operating range extends, therefore (with one qualification described below), from zero to 92,000, in dramatic contrast with the performance of systems committed to random colony growth. Sharpe and Michaud (1978) have demonstrated the linear recovery of such HGMFs over this range.

The value of 92,000 MPNGU quoted for the 10,000 grid cell HGMF (or, for that matter, any other saturation value quoted for HGMFs) must be treated with some reservation. The maximum countable state of the HGMF occurs when N-1 grid cells are positive (i.e., observed to have growth). By putting this value into Eq. (1), we see that the maximum MPNGU that may be quoted is $N \log_e N$. This must be the maximum, because when N grid cells are positive, the MPNGU is greater than $N \log_e N$ but less than infinity, and no value can be calculated. However, it must also be remembered that to filter $N \log_e N$ growth units through an HGMF would produce almost as high a probability of finding N positive grid cells as of N-1 (i.e., of not producing a quotable result). We can say that N-1 positive

TABLE 1 Most Probable Numbers and 95% Confidence Limits for MPN Systems of Different Sizes

Number of positive compartments	MPN[a]	95% Confidence limits[b] (as % of MPN Lower	Upper
	100 Compartments		
1	1	3	560
2	2	12	370
5	5	32	234
10	11	49	185
20	22	61	155
50	69	73	147
90	230	75	131
95	299	73	137
98	391	68	154
99	461	65	176
	1,000 Compartments		
1	1	1	560
2	2	15	360
10	10	48	182
100	105	81	122
200	223	87	115
500	693	91	109
900	2,302	92	109
950	2,996	91	110
998	6,214	79	137
999	6,908	75	∞
	10,000 Compartments		
1	1	0	610
10	10	50	181
500	513	92	109
1,000	1,054	94	101
5,000	6,931	98.0	102.1
9,000	23,026	99.1	102.5
9,500	29,957	97.3	102.9
99,50	52,983	95	106
9,998	85,172	85	∞
9,999	92,103	81	∞

[a]Values rounded to nearest integer.
[b]Most values rounded to nearest integer, except where a more accurate value is of interest.

grid cells represents an MPNGU of N $\log_e$ N, but the reverse is not true--we cannot be sure that N $\log_e$ N growth units will give us N-1 positive grid cells. Thus, while the value N $\log_e$ N is useful for comparing the absolute capabilities of HGMFs, the practical operating limits will be lower. It seems reasonable to set an upper limit of 95% saturation (x = 0.95N) for routine counting. For the 10,000 grid cell HGMF, then, the lower and upper 95% confidence limits with 9,500 grid cells positive are 29,100 and 30,810, respectively, with an MPNGU of 29,957--still a healthy contrast compared with compared with conventional systems.

Accompanying the increase in operating range, we also observe an increase in precision, resulting from the larger numbers it is possible to involve. HGMFs can provide data (see Table 1) for which the 95% confidence limits are as close as 1 or 2%. We gain in this way by sacrificing only information about relative colony sizes. To some, this information may be important; to others, it may appear to be less important, even misleading. It is a subjective matter.

C. Isolation and Purification

A third benefit is to be gained from trading information about relative colony sizes. This is an ability to quantify the degree to which organisms have been isolated or purified.

Many areas of microbiology require the purification of microbial species, and there is very little instrumentation available to facilitate the process. The common streaking technique used to "dilute" and "isolate" colonies may appear to work satisfactorily, manually, in that the operation usually yields a series of continuous lines, punctuated by one or two blobs which are assumed to have resulted from single CFUs. The mechanics of the transfer of microorganisms from agar to wire, and back to agar again during the streaking are, however, almost certainly extremely complex. There is a vagueness and human skill involved in the whole business, guaranteed to delight the microbiologist and make the instrument engineer tear his hair.

To attempt to simulate such a process mechanically invites several leading questions, which, although they may be asked of the manual process, are at present unanswerable and therefore easily put on one side. On viewing the serpentine strokes, for example, how does the human technician decide which of the final islands of growth are pure enough for subculture? How many times must the process be repeated before one can be reasonably certain of having a pure culture? How certain is certain--what is the residual probability of contamination? What is the probability of any island of growth being pure? How do streaking speed, agar concentration, surface viscosity, roughness of wire, deformation of the agar surface, temperature, ionic strengths, and so on, affect the kinetics of the adsorption and desorption of microorganisms on that final streak? Is the release of microorganisms merely a slowly attenuating process whereby CFU fall off steadily along the track, or is it also affected by vibration, the twitching of the technician's muscles, with CFU falling off in clumps at every twitch?

The HGMF can eliminate some of these questions and provide answers to the others. As a result, it could again be an excellent component in instruments, for isolation. After incubation:

1. Every growth on an HGMF has exactly the same probability of being (or not being) pure. That is, any grid cell containing growth of, say, the correct color, may be subcultured, being just as good for the purpose as any other grid cell in that selection category.
2. Its probability of purity (P) may be calculated easily from the formula

$$P = -\frac{N - y}{y} \log_e \frac{N - y}{N} \qquad (2)$$

   where y is the number of grid cells conforming to a given set of selection criteria.
3. The coordinate reference system embodied in the grid is ideally suited to the guidance of a mechanical subculturing device.

Table 2 shows several values obtained for an HGMF of 10,000 grid cells, using Eq. (2). Even with as many as 100 grid cells containing growth, chances are still 200:1 that subcultured growths

TABLE 2 Probabilities of Purity of Colonies on 10,000 Grid Cell HGMFs

Number of positive grid cells	Probability of purity
1	0.99995
10	0.9995
100	0.995
500	0.975
1,000	0.948

will be pure. To use the HGMF in this manner obviously requires that one or two dilutions be made. However, the quantifiability of the result may often be considered worthwhile, even in manual purification work. The ability of the HGMF to isolate growths and reduce the extent of interactions between them was used by Hendry and Sharpe (1976) to explain the apparent fourfold greater recovery of coliforms from lake waters in the presence of other organisms.

D. Shapes, Sizes, and Preparation of HGMFs

Initial studies of the properties of HGMFs were made simply by printing rectilinear wax grids on standard 47 mm circles, using etched zinc plates or roller printers of the type described below. Most common waxes gave satisfactory results, although the dental sticky waxes (e.g., Ash Model Cement) were preferred because of their solidifying properties. Using a rectilinear grid in a circular filtration unit led to some doubt about the actual number of grid cells being inoculated, and later HGMFs were printed as 60 mm squares, with heavy wax borders defining a 51 mm active area. Ink-stained HGMFs are shown in Fig. 2.

The width of the grid lines in HGMFs must be a trade between filtration speed and barrier efficiency. Grid lines of zero width would be ideal because, with no loss of filtration area, filtration speed would be unaffected; however, these would not act as barriers. Typically, line widths of 0.1 to 0.2 mm have been found adequate as barriers. The actual loss of filtration area is small enough to be

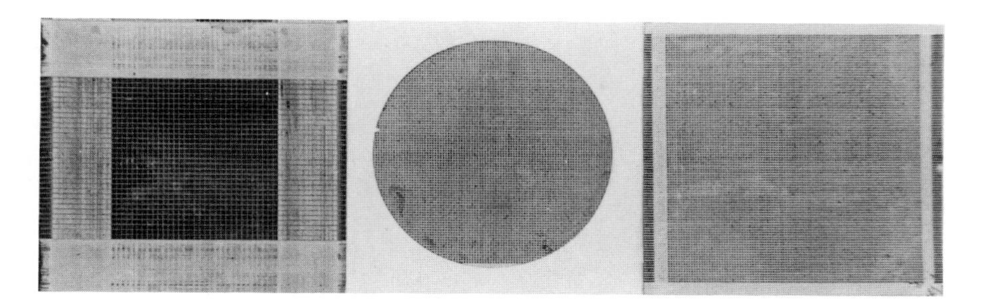

FIGURE 2 Ink-stained specimens of some of the hydrophobic grid-membrane filters used by the author. Commercially available HGMFs are somewhat prettier.

offset by the change from circular to square format. It should be noted, for example, that the active area (and therefore filtration speed) of a 60 mm^2 HGMF containing 10,000 grid cells with a line width of 0.13 mm is 50% greater than that of a conventional 47 mm circle, for which the funnel diameter is generally about 35 mm.

Ideally, barriers would rise above the filter surface, so as to present a mechanical impediment to colony growth. The printing process used so far, however, does not produce raised barriers. Indeed, the lines may sometimes be depressed slightly below the surface. The barrier effect thus results solely from the hydrophobicity of the line material. Wax typically penetrates one-fourth to one-half of the depth of the filter and, to date, has only been applied to one side of the filter.

The chosen size of grid cells is preferably somewhat smaller than the area normally occupied by colonies, so as to ensure that any growth completely fills a cell. However, this is not vital for manual counting.

By exercising a little care in handling during printing, contamination of HGMFs may be kept to such a low level that subsequent sterilization is unnecessary for most work. Should it be necessary, HGMFs could presumably be sterilized by irradiation or by ethylene or propylene oxide treatment. Autoclaving, of course, will not be possible when waxes are used for the barrier material.

FIGURE 3 Roller printer for HGMFs, constructed by I. Dudas. The width of the barriers is determined by the width of fins on the lower roller. Wax is picked up from the electrically heated compartment beneath the roller and applied to the fins by a microcloth pad. Each filter is run through beneath a 60 mm^2 blotting paper backer, turned through 90°, and run through again.

Finally, the apparatus needed to print HGMFs, for example the roller printer in Fig. 3 (Sharpe, 1975), is minimal. Printing by hand is simple enough but sufficiently tedious as to curb interest in HGMFs until recently, when they became available commercially.

III. USING HGMFs

A. Inoculation

Experimental HGMFs are printed on one side only; therefore, it is important that they be placed in the filtration apparatus with the barriers uppermost. Filters should be horizontal to ensure uniform filtration volumes over the whole surface. Similarly, the volume filtered should be sufficient to ensure a uniform probability of inoculation. For water samples this is usually no problem; however, for 1.0 ml aliquots of food suspension, it is best to first add about 10 ml diluent to the HGMF and disperse the suspension in this.

B. Removing Food Debris

It is desirable that food debris does not cover the HGMF surface. Sharpe et al. (1978) suggested that this could be accomplished after filtration, without losing bacteria, by rinsing the HGMF using a jet of sterile diluent. Later evidence (P. Entis, QA Laboratories Ltd., personal communication) indicates that the success of this method varies; it may, perhaps, depend on the MF material. However, the use of sterile fine mesh prefilters, for example, in the form of disposable pipet "tips" when pipetting the aliquot for filtration (Sharpe et al., 1979) is to be preferred. Unlike depth prefilters (paper, asbestos, etc.), mesh prefilters appear to have no effect on bacterial recovery (Peterkin and Sharpe, unpublished results).

C. Plating and Incubation

HGMFs are used in exactly the same manner as MFs. The 60 mm square HGMFs fit standard 90 mm petri dishes.

D. Counting

This is a straightforward operation, simplified somewhat by the regimentation of the growths. However, the large numbers of positive grid cells encountered may be a little daunting at first, and it is suggested that the following guidelines be combined, as appropriate, into a technique that will most efficiently fit the situation:

1. Low-order HGMFs (e.g., up to 1024 grid cells) may be counted without visual aids.
2. High-order HGMFs (e.g., up to 10,000 grid cells) should be viewed under a 5X or 10X magnifier. A simple apparatus consisting of a board on which about 10 cm of wire is stretched

between supports (e.g., nails) just high enough to clear the petri dishes makes a useful aid to keeping one's eye on the correct line of grid cells. The HGMF is moved under the wire, line by line, as counts are made.

3. Unless the utmost in accuracy is required, high-order HGMFs need not be counted completely, particularly when they are about 50% full. Sufficient lines should be counted to take in, say, 300 growths, from which the average score for the whole HGMF can be estimated. The confidence limits of the count will be similar to those from a normal plate count.

4. When most grid cells are positive, the count of negative ones should be taken instead and subtracted from the total.

E. Calculating

When only a small percentage of grid cells, say 10%, are positive, the unmodified score may be taken as the actual number of growth units (GU) in the aliquot used for inoculation. When a larger percentage are positive, the most probable number of growth units (MPNGU) must be calculated. This may be done in one of three ways:

1. A standard table of MPNGU against the number of positive grid cells (x) may be prepared beforehand for an HGMF of N grid cells. The MPNGU are then read from the table.

2. An electronic calculator may be used to compute values as required. Rearranging Eq. (1) to read

$$\text{MPNGU} = N \log_e \frac{N}{N - x}$$

yields a calculation taking but seconds to execute. A programmable calculator, of course, will provide an almost instantaneous value of MPNGU.

3. A nomogram showing MPNGU against x may be constructed from calculated values. MPNGU are simply read from the graph. This method is quick and simple, but its use is not recommended when full counts of positive grid cells have been made in the sake of maximum accuracy or when the HGMF approaches saturation.

The use of counts obtained when more than 95% of the HGMF grid cells are positive is not advised (see Sec. II.B) except when there is no alternative. Table 3 shows 95% saturation values for HGMFs of various sizes. If all the grid cells are found to be positive, no value can be calculated, and the observation that

$$\text{Level exceeds } \frac{D}{V} \times \text{ saturation value}$$

TABLE 3 Saturation and 95% Saturation Values for Typical HGMFs

Number of HGMF grid cells	Saturation value (x = N-1)	95% of grid cells (rounded)	95% Saturation value (x = 0.95N)
25	81	23	63
36	129	34	104
64	266	61	196
100	461	95	300
625	4,024	594	1,877
1,024	7,098	973	3,071
2,500	19,560	2,375	7,489
3,650	29,939	3,468	10,944
10,000	92,103	9,500	29,957

(V being the volume of suspension of dilution factor D filtered) should be recorded. Saturation values for different HGMFs are also shown in Table 3.

IV. QUANTITATIVE FOOD MICROBIOLOGY USING HGMFs

It is unusual for membrane filters to be used in quantitative food microbiology. One possible reason for this is the small numerical operating range of conventional MFs, compared with petri dishes, which would necessitate the use of an unacceptable number of dilutions in order to accommodate the range of values commonly encountered in foods. The fault lies, however, not with the MF material, but with the small growth areas normally employed. Colony densities may actually be higher on MFs than on plates; for example, the arbitrary limit of 80 colonies for total coliform counts on MFs, for which the inoculated area is about 35 mm, would increase to 590 colonies if the area inoculated equaled that of a 90 mm petri dish.

Perhaps the most important reason, however, is the apparent general belief that food suspensions immediately block the pores of MFs and are thus unfilterable. If this were true, the use of MF-based techniques in food microbiology would be impossible. It is, however, feasible to filter adequate quantities of many food suspensions through MFs and HGMFs, particularly those prepared using the Stomacher blender rather than conventional rotary blenders such as the Osterizer (Sharpe et al., 1978), so that the peculiar properties of HGMFs may have wider implications for food microbiology than would at first appear.

Sharpe et al. (1978) initially showed that 0.5 ml aliquots of 1:10 homogenates, prepared using an Osterizer blender, and 1.0 ml aliquots of higher dilutions of foods could be filtered through 1024 grid cell HGMFs in reasonable times to yield counts comparing favorably with the corresponding plate counts. A range of foods, including ground beef, sausage, ham, salmon, carrots, green beans, peas, corn, and chow mein, containing *Pseudomonas aeruginosa*, *Escherichia coli*, and *Staphylococcus aureus*, were examined. Later, Sharpe et al. (1979) prepared continuous filtration rate curves in an intensive study of many of the factors affecting membrane filtration. The study covered 58 foods, 13 brands or pore sizes of filters, homogenization procedures, concentration of food, settling periods, pressure differential, direction of flow through the membrane, temperature, and presence of surfactants. Stomached homogentates filtered more rapidly than blended ones. Some surprising effects of flow direction through the membrane were noted. Only four foods could not be filtered at the 0.1 g level, which is the maximum quantity used in conventional plating procedures. Soon after, Peterkin and Sharpe (1980) defined procedures, some of which included short incubations with proteases, allowing the satisfactory membrane filtration of dairy products.

Sharpe et al. (1979) also showed that the organisms from more than 0.1 g food could often be concentrated and counted on HGMFs. This improved the limit of detection for coliforms and *Escherichia coli* compared with conventional methods. At the same time, growth inhibitors, which occur in some foods and cause poor recovery by conventional counting procedures, were completely removed in the membrane filtration steps. In like manner, the sucrose in ice cream, which may cause false "lactose fermentation" reactions, was completely removed, and with it, most likely, the need to confirm coliforms during the analysis of this food.

The demonstrated filterability of food suspensions indicate that the great numerical range of HGMFs could be used to advantage in food microbiology. It is quite possible that many foods for which counts do not normally range greater than 10^5-10^6 organisms per gram could be examined by the preparation of a single HGMF. The use of one extra dilution and another HGMF would probably suffice for most remaining foods.

V. ELECTRONIC COUNTING AND AUTOMATION

The HGMF was actually developed as an aid to electronic colony counting, and its ability to mimic a perfect recovery system was a rather serendipitous observation. In this area, and particularly in its potential usefulness in totally automated analytical instruments, the HGMF warrants consideration.

Regardless of manufacturer's claims, electronic colony counters are not yet perfected. Although their use is increasing, there are areas, such as the food industry, where counters of today's level of sophistication are unlikely to satisfactorily simulate the counting ability of a human technician. Existing colony counters are better described as particle counters; when particles other than colonies (e.g., bubbles, food debris) are present or when colonies do not exist as well-defined particles, their performance may be poor. Recent dramatic changes in the availability of image-sensing devices, microprocessors, and so on, will perhaps soon allow sufficiently discerning counters to be produced at realistic prices. However, the following argument, which originally stimulated development of the HGMF, will always apply:

1. If colonies are allowed to develop randomly and unchecked over the growth area, any counting device must scan the whole area to make its count.

2. Since colonies are recognized from their background by, say, differences in optical properties, the counting device must reliably discern their borders.

3. At every point in its scan, therefore, it must decide whether it is looking at colony or background, on the basis of the electrooptic signal. Frequently, the signal/noise ratio is low and the error rate is high.

4. Predetermining the positions of colonies eliminates the need to locate their borders and drastically reduces the number of decisions needed. The relative counting error should thus be proportionally reduced.

This argument is well illustrated by considering the following example, in which the counting circle of an MF just fills the vertical frame height of a conventional 625 line TV frame. If we assume that the counter can be programmed so as to ignore the area outside the circle, the image can be regarded as being composed of approximately 216,000 picture points (Sharpe and Michaud, 1978) at each of which the counting circuit must "decide" whether it is looking at colony or background. Even an error rate of 10^{-4} would still represent the addition or subtraction of 22 false counts, which could be embarrassing with an MF containing only 20 colonies.

Suppose now that the MF is replaced by an HGMF capable of enumerating 80 CFU. Surprisingly, such an HGMF need contain only 25 grid cells [placing N = 25 and x = 24 in Eq. (1) yields a maximum MPNGU of 80.5]. A counter designed for use with this HGMF would need to make only 25 decisions, for we could program it to examine only the optical properties of the centers of these grid cells. Other things being equal, therefore, the use of the HGMF instead of an MF would reduce the error rate in this example by a factor of 216,000/25, or 8640. To this we might add a less quantifiable factor resulting from an improvement in signal/noise ratio, for it is not difficult to demonstrate (see Fig. 1) that when colonies are restrained from growing laterally, they tend to grow upwards instead, thereby enhancing the optical differences between themselves and their background.

To this potential improvement in accuracy should be added another factor, resulting from the possibility of removing food debris as a source of error in the count. Presumably, this is also true of MFs, although I am only aware of rinsing having been described with HGMFs. Certainly, though, rinsing conventional petri dishes to remove food debris but not bacteria is unlikely to be practical.

One of the most valuable properties expected of the HGMF, however, is an ability to operate safely under saturation conditions in an automated instrument. The performance of a combination of conventional colony/particle counter with conventional MFs or petri dishes is unacceptable for any totally automated system (i.e., one intended to operate without human intervention at the counting stage). The problem results from the conventional need to detect colonies through the existence of borders. High colony densities on the growth area (such as might result from a situation where the contamination level exceeds the designed capacity of the instrument) will yield lawns of growth; with few detectable borders such a system will record a false (and potentially serious) low result. Systems operating with HGMFs, however, would not rely on the detection of borders, and saturation of the HGMF would simply yield a maximum score (N positive grid cells). Although not allowing a true count to be obtained in this circumstance, there need be no problem in programming an instrument to indicate that the contamination level exceeds its analytical capability. Thus, the use of the HGMF would allow automation of the counting stage without danger of the occurrence of false low reports.

Semiconductor image-sensing arrays are appearing on the market in increasing numbers and are likely to replace conventional devices (e.g., videcons) for most video applications. Arrays containing 32 X 32 (1024), 50 X 50 (2500), and 100 X 100 (10,000) elements are already available. The rectilinear pattern of HGMF grid cells matches that of the elements in such arrays, so that HGMFs prepared in rigid supports for automated handling would seem to be ideally suited to counting with these devices, for example in systems where each grid cell is inspected by a corresponding semiconductor element. There are enormous possibilities for using microprocessors in conjunction with these devices to obtain a great deal of information about the organisms on a filter.

VI. THE HGMF AS A POINTER FOR FUTURE SYSTEMS

The real value of enumerating microorganisms in water or food is probably as obscure as some of the samples received for analysis. Although it is a relatively easy thing to carry out manually, my opinion is that enumeration is generally a futile and misleading exercise unless one also measures relevant metabolic activities or infective doses, multiplication rates, interactive effects of the other flora, and so on. Detection methods based on, for example, metabolism of ^{14}C-labeled sugars (Reasoner and Geldreich, 1978) may be more meaningful than counting in this respect, even when the organisms measured only indicate the possible presence of some more malicious beast. However, technicians will probably still be counting colonies on membrane filters many years from now, and if they are there will be those who will try to mechanize the process. The HGMF is an idea, but there are probably better adaptations of conventional techniques still to be developed, and hopefully so. However, a brief consideration of the source of those advantages it does possess may provide pointers for scientists/engineers wishing to mechanize enumerative microbiology.

The HGMF's advantages--extended operational range, linear recovery, suitability for automation of counting, and isolation of microorganisms--are due to its ability to control the growth of colonies in a manner impossible with conventional MFs or plates. Colonies should not be allowed to be the master of a microbiologist or an engineer. It is surely more desirable, more technologically sound, and more aesthetically satisfying to suggest to microorganisms that they fit the machine than vice versa. I believe that, if truly satisfactory instruments for enumeration of microorganisms or their colonies are to be built, they will exercise at least as much control over the permitted characteristics of colonies as do HGMFs.

The HGMF bridges the conventional colony and most probable number counting techniques. The conventional tube MPN technique has deservedly earned a reputation for inaccuracy compared with colony counting techniques, but it should be noted that this need not be so; in fact, the MPN method has within it the capability for great accuracy when we use it correctly.

Indeed, the familiar MPN technique is not *inaccurate*, although it is certainly very *imprecise*; it merely produces data having very wide confidence limits. For example, the best lower and upper 95% confidence limits of the standard three-tube/dilution MPN test are about 25 and 400% of the actual MPN, respectively. But the minimum width of the confidence limits is a function of the number of tubes (compartments, grid cells, etc.) employed; it is large in conventional MPN techniques simply because manual inoculation of more than three or five tubes for each dilution is prohibitively toilsome. If we inoculated 100 or 200 tubes for each dilution, however, our confidence limits would compare extremely favorably with those from plating, and they would improve continuously if we increased the number of tubes still further. Table 1 shows most probable number values, and lower and upper 95% confidence limits for MPN systems containing 100, 1000, and 10,000 compartments, respectively, when various proportions of the compartments are counted as being positive. With only 100 compartments, the MPN confidence limits compare well with plates or MFs, at 75-131%. With 10,000 compartments, however, the confidence limits may be as close as 99.1-102.5%, a level of precision inconceivable in plate or conventional membrane filter counting.

By treating microbial enumeration experiments as exercises in the communication of information, one can show (Sharpe, 1978) that systems of discrete growth compartments in which each compartment is capable of "transmitting" only one "bit" of information are inherently more accurate in automated counting systems than those where growth compartments may transmit a range of values. In practical terms, the binary nature of MPN systems, where each growth compartment can exist in only one of two states (0 for no growth, 1 for growth, say), is far better suited than that of plates or MFs (which may take on any one of a large number of values) to the information-gathering capabilities of electronic instruments. We have, therefore, a second indication that the MPN way may be the preferred route in automating quantitative microbiology.

There is a third indication, however, which may be more important. We see from Eq. (1) that the maximum number of organisms countable in MPN systems is greater than the number of compartments, by the factor $\log_e N$. Therefore, the numerical range of operation of the whole system increases faster than the number of compartments. Thus, with increasing numbers of compartments, the number of dilutions required to be able to handle a given range of microbial concentrations decreases rapidly, and with it the mechanical complexity of the instrument. It is a fact that one of the most difficult engineering problems in developing machinery to simulate conventional counting methods is the provision of a suitable means for repeatedly diluting the sample. But it is possible to consider the development of instruments using the MPN method, in which the numerical range of operation is so great that the need to make dilutions is eliminated altogether.

It has already been shown, for example (Sharpe and Michaud, 1978) through the preparation of HGMFs, that the inoculation of 10,000 compartments to provide a counting range of nearly 10^5 is entirely feasible. This is more than adequate for the examination of water. The mechanized inoculation of thousands of compartments is thus not unrealistic, regardless of whether it is carried out simultaneously and, as it were, passively, as would be the case with instruments using the HGMF, or step by step, as might be the case with some other more discrete forms of growth vehicle yet to be developed.

REFERENCES

American Public Health Association (1971). *Standard Methods for the Examination of Water and Waste Water,* 13th ed. American Public Health Association, Washington, D.C., p. 678.

Hendry, G. S., and Sharpe, A. N. (1976). Superior performance of hydrophobic grid-membrane filters for membrane total coliform determinations of recreational lake waters. *Can. J. Public Health 67*: 137.

Peterkin, P. I., and Sharpe, A. N. (1980). Membrane filtration of dairy products for microbiological analysis. *Appl. Environ. Microbiol. 39*: 1138-1143.

Reasoner, D. J., and Geldreich, E. E. (1978). Rapid detection of water-borne fecal coliforms by $^{14}CO_2$ release. In *Mechanizing Microbiology,* A. N. Shapre and D. S. Clark (Eds.). Thomas, Springfield, Ill., pp. 120–139.

Sharpe, A. N. (1975). Machine for printing hydrophobic grids on membrane filters. *Appl. Microbiol. 30:* 110–112.

Sharpe, A. N. (1978). Some theoretical aspects of microbiological analysis pertinent to mechanization. In *Mechanizing Microbiology,* A. N. Sharpe and D. S. Clark (Eds.). Thomas, Springfield, Ill., pp. 19–40.

Sharpe, A. N., and Michaud, G. L. (1974). Hydrophobic grid-membrane filters: new approach to microbiological enumeration. *Appl. Microbiol. 28*: 223–225.

Sharpe, A. N., and Michaud, G. L. (1975). Enumeration of high numbers of bacteria using hydrophobic grid-membrane filters. *Appl. Microbiol. 30*: 519–524.

Sharpe, A. N., and Michaud, G. L. (1978). Emuneration of bacteria using hydrophobic grid-membrane filters. In *Mechanizing Microbiology,* A. N. Sharpe and D. S. Clark (Eds.). Thomas, Springfield, Ill., pp. 140–153.

Sharpe, A. N., Diotte, M. P., Dudas, I., and Michaud, G. L. (1978). Automated food microbiology: potential for the hydrophobic grid-membrane filter (HGMF). *Appl. Environ. Microbiol. 36*: 76–80.

Sharpe, A. N., Peterkin, P. I., and Dudas, I. (1979). Membrane filtration of food suspensions. *Appl. Environ. Microbiol. 37*: 21-35.

Sharpe, A. N., Peterkin, P. I., and Malik, N. (1979). Improved detection of coliforms and *Escherichia coli* in foods by a membrane filter method. *Appl. Environ. Microbiol. 38*: 431-435.

20

ADVANCES IN MEMBRANE FILTER APPLICATIONS FOR MICROBIOLOGY

D. A. SCHIEMANN   Ontario Ministry of Health, Toronto, Ontario, Canada

I. INTRODUCTION

Within the relatively short period of about 25 years, beginning with development of methods for enumeration of coliform bacteria in water, the membrane filter has seen wide application in diverse areas of microbiology while essentially revolutionizing water bacteriology. Current investigations with membrane filters in microbiology are concentrated in three primary areas: (1) modified membrane structures and procedures for improved recovery of injured microorganisms, (2) development of appropriate media for enumeration of other groups of microorganisms by the membrane filter technique, and (3) design of sequential transfer schemes for identification of bacteria recovered on the membrane filter.

The inherent advantages of the membrane filter for concentration of organisms from large volumes of liquid, provision of quantitative results having greater precision than alternative tube dilution methods, and the relative simplicity in materials and procedures have encouraged applications of this technique in many different areas of microbiology. This chapter will present some of those applications which are either new and have not been fully evaluated or standardized, or are unusual and have not seen official or wide adoption. The principal and essential steps for each method will be presented, but the original description should be consulted for procedural details before attempting any of these techniques.

II. WATER MICROBIOLOGY

A. Coliform Bacteria

The historic development and application of the membrane filter are intimately tied to the coliform group of bacteria, which have long been used as indicators of water quality. Standard procedures for enumeration of coliforms by membrane filter techniques are described elsewhere, and the methods presented here are added only as new and unusual procedures which have not been fully investigated or widely adopted.

Guthrie and Reeder (1969) described a membrane filter-fluorescent-antibody technique for enumeration of *Escherichia coli* in water samples. Antisera against selected strains of *E. coli* are prepared in rabbits and conjugated with fluorescein isothiocyanate. The water sample is filtered through a black, gridded membrane (Millipore HABG047) that is transferred to trypticase soy agar and incubated at 35°C for 12 hr. The membrane is returned to the filter-funnel apparatus and overlayed with 1-2 ml of pooled normal rabbit serum for 5 min at room temperature. This serum is then removed under vacuum. Colonies are overlayed with specific labeled antiserum for 15-20 min. The labeled antiserum is removed under vacuum and the colonies on the membrane are washed with 10-15 ml of phosphate-buffered saline (PBS) and then glycerol mounting fluid (pH 9.5). Examination of the membrane is performed under a dissecting scope with 10X magnification and ultraviolet (UV) illumination giving green fluorescing colonies. Quantitative results on spiked water samples were found comparable to standard membrane filter and most probable number (MPN) methods. Allowing growth before staining permits low-power, large-field microscopy.

A rapid test for coliform bacteria in water based on $C^{14}O_2$ evolution was described by Levin et al. (1959). Brilliant green lactose bile broth (BGLBB) containing 0.01% C^{14}-formate is shaken for several hours before use to reduce nonmetabolic $C^{14}O_2$. The shaken medium is added to a paraffin-coated planchet. A membrane filter is used to concentrate the bacteria from the water sample, and is then placed in the planchet. The planchet is placed in an ointment jar with the lid screwed tight and incubated at 37°C for 3 hr. A second planchet containing several drops of saturated barium hydroxide solution is added to the jar and incubation is continued for an additional hour. This planchet is removed, dried over a heat lamp, and counted in an internal flow Geiger counter. Noncoliform organisms that also liberate gas from formate interfere with the method.

Scott et al. (1964) described a variation of the foregoing method in which the membrane filter (Millipore HA 1 in. diameter 0.45 µm pore size) is placed in the bottom of a planchet and 0.5 ml of C^{14}-formate Endo medium is added. A glass cover slip is placed over the liquid and the planchet is agitated for 1 min and then placed in a 35°C incubator for 3-1/2 hr. The sample planchet is then held at room temperature for 5 min. The cover slip is removed, and a second planchet containing an absorbent pad impregnated with 5 drops of a saturated barium hydroxide solution is placed over the top. The double planchet, one inverted over the other, is incubated for 30 min, then gently shaken to release all CO_2, and the bottom half discarded. The upper planchet is dried under an infrared lamp for a few minutes and then covered with 5 drops of 0.5% collodion solution and dried until caramel brown under a heat lamp. The dried pads are counted in a windowless gas flow proportional counter for 5 min. The correlation with bacterial counts was not found constant, owing possibly to interference by other bacteria capable of utilizing formate.

The Millipore Corp. (1973a) has devised a type of "dip stick" for simultaneous collection and examination of water for bacteria.

The dip-stick sampler consists of a plastic tab supporting a membrane filter bonded to an absorbent pad containing dehydrated medium. When immersed in a liquid, the pad draws up 1 ml through the membrane, retaining any bacteria on the filter surface which develop into visible colonies with incubation. The Coli-Count Sampler is designed for enumeration of total and fecal coliform bacteria. Without sample dilution, the sampler provides a counting range of less than 100 to about 6000-8000 per 100 ml of water. The 18 ml chamber with the sampler is calibrated at 1.8 ml to permit a 1:10 dilution, raising the upper counting limit by a factor of 10.

Cotton et al. (1975), who are associated with the Millipore Corp., found that their comparisons of the coliform and total bacteria samplers correlated closely with standard techniques applied to many different types of water. Hedberg and Connor (1975) evaluated recovery by the Coli-Count Sampler using laboratory cultures of coliform bacteria, and found the counts significantly lower than by standard membrane filter procedures for total and fecal coliforms. Although the sampler is a convenient device, it does not appear acceptable for the precise enumeration required for official testing, and further lacks the sensitivity required for fecal coliforms in water because no more than 1 ml of sample can be examined.

B. Total Counts

Direct microscopic counts of bacteria can be made by utilizing the membrane filter. A procedure described by Harrigan and McCance (1976) suggested staining of cells recovered on the membrane by direct addition of Loeffler's methylene blue to the filter funnel, where it is held for 5 min. The staining solution is removed under vacuum and the filter rinsed with distilled water. The filter is dried in air and then floated on a small amount of immersion oil in a petri dish to render it transparent. It is then transferred to a microscopic slide for examination under an oil objective. The bacterial density can be calculated using the microscopic factor and volume of sample filtered.

A rapid method for determining total viable bacterial density based on incorporation of p^{32} was described by MacLeod et al. (1966). The procedure was evaluated only with laboratory cultures of *Aerobacter aerogenes*. Cells and p^{32} are added to an incubation medium (KCl, 0.005 M; $MgSO_4$, 0.002 M; glucose, 10 mg/ml; p^{32} in 0.05 M Tris buffer, 1 μc/ml), which is placed in a rotary water-bath shaker at 37°C for 1 hr. One milliliter of the incubation mixture is filtered through a sterile 25 mm Millipore HA filter (or Gelman Acropor ANW filter), followed by glass-distilled water. The filter is placed in a planchet and air-dried. Radioactivity is determined with a Geiger-Müller tube attached to a Picker scaler. A number of factors were found to influence p^{32} uptake, requiring standardization for application to natural materials. In addition to providing rapid results, the method was able to detect 23 cells per milliliter, thus providing good sensitivity.

Jannasch and Jones (1959) used three variations of a membrane filter technique for counting bacteria in seawater. The macrocolony technique is completed by incubation of a 47 mm diameter membrane after sample filtration on pads soaked with nutrient medium and then staining with Loeffler's methylene blue for examination by the naked eye. The microcolony technique uses 25 mm diameter filters and microscopic examination under 430-970X magnification. The Cholodny method involves concentration over a membrane filter in a funnel equipped with a collodium stirring loop and then direct microscopic examination on a glass slide. The concentrated sample is treated with 5% formaldehyde and then dried on a slide. The preparation is stained with 2% erythrosine in 5% phenol for 5 hr before examination.

Anderson and Heffernan (1965) observed that seawater contains a number of bacteria which pass through a 0.45 μm membrane filter but not a 0.22 μm. Most of the filterable isolates belonged to the genus *Spirillum*, and others were identified as *Leucothrix*, *Flavobacterium*, *Cytophaga*, and *Vibrio*.

Snell (1968) demonstrated that bacterial spores could be counted on the membrane filter using phase optics. Spore suspensions are filtered through a 0.22 or 0.45 μm 47 mm diameter membrane without the use of a rinse. The membrane is air-dried and mounted under a cover slip in Cargille's nondrying type A immersion oil. The spores appear dark under phase microscopy. The method was applied only to pure laboratory suspensions.

A procedure for direct counts of microorganisms from small samples using a membrane filter was described by Ecker and Lockhart (1959). A plain white membrane (Millipore HA) is divided into twelve 25 mm diameter areas by stamping with a metal die smeared with a thin film of slightly warmed 10% paraffin in petroleum jelly. The membrane is placed on a fritted glass base and small amounts of sample dilutions are added to each area with a microtiter pipette. The holder is then placed in a filter flask and vacuum applied. Diluted samples can be fixed simultaneously by preparing them in 0.85% NaCl containing 1% picric acid. Where dilutions are not required, a drop of the saline-picric acid solution is placed directly on each filter area after filtration and held for about 1 min. Staining is accomplished by adding a drop of 0.1% acid fuchsin (pH 3.0) on the filtering area, holding for 1 min, and then removing under vacuum. The membrane is taped to a glass slide, dried at 37°C for 15-20 min, and examined microscopically under oil immersion. Bacterial cells appear deep red on a clear to pink background.

Various staining techniques have been described to improve the contrast of nonchromogenic bacterial colonies on membrane filters. Cooke et al. (1957) applied 5 ml of 1% blue tetrazolium in 70% ethanol, washed the filter with distilled water, and then covered the membrane with 5-10 ml of 0.001% aqueous trypan blue. Colonies of *Pasteurella pestis* and *Malleomyces pseudomallei* appear white to faint pink on a blue background. These workers also tried staining by adding 10 ml of 0.1% aqueous quinacrine hydrochloride and allowing the stain solution to drain through the membrane without suction. Vacuum was applied and the filter was then washed with distilled water and covered with 5 ml of 0.05% aqueous vital red. The solution was removed and the membrane dried under vacuum. Colonies of *P. pestis* are lemon yellow versus a clear pink to light red background.

Lumpkins and Arveson (1968) first placed the membrane filter on a pad saturated with 10.0% nitric acid for 2 min, a pretreatment that improved the stain intensity of bacterial spores. The filter is then placed on a pad saturated with 0.1% acid fuchsin for 3 min, followed by transfer to a pad with 0.02% methylene blue for 2 min. The filter is oven-dried at 37°C and cleared with immersion oil (refractive index 1.500) for examination. Bacteria appear dark red against a faint blue background. Spores appear dark red at their periphery and light red in the center.

A modified technique of that described by Francisco et al. (1973), using a fluorescent stain for improving discrimination of bacteria from other particles on membrane filters, was described by Daley and Hobbie (1975). Acridine orange (5 mg/liter) is mixed with the sample for staining the bacteria, held for 3 min, and then filtered (Sartorius black 25 mm diameter, 0.45 μm pore size). After a 5 ml rinse with distilled water, the filter is removed and placed on a drop of nonfluorescent, low-viscosity immersion oil on a glass slide. A second drop of oil is placed on the filter and a glass cover slip is added. Bacterial cells appear as green fluorescing particles under epiillumination. Fixation of samples at the time of collection with formalin at a final concentration of 2% permits storage until examination.

Early in the investigation of the membrane filter for microbiology, it was reported that all microorganisms which could be cultured on a laboratory medium could also be cultured on the membrane when it was placed on the medium (Orlando and Bolduan, 1953). Qualitatively, this may still be essentially correct, but quantitatively it certainly is not. Taylor and Geldreich (1979) recently demonstrated, for example, that nine different nutrient media used with the membrane filter failed to perform equal to the standard pour plate method for total counts on potable waters. These workers developed a new formulation designated mSPC medium (peptone, 2.0%;

gelatin, 2.5%; glycerol, 1.0%; agar, 1.5%; pH, 7.1) that gave closer agreement when used with the membrane filter compared to the standard pour plate procedure. The medium is used with 0.45 μm pore size membranes and incubated at 35°C for 48 or 72 hr.

C. Salmonellas

The membrane filter presents an attractive device for the concentration of salmonellas from the large sample volumes that are required for reliable detection of this organism in water. However, the necessity of large sample volumes presents the always restrictive problem of filter clogging, especially pronounced with surface waters and sewage effluents. Demissie (1964) found, for example, that 4-5 hr was required for filtering 3 liters of polluted river water through a 47 mm diameter, 0.45 μm pore size membrane.

Kenner et al. (1957) used a preenrichment of 1 ml of water sample in double-strength selenite-brilliant green broth with and without sodium sulfapyridine at 35°C for 4-6 hr. The sample is then suspended in 30 ml of sterile dilution water and filtered through a 47 mm diameter membrane followed by two rinses of 30-35 ml each. The membrane is placed on a pad containing 2.0-2.2 ml of single-strength medium, inverted, and incubated at 35°C for 12-18 hr. Flat, dark orange colonies are selected for confirmation.

Tompkin et al. (1963) used the membrane filter (Millipore HA) for recovering *Salmonella* and *Shigella* from farm pond water, but clogging necessitated the use of several membranes for a single sample. After filtration, the membrane is immersed in selenite enrichment broth in a test tube with no more than two membranes per tube for enrichment.

Presnell and Andrews (1976) developed a technique that permitted analysis of larger volumes of water for both indicator bacteria and *Salmonella*. A filter aid, Celite (Johns-Manville), is added to the sample at a concentration of 0.1%. The sample is filtered through a 142 mm diameter, 0.45 μm pore size membrane (Millipore HA). The filter is blended with 100 ml of sterile 0.1% peptone water and the homogenate inoculated into Kauffman's tetrathionate broth with brilliant green at 0.00001% and into selenite broth. After 24 hr at 35°C, the enrichment broths are streaked onto selective agar media.

Kenner and Clark (1974) evaluated several glass fiber filters for concentrating *Salmonella* from water, and chose the Reeve Angel Corp. 984H 47 mm diameter filter for use. A newer glass filter, the Whatman GF/F, showed slightly better retentive properties than the 984H.

We have used in our laboratory both a large-diameter membrane filter (Sartorius 11306 0.45 μm pore size, 142 mm diameter) and the Whatman GF/F glass fiber filter (4.7 cm diameter) for concentration of salmonellae from water, sewage effluents, and sediment washings. A limited comparison of the two techniques found them equivalent for recovery of salmonellae from sewage effluent (Schiemann et al., 1978). The Gelman membrane filter tends to clog easier than the GF/F, and also adsorbs the brilliant green dye from the tetrathionate-brilliant green enrichment medium. It is usually possible to pass up to 1 liter of secondary sewage effluent through the GF/F filter without serious clogging. As Kenner and Clark (1974) previously showed, not all glass fiber filters have equal retentive properties. We found, for example, that the Gelman A-E glass fiber filter allowed passage of salmonellae, whereas the GF/F did not.

D. Staphylococci

Staphylococci in water have been of particular interest as indicators of swimming pool disinfection. A variety of media have been used with the membrane for recovery of staphylococci. Standard Methods (American Public Health Association, 1976) includes Chapman-Stone agar in a tentative membrane filter method for staphylococci in water. Favero et al. (1964) found Chapman-Stone agar slightly superior to phenol red mannitol salt agar and Vogel-Johnson agar for enumeration of staphylococci in swimming pool water. The confirmation rate for *Staphylococcus aureus* was highest with Vogel-Johnson agar, but total staphylococci counts were lower on this medium. Crone and Tee (1974) used staphylococcal medium 110(Oxoid) with addition of bromthymol blue and incubation at 30°C for 48 hr for recovery of staphylococci from swimming pool water. Alico and Palenchar (1975) found that the type of membrane filter used with m-Staphylylococcus broth plus 1.7% agar had a distinct influence on recovery, with the Nuclepore filter giving consistently lower results. These workers also pointed out that reliance on yellow-gold pigmentation with m-Staphylococcus broth for identification of *S. aureus* is not reliable.

Our own experience with Chapman-Stone agar showed several disadvantages for recovery of staphylococci from water by membrane filtration. The medium is not highly selective; colonies tend to spread and show no differential characteristics or contrast with the membrane. The technique employed by our laboratory uses tellurite-polymyxin egg yolk (TPEY) agar. The medium is poured in 15 x 60 mm petri dishes which are inverted for incubation in a sealed chamber containing moist paper towels at 35°C for 48 hr. Colonies of *Staphylococcus* are deep black in color and are smaller and not mucoid like those of *Bacillus*, which also grows on this medium. Confirmation is usually limited to a catalase test performed by direct addition of 3% hydrogen peroxide to the colony on the membrane. If more reliable confirmation is desired, the colony is fished to brain heart infusion agar and then examined after overnight incubation for catalase, Gram's stain, and OF dextrose. A test for coagulase or thermonuclease is completed if identification as *S. aureus* is desired. In studies involving a number of different swimming pools, a total of 1834 colonies were selected from membrane filters on TPEY agar for confirmation and 1555 or 84.8% confirmed as *Staphylococcus* (i.e., gram-positive, catalase-positive cocci).

Our laboratory has also developed a technique for recovery of staphylococci from the surface film of pool water. A dry sterile Nuclepore filter (Nuclepore Corp. No. 1111-7, 47 mm diameter, 0.4 μm pore size) is placed with the aid of a forceps on the surface of the water and picked up after a few seconds of contact. The filter is immediately immersed in 25 ml of Butterfield's phosphate buffer (pH 7.2) with 0.85% saline for transport to the laboratory. Analy-

TABLE 1 Recovery of Staphylococci from Pool Water and Surface Film

	Water	Surface film
Number of comparisons	170	170
Number samples positive[a]	94 (55.3%)	157 (92.4%)
Median count	2/100 ml	50/filter area
90th percentile count	70/100 ml	>400/filter area[b]

[a]≥2/100 ml for water and ≥5/filter area for surface film.
[b]33 counts (19.4%) >400/filter area.

sis begins with mechanical shaking for 15 min, after which the membrane is removed and discarded. The buffer or dilutions are filtered through a standard 0.45 μm membrane which is placed on TPEY agar for incubation. Table 1 shows the results for 170 comparisons between staphylococci densities in the pool water vs. the surface film. Considering the small water volume actually sampled by the Nuclepore filter, which covers a surface area of 17.3 cm^2, the concentration of staphylococci in the surface film is many times that in the subsurface water.

E. Clostridium perfringens

Taylor and Burman (1964) have suggested that spores of *Clostridium perfringens* in water can be counted on membrane filters using Wilson and Blair glucose-sulfite medium. After filtration of the heated sample, the membrane is placed face down on the surface of the medium and then covered with a layer of nutrient agar. After incubation at 44 or 45°C for 24 hr, *C. perfringens* will appear as black colonies with large black halos.

Cabelli (1977) has described a membrane filter method for isolation of *C. perfringens* spores from water. Ten milliliters of sterile nutrient agar are poured into a 15 x 100 mm petri dish, which is dried at 37°C for 30 min just before use. The water sample is heated to 75°C for 20 min and passed through a 0.45 μm membrane filter. The filter is placed on the nutrient agar and dried for 30 min at 37°C. An overlay of 18 ml of sulfite-glucose iron agar (per liter nutrient agar: glucose, 20 g; 10% $MaSO_3$, 1 ml; 8%

$FeSO_4$, 5 drops; adjusted to pH 7.6 and autoclaved at 121°C for 20 min) is carefully added. After the overlay has hardened, an additional 15 ml of nutrient agar is added. After hardening of this third layer, the plates are incubated anaerobically at 37°C for 24-44 hr and all black colonies are counted. Confirmations, which were not described by Cabelli, could easily be performed by stabbing the subsurface colony with a needle and inoculating tubes of lactose-gelatin (Hauschild and Hilsheimer, 1974a) and nitrate-motility agars (Hauschild and Hilsheimer, 1974b).

Bisson and Cabelli (1979) have recently reported another membrane filter method for rapid quantitation of *C. perfringens* in water. The medium (mCP) contains (g/100 ml): tryptose, 3.0; yeast extract, 2.0; sucrose, 0.5; L-cysteine, 0.1; $mGSO_4 \cdot 7H_2O$, 0.01; bromcresol purple, 0.004; agar, 1.5. The pH is adjusted to 7.6 before autoclaving. After cooling the following are added per 100 ml: D-cycloserine, 40 mg; polymyxin-B sulfate, 2.5 mg; indoxyl-β-D-glucoside, 60 mg; phenolphthalein, 2.0 ml of a 0.5% filter-sterilized solution; $FeCl_3 \cdot 6H_2O$, 0.2 ml of a 4.5% filter-sterilized solution. Incubation is completed at 45°C for 18-24 hr. Differential characteristics for confirmation include fermentation of sucrose, production of acid phosphatase, and absence of β-D-glucosidase activity.

F. Tracer Bacteria

Tracer bacteria, especially *Serratia* (Robson, 1956), have been used for the purpose of identifying the source of water contamination or describing dispersion patterns of pollution in the environment. Berlin and Rylander (1963) described a tracer method using radioactive bacteria and an autoradiogram of a membrane filter. The bacterium, *E. coli* B, is tagged by growth in a medium containing S^{35}. The bacteria are recovered by diluting 1 ml of water sample in 500 ml of 0.9% NaCl and filtering through a 0.45 μm membrane (Millipore HA). The filter is dried at 50°C for 20 min and applied to X-ray film. After 10 days of exposure, the film is developed. Black spots on the film representing bacteria are counted under 10X magnification. The method was successful with heavily polluted water and showed no loss with storage that resulted in a decline of viable cell counts. "Background spots" produced by normal waters limit the sensitivity of the method.

G. Iron Bacteria

The membrane filter provides a convenient device for concentrating iron bacteria from water compared to older techniques such as centrifugation. Lueschow and Mackenthun (1962) used a 25-mm membrane filter with a 0.45 μm pore size for this purpose. Five milliliters of sterile water are placed on the filter, the sample is added and vacuum applied. The filter is dried in an oven at 100°C and placed on a slide on top of 2 drops of immersion oil. Additional oil is added on top of the filter and a cover slip provided for microscopic examination.

We have found in our laboratory that the membrane filter, cleared in immersion oil after drying, can be easily examined for iron bacteria under dark-field microscopy.

H. Yersiniae

There have been three reports during the past few years implicating water as the vehicle of transmission in outbreaks of human yersiniosis (Eden et al., 1977; Keet, 1974; Lassen, 1972). Highsmith et al. (1977) have recommended a membrane filter procedure for recovery of *Yersinia enterocolitica* from water. Two aliquots of the water sample are filtered, one membrane placed on m-Endo broth and incubated at 25°C for 3 days, the other immersed in cooked meat broth and incubated at 4°C for 21 days. Colonies from m-Endo broth are streaked on MacConkey agar, and 0.1 ml of the cooked meat cold enrichment is plated on MacConkey agar after 7, 14, and 21 days of incubation. Biochemical characterization of suspect isolates is required for definitive identification.

We have used a membrane filter technique in our laboratory for isolating *Y. enterocolitica* from water, which begins with the standard membrane filter method for coliform bacteria. After incubation of the membrane on m-Endo agar LES at 35°C for 24 hr, the plate is held an additional 24 hr at room temperature. Colonies of *Y. enterocolitica* on the membrane are usually small and clear to pinkish in color, often appearing as small water drops on the filter. Although we found this technique superior to enrichment of membranes at 4°C in Christenson's buffer (K_2HPO_4, 0.2%, NaCl, 0.85%; mannitol, 1.0%; pH 7.3), the membrane is frequently overgrown with coliform or other flora, especially with surface water samples, so that colonies of *Y. enterocolitica* cannot be detected (Schiemann, 1978).

I. Bacteriophages

Bacteriophages, especially coliphages, have periodically been proposed as potential indicators of water pollution and suggested as more reliable indicators of virus contamination. Loehr and Schwegler (1965) described a membrane filter technique for detecting coliphage using laboratory cultures of *E. coli* B as the host and T2 phage suspensions. The phage is mixed with a suspension of the host and mixed at 45 rpm for 3 min to improve phage adsorption on host cells. Samples of the mixture are filtered through a sterile membrane. The filter is placed on an absorbent pad saturated with m-Endo broth held in a petri dish. The dish is inverted and incubated at 37°C for 24 hr, at which time the phage plaques are counted and the titer of the phage solution is calculated. The optimal quantity of host bacteria in a filtered sample is 10^9-10^{10}. This filtration method showed lower recovery of phage than the soft-agar method. It does have, however, the benefit of a much greater sensitivity, in that larger volumes of sample can be examined. It is undoubtedly the low sensitivity of usual methods for detecting phage, such as the soft-agar overlay technique, and the lack of a universal host for all types of phages, that have discouraged more serious consideration of bacteriophages as indicators of water pollution.

Grant (1973) suggested a membrane filter technique for isolation of different bacteriophages by a series of "bacterial sandwiches." The sample material is first prepared by removal of debris with centrifugation and filtration through a 0.45 μm membrane filter to remove bacteria. The bacterial sandwich is prepared with a logarithmic-phase host bacterial culture which is layered on a Nucleopore polycarbonate membrane. A second membrane is layered on the organisms followed by a sterile grid support (General Electric Co., Lexan). A second and third layer of different host cells can be stacked on top of the first. The sample material is then passed slowly through the layers in series. The organisms are removed by washing into growth medium, to which is added a logarithmic-phase subculture of the host organism for bacteriophage production. The method is, therefore, more a qualitative recovery-propagation technique than a method for quantitative enumeration of bacteriophage.

III. MISCELLANEOUS APPLICATIONS

A. Aerobiology

The membrane filter has been employed in both direct and indirect methods for recovery of microorganisms from air. Gordon and Boyd (1953) pointed out that the membrane was a convenient way for concentrating bacteria from the large volumes of liquid required in a glass venturi scrubber which operates at 25 ft^3/min. Gordon et al. (1954) used such a sampling device as well as a Shipe tangential jet sampler for recovering anthrax bacilli from air. After passing the scrubbing fluid through a membrane filter, the trapped material is washed off with a small quantity of nutrient broth. The broth is used to streak agar plates and inoculate mice for recovery of the organism.

Agosti et al. (1965) used membrane filters (Millipore 0.45 μm pore size) for recovering *S. aureus* and *Pseudomonas aeruginosa* from air in a tuberculosis sanatorium. Chapman agar was used for *Staphylococcus* and meat infusion agar with 500 μg/ml of nitrofurantoin for isolation of *Pseudomonas*.

Vlodavec (1963) collected airborne *Serratia marcescens* on a membrane filter and then stained the membrane with 3% carbol-erythrosin followed by immersion oil for direct microscopic examination. Jost and Fey (1970) used a membrane filter-fluorescent antibody technique for recovery of bacteria from experimentally created aerosols. The air is drawn through a dry membrane (Millipore HABP 047) held in an aerosol filter holder (Millipore Corp.). The filter is washed before use with 50 ml of PBS. After sampling, 30 mm^2 pieces are stamped out and placed on a slide. A drop of conjugate is added to each piece and the slide placed in a humid chamber for 50 min. The filter pieces are washed slowly with 100 ml of PBS and then mounted in carbonate-bicarbonate buffered glycerol (pH 9.0) under a cover slip for fluorescence microscopy. Especially critical in the procedure is the quality of the conjugate. Several problems remain unsolved with the method, including background fluorescence, penetration of cells into the filter, fading of the preparation, and cell autolysis.

Fields et al. (1974) examined the use of membrane filter field monitors for recovery of bacteria from air. Six monitors are placed on a manifold and air is drawn through at 2 ft^3/min (about 0.056^3/min). The membrane is removed from the monitor and placed directly on trypticase soy agar for incubation at 35°C for 24, 48, or 72 hr. A membrane with a pore size of 0.8 μm recovered higher numbers of microorganisms, but differences were not statistically significant when compared to other pore sizes. The membrane recovered 79% of the number of organisms detected by Reynier slit samplers. Although this recovery was lower, the correlation in methods was high.

The Millipore Corp. (1972) has devised an air sampling impinger apparatus which is assembled with the Sterifil filter system. Twenty milliliters of impingement fluid (gelatin, 0.2%; disodium phosphate, 0.4%; brain heart infusion, 3.7%; octyl alcohol, 0.1 ml/liter) is placed in the Sterifil funnel previously assembled with a membrane filter and rubber cap on the outlet and sterilized. A piece of tubing is attached to the underside of the funnel cover and cut off so that it ends just above the fluid surface. Vacuum is applied to a threaded hose connector holding a limiting orifice and attached to the top of the funnel cover. After sampling, the impingement fluid is drawn through the membrane filter under vacuum and the filter is removed from the assembly to place on a suitable culture medium.

Fincher and Mallison (1967) have cautioned that direct recovery of bacteria from air on the membrane filter is usually quantitatively lower than other sampling techniques owing to desiccation during collection. Sartorius (SM-Report 3) has attempted to overcome this problem by development of a gelatin membrane (SM12602, 3 μm pore size). The gelatin membrane can be placed directly on nutrient media or can be dissolved by stirring in 30 ml of normal saline solution warmed to 30-35°C. Dissolution allows the preparation of dilutions and differential analysis by membrane filtration and suitable culture media.

B. Disinfectant Testing

A promising method using the membrane filter, which has several advantages over the standard AOAC use-dilution method, for evaluation of bactericidal activity of disinfectants was described by Ko and Vanderwyk (1968). Five milliliters of sterile water is placed in the funnel of a filtering apparatus previously assembled with a membrane filter (Millipore DAWB04700) and steam-sterilized. Two milliliters of a brain heart infusion suspension with 2×10^4 to 4×10^4 cells of the test organism per milliliter is added, the contents swirled, and vacuum applied. Five such membranes are prepared and placed on trypticase soy agar incubated at 37°C for 24 hr to obtain the suspension density used for calculating percent kill. A neutralizer control is provided by filtering 5 ml of the disinfectant solution, then 10 ml of neutralizer and 25 ml of sterile rinse water. Two milliliters of the organism suspension is then filtered and the membrane placed on medium to obtain a colony count. To test the disinfectant, 2 ml of the organism suspension is filtered and the vacuum is shut off. Five milliliters of the disinfectant solution is placed on the membrane and after exactly 60 min contact (or other selected contact time) 10 ml of neutralizer are added and the mixture is filtered rapidly. The membrane is immediately flushed with 25 ml of sterile water. Treated membranes are placed on trypticase soy agar and incubated at 37°C for 24 hr. Colonies are counted and the percent reduction is calculated. This method has the distinct advantage over the carrier technique of not being subject to loss of cells by washing off, providing a more uniform distribution on a surface, and eliminating the possibility of carryover to the growth medium.

C. Bacteremia

Winn et al. (1966) described a membrane filter method for diagnosis of bacteremia which had the advantage of rapid identification of the agent over conventional blood culture methods. Erythrocytes are sedimented by 3% dextran in distilled water containing 0.6% NaCl and 1% sodium citrate. The leucocytes are centrifuged at 1500 rpm for 5 min, washed in 0.6% NaCl plus 1% sodium citrate, and recentrifuged (85 rev/min for 5 min) to separate them from extracellular bacteria and plasma. The supernatant plasma and the supernatant fluid from the leucocyte washing are separately filtered through a 0.45 μm, 47 mm diameter membrane. The filter is washed with an isotonic solution (0.6% NaCl plus 1.0% sodium citrate, pH 7.4) and placed on blood agar and eosin methylene blue agar, which are incubated at 37°C. Other special media, such as chocolate agar or Lowenstein-Jensen, and special incubation conditions, including CO_2 or anaerobiosis, can be used for isolation of particular organisms. Field monitors (Millipore Corp.) can be used in a closed system attached to a sample tube, wash solution reservoir, and a filtrate reservoir that is under vacuum. The filter is removed and cut in half for placement on blood and eosin methylene blue agars.

Braun and Kelsh (1954) used a membrane filter technique to recover *Brucella* from blood which had several advantages over available blood culture methods, including provision of a quantitative estimation of the extent of the bacteremia. One milliliter of freshly drawn blood is transferred to a sterile tube containing 0.1 ml of heparin solution (100 U.S.P. units). Three milliliters of tryptose saline is used to wash down the sides of the tube. The mixture is centrifuged at 2500 rpm for 30 min. The supernate is drawn off and the cells suspended in 3 ml of sterile distilled water and allowed to stand for 30 min. The sample is then filtered through a sterile membrane under vacuum, which requires about 30-40 min. The membrane is placed on the surface of a solid medium for *Brucella* and incubated. Colonies appear within 3-4 days and can be detected easier by incorporation of 0.5% triphenyltetrazolium chloride (TTC) in the medium.

Stanaszek (1969) found that yeasts could be isolated from blood by membrane filtration within 48 hr compared to 3-6 days for standard methods. She also found that diagnosis of bacteremia using membrane filtration was equally reliable to blood culture systems and gave more rapid results.

D. PPLO Organisms (Mycoplasma)

Barile (1962) described a technique for cultivation of pleuropneumonia-like organisms (PPLO, now *Mycoplasma*) using the membrane filter, which had certain advantages over growth on agar where the colonies embed themselves in an agar matrix. A suspension of the organism is inoculated onto a 47 mm diameter membrane (Millipore DA, HA, or PH) previously placed on PPLO agar medium. The inverted plate is sealed with tape or enclosed in a jar and incubated at 37°C. PPLO colonies appear on the membrane in 5-10 days. The filter is removed from the medium and fixed by air drying or isopropyl alcohol at 4°C for 30 min. PPLO colonies are stained by immersion in aqueous methylene blue-azure stain or crystal violet in isopropyl alcohol for about 10 min. The stained membrane is washed three times in PBS (pH 7.4), rinsed with water, and air-dried on a blotter at 37°C for 1 hr. The membrane is cut into 2 cm sections and made transparent with low-viscosity oil (index of refraction 1.515) placed under the membrane to prevent air bubbles. The section of membrane is placed on a slide containing a drop of oil, a cover glass is provided, and another drop of oil added for microscopic examination.

E. Vibrio fetus in Bovine Semen

Shepler et al. (1963) found that the membrane filter with a 0.50 or 0.60 μm pore size was useful for removing interfering bacteria in the isolation of *Vibrio fetus* from bovine preputial fluid. The filters are used in a Swinny adapter attached to a Luer-Lok glass syringe. The filtrate is collected in sterile tubes and 0.08 ml is streaked onto brain heart infusion agar containing 10% bovine blood. The developers found some loss of *V. fetus* cells with filtration, and recommended the method only as an adjunct to direct culture on antibiotic media.

Winter et al. (1965) further evaluated the membrane filter technique for recovery of *V. fetus* from bovine semen. One-half milliliter of semen is first diluted with 4.5 ml of thiol broth and centrifuged at 22°C for 10 min at 150 g. Three milliliters of supernatant is drawn into a syringe and filtered through a 0.60 μm filter held in a Swinny adapter. The first milliliter of filtrate is discarded and 0.1 ml of the remainder is spread onto a blood agar plate that is incubated at 37°C in an atmosphere of 85% nitrogen, 10% oxygen, and 5% carbon dioxide. The loss of cells with filtration decreased the recovery rate when the concentration of *V. fetus* in the semen was low. The membrane filter method was recommended only when initial isolation by direct culture failed because of overgrowth by bacterial contaminants.

F. Mycobacteria

Rogers et al. (1955) examined the usefulness of the membrane filter for recovering tubercle bacilli from respiratory tract secretions. A mouth wash sample is first digested to reduce viscosity and eliminate normal flora. The mixture is then passed through a membrane filter, which is placed on Dubos oleic acid-albumin agar with trypan blue. Plates are inverted and sealed in cans with moistened cotton for high humidity during incubation. Growth is slower on the membrane than on medium alone with 21-30 days of incubation required compared to 16-21 days.

Sato and Fieldsteel (1977) used a membrane filter technique for enumeration of *Mycobacterium leprae*. The supernatant from a tissue preparation is drawn into a syringe and filtered through a polycarbonate membrane (Nucleopore 25 mm diameter, 0.4 μm pore size) held in a stainless steel microsyringe holder (Millipore Corp.). The filter assembly is flushed with air to force any remaining fluid through the filter. The filter is removed and placed in the center of a glass slide and allowed to dry. It is then fixed in formalin fumes and stained by a standard acid-fast stain. Chloroform is used to increase transparency of the filter for microscopic examination.

A membrane filter method for detection of *Mycobacterium tuberculosis* in sputum was developed by the Millipore Corp. (Application Note AN-232). The procedure begins with addition of 5 ml of 4% sodium hydroxide to 10 ml of sputum followed by 15 min of contact at room temperature. Twenty milliliters of Sputolysin (Calbiochem) is added for digestion. The sample is agitated for 1 min and then filtered through a 0.8 μm membrane followed by three 50 ml saline rinses. The membrane is then transferred to Middlebrook 7-H-10 medium and incubated in 10% carbon dioxide at 35°C for 6-8 days.

Wright and Mallmann (1963) described a method for staining acid-fast bacilli on the membrane filter. One to two milliliters of basic fuchsin solution is added to the wet membrane and allowed to stand for 30 min. The solution is removed under vacuum and the filter rinsed with decolorizing solution (3% concentrated HCl in 95% ethanol, filtered). Five milliliters of decolorizing solution is added and allowed to stand for 2 min. About 5 ml of distilled water is added and the mixture is filtered. The filter is removed and placed between two filter pads and weighted down firmly on a flat surface to dry for 10-15 min. The dried filter is immersed in a beaker containing light green counterstain (light green SF yellowish in 95% ethanol, 1:10,000) for 5-6 min. The filter is rinsed with distilled water until all excess stain is removed, and then dried between two filter pads. The dried filter is dipped in xylene and placed on a slide with permount. Permount is added to the surface and a cover slip provided. The mounted filter is dried at 37°C for 30 min before examination. Acid-fast bacilli appear dark red on a light green background.

G. Streptoccoccus mutans

Streptococcus mutans has been associated with the development of dental caries. The Millipore Corp. has conducted some studies on a membrane filter procedure for isolation of this organism (Millipore Corp., unpublished data). Preliminary results indicate that the membrane filter (Millipore HAWG047S1) gives equivalent recovery to spread plates with mitis-salivarius agar incubated at 37°C for 24-48 hr in an atmosphere of 90% nitrogen and 10% carbon dioxide.

H. Oral Treponemes

Loesche and Socransky (1962) used the ability of spirochetes to penetrate membrane filters to separate oral treponemes from other bacteria. A sterile filter of 0.10, 0.05, or 0.01 μm pore size is placed on the surface of PPLO medium containing 1.2% agar. Sample material from the mouth is placed in 2 ml of PPLO broth or phosphate buffer (pH 7.0) and mixed. The mixture is added with a Pasteur pipette to the filter surface. The inoculated filter plates are incubated in an upright position in a Brewer jar with an atmosphere of 95% hydrogen and 5% carbon dioxide at 37°C for 5-8 days.

I. Sulfate-Reducing Bacteria

Tsuneishi and Goetz (1958) demonstrated that the sulfur-reducing bacterium *Desulfovibrio aestuarii* could be recovered on membrane filters. After concentration from the water sample by filtration, the membrane is cut in two and one section is placed downward and the other upward in a thin layer (about 1/8 in.) of nutrient agar (0.75% agar). A thick layer (about 3/8 in.) of nutrient agar (2% agar), heavily inoculated with a 24-hr culture of *Serratia marcescens*, is poured on top of the first layer and sections of filter. This aerobic organism serves as an oxygen scavenger and provides a surface barrier to oxygen diffusion into the medium. Anaerobes, distinguished by black coloration in the area of the colony, will grow best on the inverted filter within 24-36 hr at 30°C. The composition of the nutrient medium is (per liter): K_2PHO_4, 0.6 g; $MgSO_4$, 0.3 g; $Fe(NH_4)_2$ (SO_4), 0.2 g; sodium lactate, 18.0 g; ascorbic acid, 0.3 g; sodium thioglycollate, 0.3 g; yeast extract, 3.0 g; NaCl, 30.0 g; Bacto-peptone, 3.0 g; pH 7.2. A 1% solution of Fe $(NH_4)_2(SO_4)_2$ is prepared in sterile water and added at a rate of 1 ml/100 ml of medium to liquid agar just prior to pouring the plates.

J. Sequential Transfer

Probably one of the most significant new developments in use of membrane filters, and one that will no doubt see further applications, is the use of sequential transfer for completing diagnostic biochemical tests on recovered organisms. This technique has been used in a method for recovery of *E. coli* from foods where the membrane is transferred after incubation to a pad saturated with reagent for the indole test. Dufour and Cabelli (1975), using a specially formulated medium for initial recovery of coliform bacteria on the membrane filter, designed a sequential transfer scheme for completing tests for urease, oxidase, and indole. The filter with lactose-positive colonies (blue) is removed from the growth medium and placed on a pad saturated with urea substrate (urea, 2 g; phenol red, 10 mg; distilled water, 100 ml; pH 5.0 ±0.2). Lactose-positive colonies are first marked by piercing the filter adjacent to the colony with a needle. After 15-20 min at room temperature, urea-positive colonies are identifiable by their magenta color, and these colonies are marked a second time. The filter is then transferred to a pad saturated with 1% aqueous N,N,N´,N´-tetramethyl-p-phenylenediamine dihydrochloride. Oxidase-positive colonies show a deep purple color within 10 sec and are obliterated with a wire loop. The filter is then transferred immediately to another pad saturated with tryptophan substrate (trypticase, 0.5 g; tryptophan, 0.2 g; distilled water, 100 ml; pH 7.2; sterilized by filtration).

After 20 min at room temperature, the filter is again transferred to a pad saturated with indole reagent (p-dimethylaminobenzaldehyde, 5.0 g; ethanol, 90 ml; concentrated HCl, 10 ml). Indole-positive colonies show a red color or red halo within 10 min.

Watkins et al. (1976) also used a sequential transfer scheme for biochemical identification of *Vibrio parahaemolyticus* recovered on the membrane filter with mVP agar. Typical colonies of *V. parahaemolyticus* on this medium are circular, 0.7-2.5 mm in diameter, convex or raised, and green in color. Each typical colony on the filter is marked by piercing the membrane adjacent to the colony with a needle. The filter is removed and placed on galactose medium (polypeptone peptone, 1.0 g; yeast extract, 0.7 g; galactose, 1.0 g; NaCl, 3.0 g; bromthymol blue, 0.0025 g; agar, 1.5 g; distilled water, 100 ml; boiled and cooled, sodium cholate, 0.15 g added; pH adjusted to 8.0 ±0.2). After 2 hr at 41°C, galactose-positive colonies are a yellow color, and are again marked by piercing the membrane with a needle. The filter is then transferred to a pad saturated with PBS (pH 7.6) for 10 min at room temperature to neutralize the acid produced from galactose fermentation. The filter is then moved to a pad containing sucrose medium (same formulation as galactose medium with sucrose instead of galactose). After 2 hr at 41°C sucrose-positive colonies are yellow or have yellow halos that can be observed from the reverse side of the plate. The filter is then transferred to a pad saturated with oxidase reagent, producing a dark purple halo within 20-30 sec for oxidate-positive colonies. *V. parahaemolyticus* is identified as typical colonies that are galactose-positive, sucrose-negative, and oxidase-positive. The developers of this procedure state that they have also devised in situ tests for lactose and mannitol fermentation, starch hydrolysis, and hemolysis.

IV. CONCLUDING REMARKS

The membrane filter is a versatile tool that will undoubtedly remain a permanent part of microbiology. Although some early applications may have been a bit hasty in retrospect, the potential for correcting deficiencies--such as the inability to recover injured bacteria--appears promising, and it is not unrealistic to expect that most of these shortcomings will eventually be overcome. The alternative would be to return to methods used prior to the availability of the membrane, which most microbiologists would consider an undesirable regression.

The versatility of the membrane filter is apparent from the many types of applications described in this review, that has in fact been confined to a relatively narrow subject area. The variety of techniques in which the membrane filter has been used testifies further to the ingenuity of microbiologists, which has hopefully not diminished. If this compilation of methods will spark one new idea, or stimulate further investigation and development of techniques that have been suggested by others, it could serve no greater purpose.

REFERENCES

Agosti, E., Bernardini, V., and Baronti, A. (1965). Use of membrane filters to isolate *Staphylococcus aureus* and *Pseudomonas aeruginosa* from the air in a tuberculous sanatorium environment (Italian). *Ann. Med. Sondalo 13*: 287—299.

Alico, R. K., and Palenchar, C. A. (1975). *Staphylococcus aureus* recoveries on various brands of membrane filters. *Health Lab. Sci. 12*: 341—346.

American Public Health Association (1976). *Standard Methods for the Examination of Water and Wastewater*, 14th ed. American Public Health Association, Washington, D.C.

Anderson, J. I. W., and Heffernan, W. P. (1965). Isolation and characterization of filterable marine bacteria. *J. Bacteriol. 90*: 1713—1718.

Barile, M. (1962). Cultivation of pleuropneumonia-like organisms on membrane discs for microscopic examination. *J. Bacteriol. 83*: 430—432.

Berlin, M., and Rylander, R. (1963). Autoradiographic detection of radiactive bacteria introduced into sea water and sewage. *J. Hyg. Camb. 61*: 307—315.

Bisson, J. W., and Cabelli, V. J. (1979). Membrane filter enumeration method for *Clostridium perfringens*. *Appl. Environ. Microbiol. 37:* 55-66.

Braun, W., and Kelsh, J. (1954). Improved method for cultivation of *Brucella* from the blood. *Proc. Soc. Exp. Biol. Med. 85*: 154–155.

Cabelli, V. J. (1977). *Clostridium perfringens* as a water quality indicator. In *Bacterial Hazards/Health Hazards Associated with Water,* A. W. Hoadley and B. J. Dutka (Eds.). American Society for Testing and Materials, Philadelphia, pp. 65–79.

Cooke, G. M., Duffy, E., and Wolochow, H. (1957). Staining of bacterial colonies on the Millipore filter to improve contrast. *Stain Technol. 32*: 63–66.

Cotton, R. A., Sladek, K. J., and Sohn, B. I. (1975). Evaluation of a single-step bacterial pollution monitor. *J. Am. Water Works Assoc. 67*: 449–451.

Crone, P. B., and Tee, G. H. (1974). Staphylococci in swimming pool water. *J. Hyg. Camb. 73*: 213–220.

Daley, R. J., and Hobbie, J. E. (1975). Direct counts of aquatic bacteria by a modified epifluorescence technique. *Limnol. Oceanogr. 20*: 875–882.

Demissie, A. (1964). The isolation of *Salmonella* in a Swedish water course (the River Fyris). *Acta Pathol. Microbiol. Scand. 62*: 409–416.

Dufour, A. P., and Cabelli, V. J. (1975). Membrane filter procedure for enumerating the component genera of the coliform group in seawater. *Appl. Microbiol. 29*: 826–833.

Ecker, R. E., and Lockhart, W. R. (1959). A rapid membrane filter method for direct counts of microorganisms from small samples. *J. Bacteriol. 77*: 173–176.

Eden, K. V., Rosenberg, M. L., Stoopler, M., Wood, B., Highsmith, A. K., Skaliy, P., Wells, J., and Feeley, J. C. (1977). Waterborne gastroenteritis at a ski resort associated with isolation of *Yersinia enterocolitica*. *Public Health Rep. 92*: 245–250.

Favero, M. S., Drake, C. H., and Randall, G. B. (1964). Use of staphylococci as indicators of swimming pool pollution. *Public Health Rep. 79*: 61–70.

Fields, N. D., Oxborrow, G. S., Puleo, J. R., and Herring, C. M. (1974). Evaluation of membrane filter field monitors for microbiological air sampling. *Appl. Microbiol. 27*: 517–520.

Fincher, E. L., and Mallison, G. F. (1967). Intramural sampling of airborne microorganisms. In *Air Sampling Instruments Manual,* 3rd ed. Am. Conf. Govt. Ind. Hyg., Cincinnati, Ohio.

Francisco, D. E., Mah, R. A., and Rabin, A. C. (1973). Acridine orange-epifluorescence technique for counting bacteria in natural waters. *Trans. Am. Microsc. Soc. 92*: 416–421.

Gordon, M. A., and Boyd, F. M. (1953). Large-scale biological air sampling with the Venturi scrubber. *Bacteriol. Proc. G5*: 26.

Gordon, M. A., Moody, M. D., Barton, A. M., and Boyd, F. M. (1954). Industrial air-sampling for anthrax bacteria. *Arch. Ind. Hyg. Occup. Med. 54*: 16–22.

Grant, J. (1973). Membrane filter techniques for the study of bacteriophage isolation and replication kinetics. *Am. Lab. 5*: 17–23.

Guthrie, R. K., and Reeder, D. J. (1969). Membrane filter-fluorescent-antibody method for detection and enumeration of bacteria in water. *Appl. Microbiol. 17*: 399–401.

Harrigan, W. F., and McCance, M. E. (1976). *Laboratory Methods in Food and Dairy Microbiology*. Academic Press, London, pp. 43–45.

Hauschild, A. H. W., and Hilsheimer, R. (1974a). Enumeration of food-borne *Clostridium perfringens* in egg yolk-free tryptose-sulfite-cycloserine agar. *Appl. Microbiol. 27*: 78–82.

Hauschild, A. H. W., and Hilsheimer, R. (1974b). Evaluation and modification of media for enumeration of *Clostridium perfringens*. *Appl. Microbiol. 27:* 78-82.

Hedberg, M., and Connor, D. A. (1975). Evaluation of the Coli-Count Sampler for possible use in standard counting of total and fecal coliforms in recreational waters. *Appl. Microbiol. 30:* 881–883.

Highsmith, A. K., Feeley, J. C., and Morris, G. K. (1977). *Yersinia enterocolitica:* a review of the bacterium and recommended laboratory methodology. *Health Lab. Sci. 14*: 253–260.

Jannasch, H. W., and Jones, G. E. (1959). Bacterial populations in sea water as determined by different methods of enumeration. *Limnol. Oceanogr. 4*: 128–139.

Jost, R., and Fey, H. (1970). Rapid detection of small numbers of airborne bacteria by a membrane fluorescent-antibody technique. *Appl. Microbiol. 20*: 861–865.

Keet, E. E. (1974). *Yersinia enterocolitica* septicemia. *N.Y. State J. Med. 74*: 2226–2230.

Kenner, B. A., and Clark, H. P. (1974). Detection and enumeration of *Salmonella* and *Pseudomonas aeruginosa*. *J. Water Pollut. Control. Fed. 46:* 2163–2171.

Kenner, B. A., Rockwood, S. W., and Kabler, P. W. (1957). Isolation of members of the genus *Salmonella* by membrane filter procedures. *Appl. Microbiol. 5*: 305–307.

Ko, J. H., and Vanderwyk, R. W. (1968). Membrane filter technique for testing the bactericidal activity of iodophors. *J. Pharm. Sci. 57*: 2013–2015.

Lassen, J. (1972). *Yersinia enterocolitica* in drinking water. *Scand. J. Infect. Dis. 4*: 125–127.

Levin, G. V., Harrison, V. R., Hess, W. C., Heim, A. H., and Stauss, V. L. (1959). Rapid, radioactive test for coliform organisms. *J. Am. Water Works Assoc. 51:* 101–104.

Loehr, R. C., and Schwegler, D. T. (1965). Filtration method for bacteriophage detection. *Appl. Microbiol. 13*: 1005–1009.

Loesche, W. J., and Socransky, S. S. (1962). Defect in small Millipore filters disclosed by new technique for isolating oral treponemes. *Science 138*: 139–140.

Lueschow, L. A., and Mackenthun, K. M. (1962). Detection and enumeration of iron bacteria in municipal water supplies. *J. Am. Water Works Assoc. 54*: 751–756.

Lumpkins, E. D., and Arveson, J. S. (1968). Improved technique for staining bacteria on membrane filters. *Appl. Microbiol. 16:* 433–434.

MacLeod, R. A., Light, M., White, L. A., and Currie, J. F. (1966). Sensitive rapid detection method for viable bacterial cells. *Appl. Microbiol. 14*: 979–984.

Millipore Corp. (1972). Detecting microorganisms in air. Application Procedure AP309. Bedford, Mass.

Millipore Corp. (1973a). Microbiological water testers. Bulletin MB407. Bedford, Mass.

Millipore Corp. (1973b). Simplified quick-count method for microbiological control in food processing. Application Procedure AP601. Bedford, Mass.

Millipore Corp. (1973c) Microbiological analysis of beverages. AB601. Bedford, Mass.

Orlando, M. D., and Bolduan, O. E. A. (1953). The application of the membrane filter to a variety of pathogenic bacteria and fungi imperfecti. *Bacteriol. Proc.*, pp. 26–27.

Presnell, M. W., and Andrews, W. H. (1976). Use of the membrane filter and a filter aid for concentrating and enumerating indicator bacteria and *Salmonella* from estuarine waters. *Water Res. 10:* 549–554.

Robson, J. (1956). A bacterial method for tracing sewage pollution. *J. Appl. Bacteriol. 19*: 243–246.

Rogers, D. E., Cooke, G. M., and Meyers, C. E. (1955). The detection of tubercle bacilli in mouth wash specimens by the use of membrane filter cultures. *Am. Rev. Ruberc. Pulm. Dis. 71*: 371–381.

Sartorius-Membranefilter GmbH. *The Determination of the Microorganism Content of the Air in the Food, Drink and Packaging Industries*. SM-Report 3.

Sato, N., and Fieldsteel, A. H. (1977). New method for concentration and quantitation of *Mycobacterium leprae*. *J. Clin. Microbiol. 5*: 326–328.

Schiemann, D. A. (1978). Isolation of *Yersinia enterocolitica* from surface and well waters in Ontario. *Can. J. Microbiol. 24:* 1048-1052.

Schiemann, D. A., Brodsky, M. H., and Ciebin, B. W. (1978). *Salmonella* and bacterial indicators in ozonated and chlorine dioxide-disinfected effluent. *J. Water Pollut. Control Fed. 50*: 158–162.

Scott, R. M., Seiz, D., and Shaughnessy, H. J. (1964). I. Rapid carbon^{14} test for coliform bacteria in water. *Am. J. Public Health 54*: 827–833.

Shepler, V. M., Plumer, G. J., and Faber, J. E. (1963). Isolation of *Vibrio fetus* from bovine preputial fluid, using Millipore filters and an antibiotic medium. *Am. J. Vet. Res. 24*: 749–755.

Snell, N. (1968). Direct counts of bacterial spores on membrane filters under phase optics. *Appl. Microbiol. 16*: 346.

Stanaszek, P. M. (1969). A new rapid and practical laboratory procedure utilizing differential membrane filtration in the rapid diagnosis of bactermia. Master's thesis, Ohio State University, Columbus, Ohio.

Taylor, E. W., and Burman, N. P. (1964). The application of membrane filtration techniques to the bacteriological examination of water. *J. Appl. Bacteriol. 27*: 294–303.

Taylor, R. H., and Geldreich, E. E. (1979). A new membrane filter procedure for bacterial counts in potable water and swimming pool samples. *J. Am. Water Works Assoc. 71:* 402-405.

Tompkin, R. B., Weiser, H. H., and Malaney, G. W. (1963). *Salmonella* and *Shigella* organisms in untreated pond water. *J. Am. Water Works Assoc. 55*: 592–596.

Tsuneishi, N., and Goetz, A. (1958). A method for the rapid cultivation of *Desulfovibrio aestuarii* on filter membranes. *Appl. Microbiol. 6*: 42–44.

Vlodavec, V. V. (1963). Nachweis der Lebensfähigkeit von Bakterien im Aerosol. *Zh. Mikrobiol. Epidemiol. Immunobiol. 4*: 46–49.

Watkins, W. D., Thomas, C. D., and Cabelli, V. J. (1976). Membrane filter procedure for enumeration of *Vibrio parahaemolyticus*. *Appl. Environ. Microbiol. 32*: 679–684.

Winn, W. R., White, M. L., Cater, W. T., Miller, A. B., and Finegold, S. M. (1966). Rapid diagnosis of bacteremia with quantitative differential-membrane filtration culture. *J. Am. Med. Assoc. 197*: 539–548.

Winter, A. J., Burda, K., and Dunn, H. O. (1965). An evaluation of cultural technics for the detection of *Vibrio fetus* in bovine semen. *Cornell Vet. 54–55*: 431–444.

Wright, G. L., Jr., and Mallmann, W. L. (1963). Method for staining acid-fast bacilli on the membrane filter. *Am. Rev. Respir. Dis. 87*: 432–434.

21

MEMBRANE FILTERS IN INDUSTRY

*CLIFFORD F. FRITH, JR.** *Millipore Corporation, Bedford, Massachusetts*

I. INTRODUCTION

The development of membrane filters in the 1950s for commercial applications initiated a new fluid separations technology. For centuries, filtration as a unit process has been a standard method for separating material. However, the advancement provided by a truly absolute mechanism of separating discrete matter from liquids and gases has benefited both the analytical methodologies and process industries.

The age of "microstate technology" in the twentieth century demanded new and advanced methods for analyzing and controlling infinitesimally small substances. The membrane filter technology is a vital part of the success thus far achieved. The potential membrane market, estimated to triple in the next 5 years, is indicative of the requirements for absolute filtration technology. The substances that may be monitored and controlled by membrane filters range from the virus organism (0.010 μm) to the atmospheric dust in a space capsule (100 μm). Figure 1 illustrates the range of filters available and their relationship to discrete particles. Selection of the

*Present affiliation: Vaponics, Inc., Plymouth, Massachusetts

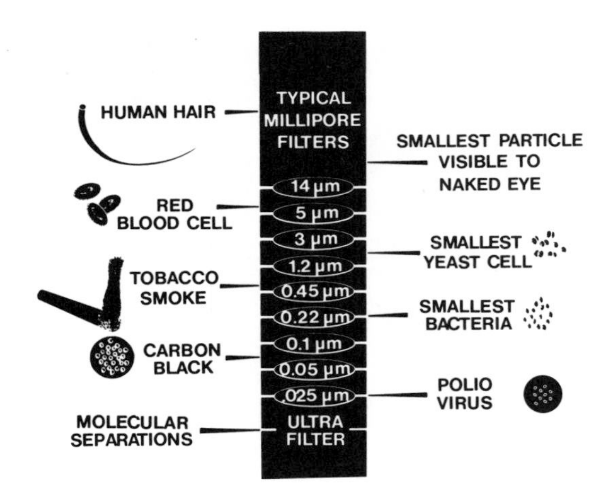

FIGURE 1 Comparison of particle size and pore size of some membrane filters.

appropriate filter depends on the dimensions of particles to be removed.

The advanced methods of production have provided membrane filters in all sizes, shapes, absolute retention characteristics, porosity, and structure morphology. The materials from which membrane filters are produced are mixed esters of cellulose, cellulose acetate, polyvinyl chloride, nylon, polycarbonate, and polytetrafluoroethylene. Additional proprietary materials are also available and are recommended for certain critical applications. Physical, chemical, microbiological, and thermal resistance properties of the various polymers are important for the applications.

The membrane may be fabricated into cartridge configurations to compact as much filtration material per unit volume as possible. Sterile, disposable filtration devices have been developed and are an integral part of medical technologies. Special applications for both analysis and control required fabricated membrane products, which were developed, designed, and manufactured to specific requirements.

The membrane filter has utility in almost every industry and may affect the lives of each living being. It is difficult to find applications in our microstate technology age that are not influenced by an analysis or control using membrane filters.

Previous chapters of the book discuss membrane filter procedures for detecting microbiological organisms in fluids. Chapter 2 details the properties of the filter. Two properties, absolute filtration of all particles larger than the pore openings and filtration at the surface interface, are essential for reliable analytical methods. Process filtration may not require the latter, but the same membrane is used for both analytical and process control. This provides an economical method for production and marketing.

Historically, new applications were researched and developed using the membrane as an analytical tool. Once established as a method for detecting and enumerating contaminants and/or foreign substances, the membrane was rapidly accepted as a process tool. An example of this two-stage event is the microelectronics semiconductor. The filter was employed to test for particulates, including microorganisms, in the production process. Monitoring such fluids as high-purity water, ultraclean solvents, photoresists and developers, etchants, gases, and the environmentally controlled workspace was the first step. The most reliable and economical method of eliminating contaminants detected employed the membrane filter in the process. This was the second step. The basic difference between the first and second steps was the physical size and configuration of the membrane filter.

A. Filtration Mechanism

The mechanics of removing discrete particles employing filtration membranes is defined as sieving. Substances suspended in the fluid stream are collected on or near the receiving surface. Particles equal to or larger than the pore openings are retained (Fig. 2). Depending on the fluid state, liquid or gas, additional forces may

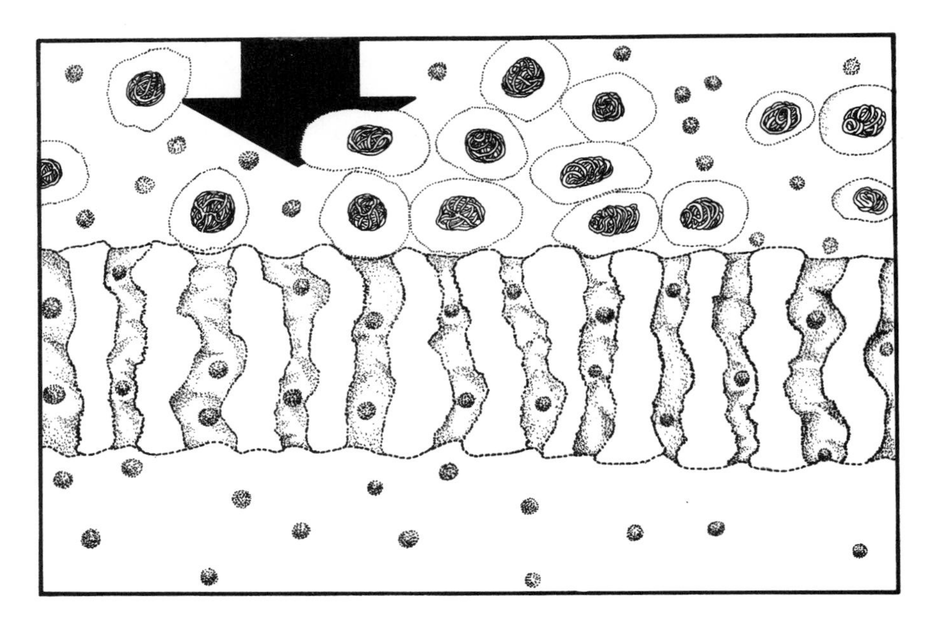

FIGURE 2 Absolute retention of all particles larger than pore openings.

increase the efficiency of the filter. Electrostatic charge in the case of gas filtration and van der Waals forces in liquid systems are primarily responsible for improved retention characteristics. For instance, cigarette smoke particles (0.5 μm) are efficiently removed by a 0.8 μm membrane. Both the charge of the filter matrix and the particle charge play a role in the separation.

The open membrane structure is not a straight-through channel but one of a tortuous path. This also results in the collection of particles smaller than the pore size by random entrapment within the matrix. Therefore, we will label the process as sieving with one precaution--the membranes do not serve as a method for classifying particles into dimensional categories. The traditional term "sieving," a civil engineering method for separating and classifying particles, is not applicable to membrane filter analysis.

B. Integrity Test

A major benefit for process filtration is the unique, simple, reliable integrity test. This provides assurance that the membrane is the correct pore size, without imperfections, and that the filtration system is qualified to prevent bypass or leakage. This test, in the case of liquid processes, employs the fluid being filtered. The bubble point test is now considered as a standard operating procedure for all applications. The test follows the principle of capillary action, as the pores in the matrix resemble small capillaries. These pores are filled and wetted with the liquid to be filtered, then gas is forced against the membrane to expel the liquid from the pores. Depending on the surface tension of the liquid, the size of the pore opening, and the amount of gas pressure required to force the liquid out of the matrix, the membranes

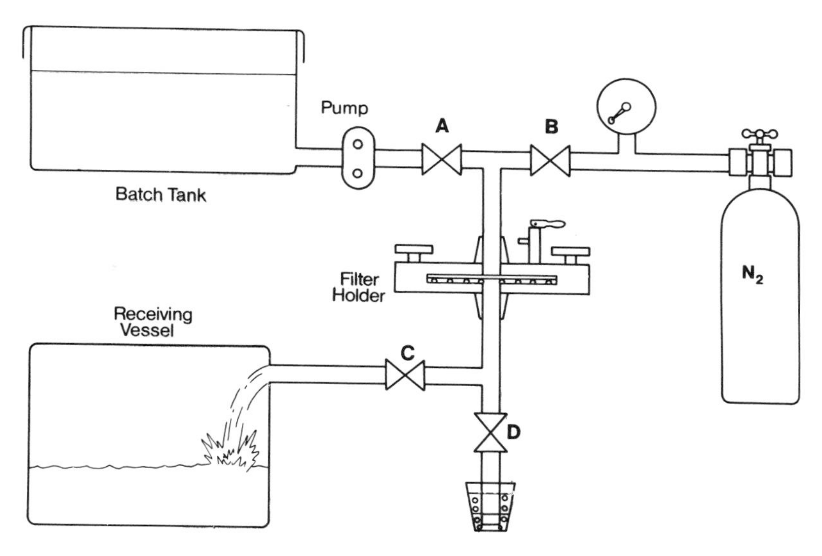

FIGURE 3 Schematic of in-line bubble point test set up for typical process filtration.

largest pore can be determined. The capillary action equation states that the smaller the capillaries, the more pressure that is required to expel the liquid and allow the gas to pass downstream.

The procedure is a noncontaminating, nondestructive test and may be used before, during, and after the process. Figure 3 shows a typical setup for testing the integrity of the system. Valves A and C are opened to allow for filling the system with the liquid to be processed until the membrane pores are completely wetted. Then these values are closed and valves B and D are opened. Gas (nitrogen) is introduced to force the liquid from the pores. The pressure of the gas is gradually increased until bubbles are observed downstream of the membrane holder. To illustrate the selectivity of the test, note the difference in pressure required for a 0.45 μm and a 0.22 μm membrane filter. With water as the

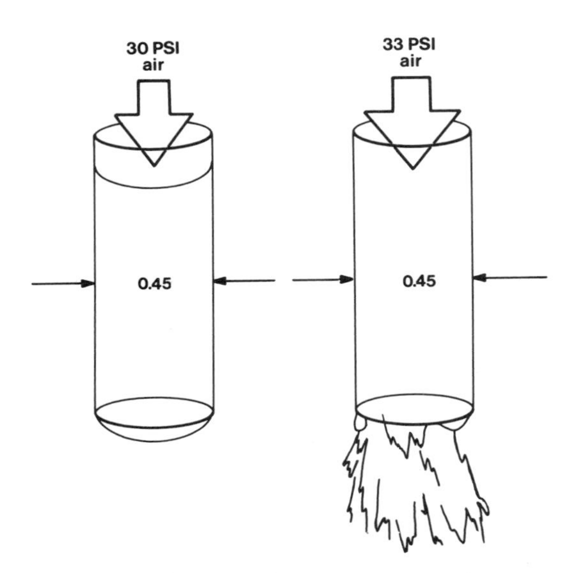

FIGURE 4(a) A 0.45 μm membrane filter requires a bubble point pressure in excess of 30 psi.

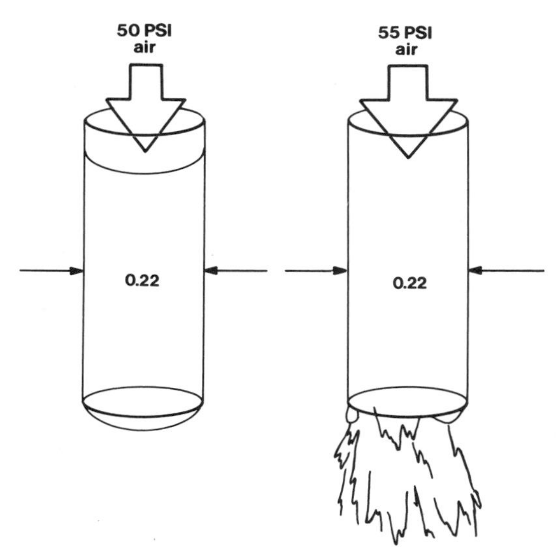

FIGURE 4(b) A 0.22 μm membrane filter requires a bubble point pressure in excess of 50 psi.

test liquid, the 0.45 μm membrane will require a 33 psi differential challenge pressure (Fig. 4a), compared to 55 psi differential with the 0.22 μm membrane (Fig. 4b).

II. ANALYTICAL TECHNIQUES

The value of the membrane filter for analytical procedures increases with the advancements in monitoring methodologies (Dwyer, 1966). Its ability to retain, segregate, and concentrate foreign materials from a fluid stream can best be illustrated by reviewing the following methods of analysis. Most of these methods are published as standards by organizations such as the federal government, American Society for Testing and Materials (ASTM), Society of Automotive Engineers (SAE), American Public Health Association (APHA). American Water Works Association (AWWA), Water Pollution Control Federation (WPCF), *United States Pharmacopoeia* (USP), and National Committee for Clinical Laboratory Standards (NCCLS).

A. Microbiological Analysis

Throughout the rest of this book numerous references are cited for collecting and enumerating the microbiological population of a fluid. The efficacy to retain the organisms for subsequent culture and direct enumeration as well as to analyze large volumes where few organisms exist have made the membrane procedure the method of choice. This method is more favorable than the most probable number (MPN) technique, with its statistically inferred result, or plate counts (streak or pour), which limit the sample volume.

Virology was limited for many years, owing to the small virus population in a large volume of liquid. The ability to concentrate this population by filtering large volumes through a small surface area membrane initiated improved and simple virus detection methods (Hill et al.,1972).

Fabricated devices, including the filter, absorbent pads, and in some cases, dehydrated media, are readily available for field survey procedures. In one unit, as illustrated in Fig. 5(a), the filter, absorbent pad, and dehydrated media are preassembled and sterilized ready for use. The unit [Fig. 5(b)], absorbs a given volume of water through the 0.45 μm membrane and rehydrates the

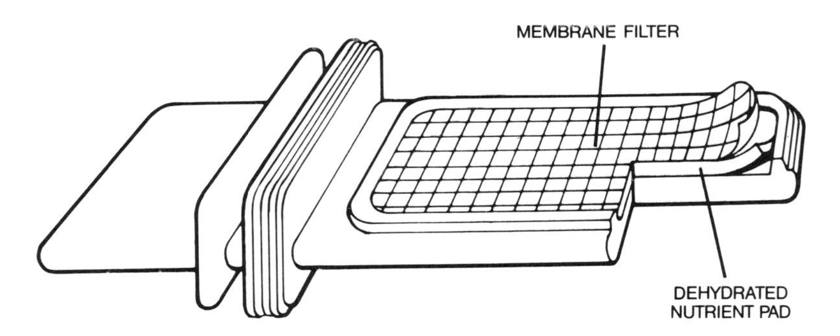

FIGURE 5(a) Schematic of a Millipore Sampler, a complete preassembled, presterilized water test kit.

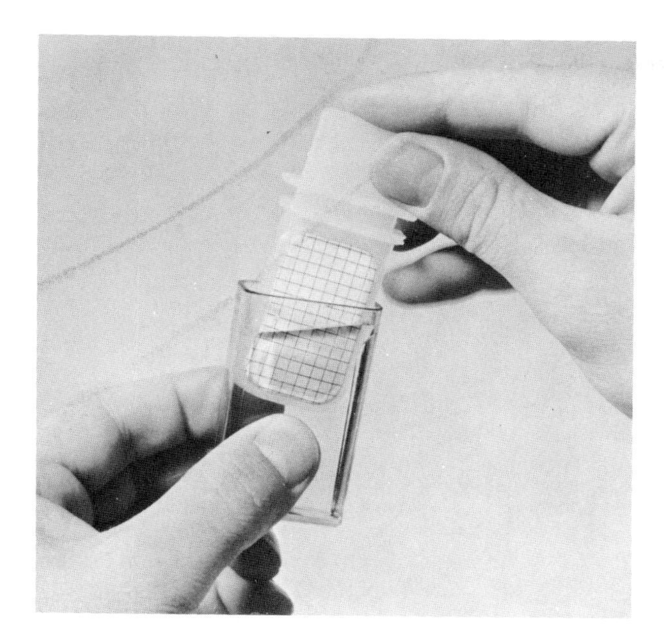

FIGURE 5(b) Testing for coliform bacteria in an untreated water supply with the Coli-Count Sampler.

nutrient media for subsequent culturing (ASTM, 1978). Such devices are completely disposable and save the preparation time and cleanup. For drinking water analysis, these units have not been accepted. However, for industrial monitoring and environmental surveys, such products cannot be discounted.

Kits for portable laboratory analysis were developed in the early 1960s for Department of Defense mobile field stations and remote installations (Fig. 6). Environmental agencies at the state and federal levels are employing these units. The sample can be processed easily in the field using the membrane filter methods. Incubators in the portable kits are operated from the generators of the mobile units or batteries from the transport vehicles.

FIGURE 6 Complete membrane filter test being performed in the field with a portable kit.

B. Sterility Testing

The demands for testing all types of liquids to ensure the absence of organisms has increased at a phenomenal rate. Since the enforcement of Food, Drug and Cosmetic Act regulations, pharmaceutical and medical device manufacturers are performing hundreds of thousands of sterility tests (USPXX, 1980). The technique of filtering large sample volumes to detect extremely low levels of living organisms has greatly improved the sensitivity and reliability of the sterility test.

C. Airborne Microbiology

In the early days of developing analytical methods, large volumes of gas were passed through the filter to collect all living matter.

The filter was then placed onto a nutrient pad and the media was absorbed into the filter matrix. Subsequent incubation provided a direct enumeration of the airborne microorganisms. The efficiency was poor compared to media impaction plates such as the slit sampler. Additional development efforts showed that the membrane surface was not the proper medium for collecting organisms. The shock treatment due to high-velocity gases desiccating the cell walls actually destroyed all but the most resistant cells. Further research provided the method of combining an impinger-filtration procedure to protect the organisms. The gas sample is drawn by vacuum through an impinger containing an appropriate growth media. The particles and organisms are removed from the gas stream in a bathing action. The media containing the particles is then filtered through the 0.45 μm membrane. Figure 7 shows the impinger with a preassembled, sterile Millipore 0.45 μm Clinical Monitor. After incubation, colonies are counted and calculations made to determine organisms per unit volume of gas sampled (Millipore, 1976a).

D. Microscopic Examination

The properties of the membrane filter make it an ideal collecting medium for particles smaller than are visible to the eye (less than 50 μm). Both the collection of particles on the surface and the ability to detect low levels of particles due to the large volumes of fluids that can be filtered through the membrane in a short time period make this the preferred method. Techniques for counting particles as small as 5 μm are standard and employ incident lighting and a microscope with 100X magnification (ASTM, 1968; SAE, 1966). The enumeration and identification of the particles are easily performed using filters with a grid pattern. As an example, the 47 mm diameter membrane has an effective filtration area with standard

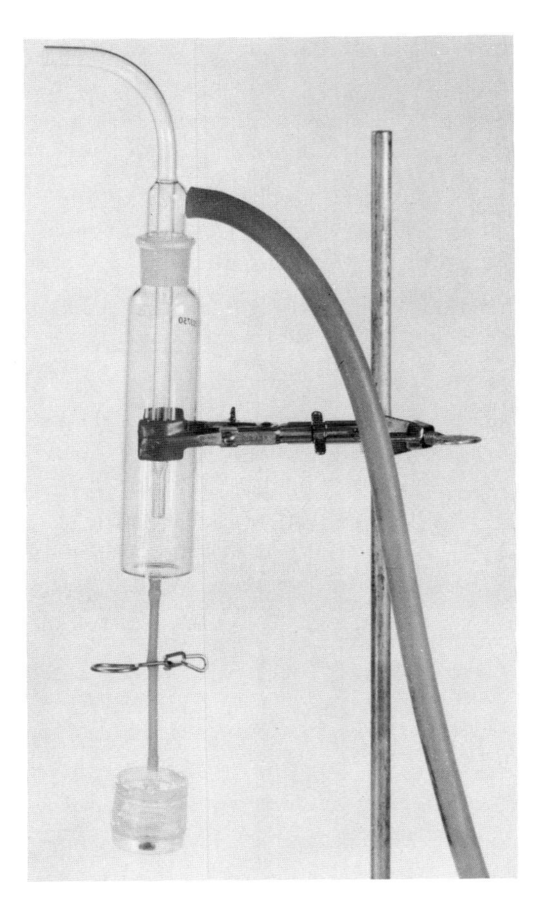

FIGURE 7 Impinger and membrane filter monitor for sampling airborne microorganisms.

analysis filter holders of 960 mm^2. The grids are imprinted as 9.6 mm^2 squares. Statistical counting methods have been developed so as to count only a portion of the entire surface to obtain valid results. The calculated number of particles on the entire surface and the total volume of fluid filtered are entered into an equation and provide the analyst with the number of particles per unit volume.

A size distribution curve may be developed from any sample using the foregoing procedure and counting particles in various size ranges (U.S. Air Force, 1965). Plotting the results on a graph with the total number of particles greater than a measured size on one axis and the measured size on another axis provides a curve. Using log-log graphical presentation, a line can be drawn through the points on the graph and further information obtained by extrapolating to other size ranges not actually counted. This method is frequently used to calibrate automatic particle analyzers.

Additional methods have been developed using transmitted light microscropy for detecting particles in the 0.5-5.0 μm range (Millipore, 1977). The mixed esters membrane is a matrix of approximately 80% air. Filling the pores with an immersion oil having a refractive index of 1.5 renders the filter transparent. The higher level of resolution is a major advantage for analyzing ultraclean fluids.

E. Gravimetric Analysis

The membrane filter collection efficiency for all particles greater than the pore openings is absolute. This screen, in fact, removes a high percentage of particulate matter smaller than the pore opening, owing to physical, chemical, and mechanical phenomona of the membrane composition, particle properties, and the fluid being sampled. The fact that the channels in most filters are not straight-through holes allows random entrapment of the smaller particles. This fraction of the sample may not be significant to the results. However, the technician is assured of high-efficiency collection (Megaw and Wiffen, 1963; Liu and Lee, 1976), and the reproducibility and accuracy of the method are excellent.

The concentration of large sample volumes through a small filter disk is an outstanding advantage. In comparison, evaporation and centrifugation techniques are inefficient and quite laborous. With the filter technique, the sample may be forced through by either vacuum or pressure (ASTM, 1973; SAE, 1963). Depending on the concentration of material to be removed, the volume filtered can be selected so as to allow the technician the ability to collect enough sample to eliminate many of the variables associated with other methods. Ultraclean fluids are not analyzed by gravimetric techniques. However, as a practical matter, the filtration membrane used in the process as a final step of the cleaning system to assure the quality of the fluid may be analyzed gravimetrically.

A control filter is always recommended for improving the accuracy of the gravimetric test. Errors due to the fluid, types of contamination, filter hygroscopicity, or balance changes are possible. The standard procedure with the control proves beneficial in correcting these possible errors. The analytical membrane (mixed esters of cellulose) contain a water-extractable component. Therefore, in sampling water or polar liquids, the control filter eliminates any weight loss error, as the second filter undergoes the same extraction process (Frith, 1967).

To assist the technician, membrane filters specially prepared for gravimetric analysis are commercially available. This product is defined as a "matched weight" membrane (Millipore, 1978). Two identical filters have been preweighed to within ±0.1 mg and matched as a pair. The filters are used in series for the test. The top filter collects the contamination and the bottom filter serves as a control (Fig. 8). Since both are treated identically in the procedure, variables in the technique are experienced by each and therefore reduce errors. Additionally, the time to run the test is greatly reduced because the preliminary "tare" procedure is eliminated.

The organic membrane filter is used in the gravimetric procedure to identify the percentage inorganic vs. organic matter of the

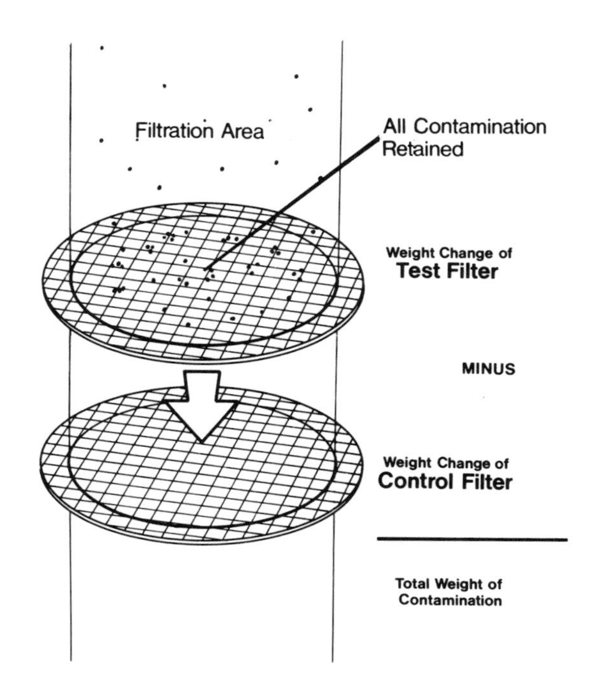

FIGURE 8 Control filter gravimetric analysis technique; utilizes two filters for improved accuracy.

sample. After sample collection and weighing, the total contamination is calculated. Then the filter is ashed in a muffle furnace at 600°C. The organic fraction plus the filter are eliminated and the weighable material remaining is the inorganic matter. The ashed residue of the pure membrane adds less than 0.0001 mg inorganic, which is two orders of magnitude less than the sensitivity of the gravimetric technique.

F. Color Comparison

Membrane filter "patch test" techniques are available which provides visual color comparisons (Millipore, 1975a). This is used for testing pure water in supercritical boilers employed for efficient generation of power by steam turbines. The patch test technique is extensively employed for evaluating the concentration of contaminants in aircraft jet fuel. Known standards of contaminants are imposed on the filter surface using gravimetric techniques to establish color comparison charts. The quantitative charts are used by the technician to qualify the unknown concentration of matter in the fluid system. A known volume of the fluid is passed through the filter and matched to the closest visual standard to determine the relative concentration of contamination. Both preselected charts and charts made for the individual application are recognized as in-process test methods, and in some cases determine the "go-no go" of the process. In the case of the utility industry, this method is recognized as an extremely valuable analytical tool to prevent the destruction of steam turbines and loss of revenue. Figure 9 is a typical chart used in monitoring boiler feed water. Additionally, the U.S. Department of Defense and commercial airlines include the patch test as a requirement for the qualification of both hydraulic oils and fuels (Fig. 10) in high-performance jet aircraft. The requirements for clean hydraulic oils in aerospace components and military hardware include color comparison membranes as a test specification.

The manufacturer of integrated circuits require that the purity of the water used as a solvent be of the highest degree. The bacterial population does not produce a visible color; however, this microscopic matter can destroy integrated circuits in a very efficient manner. A simple patch testing procedure requires filtration of large volumes of water through the membrane and then a subsequent staining in a Ponceu-S dye to detect the presence of proteinaceous materials (Frith, 1971). The degree of red stain of the test filter is a semiquantative indication for the level of microorganisms. This technique is being used widely to measure the quality of the high-purity water and provides the operator with information in a timely manner as compared to the more time-dependent culture methods.

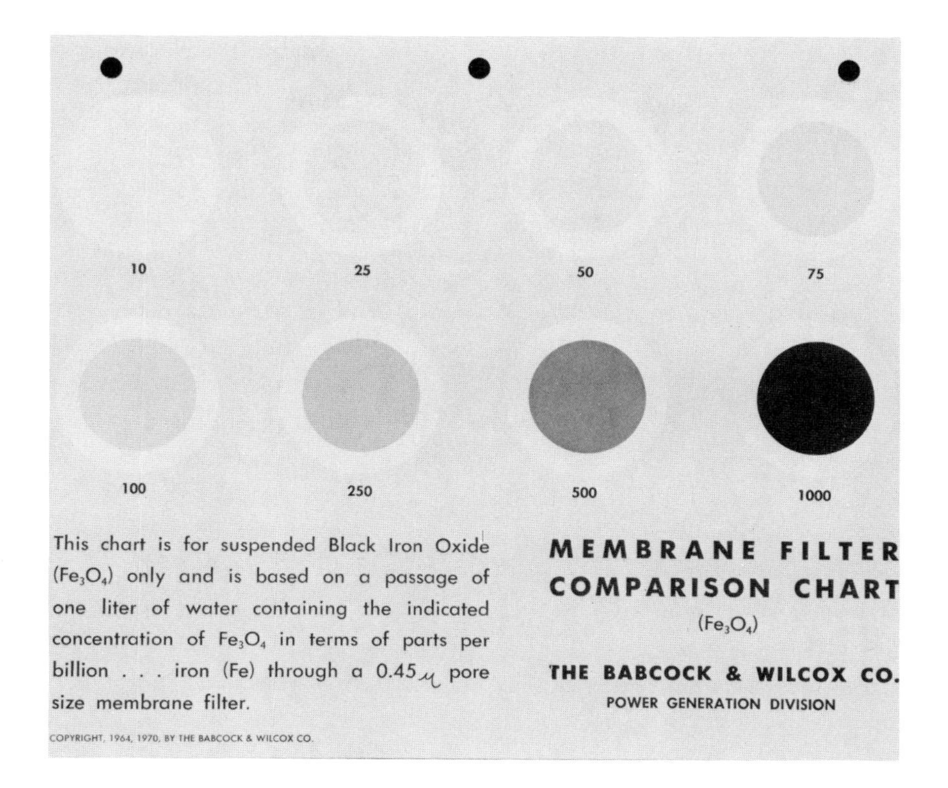

FIGURE 9 Published standards for determining levels of hydrated iron oxide in water.

G. Fluid Flow Restriction

The screen characteristics of the membrane surface provide a unique analytical method known as fluid flow decay. This analytical tool can be used in the unit process as an in-line indicator of the level of contamination in the fluid stream, or samples can be tested in the laboratory. The technique employs the clogging tendency of the membrane surface due to the size and nature of the contaminants. There are two methods that may be employed. In the first, the fluid is processed at a constant pressure and the rate of fluid flow

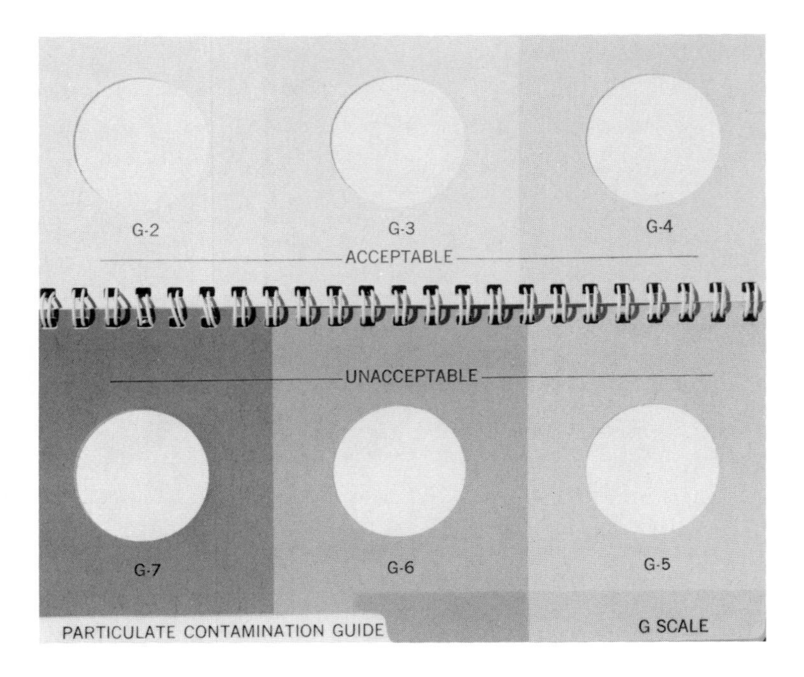

FIGURE 10 Aviation turbine fuel contamination standards.

through the membrane is monitored. The decay of the flow rate is dependent on the concentration of contaminants. In the second technique, the fluid is processed at a constant flow rate and the pressure differential developed across the filter is monitored. This pressure differential can be compared to that from a properly pretreated standard solution. There are various names for these analytical procedures, but the most common are the Silting Index, Fouling Index, Plugging Index, and Filterability Index (ASTM, 1969; SAE, 1963b).

As described earlier, there are various pore size filters available and, depending on the limit of contamination control desired, the proper pore size filter may be selected. As the state of the art for many industries reduces product size and increases the level of product sophistication, submicron pore size filters are becoming the standard for these tests.

H. Chemical Reactivity

The contamination collected on the surface of the membrane may be individual particles or a continuous mat. Microchemical techniques have been developed for identifying the classes of contaminants as well as individual particles (ASTM, 1968b). Direct detection methods are employed on the membrane surface, or the sample and membrane may be ashed if only the inorganic fraction is to be qualatatively assayed.

In recent years, membranes have been pretreated with various chemical reagents to provide colorimetric reactions. The membrane with reagent in the presence of certain particulates or gases produces a color reaction that is visible to the unaided eye. Again, the value of the membrane lies in the large amount of fluid that can be sampled to detect trace amounts of contaminants.

I. Instrumental Analysis

The modern laboratory depends on automated and instrumental analytical procedures for sensitivity and reliability. Both direct and indirect procedures have been developed for sophisticated instruments utilizing the membrane filter as the sample collection mechanism. The matrix of the filter is insignificant in most instrumented analysis because it does not present interference and in certain methods, the background "noise" is eliminated. The absolute filtration rating provides the analyst with assurance that the sample has been concentrated efficiently to qualify and quantify the level of contamination. The membrane filters are inert and do not contribute contaminants. The retentate may be digestible in various solvents and acids. The analyst may dissolve the material in and on the membrane or extract it.

In many cases the amount of substance to be detected is in microquantities and the sensitivity and reproducibility of the results would be quite difficult if the unconcentrated sample was used (Dwyer and Frith, 1967). Large volumes passed through a relatively small filter area allow the analyst a rapid and reliable method for sample preparation.

Infrared and ultraviolet spectrophotometry and colorimetry are methods that employ absorption by chemical substances to provide qualitative and quantitative results. The filter as a sampling mechanism may be employed in a direct or indirect method.

Emission spectroscopy techniques in which the materials to be identified emit light, include flame photometry, arc emission spectroscopy, X-ray florescence and X-ray defraction. All these methods have various levels of sensitivity and in most cases require very expensive and elaborate instrumentation. On the other hand, their increased use in laboratories requires rapid sample preparation and highlights the value of the membrane filter.

J. Soluble vs. Insoluble

Chemical analysis procedures for a given sample often require the measurement of dissolved and suspended components separately. Several reasons may be given: (1) insoluble materials may interfere with the standard chemical reaction; (2) the insoluble fraction will not completely dissolve during the procedure, so that the desired element is inaccurately measured; and (3) the analyst may be required to report separately the soluble and insoluble components. The 0.45 μm membrane filter has been accepted as the standard method for preparing samples that require this division. As the precision and accuracy of trace chemical analysis is improved by sophisticated instruments, this step in sample preparation is becoming routine, thus advancing our knowledge of collodial suspensions. A recent example of this is the development of ion-exchange resins with macroporous structures for removing colloidal iron and silica. The membrane filter was employed in the R&D phase of these resins (Rohm and Haas Company, 1966). It is now recommended as a test to detect colloidal particles in high-purity water prior to installation of these special resins.

III. PROCESS FILTRATION

Membrane process filtration can be divided into two distinct types, both of which use the absolute retention properties of membranes. The first is the removal of unwanted discrete particulate matter, whether the particles are viable or nonviable. The second is that of reclaiming valuable materials and the subsequent digestion or elimination of the filter matrix.

Contamination control has developed over recent years to a dynamic technology. Almost any fluid used in industry today can be processed by a defined filter matrix. The microbiological, physical, and chemical properties of the filter are extremely important and have been characterized for almost every application.

The unit process "final filtration" serves a valuable function as an integral part of other purification systems, for example the production of ultrapure water. The four major types of contaminants found in all waters are inorganics, organics, particulates, and microorganisms. In the design of a water treatment system, the engineer will include equipment for organic adsorption (activated carbon), deionization of charged inorganic ions (ion exchange), and absolute membrane filtration (Frith, 1969). The roll of the membrane has three very important aspects: (1) elimination of particulate and colloidal matter in the feedwater supply; (2) elimination of particles generated from the carbon bed, or ion-exchange columns; and (3) elimination of any microorganisms in the feed, as well as those growing within the carbon, ion-exchange, and distribution system.

Many examples can be cited where the unit process of "filtration" is dependent upon the absolute retention of discrete particles and microorganisms as well as the requirement that the filter system not generate undesirable material. There are many final filtration units in common use that were improperly selected and placed within the purification system. They may produce contaminants in the final product that exceed the level entering with the untreated fluids or fail to remove the particulate matter to the standard desired for the application. These are not true membrane filter systems.

IV. INDUSTRIAL APPLICATIONS

The variations of fabricated products, together with different types of membrane materials, have created a variety of filtration applications. A significant part of the development and marketing strategy of the membrane filter industry has been to provide standard products that could be used in many different applications, regardless of industry. The basic technical information, including installation, operation, and maintenance instructions, is identical.

Table 1 is a summary of the important applications for membrane filters in various industries. From this table, the scope of membrane technology and its value to both analytical methods and process filtration are evident. It is interesting to note that one industry will often solve a problem or advance its technology using membrane filters, and an unrelated industry will satisfy one of its needs by employing the same method and equipment. An example is the use of aviation fuels "patch test" in the utility industry. The same membranes, equipment, and method used for jet aircraft fuel are employed to test the condensate for boiler makeup water. Only the results will be different, and appropriate reference standards are available for the application.

A. Public Health Industries

The major applications in the public health industry are analytical techniques which trace their origin back to the original development of the membrane. Throughout this book details of microbiological techniques have been presented. Other areas of significance are virology, ecological problems such as plankton or algae, and dairy-related methods such as the detection of bovine mastitis (Millipore, 1976b).

TABLE 1 Industry-Related Applications

	Public health	Industrial hygiene	Food	Beverage	Medical	Pharmaceutical	Cosmetic	Electronic	Chemical	Energy/utilities	Aerospace/transportation
Analytical testing											
Microbiological	x	x	x	x	x	x	x	x	x	x	x
Sterility	x				x	x					
Particle counting	x	x			x	x	x	x	x	x	x
Particle sizing	x	x			x	x	x	x	x	x	x
Gravimetric	x	x						x	x	x	x
Color comparison		x						x	x	x	x
Flow decay			x	x		x	x	x	x	x	x
Microchemical	x	x			x	x		x	x	x	x
Instrumental	x	x			x	x	x	x	x	x	x
Process filtration											
Contamination control											
Liquids			x	x	x	x	x	x	x	x	x
Gases			x	x	x	x	x	x	x	x	x
Material recovery								x	x	x	x

B. Industrial Hygiene Industries

The work environment as well as atmospheric conditions for the population have received high priorities in pollution control. Steps have been taken by federal, state, and local agencies to determine the origin of the pollutants and to take regulatory action to stop or reduce their generation. The most common problem is that of air pollution since it is visible to the naked eye in terms of both atmospheric conditions and the resulting surface depositions.

Atmospheric pollution, whether from smoke stacks, industrial operations, or transportation emissions, has been brought under control in recent years, and filter monitoring procedures are among the standard methods being used.

The current impact of protection techniques for the worker has probably been most notable in the mining industry. In recent years black lung disease has been classified as a hazard to coal miners. Major efforts have also been directed at various industries that mine, process, or fabricate such materials as asbestos, lead containing compounds, beryllium, and textile fibers. The membrane technique is approved for testing the environment to assure that levels do not exceed limits for operator safety.

One of the earliest applications recognized by the U.S. government when it first classified the membrane as a secret technology was for the detection of radiological and biological warfare agents. More recently, high-altitude sampling methods used to monitor radioactive fallout and the environment around industries where radioactive material is present employ the membrane filter. The radioactivity is in the form of discrete particles from either the active material or airborne particles that become charged due to radiation in the environment.

C. Food Industries

The control of food processing and distribution are regulated for public health requirements. Adequate control of incoming raw materials, on-line processing, and distribution offer many new opportunities for filtration. The economic impact for a well-controlled process at each step has been documented and the payback is quite large. In supermarket operations, the swab sampling and membrane filter techniques are used to indicate levels of cleanliness within the meat cutting and packing area. This is a location where economic justification is easily assured. Another often neglected area involving potential contamination is that of frozen foods, including desserts. Simple and rapid techniques have been developed and are recommended as screening methods.

Enzymatic and antibiotic processes which involve high concentrations of sugar and protein mixtures are monitored for undesirable organisms. The existence of one foreign organism can spoil large volumes of product. Here, the raw materials are well controlled and processed through high-volume membrane filtration systems (Russell, 1971). The venting of tanks is most critical, and absolute assurance that microorganisms do not enter through these vents must be guaranteed. Properties of membrane filters offer the operator this assurance.

D. Beverage Industries

The most common application recognized internationally for the membrane filter relates to the stabilization of beer and wine. This "cold pasteurization" process is the key in producing draft beer in cans. In the mid-1960s much effort was devoted to the membrane processing of alcoholic beverages. Before the recognition of large-volume process filtration, analytical chemists and microbiologists had established the membrane filter as a standard laboratory method for detecting spoilage organisms (Millipore, 1969). The simplicity of the detection techniques has provided a method of on-line control to ensure the validity of the process.

The bottled water industry, a rapidly growing market, learned from the brewing industry that process membrane filters are a good insurance policy for protecting packaged products. The American Bottle Water Association has been a leader in establishing product quality standards without strong influences from regulatory agencies. Technicians also utilize the membrane filter for detecting total bacteria population to ensure the efficiency of the treatment process prior to packaging. Most manufacturers will readily admit that producing low count bottle water is not an easy task. In Europe the volume requirements for mineral water for drinking purposes are large and the cost of producing the product can be expensive, depending on the quality of the source water. The membrane filter has provided assurance of the stability and quality of the product.

The quality control of soft drink production is a relatively new technology for the membrane filter. This does not imply a lack of previous microbiological analytical techniques, but with the advancements of simple and rapid procedures, such raw materials as the syrups and makeup waters are now continuously tested using the membrane filter. The employment of a swab and membrane combination also allows the technician to monitor the condition of the filling machine and other contact surfaces that could be contaminated and cause product spoilage. The compositions of these products provide an excellent nutrient for growth of spoilage organisms that are common to the environment. This makes the process difficult without adequate control.

The analysis and control of packaging gases such as carbon dioxide, used for fluid transfer and filling operations, has received much attention. The assurance that microbiological contaminants are eliminated from the gas stream is very important. Simple filtration systems that can be steam-sterilized in situ are becoming a standard component in the filling equipment (Millipore, 1975b).

Many of the gases common to food and beverage processing are also used in the fermentation industry and require the same contamination control.

E. Medical Industries

The medical research and hospital environments are the fastest-growing areas of technology in the world today. Both the degree of sophistication for detecting medical problems and methods for curing diseases have made great progress in the past two decades. Researchers have used various techniques for isolating and identifying the cause of various illnesses. Here again, the membrane filter has provided a tool for concentrating the specimen and improving the ability to identify and confirm the problem.

In the hospital, the treatment of patients uses fluids that have to be sterile and nonreactive to the patients. There are thousands of materials required in patient care. Most of these solutions, ointments, and preparations are purchased from companies that have used proper techniques for assuring the quality of their products. The pharmaceutical industry has advanced membrane filter technologies for preparing such products. One of the latest developments that provides a significant application for membrane filters is the final filter used on intravenous administration sets (Rapp et al., 1975). The number and size of particulates that can be injected during administration has been a major concern, and now the final-filter concept is a guarantee against particulate and microbial contamination being introduced to the patient. In addition, such products as the Millipore Ivex-2, distributed by Abbott Laboratories, include a hydrophobic membrane to eliminate air during intravenous feeding. The Ivex-2 (Fig. 11) is unique because the 0.22 μm hydrophillic membrane allows the intravenous fluid to pass but eliminates any chance of air embolism. The entrapped air collected upstream of the hydrophilic filter is discharged through the hydrophobic membrane, thus preventing flow stoppage due to entrapped air (Rapp et al., 1974).

Contamination control of intravenous solutions injected in infants and small children is extremely important. The practitioner

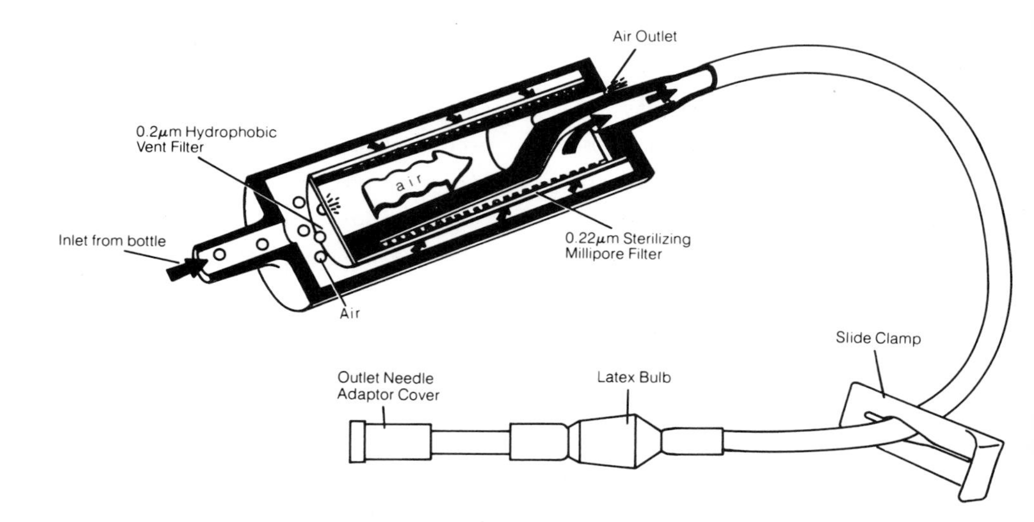

FIGURE 11 Ivex-2, 0.22 μm Filterset, an in-line final filter and extension set for connection to any i.v. administration set.

prescribes a guard by using the membrane filter "in line" of the parenteral alimentation administration set. Particles and any microorganisms are eliminated as well as air embolisms.

Because of the nature of many of the solutions and preparations, materials must be prepared in situ or within a reasonable time prior to use. This requires that a hospital pharmacy be equipped to produce sterile materials such as intravenous additives, radiopharmaceuticals, ointments, and antibiotics (Millipore, 1976c). These extemporaneously prepared materials use presterilized membrane filters (Fig. 12).

Ophthalmic surgery requires that bacteria and microscopic particles be eliminated during the procedure. The use of a presterilized filtration unit on the hyperdermic syringe (Fig. 13) provides the surgeon with a particle-free sterile rinse. The membrane filter is used successfully as a microporous bandage in nerve tissue regeneration and bone graft operations. The appropriate materials pass

FIGURE 12 Millipore Millex being used to sterilize and clarify i.v. additive in the hospital pharmacy.

through the membrane bandage, with microorganisms and particles being rejected to keep the sensitive area free of contaminants.

The irrigating fluids for peritoneal dialysis must be controlled to eliminate microorganisms which would cause further complications. Filters serve this function and provide the patient with the proper conditioned solution.

Diagnosis and prevention of contagious diseases are the responsibility of the practitioner. Technicians perform many chemical and microbiological techniques to provide information for prescribing drugs, verifying the need for surgery and determining when the patient is cured. Body fluids are analyzed by electrophoresis

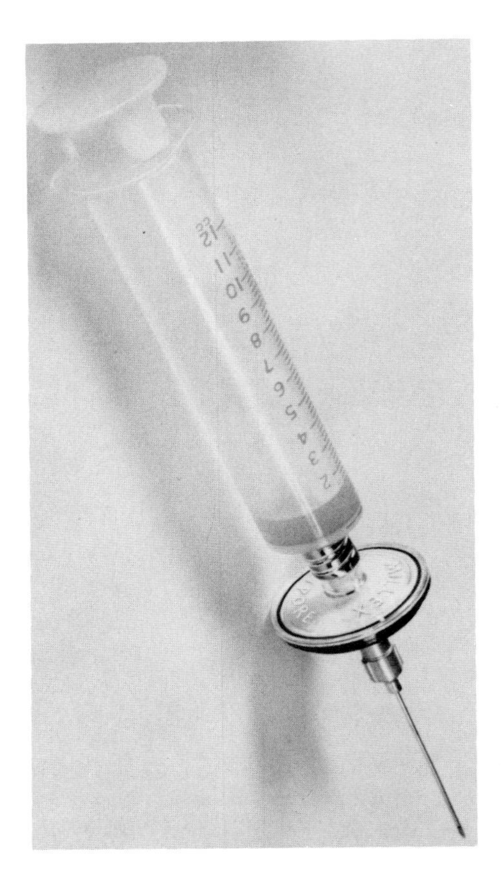

FIGURE 13 Simplest and easiest technique for sterilizing or ultracleaning small volumes of fluids.

techniques using membrane filters. The sample is separated for analysis by the use of electrophoretic mobility and indicates if there is an abnormality of the body fluid.

In cancer cell identification procedures, use of a 5 μm pore size membrane allows the cells to be isolated from other cells for staining and microscopy. More recently, research efforts for better understanding of sickle cell anemia factors have been greatly advanced with the use of the membrane filters.

In the laboratory, reagent-grade water requires final membrane filtration for removal of particles and microorganiams. Previously, distillation was the standard production technique. However, with the advancements in analytical methodologies, the specifications have been increased to provide a more uniform reagent water and purer-than-distillation capabilities (NCCLS, 1977).

Hospital environments are controlled to prevent the spread of disease. Surface and airborne monitoring are standard practices that use the membrane filter. In recent years, such areas as inhalation therapy have been suspect, and both monitoring and filtration of gases incorporate the membrane. In the case where nonviable particles in the atmosphere may aggravate a severely burned patient, particle counting is instituted to determine the need for air purification equipment.

Medical research efforts employ membrane filters for electron microscopy, radiochemistry, radiomicrobiology, virology, and parasitology. In the field of radio tracers, for example, the cell metabolism, receptor binding, and genetic composition studies rely on membrane filter properties.

F. Pharmaceutical and Cosmetics Industries

The pharmaceutical industry is closely regulated because it is related to health care and the prevention and elimination of disease. The elimination of particulate matter and the production of sterile products both have a high priority. The manufacturing and distribution of pharmaceutical products must emphasize contamination control.

Water for injection (WFI) manufacturing and packaging operations are dedicated to the preparation of sterile, chemically pure water (Frith et al., 1975). The pharmaceutical industry has recognized the requirements and has imposed self-regulations that are quite stringent. One standard for filtration eliminates asbestos and fiber-releasing materials and recommends use of the filter membrane.

Use of the membrane filter for verifying sterility and the absence of particles is an accepted procedure. In addition, to ensure that the product will pass the rigid final inspection, membrane filters are used in the process. Heat-sensitive or heat-labile solutions must be "cold"-sterilized. Antibiotics, allergenic extracts, and virus vaccines are typical solutions that are filtered. Production employs the membrane for final product filtration as well as to prepare raw materials, produce sterile gases, and vent makeup air. Serum filtration presents very difficult operating problems. Therefore, serial filtration procedures are required. A depth prefiltration stage is placed upstream of a series of membrane filters and this is followed by a final 0.22 μm sterilizing filter unit (Millipore, 1975c).

Toiletries, cosmetics, and oral products utilize membranes for quality control analysis. New government regulations require that the production of products be well controlled, and therefore an opportunity for final process filtration is being recognized.

G. Electronics

The membrane filter in analytical and process applications has been a major factor in the development of microstate technology. Devices for the semiconductor industry require that only the materials designated as components be fabricated and packaged; therefore, the environment, materials, production process, and workers are rigidly controlled.

The measurement of atmospheric contaminants within a clean-room environment and the particles on workers' garments (ASTM, 1968c), are of prime concern. Foreign matter (Fig. 14) can easily damage valuable microcircuit functions,and thus provides justification for the use of high-volume process filtration equipment (Fig. 15). Even though the membrane filter is not essential for purifying the clean-room environment, it is used to monitor the controlled facility (ASTM, 1968d).

The most widely used universal solvent in the electronics industry is water. Ultrapure specifications demand the most advanced technologies possible for the purification process (ASTM, 1977). Of the four major types of contaminants (organics, inorganics, particles, and microorganisms), the membrane filtration unit eliminates

FIGURE 14 Particles removed from a fluid in a microelectronics production process. (For particle size reference, the grid lines on the filter are 150 μm.)

two: particles and microorganisms (Frith, 1969). In addition, the filtration unit is located as the final step to remove any of the adsorbent or deionization materials used to eliminate the soluble organics and inorganics. The fact that the inert membrane does not impart additional contamination is an important factor for the fabrication of these microelectronic elements.

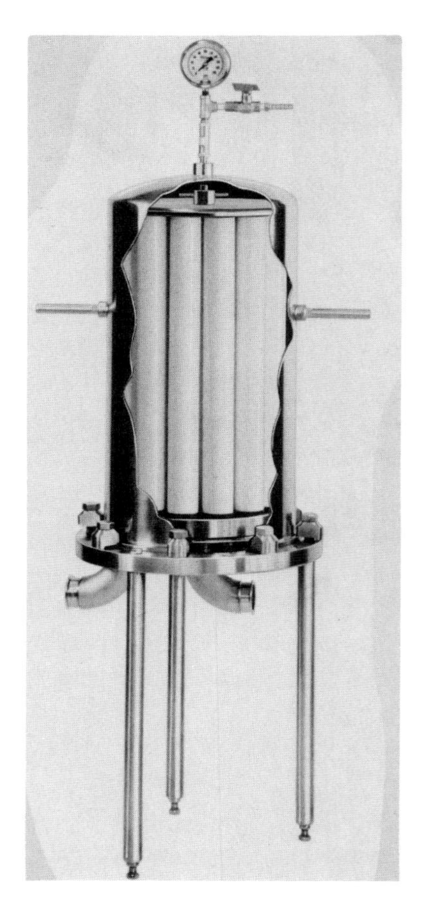

FIGURE 15 Cutaway view showing a Millipore Multitube Filter Holder.

In a production environment, all types of gases are required for cleaning, purging, and filling. The elimination of foreign matter is important. Solutions such as solvents, acids, photoresists, developers, and etchants must be ultraclean because of the narrow limits of resolution (Lafeuille et al., 1975).

Advancements in the data processing industry have become so sophisticated that the environment and components for the computer systems must be rigidly controlled. The 0.2 μm membrane filter is standard in most applications.

H. Energy/Utilities

The ever-increasing demands for energy require both that efficient sources be developed and current systems be improved. Analytical methods for contamination control are important whether it be for a supercritical boiler/turbine system or the cooling water for a nuclear generator (Millipore, 1972).

Because of the operating requirements for power-generating equipment, the specifications for contamination are extremely low. Therefore, concentrating large sample volumes through the membrane filter for detection and identification is a major advantage. Many tests may be performed on a sample retained on the filter, such as chemical, sophisticated instrumental methods, and physical examinations.

In the nuclear reactor, small particles in the cooling water shield may become contaminated due to radiation from the active elements. Membrane filtration is an advantage because particles in the colloidal range and larger can be eliminated in the pretreatment process. Optical clarity of this protective shield is also provided by the removal of the small particles.

I. Chemical Industry

The analytical methods used in the chemical industry are similar to those of other industries. Unique analytical applications exist for membranes, such as monitoring the high-pressure water and steam employed for secondary oil well recovery operations. Deionized water is a universal solvent and membranes are required, as noted earlier. Also, specialty chemicals with high quality levels are marketed to other industries, such as aerospace, electronics, and pharmaceuticals. High-volume membrane filtration systems are used prior to packaging these solvents (Frith, 1968).

High-resolution developers in photography have extremely rigid specifications and the membrane filter is a production process tool. For instance, in photoreconnaissance, pictures are photographed at heights in excess of 50,000 feet. The presence of submicron particles in large quantities can cause the developing solution to haze or obscure an important detail. The same is true for precision plating solutions. The presence of microscopic and colloidal matter can contaminate the solution and the surface interface of the component to be plated and cause a reliability problem. Also, filtration of polymers is a growing field because of requirements in the production of contact lenses and optical lenses.

In the field of chemical reclamation, the absolute filtration characteristics allow the collection of precious metals in the waste product. Once the material has been collected, it is very simple to ignite the filter and recover the retentate. This is a simple and rapid separations process, and large-scale membrane filter equipment satisfies the ever-increasing demands of this industry. The economic impact is quite substantial and it is anticipated that other, less expensive materials will be reclaimed in the future.

J. Aerospace/Transportation

Perhaps the timely development of the membrane filter for commercial use and the decision that aerospace exploration was essential can be considered a perfect marriage. From the beginning, such subjects as ultrapure solutions and clean-room environments were important in the rapid growth of this industry. The success of the "Man on the Moon" program, and further developments such as Sky Lab, are having repercussions in the transportation industry.

The air and water pollution problems that are common to other industries are very significant in the aerospace industry. The allowable levels of contaminants are many orders of magnitude lower. Where a normal earth environment may contain millions of submicron particles and still be considered safe, the demands of high-precision technology in space may reject surroundings having 100 to 1000 submicron particles per cubic foot.

The reliability factor demanded of high-performance jet aircraft provided new applications for membrane filter techniques.

The jet fuel is monitored for particulates and bacteria and the electromechanical systems require extremely clean oils and lubricants for proper operation. These petroleum products are tested with membrane filters.

The monitoring of cryogenic materials is a unique application and one to which membrane filter materials are readily adapted. Auto emission testing is another relatively new procedure but is gaining recognition as government agencies enforce regulations for clean air. The filter is used to collect particulate lead which is emitted from the various types of engines used in transportation.

The low levels of contamination acceptable to the aerospace industry for extremely clean lubricants, fuels, solvents, and gases justify the absolute retention properties of the membrane filter (Millipore, 1968). The wide range of membrane materials allows the technician to filter almost any liquid or gas common to the industry.

Missile components rely on fuels and hydraulic oils that have extremely tight particle limit specifications. In certain cases, a continuous filtration process is designed into the equipment. This is similar to the oil filter in the automobile. However, the assurance that particles in the submicron range are eliminated is much more critical. An absolute filter is essential for missiles but would not be required for automobiles.

V. SUMMARY

The membrane filter has applications in every industry and is improving environmental conditions and manufacturing processes. The value of such technology is recognized when one studies the growth of the membrane filter industry in developing, producing, and distributing the technologies and products. Each year, hundreds of new applications are provided to industry, and sufficient material is available to devote an entire book to the many analytical procedures and filtration processes. The future microminiaturization and improved quality of our life-style will demand that industries include the membrane filter as a valuable analytical and process tool.

REFERENCES

American Society for Testing and Materials (1968a). Sizing and counting of airborne particulate contamination in clean rooms and other dust-controlled areas designed for electronic and similar applications. F25-68 (1973). Philadelphia.

American Society for Testing and Materials (1968b). Identification of metal particulate contamination found in electronic and microelectronic components and systems using the ring oven techniques with spot tests. F59-68 (1973). Philadelphia.

American Society for Testing and Materials (1968c). Sizing and counting particulate contaminants in and on clean room garments. F51-68 (1973). Philadelphia.

American Society for Testing and Materials (1968d). Detection and enumeration of microbiological contaminants in water used for processing electron and microelectronic devices. F60-68 (1973). Philadelphia.

American Society for Testing and Materials (1969). Test for silting index of fluids for processing electron and microelectronic devices. F52-69 (1973). Philadelphia.

American Society for Testing and Materials (1973). Tests for particulate contaminant in aviation turbine fuels. D2276-73. Philadelphia.

American Society for Testing and Materials (1977). Specifications for reagent water. D1193-77. Philadelphia.

American Society for Testing and Materials (1978). Test for total bacteria count in water used for processing electron and microelectronic devices. F488-78. Philadelphia.

Dwyer, J. L. (1966). *Contamination Analysis and Control*. Reinhold, New York.

Dwyer, J. L., and Frith, C. F. (1967). New analytical techniques for determination of colloidal contamination in high purity steam generating systems. *Proc. Am. Power Conf.*, Chicago.

Frith, Clifford F. (1967). A new gravimetric analysis technique for suspended contamination in high purity water systems. Paper presented to the International Water Conference, Pittsburg, Pa., October.

Frith, Clifford F. (1968). Sterile water and industry. Paper presented for Liberty Bell Corrosion Course, Drexel Institute of Technology, Philadelphia, September.

Frith, C. F. (1969a). Technology of producing ultra-pure water. *Proc. 8th Annu. Tech. Meet., Am. Assoc. Contam. Control*, New York, May.

Frith, C. F. (1969b). Sterile water for use in microelectronics production. *Circuits Manuf.* *9*(9).

Frith, C. F. (1971). The membrane filter: its role in water analysis. Liberty Bell Corrosion Course, Drexel Institute, Philadelphia, 1971.

Frith, C. F., Dawson, F. W., and Sampson, R. L. (1975). Water for injection, USPXIX, by reverse osmosis. *Proc. Annu. Meet. Parenteral Drug Assoc.*, New York, October.

Hill, William F., et al. (1972). Virus in water: evaluation of membrane cartridge filters for recovering low multiplicities of poliovirus from water. *Appl. Microb.* *23*(5): 880–888.

Lafeuille, D., et al. (1975). Purity of chemicals for semiconductor processing. *Solid State Technol.*, January.

Liu, Y. H., and Lee, K. W. (1976). Efficiency of membrane and nuclepore filters for submicrometer aerosols. *Environ. Sci. Technol.* *10*(4): 345–350.

Megaw, W. J., and Wiffen, R. D. (1963). The efficiency of membrane filters. *Int. J. Water Pollut.* *7*: 501–509.

Millipore Corp. (1968). Ultracleaning of fluids and systems. ADM60. Bedford, Mass.

Millipore Corp. (1969). Microbiological control in the brewery. AR71. Bedford, Mass.

Millipore Corp. (1972). Analysis and control of particulate contamination in electric power plants. AG3. Bedford, Mass.

Millipore Corp. (1975a). Patch testing for particulate contamination. AB417. Bedford, Mass.

Millipore Corp. (1975b). Sterilization of gases by filtration. AB315. Bedford, Mass.

Millipore Corp. (1975c). Filtration of serum products. AB719. Bedford, Mass.

Millipore Corp. (1976a). Sampling Microorganisms in air. TS001. Bedford, Mass.

Millipore Corp. (1976b). The MF-DNA somata count test for bovine mastitis. AB815. Bedford, Mass.

Millipore Corp. (1976c). Preparing sterile particle-free fluids in the hospital pharmacy. AM303. Bedford, Mass.

Millipore Corp. (1977). Filter clearing procedure for particle counting using transmitted light. TS018. Bedford, Mass.

Millipore Corp. (1978). Matched-weight aerosol monitors for gravimetric analysis. TS030. Bedford, Mass.

National Committee for Clinical Laboratory Standards (1978). Specifications for reagent water used in the clinical laboratory. Tentative Standard: PSC-3.

Rapp, R., et al. (1974). Evaluation of a prototype air-venting in-line intravenous filter set. *Am. J. Hosp. Pharm. 32*: 1253–1259.

Rapp, R., et al. (1975). In-line filtration of I.V. fluids and drugs. *Am. J. I. V. Ther.*, April-May, 18–23.

Rohm and Haas Company (1966). Amberlite XE-238: preliminary notes.

Russell, J. H. (1978). High-volume air sterilization. *Process Biochem.*, September, 25–28.

Society of Automotive Engineers (1963a). Procedure for the determination of particulate contamination in hydraulic fluids by the control filter gravimetric procedure. ARP785.

Society of Automotive Engineers (1963b). Procedure for the determination of the silting index of a fluid. ARP788.

Society of Automotive Engineers (1966). Procedure for the determination of particulate contamination of air in dust controlled spaces by the particle count method. ARP-743A.

United States Air Force (1965). Technical order-standards and guidelines for the design and operation of clean rooms and clean work stations. T.O. 00-25-203, July.

United States Pharmacopoeial Convention, Inc. (1980). *The United States Pharmacopoeia, Twentieth Revision*. Mack Printing Co., Easton, Pa.

INDEX